AFTER CALCULUS: ANALYSIS

AFTER CALCULUS: ANALYSIS

DAVID J. FOULIS
UNIVERSITY OF MASSACHUSETTS

MUSTAFA A. MUNEM
MACOMB COLLEGE

DELLEN PUBLISHING COMPANY
SAN FRANCISCO

COLLIER MACMILLAN PUBLISHERS
LONDON

DIVISIONS OF
MACMILLAN, INC.

On the cover: The cover is an untitled 1986 monoprint by Sam Tchakalian. The artist's "attitude," rather than formalized structural elements, becomes the content of his work. Tchakalian's work may be seen at the San Francisco Museum of Modern Art; the Oakland Museum, Oakland, California; the Albright Knox Museum, Buffalo, New York; and the University Art Museum of the University of California, Berkeley. Tchakalian is represented by Modernism in San Francisco, California.

Permissions: Dellen Publishing Company
400 Pacific Avenue
San Francisco, California 94133

Orders: Dellen Publishing Company
c/o Macmillan Publishing Company
Front and Brown Streets
Riverside, New Jersey 08075

Collier Macmillan Canada, Inc.

LIBRARY OF CONGRESS CATALOGING-IN-PUBLICATION DATA

Foulis, David J.,
    After calculus—analysis/David J. Foulis, Mustafa A. Munem.
        p.   cm.
    Includes index.
    ISBN 0-02-339130-8
    1. Mathematical analysis.   I. Munem, Mustafa A.   II. Title.
QA300.F678   1989
515—dc 19

89-1579
CIP

Printing: 1 2 3 4 5 6 7 8 9   *Year:* 9 0 1 2 3
ISBN 0-02-339130-8

Dedicated to the memory of our colleague and friend,
Charles H. Randall 1928–1987

CONTENTS

1

INTRODUCTION TO LOGIC AND SETS   1

2

RELATIONS AND FUNCTIONS   83

6

INFINITE SERIES   435

PREFACE FOR THE INSTRUCTOR

This textbook is written for students who have completed at least one semester of calculus and who intend to take more advanced courses in mathematics. It is designed to provide a gentle transition from "cookbook" mathematics to the more rigorous and sophisticated material in post-calculus courses. Our primary concern is to acquaint the student with the language and general methods of contemporary mathematics; however, since this cannot be done in a vacuum, we have chosen to present this material in the context of real analysis.

This is not a cultural survey for liberal arts students, nor is it a vehicle for mathematical proselytization. We assume that our readers already have a strong attraction to mathematics, or are going into a field in which more advanced mathematical techniques are used. We believe that these readers are best served by a straightforward exposition without frills. Therefore, although we do include a few historical vignettes and annotated bibliographies for suggested further reading, we have indulged in a minimum of motivational rhetoric.

There are no prerequisites for an intelligent reading of this textbook other than competence in precalculus mathematics and a working computational knowledge of the basic techniques of differentiation and integration. In our writing, we have used the so-called spiral approach. Thus, we assume at the outset that the reader is familiar with some elementary facts about calculating, counting, and dealing with the elementary functions encountered in basic calculus courses; later, we return to these concepts and show how they can be handled more carefully. For instance, in Section 5 of Chapter 1, we assume that the reader knows what is meant by the number of elements in a finite set, and we make free use of this idea; later, in Section 6 of Chapter 4, we reexamine this concept, make it more rigorous, and extend it to infinite sets.

This textbook includes well over 1200 problems, ranging from simple drill-type exercises to problems that probe for deep understanding and insight. The odd-numbered problems at the end of each section are usually quite straightforward and should present little difficulty for the student who has carefully read the section. Numerous examples, given in an example–solution format, show the student how these problems are to be handled; answers or hints for selected odd-numbered problems can be found at the end of the book. Some of the even-numbered problems— particularly those toward the ends of the problem sets—are considerably more challenging. In general, *even-numbered problems should not be assigned without careful prior consideration* since the more difficult of these problems can be frustrating for all but the best students.

In those cases for which there is no universal agreement regarding notation or terminology, we have made our choice (when possible) on the basis of what seems to be most common in the standard literature. In those instances for which there is no clear basis for a decision, we have made an arbitrary choice, but we caution our readers about the alternatives used by other authors.

Since even students with otherwise strong backgrounds in mathematics often have difficulty with mathematical induction, we have included a *review of mathematical induction* in an appendix. Some instructors may prefer to organize their first lesson around this appendix; others may suggest that students refer to the appendix for help with problems that require this technique.

Other special features of the book are as follows:

1. Propositional calculus and predicate calculus are presented informally in the first two sections of Chapter 1. It is emphasized that this machinery will be used only when necessary to help clarify portions of mathematical arguments.

2. In the third section of Chapter 1, we discuss the notion of a mathematical proof based on axioms, definitions, and rules of procedure, and we explain the idea of an indirect proof.

3. In the last three sections of Chapter 1, we consider the basic set-theoretic machinery that will be needed in the sequel.

4. In Chapter 2, relations are defined as sets of ordered pairs, and functions are introduced as special kinds of relations. Order relations and the affiliated concepts of infimum and supremum are discussed here. The chapter includes an optional final section on partitions and equivalence relations.

5. Chapter 3 is devoted to a study of the real number system $\mathbb{R}$ and its subsystems $\mathbb{N}$, $\mathbb{Z}$, and $\mathbb{Q}$. To lay the groundwork, groups, rings, and fields are introduced in the first two sections of the chapter. In the third section, we study ordered fields and postulate the existence of a

complete ordered field $\mathbb{R}$. In the last two sections of Chapter 3, the number systems $\mathbb{N}$, $\mathbb{Z}$, and $\mathbb{Q}$ are extracted from $\mathbb{R}$, and their basic properties are established. We feel that this "top-down" approach provides an excellent preparation for those students who go on to study the constructive development of $\mathbb{R}$.

6. In Chapter 4, the basic topological features of the real line are developed and exploited. Although we confine our attention to $\mathbb{R}$, all concepts are presented in such a way that the natural extension to $\mathbb{R}^n$, metric spaces, and topological spaces in general, is almost obvious. The intermediate value theorem is obtained as a consequence of the fact that continuous functions preserve connected sets, and the extreme value theorem is shown to follow from the fact that a continuous image of a compact set is compact.

7. The unifying concept in Chapter 5 is the idea of approximation. The derivative solves the problem of local linear approximation, Taylor's theorem solves the problem of local polynomial approximation, and the integral is a limit of approximating sums. Again, this approach prepares the way for generalization to higher-dimensional spaces. Both the Darboux and Riemann definitions of the integral are presented, and it is shown that these definitions are equivalent.

8. Sequences and series are introduced and studied in the final chapter of the book. The theme of approximation continues to unify the presentation of this material.

Although this textbook is primarily designed for use in a one-term course at the undergraduate level, it is probably unrealistic to attempt a thorough coverage of the entire book in such a course. In planning the course, it is suggested that the instructor keep in mind that Chapters 4 and 5 comprise the heart of the book, and that this is where the emphasis belongs. If time constraints make it necessary, the first chapter can be treated with dispatch, stressing only the peculiarities of the implication connective, the use of quantifiers, indirect proof, and the elements of set algebra. The final section of Chapter 2 can be omitted without loss of generality. For students with a strong background, the course could begin with Chapter 3, or even with Chapter 4.

It is a pleasure to acknowledge the valuable advice of the following reviewers of our manuscript:

David Barnette, University of California, Davis

Sherralyn Craven, Central Missouri State University

David Ellenbogen, University of Vermont

Michael Evans, North Carolina State University

Robert Gamble, Winthrop College

Juan Gatica, University of Iowa

Robert Hunter, Pennsylvania State University

Jack Johnson, Brigham Young University, Hawaii Campus

Daniel Kocan, Potsdam College of the State University of New York

Joseph H. Oppenheim, San Francisco State University

Robert Piziak, Baylor University

Daniel Sweet, University of Maryland

Howard Taylor, West Georgia College

We are grateful for the superb work of our editor, Phyllis Niklas, and to Donald Dellen for shepherding the book through the various stages of editing and production. Special thanks are also given to Gerald L. Alexanderson of Santa Clara University and Al Bednarek for providing the annotated bibliographies and the historical notes. Finally, we would like to express our gratitude to Steve Fasbinder and Hyla Gold Foulis for assisting with the proofreading of the pages.

David J. Foulis
Mustafa A. Munem

INTRODUCTION

From ancient times until nearly the end of the nineteenth century, mathematics was generally viewed as the study of number (algebra) and form (geometry). This view was reinforced in the seventeenth century by the great synthesis of algebra and geometry (analytic geometry) created by the French mathematicians René Descartes and Pierre de Fermat. By the early part of the twentieth century, however, it had become clear that this perception of mathematics was overly restrictive, that the study of algebraic and geometric structure is only part of the mathematical enterprise, and that the true concern of mathematics is the study of abstract structure in general. It is because sets are natural carriers of this structure that set theory has played such a central role in twentieth century mathematics.

The unification effected by the modern structural view of mathematics has been even more profound than the synthesis wrought by Descartes and Fermat; and, as a consequence, all branches of mathematics now share a common core of ideas and methods. The purpose of this textbook is to acquaint you with these fundamental mathematical concepts and techniques. In your study of this material, you may find the following suggestions helpful:

1. Your previous mathematical experience may have been with "cookbook" courses requiring only that you memorize formulas and procedures and apply them to solving problems, usually by simply "plugging in" suitable values for variables. If so, be forewarned that the material in this texbook is a departure from this approach. You will find that this material has some of the flavor of a foreign language, with its own alphabet, vocabulary, orthography, grammar, and idiomatic expressions. Indeed, you will be learning the language used by mathematicians in communicating their ideas and discoveries.

2. Try to understand every word in every sentence in every paragraph in every section of this textbook. In mathematics, new ideas almost always depend strongly on previously introduced concepts, and a lack of understanding of earlier material often interferes with comprehension of

later ideas. Since you probably will not grasp everything the first time around, we recommend a spiral approach, in which you repeatedly reconsider earlier ideas after acquiring a better notion of where they lead.

3. In reading this textbook (or mathematics in general) it is *imperative* that you have a pencil and paper at hand. You must *write out* the definitions, formulas, theorems, and key ideas that you encounter as you go along. Some students like to use a highlighting pen to color this material in the text, but this is not nearly as effective as the act of transcribing them to a sheet of paper. *Doing the writing* focuses your attention. Considerable effort is required to learn mathematics, and it is impossible to do so sitting in an easy chair, reading your textbook as if it were a novel.

4. Draw pictures and diagrams when possible to illustrate the concepts under consideration.

5. The problem sets at the ends of the sections are integral parts of this textbook. Although some of the even-numbered problems, particularly those toward the ends of the problem sets, are quite challenging, most of the odd-numbered problems are straightforward, and working them will enhance your grasp of the material. An odd-numbered problem with which you cannot cope may indicate a lack of understanding of the concepts involved. Answers and hints for most of the odd-numbered problems are provided at the end of the book.

6. In working the problems (and in writing mathematics in general) cultivate the habit of writing complete sentences. An isolated expression such as $x^2 + 2x + 1$ means very little. What about it? Do you want to find the values of x for which the expression is zero? If so, say so. Are you writing this as an example of a polynomial with integer coefficients? If so, say so.

7. Every mathematical concept has to be understood on two levels: the *formal* level and the *intuitive* level. To understand the concept on the formal level, you need to know its mathematical definition. Although it might be possible to get through a beginning calculus course without knowing the definition of, say, the Riemann integral, it is inconceivable to do more advanced mathematics without paying careful attention to definitions. Some people prefer to memorize definitions outright; others would rather look them up as they need them in working problems; this is a matter of individual preference. An intuitive grasp of a mathematical concept may be more difficult to come by; it generally develops only after you deal with the concept over a period of time. To build your intuition, you might begin by collecting pertinent examples of the concept. Perhaps, as you go along, you will want to keep a notebook of interesting and suggestive examples. After you have a suitable stock

of examples at hand, you should begin to use the concept, and this is where the problems are of importance. Finally, you should think about the concept, mull it over, see how it relates to other ideas, and see if it helps to clarify notions that might otherwise be hazy or imprecise.

8. From time to time you will encounter an idea that you find exciting. When you do, why not decide to become a "local expert" on this idea and its ramifications? Go to the library and read everything you can find that pertains to the idea. Start a notebook for your own use, devoted to this and related ideas. Collect pertinent examples, exercises, definitions, theorems, diagrams, remarks, historical lore—anything you can find that is germane to the idea. Explain the idea to your friends and tell them why you find it so fascinating. If you persist in this, you will come to the realization that mathematics is remarkably holistic and that, in learning all you can about one apparently isolated concept, you will automatically learn considerably more mathematics than you might expect.

This textbook is organized into six chapters, and each chapter is divided into numbered sections. Problem Set 3.2, for instance, refers to the problem set at the end of Section 2 in Chapter 3. Major definitions, theorems, and examples are numbered consecutively through each section; for instance, Theorem 2.13 refers to the thirteenth numbered item (i.e., definition, theorem, or example) in the second section of a chapter. If no chapter is mentioned, it is understood that the item appears in the current chapter; otherwise, the chapter is always indicated.

The most important definitions and theorems are in boxes, and all major theorems have names (as well as numbers) for ease of reference. Important words, phrases, and remarks are *italicized*, and terms are written in **boldface** type when they are defined for the first time. Separate, but related, parts of definitions or theorems are labeled with small Roman numerals (i), (ii), (iii), and so on; separate, but related, parts of problems are labeled with small letters (a), (b), (c), and so on. The symbol □ is used to signal the end of a proof or solution.

1

INTRODUCTION TO LOGIC AND SETS

In 1666, the German mathematician and philosopher G. W. Leibniz (one of the codiscoverers of the infinitesimal calculus) conceived of a "calculus of reasoning" that could be used to make complicated logical deductions merely by mechanically manipulating suitable symbols. The formulas of such a **symbolic logic** would be handled in much the same way as the formulas of ordinary algebra; and logical errors, which would arise only through mistakes in calculation, could quickly be detected and corrected. Reasoning would be reduced to computation!

Between 1666 and the 1690s, Leibniz himself made an effort to lay the foundations for symbolic logic, but it remained for a nineteenth-century British mathematician, George Boole, to take the first really decisive steps in this direction. In the hands of Boole and his successors, symbolic logic developed into a powerful tool that not only helped to clarify mathematical thought but also had a number of unforeseen applications. For instance, in the late 1930s, the American electrical engineer Claude Shannon showed that Boolean algebra, a branch of symbolic logic, could be used to analyze and design electrical switching circuits.

In relation to mathematics, symbolic logic plays a dual role: First, it functions as a precise *language* that is tailor-made for clarifying complex logical deductions; second, in connection with a discipline called *metamathematics*, it is a tool for studying the potentialities and limitations of mathematics itself. However, symbolic logic is not a miraculous machine for grinding out mathematical theorems; and the would-be provers of such theorems must be prepared for some hard work no matter how much symbolic logic they happen to know.

Although it is possible to approach symbolic logic in a formal axiomatic way, we propose to present it informally as a language with its own alphabet, grammar, and idioms. You should expect to learn this material in much the same way that you would learn any foreign language; however, you will find that it is a language capable of exquisite precision.

Although most of our mathematical deliberations will be conducted in ordinary English, we intend to use symbolic logic when necessary to clear up arguments that might otherwise cause difficulty.

1

ELEMENTS OF PROPOSITIONAL CALCULUS

The Latin word *calculus* refers to a small stone or pebble. In ancient times, pebbles were used as counters to aid in doing arithmetic, and so the word calculus has come to be used in mathematics to refer to a systematic method of calculating or reasoning. Thus, *differential calculus* is the study of calculating with differentials or derivatives, whereas *integral calculus* is the study of calculating with integrals. One of the important branches of symbolic logic is **propositional calculus**, a systematic method of reasoning or "computing" with statements or propositions. We shall launch our study of propositional calculus with the following basic definition.

1.1 DEFINITION

> **PROPOSITION** By a **proposition** we mean a complete declarative sentence that is either true or false. If a proposition is true, we say that its **truth value** is 1, and if it is false, we say that its truth value is 0.

Some people use the letters T (true) and F (false) to denote the truth values of propositions. However, since most computer languages denote truth values by 1 and 0, we prefer to use 1 and 0 rather than T and F.

Our definition of a proposition is not intended as a formal mathematical definition but merely as an informal working definition that will serve our present purposes. We often use the word **statement** as a synonym for a proposition. Note that there is no condition on the language in which a proposition is expressed; it may be, for instance, English, German, Esperanto, or mathematical symbolism. Thus, *two is smaller than three, Zwei ist kleiner als Drei*, and *2 < 3* are propositions. Also, in testing to see whether an expression is a proposition, you are not required to determine its truth value. Thus, the statement, *the trillionth digit after the decimal point in the decimal expansion of pi is 7* is a proposition even though no one may know whether it is true or false.

1.2 Example   Which of the following are propositions?

(a) Hoping to evaluate the integral
(b) $2 + 2 = 4$
(c) Is pi equal to 22/7?
(d) Please go to the blackboard and solve the equation $x^2 + 1 = 0$

SOLUTION Only (b) is a proposition; (a) is not a complete sentence; (c) is a question, not a declarative sentence; and (d) is a request, not a statement of fact. ☐

In ordinary algebra, we use symbols such as $x, y, z, \ldots$, called *number variables* to denote arbitrary numbers. Likewise, in propositional calculus, we use symbols such as $P, Q, R, \ldots$, called **propositional variables** to stand for arbitrary propositions. In ordinary algebra, we have operations such as addition, subtraction, multiplication, division, negation, and reciprocation, which can be used with number variables to form new number expressions $x + y$, $x - y$, $x \cdot y$, $x \div y$, $-x$, and x^{-1}. *Propositional connectives*, which we now define, play a similar role in propositional calculus.

1.3 DEFINITION   **BINARY PROPOSITIONAL CONNECTIVES**   A **binary propositional connective** is an operation that combines two propositions to yield a new proposition whose truth value depends only on the truth values of the two original propositions.

The English word *and* provides a good example of a binary propositional connective. Indeed, if P, Q are propositions, then

P and Q

is a new proposition declaring that both P and Q are true. Its truth value is 1 (true) if and only if both P and Q have truth value 1; otherwise, its truth value is 0 (false).

More generally, if the symbol c represents some binary propositional connective, we write

P c Q

for the new proposition obtained by combining the propositions P and Q with the use of the connective c. It is important to understand that the truth value of the new proposition $P \, c \, Q$ does not depend on the *content* or *meaning* of the original propositions P and Q, but *only on their truth values*. For instance, to determine the truth value of

P and Q

you do not need to know what either P or Q happens to say; you need only to know their truth values. Propositions built up by combining propositional variables with the aid of propositional connectives are called **compound** propositions and statements.

1.4 DEFINITION   **TRUTH COMBINATION**   Let $P, Q, R, \ldots$ be propositional variables. Any definite assignment of truth values to each of these propositional variables is called a **truth combination** for $P, Q, R, \ldots$.

For instance, one possible truth combination for the triple P, Q, R of propositional variables would be obtained by assigning, say, the truth value 1 to P, the truth value 0 to Q, and the truth value 1 to R. This particular truth combination is displayed in the following table:

P	Q	R
1	0	1

How many different truth combinations can be formed for n different propositional variables

$$P_1, P_2, P_3, \ldots, P_n?$$

Any one of the two possible truth values (0 or 1) can be assigned to P_1; then, any one of two possible truth values (0 or 1) can be assigned to P_2, for a total of $2 \times 2 = 4$ different ways of assigning truth combinations to P_1 and P_2. Continuing in this way, we double the number of truth combinations for each additional propositional variable, and it follows that there are

$$\overbrace{2 \times 2 \times \cdots \times 2}^{n \text{ factors}} = 2^n$$

different ways of assigning truth combinations to

$$P_1, P_2, P_3, \ldots, P_n.$$

Thus:

For n different propositional variables, there are exactly 2^n different truth combinations.

1.5 Example   List in tabular form all of the $2^3 = 8$ possible truth combinations for the three propositional variables P, Q, and R.

SOLUTION

P	Q	R
1	1	1
0	1	1
1	0	1
0	0	1
1	1	0
0	1	0
1	0	0
0	0	0

Notice that we have obtained all possible truth combinations by alternating 1's and 0's in a systematic way: singly in the first column, in pairs in the second column, and in sets of four in the third column.   □

Now, let's use a table to show exactly how the truth value of the compound proposition

P and Q

depends on the truth values of the two propositional variables P, Q. There are $2^2 = 4$ different truth combinations for these two propositional variables; and, for each such truth combination, the compound proposition

P and Q

will have the corresponding truth value, shown in the following table:

P	Q	P and Q
1	1	1
0	1	0
1	0	0
0	0	0

As this table shows, the compound proposition

P and Q

is true only if both P and Q are true; otherwise, it is false.

More generally, we can make a **truth table** for any binary propositional connective c by showing the truth values of $P c Q$ that correspond to the four possible truth combinations for the two propositional variables P and Q:

P	Q	$P c Q$
1	1	v_1
0	1	v_2
1	0	v_3
0	0	v_4

In this table the entries v_1, v_2, v_3, and v_4 represent the truth values of the compound proposition $P c Q$ corresponding to the four possible truth combinations for P and Q. Thus, v_1 is the truth value of $P c Q$ when P and Q are both true, v_2 is the truth value of $P c Q$ when P is false and Q is true, and so forth. Notice that a *propositional connective is completely determined by its truth table*. Since each of the entries v_1, v_2, v_3, and v_4 has two possible values (1 or 0), it follows that there are only 16 possible ways to assign these values (see Problem 43). Therefore, *there are only 16 possible binary propositional connectives*.

We are going to focus our attention on four of the sixteen possible binary propositional connectives. Special names and symbols are used for these four connectives.

1.6 DEFINITION

∧, ∨, ⇒, **AND** ⇔   The binary propositional connectives $\wedge$, $\vee$, $\Rightarrow$, and $\Leftrightarrow$, called **conjunction, disjunction, implication**, and **equivalence**, respectively, are defined by the following truth tables:

P	Q	$P \wedge Q$
1	1	1
0	1	0
1	0	0
0	0	0

P	Q	$P \vee Q$
1	1	1
0	1	1
1	0	1
0	0	0

P	Q	$P \Rightarrow Q$
1	1	1
0	1	1
1	0	0
0	0	1

P	Q	$P \Leftrightarrow Q$
1	1	1
0	1	0
1	0	0
0	0	1

Notice that the conjunction connective $\wedge$ is just alternative notation for the connective *and*. These four connectives are usually read (in English) as follows:

$P \wedge Q$ is read "*P and Q*,"

$P \vee Q$ is read "*P or Q*,"

$P \Rightarrow Q$ is read "*P implies Q*,"

and

$P \Leftrightarrow Q$ is read "*P is equivalent to Q*."

Sometimes,

$P \Rightarrow Q$ is read "*if P then Q*,"

and sometimes,

$P \Leftrightarrow Q$ is read "*P if and only if Q*."

Although the connective $\wedge$ is used in propositional calculus in exactly the same way the conjunction *and* is used in English, the connective $\vee$ is not always used in the same way as the English word *or*. Notice that, in the statement

John will marry Marsha or Dolores,

the word *or* is used in the **exclusive** sense of *one or the other but not both*. On the other hand, in the statement

Most families in our town have a cat or a dog,

the word *or* is used in the **inclusive** sense of *at least one and possibly both*.

This ambiguity in the meaning of *or* normally does not cause any difficulty because we can tell the intended meaning from the context; however, to avoid any possible misunderstandings in carefully written documents, *and/or* is often used for *or* in the inclusive sense. As you can see from the table for ∨ in Definition 1.6:

> *The connective* ∨ *is used in mathematics as a symbol for* or *in the inclusive sense of* and/or.

Thus, in mathematical writing, to say that $P \lor Q$ is true means that *at least one and possibly both* of the propositions P, Q are true.

The implication connective ⇒ is usually a source of difficulty for beginning students of propositional calculus. As the truth table for ⇒ in Definition 1.6 shows, *whenever the proposition P is false, the proposition $P \Rightarrow Q$ is true!* Thus, for instance, the compound proposition

$1 = 2 \Rightarrow$ *all pigs have wings*

is true because $1 = 2$ is false. This situation understandably upsets many people; they feel that the propositional connective ⇒ does not correspond to their intuitive idea of logical implication. Nevertheless, it has been found that the connective ⇒, as defined by the truth table in Definition 1.6, is precisely the version of logical implication appropriate for the construction of mathematical proofs. After studying Section 1.3, you will understand why this is so.

If you look at the truth table for the equivalence connective ⇔ in Definition 1.6, you will see that the compound proposition $P \Leftrightarrow Q$ is true precisely when the two propositions P, Q have the same truth value—both true or both false. For instance, the compound proposition

$1 = 2 \Leftrightarrow$ *all pigs have wings*

is true because both of the propositions from which it is formed have the same truth value (namely, 0).

In ordinary conversation, we often have occasion to deny that something is true. The same thing is accomplished in propositional calculus by the use of the *negation connective.*

1.7 DEFINITION

NEGATION CONNECTIVE   If P is a propositional variable, we define the expression $\sim P$ to be a new proposition that is true if P is false and false if P is true.

The expression $\sim P$ is usually read (in English) as *not P* and the symbol $\sim$ is called the **denial connective** or the **negation connective**. Note that $\sim$ is a **unary** connective rather than a binary connective; that is, it "acts on" a *single* propositional variable rather than on a pair of propositional variables. The truth table for $\sim$ is simple enough:

P	$\sim P$
1	0
0	1

By a *formula* of propositional calculus we mean, roughly, any expression built up in a meaningful way with the use of a finite number of propositional variables and the five connectives $\wedge$, $\vee$, $\sim$, $\Rightarrow$, and $\Leftrightarrow$. More precisely:

1.8 DEFINITION

> **FORMULA**   Propositional variables are defined to be **formulas**. If $\mathscr{F}$ and $\mathscr{G}$ are formulas, then the expressions
>
> $$\mathscr{F} \wedge \mathscr{G}, \mathscr{F} \vee \mathscr{G}, \sim \mathscr{F}, \mathscr{F} \Rightarrow \mathscr{G}, \quad \text{and} \quad \mathscr{F} \Leftrightarrow \mathscr{G}$$
>
> are **formulas**. Only those expressions that can be obtained in these ways are **formulas**.

In building up formulas by the use of this definition, it may be necessary to enclose previously constructed formulas in parentheses. Formulas are also called **compound** propositions or statements.

1.9 Example   Show that the expression $(P \vee Q) \Leftrightarrow (Q \Rightarrow (\sim R))$ is a formula.

SOLUTION   The propositional variables P and Q are formulas; hence, $P \vee Q$ is a formula. The propositional variable R is a formula; so $\sim R$ is a formula. Because Q and $\sim R$ are formulas, it follows that $Q \Rightarrow (\sim R)$ is a formula. Finally, because $P \vee Q$ and $Q \Rightarrow (\sim R)$ are formulas, $(P \vee Q) \Leftrightarrow (Q \Rightarrow (\sim R))$ is a formula. Notice that we enclosed the previously constructed formulas $\sim R$, $P \vee Q$, and $Q \Rightarrow (\sim R)$ in parentheses before forming the final compound proposition.   □

Evidently:

> *Every formula of propositional calculus is a proposition whose truth value depends in a definite way on the truth values of the propositional variables from which it is formed.*

The dependence of the truth value of a formula on its propositional variables can be shown by means of a truth table.

1.10 Example   Construct the truth table for the formula

$$(P \lor Q) \Leftrightarrow (Q \Rightarrow (\sim R)).$$

SOLUTION   Because there are three propositional variables, P, Q, and R, involved in this formula, we need a truth table showing all of the $2^3 = 8$ truth combinations for these variables:

P	Q	R	$(P \lor Q) \Leftrightarrow (Q \Rightarrow (\sim R))$
1	1	1	1 1 1 **0** 1 0 0
0	1	1	0 1 1 **0** 1 0 0
1	0	1	1 1 0 **1** 0 1 0
0	0	1	0 0 0 **0** 0 1 0
1	1	0	1 1 1 **1** 1 1 1
0	1	0	0 1 1 **1** 1 1 1
1	0	0	1 1 0 **1** 0 1 1
0	0	0	0 0 0 **0** 0 1 1

In forming this truth table, we work from the "inside out." For instance, in the first row, we begin with $P \lor Q$, noting that when P and Q both have truth value 1, $P \lor Q$ has truth value 1. Thus, in the first row, we place a 1 under the symbol $\lor$ in the middle of $P \lor Q$. Likewise, in the first row, R has truth value 1, so $\sim R$ has truth value 0; and since Q has truth value 1, the implication $Q \Rightarrow (\sim R)$ has truth value 0, which we indicate by placing a 0 under the symbol $\Rightarrow$ in the middle of $Q \Rightarrow (\sim R)$. Finally, since the propositions $P \lor Q$ and $Q \Rightarrow \sim R$ on both sides of the equivalence $\Leftrightarrow$ have opposite truth values (0 and 1), we place a 0 under the symbol $\Leftrightarrow$ in the first row. The 0 under the symbol $\Leftrightarrow$ denotes the truth value (false) of the compound statement $(P \lor Q) \Leftrightarrow (Q \Rightarrow (\sim R))$ for the case in which P, Q, and R are all true. Continuing in this way, we work out the rest of the truth table as shown.   □

1.11 DEFINITION

> **TAUTOLOGY**   Let $\mathcal{T}$ be a formula built up from the propositional variables $P, Q, \ldots, R$, and suppose that $\mathcal{T}$ is always true for every possible truth combination for these variables. Then we call $\mathcal{T}$ a **tautology**.

Tautologies are the theorems of propositional calculus. It is an easy matter to decide whether a formula of propositional calculus is a tautology: Just construct its truth table and look to see if an unbroken column of 1's appears beneath the formula.

1.12 Example   Show that $P \vee (\sim P)$ is a tautology.

SOLUTION   The truth table

P	$P \vee (\sim P)$
1	1 **1** 0
0	0 **1** 1

shows that $P \vee (\sim P)$ is a tautology.   □

The formula $P \vee (\sim P)$, which is perhaps the simplest of all tautologies, is called the **law of the excluded middle** because it states *P or not P*, that is, P is either true or false—there is no "middle" possibility. The formula

$$((\sim P) \vee Q) \Leftrightarrow (P \Rightarrow Q)$$

is a somewhat less trivial example of a tautology.

1.13 Example   Show that $((\sim P) \vee Q) \Leftrightarrow (P \Rightarrow Q)$ is a tautology.

SOLUTION

P	Q	$((\sim P) \vee Q) \Leftrightarrow (P \Rightarrow Q)$
1	1	0 1 1 **1** 1 1 1
0	1	1 1 1 **1** 0 1 1
1	0	0 0 0 **1** 1 0 0
0	0	1 1 0 **1** 0 1 0

□

Here is a list of some of the more useful tautologies:

1. $P \Leftrightarrow \sim(\sim P)$ Double denial
2. $(P \wedge Q) \Leftrightarrow (Q \wedge P)$ Commutative law for conjunction
3. $(P \vee Q) \Leftrightarrow (Q \vee P)$ Commutative law for disjunction
4. $(P \wedge Q) \wedge R \Leftrightarrow P \wedge (Q \wedge R)$ Associative law for conjunction
5. $(P \vee Q) \vee R \Leftrightarrow P \vee (Q \vee R)$ Associative law for disjunction
6. $P \vee (\sim P)$ Law of the excluded middle
7. $\sim(P \wedge (\sim P))$ Law of noncontradiction
8. $P \wedge (Q \vee R) \Leftrightarrow (P \wedge Q) \vee (P \wedge R)$ Distributive law for $\wedge$ over $\vee$
9. $P \vee (Q \wedge R) \Leftrightarrow (P \vee Q) \wedge (P \vee R)$ Distributive law for $\vee$ over $\wedge$
10. $\sim(P \wedge Q) \Leftrightarrow (\sim P) \vee (\sim Q)$ De Morgan's law for $\wedge$
11. $\sim(P \vee Q) \Leftrightarrow (\sim P) \wedge (\sim Q)$ De Morgan's law for $\vee$
12. $(P \Rightarrow Q) \Leftrightarrow ((\sim Q) \Rightarrow (\sim P))$ Law of contraposition
13. $P \Leftrightarrow P$ Reflexivity of equivalence
14. $(P \Leftrightarrow Q) \Leftrightarrow (Q \Leftrightarrow P)$ Symmetry of equivalence
15. $(P \Leftrightarrow Q) \wedge (Q \Leftrightarrow R) \Rightarrow (P \Leftrightarrow R)$ Transitivity of equivalence
16. $(P \Rightarrow Q) \wedge (Q \Rightarrow R) \Rightarrow (P \Rightarrow R)$ Transitivity of implication
17. $((\sim P) \Rightarrow (Q \wedge (\sim Q))) \Rightarrow P$ Reductio ad absurdum
18. $(P \wedge Q) \Rightarrow P$ Simplification
19. $P \Rightarrow (Q \Rightarrow (P \wedge Q))$ Conjunction
20. $P \Rightarrow (P \vee Q)$ Disjunction

21. $(P \Rightarrow Q) \Leftrightarrow ((\sim P) \vee Q)$

22. $(P \Leftrightarrow Q) \Leftrightarrow ((P \Rightarrow Q) \wedge (Q \Rightarrow P))$

23. $Q \Rightarrow (P \Rightarrow Q)$

24. $(\sim P) \Rightarrow (P \Rightarrow Q)$

25. $(P \Rightarrow (Q \Rightarrow R)) \Leftrightarrow ((P \wedge Q) \Rightarrow R)$

26. $((P \Rightarrow R) \wedge (Q \Rightarrow R)) \Leftrightarrow ((P \vee Q) \Rightarrow R)$

27. $((P \wedge Q) \Rightarrow R) \Leftrightarrow ((P \wedge (\sim R)) \Rightarrow (\sim Q))$

28. $(P \Leftrightarrow Q) \Leftrightarrow ((\sim P) \Leftrightarrow (\sim Q))$

29. $(P \wedge (Q \Rightarrow R)) \Rightarrow ((P \wedge Q) \Rightarrow (P \wedge R))$

30. $(P \Rightarrow Q) \vee (Q \Rightarrow P)$

If $\mathscr{T}$ is a tautology, then $\sim \mathscr{T}$ is a formula whose truth value is always 0 (false), no matter what truth combination is assigned to its propositional variables. Such a formula is called an **antitautology**. The simplest example of an antitautology is the formula

$$P \wedge (\sim P)$$

which says, in words, that P is both true and at the same time false—an obvious impossibility. A compound statement of the form

$$P \wedge (\sim P)$$

is called a **contradiction**. Note that tautology 17 above says, in words, that *if the denial of P leads to a contradiction, then P must be true.* This is an important tautology, for it supplies the basis for the indirect proofs to be discussed in Section 1.3.

In the following definition, we set forth some terminology often used in connection with propositions of the form $P \Rightarrow Q$.

1.14 DEFINITION

HYPOTHESIS, CONCLUSION, CONVERSE, CONTRAPOSITIVE In an implication of the form

$$P \Rightarrow Q,$$

the proposition P is called the **hypothesis** or the **antecedent**, and the proposition Q is called the **conclusion** or the **consequent**. By interchanging the hypothesis and conclusion, we obtain a new implication

$$Q \Rightarrow P$$

called the **converse** of the original implication. By negating both the hypothesis and the conclusion in this converse implication, we obtain yet another implication

$$(\sim Q) \Rightarrow (\sim P)$$

called the **contrapositive** of the original implication.

According to the truth table

P	Q	$P \Rightarrow Q$
1	1	1
0	1	1
1	0	0
0	0	1

we have the following:

> *If the hypothesis of an implication is false, or if its conclusion is true, then the implication is true. Indeed, an implication is false only in the case in which its hypothesis is true and its conclusion is false.*

As a consequence of tautology 12 in the list above:

> *An implication is always logically equivalent to its own contrapositive implication.*

For instance, if P is the proposition *today is Thanksgiving* and Q is the proposition *tomorrow is Friday*, then the implication

$$P \Rightarrow Q$$

is certainly true. Indeed, if today is Thanksgiving, then tomorrow is Friday. The contrapositive implication

$$(\sim Q) \Rightarrow (\sim P)$$

is also true, because, if tomorrow is not Friday, then today cannot possibly be Thanksgiving. However, the converse of $P \Rightarrow Q$, the implication

$$Q \Rightarrow P,$$

claims that, if tomorrow is Friday, then today is Thanksgiving! Because not every Thursday is Thanksgiving (a pity, isn't it?), the converse implication $Q \Rightarrow P$ is false.

> *From the fact that an implication $P \Rightarrow Q$ is true, we can draw no conclusion about the truth or falsity of its converse implication $Q \Rightarrow P$.*

When a mathematician proves a theorem having the form $P \Rightarrow Q$, the question of whether the converse $Q \Rightarrow P$ is also true naturally presents itself. If the converse implication also proves to be true, then tautology 22 can be used to conclude that $P \Leftrightarrow Q$; that is, P and Q are logically equivalent (have the same truth value).

Advances in our mathematical knowledge often begin with educated guesses or conjectures concerning the truth value of various mathematical propositions. Suppose that a mathematician is somehow led to conjecture that a certain implication $P \Rightarrow Q$ is true. Such a conjecture could be defeated by finding a particular example for which the hypothesis P is true and the conclusion Q is false. An example used to show that a guess or conjecture is false is called a **counterexample**.

1.15 Example   Let x, y, and z represent arbitrary real numbers. Show by means of a counterexample that the implication

$$xy = xz \Rightarrow y = z$$

is false.

SOLUTION   At first glance, this implication looks plausible enough: It seems that the conclusion $y = z$ follows from the hypothesis $xy = xz$ if we just "cancel the x on both sides." However, consider the case in which $x = 0$, $y = 1$, and $z = 2$. In this case, the hypothesis $xy = xz$ is true (it just says that $0 = 0$), whereas the conclusion $y = z$ is false (it says that $1 = 2$).   □

In working with mathematical statements, it is often necessary to find a formula that is equivalent to the denial of a given formula, but in which no denial symbol appears in front of a compound proposition. For instance, by De Morgan's law for $\wedge$ (tautology 10), the denial of the formula $P \wedge Q$ is equivalent to the formula $(\sim P) \vee (\sim Q)$. Since every formula is built up from our five basic connectives $\wedge$, $\vee$, $\Rightarrow$, $\Leftrightarrow$, and $\sim$, it suffices to know how to deny each of the formulas $P \wedge Q$, $P \vee Q$, $P \Rightarrow Q$, $P \Leftrightarrow Q$, and $\sim P$. The following tautologies show how this is done:

(i) $\sim (P \wedge Q) \Leftrightarrow (\sim P) \vee (\sim Q)$
(ii) $\sim (P \vee Q) \Leftrightarrow (\sim P) \wedge (\sim Q)$
(iii) $\sim (P \Rightarrow Q) \Leftrightarrow P \wedge (\sim Q)$
(iv) $\sim (P \Leftrightarrow Q) \Leftrightarrow (P \wedge (\sim Q)) \vee (Q \wedge (\sim P))$
(v) $\sim (\sim P) \Leftrightarrow P$

Tautologies (i) and (ii) are De Morgan's laws (tautologies 10 and 11), and (v) is the double denial tautology (tautology 1). We leave it to you to verify (iii) and (iv) by means of truth tables (Problem 36).

1.16 Example   Write out, in English, the denial of the following statement:

If Alfie is taking chemistry, then he is smart.

SOLUTION   Let P be the statement *Alfie is taking chemistry*, and let Q be the statement *Alfie is smart*. The statement to be denied has the form $P \Rightarrow Q$; hence, by (iii) above, its denial is equivalent to $P \wedge (\sim Q)$, that is:

Alfie is taking chemistry and he is not smart.

Note that in translating from English to the symbolism of propositional calculus and vice versa, we have freely interchanged the pronoun *he* and the noun *Alfie* to which it refers. □

You will find that the following "dictionary" is handy for translating certain English expressions into the formulas of propositional calculus.

English expression	Translation
1. P or Q or both	$P \vee Q$
2. P and Q	$P \wedge Q$
3. P, but Q	$P \wedge Q$
4. P, however Q	$P \wedge Q$
5. P is not true.	$\sim P$
6. P implies Q.	$P \Rightarrow Q$
7. If P, then Q	$P \Rightarrow Q$
8. P is a sufficient condition for Q.	$P \Rightarrow Q$
9. Q if P	$P \Rightarrow Q$
10. P is a necessary condition for Q.	$Q \Rightarrow P$
11. Q only if P	$Q \Rightarrow P$
12. P is implied by Q.	$Q \Rightarrow P$
13. P if and only if Q	$P \Leftrightarrow Q$
14. P is a necessary and sufficient condition for Q.	$P \Leftrightarrow Q$
15. P is equivalent to Q.	$P \Leftrightarrow Q$
16. Either P or Q, but not both	$\sim (P \Leftrightarrow Q)$
17. P, but not Q	$P \wedge (\sim Q)$
18. P unless Q	$(\sim Q) \Rightarrow P$

PROBLEM SET 1.1

1. Which of the following expressions are propositions? [Note: Our definition of *proposition* is not sufficiently precise to allow for clear-cut answers in every case, so you might want to label some cases as indeterminate.]
    (a) If only I could make an A in this course.
    (b) Statement (b) in the present list is false.
    (c) Somewhere in the decimal expansion of pi the sequence of digits 123456789 occurs.
    (d) If p is a prime number, then $2^p - 1$ is also a prime number.
    (e) $57x^2 + \sin x - x^{-1}$.

    (f) The present king of France is bald.
    (g) How many solutions does a quadratic equation have?
    (h) $57x^2 + \sin x - x^{-1} = 243$.
    (i) If pigs have wings, then some winged animals are good to eat.
    (j) Every boojum is a snark.
    (k) It is raining.
    (l) If a function has a derivative, then it is continuous.
    (m) All squares are rectangles.
    (n) All rectangles are squares.
    (o) Let $x < 2$.

2. List in tabular form all of the $2^4 = 16$ possible truth combinations for the four propositional variables P, Q, R, and S.

3. Suppose P, Q, and R are propositions with $P \Leftrightarrow Q$ and $Q \Leftrightarrow R$.
    (a) Explain in words why it follows that $P \Leftrightarrow R$.
    (b) Which tautology in the list on pages 10–11 confirms the conclusion in Part (a).

4. Let P be the proposition $4 > 3$, and let Q be the proposition $4 > 2$. Find the truth value of each of the following:
    (a) $P \vee Q$ (b) $P \wedge Q$ (c) $P \Rightarrow Q$ (d) $P \Leftrightarrow Q$ (e) $P \vee (\sim Q)$

5. Suppose that two baseball teams A and B are scheduled to play each other with the understanding that extra innings will be used, if necessary, to break a tie. Let P be the proposition *team A will win the game*, and let Q be the proposition *team B will win the game*. Find the truth value of each of the following:
    (a) $P \vee Q$ (b) $P \wedge Q$ (c) $P \Leftrightarrow Q$
    (d) $(\sim P) \Rightarrow Q$ (e) $P \Rightarrow (\sim Q)$ (f) $(\sim P) \Leftrightarrow (\sim Q)$

6. Professor Grumbles promises Alfie that, if he makes a C on the final exam in calculus, then he will pass the course. Let P be the statement *Alfie passes the course*, and let Q be the statement *Alfie makes a C on the final exam in calculus*.
    (a) Express Professor Grumbles' promise in symbolic form as an implication.
    (b) For each of the possible truth combinations for P and Q, determine whether or not Professor Grumbles' promise is kept.

7. Let P be the proposition *interest rates are high*, and let Q be the proposition *the inflation rate is low*. Interpret the word *low* to mean *not high*. Write each of the following propositions in symbolic form:
    (a) Interest rates are high, and the inflation rate is high.
    (b) Interest rates are low, and the inflation rate is low.
    (c) If the inflation rate is low, then interest rates are low.
    (d) If the inflation rate is high, then interest rates are high.
    (e) Neither the inflation rate nor the interest rate is high.

8. If P and Q are propositions, let $P \dashv Q$ denote the proposition asserting that statement Q can be deduced from statement P. Is $\dashv$ a propositional connective? Why or why not?

9. Using De Morgan's laws, write out (in English) the denial of each of the following statements:
    (a) This equation has a solution, or I am stupid.
    (b) Alfie is studying hard and doing well in his classes.

10. Write out (in English) the denial of the statement *if it is not raining, then it is snowing.*

11. Let P, Q, and R denote the statements *it is snowing, I am cold,* and *I am going home,* respectively. Write out in English the propositions represented symbolically by the following:
    (a) $\sim Q$ (b) $Q \Rightarrow (P \wedge (\sim R))$ (c) $P \wedge (\sim R)$
    (d) $Q \Leftrightarrow R$ (e) $R \Rightarrow Q$ (f) $\sim (Q \Leftrightarrow R)$
    (g) $(P \vee Q) \Rightarrow R$ (h) $\sim (R \Rightarrow Q)$

12. Complete the tautology $(P \Rightarrow Q) \Leftrightarrow ?$ by replacing the question mark with a formula involving only P, Q, and the connectives $\sim$ and $\wedge$.

13. Continuing with the notation of Problem 11, translate each of the following statements into the symbolism of the propositional calculus:
    (a) If it is snowing, then I am going home.
    (b) If it is snowing and I am cold, then I am going home.
    (c) I am not going home unless I am cold.
    (d) A necessary condition that I go home is that I am cold.
    (e) If I am going home, then it is snowing or I am cold.
    (f) It is either snowing or I am cold, but not both.

14. Complete the tautology $(P \Leftrightarrow Q) \Leftrightarrow ?$ by replacing the question mark with a formula involving only P, Q, and the connectives $\sim$ and $\wedge$.

15. Complete the following truth tables:

P	Q	R	$P \vee R$	$Q \vee R$	$P \wedge R$	$Q \wedge R$	$P \wedge (Q \wedge R)$	$P \wedge (Q \vee R)$	$P \vee (Q \wedge R)$
1	1	1							
0	1	1							
1	0	1							
0	0	1							
1	1	0							
0	1	0							
1	0	0							
0	0	0							

16. Show that the eight truth combinations for the three propositional variables P, Q, and R correspond in a natural way to the representations of the numbers 0, 1, 2, 3, 4, 5, 6, and 7 in the binary (base 2) system.

17. Complete the truth table:

P	Q	R	$(P \wedge ((\sim Q) \vee R)) \vee Q$
1	1	1	
0	1	1	
1	0	1	
0	0	1	
1	1	0	
0	1	0	
1	0	0	
0	0	0	

18. Construct a truth table for the formula $(P \Rightarrow Q) \Rightarrow (Q \Rightarrow P)$.

19. Check tautology 1 (double denial) in the list on pages 10–11 by a truth table.

20. Professor Grumbles says, "I can't afford not to do without a new car."
(a) Can he afford the new car?
(b) Justify your reasoning in Part (a) by forming a suitable tautology.

21. Check the commutative and associative laws (tautologies 2, 3, 4, and 5) by truth tables.

22. Check each of the following tautologies by truth tables:
(a) The law of noncontradiction (tautology 7).
(b) The reductio ad absurdum tautology (tautology 17).

23. Check the distributive laws (tautologies 8 and 9) by truth tables.

24. Check tautologies 13 and 14 by truth tables.

25. Check De Morgan's laws (tautologies 10 and 11) by truth tables.

26. Check tautologies 15 and 16 by truth tables.

27. Check the law of contraposition (tautology 12) by a truth table.

28. Check tautologies 23 and 24 by truth tables, and give verbal explanations of their meanings.

29. Check tautologies 21 and 22 by truth tables.

30. Check tautologies 25 through 30 by truth tables.

31. Check the simplification, conjunction, and disjunction tautologies (tautologies 18, 19, and 20) by truth tables.

32. Make up a tautology that is not in the list on pages 10–11 and check it by a truth table.

33. The tautology $(P \wedge (P \Rightarrow Q)) \Rightarrow Q$ is called *modus ponens*. Check this tautology by a truth table.

34. (a) Without using a truth table, explain why you would expect the formula $((\sim P) \wedge (P \vee Q)) \Rightarrow Q$ to be a tautology.
(b) Check that this formula is a tautology by using a truth table.

35. Write the contrapositive of each implication:
(a) If n is an integer, then $2n + 1$ is an odd integer.
(b) You can take Math 212 only if you passed Math 211.
(c) If the series $\sum_{k=1}^{\infty} a_k$ converges, then $\lim_{n \to \infty} a_n = 0$.
(d) If a citizen is of age 18 or over, then he or she can vote.

36. Check tautologies (iii) and (iv) on page 13 by truth tables.

37. Show by means of a counterexample that the following statement is false:
If p is a prime number, then p is odd.

38. Show by means of a counterexample that the following statement is false:
If a series converges, then it converges absolutely.

39. True or false: If $x^2 > 9$, then $x > 3$?

40. Give truth tables for each of the 16 possible binary connectives, calling them $c_1, c_2, c_3, \ldots, c_{16}$. For each of these binary connectives $c_j, j = 1, 2, 3, \ldots, 16$, find a formula $\mathscr{F}_j$ involving only the propositional variables P and Q such that $P\, c_j\, Q$ is equivalent to $\mathscr{F}_j$.

41. Let $P_1, P_2, P_3, \ldots, P_n$ represent n different propositional variables. Using mathematical induction (see the Appendix on Mathematical Induction), prove that there are exactly 2^n different truth combinations for these n propositional variables.

42. Explain why any formula $\mathscr{F}$ of the propositional calculus is equivalent to a formula $\mathscr{G}$ of the propositional calculus in which $\mathscr{G}$ involves the same propositional variables as $\mathscr{F}$ but only the propositional connectives $\sim$ and $\vee$. [Hint: Note that $P \wedge Q$ is equivalent to $\sim((\sim P) \vee (\sim Q))$, $P \Rightarrow Q$ is equivalent to $(\sim P) \vee Q$, and so on.]

43. Using mathematical induction (see the Appendix on Mathematical Induction), give a careful proof showing that there are only 16 different possible binary connectives.

44. Let $P_1, P_2, P_3, \ldots, P_n$ represent n different propositional variables. Show that a truth table exhibiting all of the 2^n possible truth combinations for these variables can be constructed as follows: Head n columns with the symbols $P_1, P_2, P_3, \ldots,$ and P_n. Alternate 1's and 0's in the first column (under P_1) until 2^n rows are used. Alternate 1's and 0's in pairs in the second column (under P_2) until 2^n rows are used. Continue in this way, alternating 1's and 0's in the jth column in sets of 2^{j-1} for $j = 1, 2, 3, \ldots, n$.

2

QUANTIFIERS

Phrases of the form *for every x* and *there exists an x such that* are called **quantifiers**. The expression

for every x

is the **universal quantifier**, and the expression

there exists an x such that

is the **existential quantifier**. Although these quantifiers are indispensable in advanced mathematics, they are not often used in elementary algebra and geometry. Why? One reason is that in elementary textbooks the universal quantifier is often suppressed in the interest of brevity. For instance, if you read

$$(x + 1)^2 = x^2 + 2x + 1$$

in an elementary algebra book, you are supposed to understand that this equation holds for every value of x. Another reason is that theorems asserting the existence of something (*existence theorems*) are usually too difficult to be included in elementary mathematics courses. For instance, in elementary algebra, you learned how to handle square roots of positive numbers, but there probably was little, if any, discussion of whether these square roots actually exist.

In more advanced mathematics courses, the suppression of universal quantifiers can lead to hopeless confusion. Also, some of the more important theorems of advanced mathematics are existence theorems. Thus, to progress from elementary to more advanced mathematics, it is necessary to learn to understand and deal with quantifiers.

In any mathematical discussion, we usually have in mind some particular "universe" or **domain of discourse** that consists of those things whose properties and relationships are under consideration. For instance, in elementary algebra our domain of discourse is the set of all real numbers; in elementary geometry it is the set of all points in the plane.

In what follows, U represents some particular domain of discourse. Assuming that there is at least one object in U, we introduce symbols

$$x, y, z, \ldots$$

to stand for arbitrary objects in U, and we call these symbols **object variables**. For instance, in elementary algebra, U is the set of all real numbers, and the object variables $x, y, z, \ldots$ are just the usual numerical variables or "unknowns."

2.1 DEFINITION

PREDICATE OR PROPOSITIONAL FUNCTION   By a **predicate** or **propositional function**, we mean a complete declarative sentence $P(x)$ that makes a statement about the object variable x. We call x the **argument** of $P(x)$. Thus, if x is assigned a particular value in U, then $P(x)$ becomes a proposition with a definite truth value.

2.2 Example Give several examples of propositional functions in elementary algebra.

SOLUTION We take the domain of discourse U to be the set of all real numbers. Here are some examples of propositional functions:

(a) Let $P(x)$ be the statement $x > 0$.
(b) Let $Q(x)$ be the statement $x^2 - 5x + 6 = 0$.
(c) Let $W(x)$ be the statement x *is a whole number.*

In this example, for instance, $P(3)$ is true, $P(-2)$ is false, $Q(-1)$ is false, $Q(2)$ is true, $W(7)$ is true, and $W(\pi)$ is false.   □

We are now ready to introduce some useful symbolism for universal and existential quantifiers.

2.3 DEFINITION

> **UNIVERSAL QUANTIFIER**   Let $P(x)$ be a propositional function whose argument x ranges over the domain of discourse U. Then the expression
>
> $(\forall x)(P(x))$
>
> stands for the proposition asserting that, no matter what value (in U) is assigned to x, the resulting proposition $P(x)$ is true.

The proposition $(\forall x)(P(x))$ is read in English as

for all x, P(x)

or

for every x, P(x),

and the symbol $\forall x$ denotes the universal quantifier. We note that if y is an object in U, then

> $(\forall x)(P(x)) \Rightarrow P(y).$

2.4 Example If U is the set of all real numbers, determine the truth value of each of the following quantified propositions:

(a) $(\forall x)((x + 1)^2 = x^2 + 2x + 1)$
(b) $(\forall x)(x > 2)$

SOLUTION Proposition (a) is true because, indeed, $(x + 1)^2 = x^2 + 2x + 1$ is a true statement, no matter what real number is put in place of x. However, (b) is false because not every real number is greater than 2; for instance, if the value 1 is assigned to x, the resulting statement $1 > 2$ is false.   □

2.5 DEFINITION

> **EXISTENTIAL QUANTIFIER**   Let $P(x)$ be a propositional function whose argument x ranges over the domain of discourse U. Then the expression
>
> $(\exists x)(P(x))$
>
> will stand for the proposition asserting that there exists at least one value of x (in U) for which the resulting proposition $P(x)$ is true.

The proposition $(\exists x)(P(x))$ is read in English as

there exists an x such that P(x)

or

for some x, P(x),

and the symbol $\exists x$ denotes the existential quantifier. We note that if y is an object in U, then

$$P(y) \Rightarrow (\exists x)(P(x)).$$

2.6 Example   If U is the set of all real numbers, determine the truth value of each of the following quantified propositions:

(a) $(\exists x)(x^2 - 5x + 6 = 0)$
(b) $(\exists x)(x^2 + 2x + 6 = 0)$

SOLUTION   Proposition (a) is true because there does exist at least one real number x (for instance, $x = 2$ or $x = 3$) such that $x^2 - 5x + 6 = 0$. However, (b) is false because the discriminant of the quadratic equation $x^2 + 2x + 6 = 0$ is negative; so this equation has no real solution.   □

If $P(x)$ is a propositional function, then $(\forall x)(P(x))$ and $(\exists x)(P(x))$ are actually propositions, *not propositional functions*. Thus, even though the expressions $(\forall x)(P(x))$ and $(\exists x)(P(x))$ contain the object variable x, their truth value does not depend in any way on this variable, but only on the propositional function $P(x)$ and the domain of discourse U. For instance, if U is the set of all real numbers, then

$(\exists x)(x > 2)$

merely says that there is a real number greater than 2; hence, it is a true proposition having to do only with real numbers and the relation of being greater than. It really has nothing to do with x at all! For this reason, we say that the symbol x that occurs in the quantified expressions $(\forall x)(P(x))$ and $(\exists x)(P(x))$ is a **dummy variable** or an **apparent variable** because it only *appears* (to the uninformed) that these propositions depend on x in some way. Logicians describe this by saying that the variable x occurs as a **free variable** in the propositional function $P(x)$, so we can freely assign it any value in U; but it is a **bound variable** in the quantified expressions $(\forall x)(P(x))$ and $(\exists x)(P(x))$ and, as such, is no longer available for the assignment of values.

For ordinary numerical-valued functions $f(x)$, a similar situation occurs in the integral calculus. Indeed, we can assign x any value in the domain of such a function and the result is a real number, for instance $f(-7)$ or $f(1.3)$ or $f(\pi)$. However, the definite integral

$$\int_a^b f(x)\,dx,$$

which appears (again to the uninformed) to depend in some way on x, is actually a number depending only on the function being integrated and the interval $[a, b]$ over which the integration takes place. In the expression for the definite integral, x is a dummy variable which cannot be assigned particular numerical values.

The fact that x is a dummy variable in the definite integral can be expressed by the equation

$$\int_a^b f(x)\,dx = \int_a^b f(y)\,dy.$$

Likewise, the fact that x is a dummy variable in the quantified expressions $(\forall x)(P(x))$ and $(\exists x)(P(x))$ can be expressed by the rules

$$(\forall x)(P(x)) \Leftrightarrow (\forall y)(P(y))$$

and

$$(\exists x)(P(x)) \Leftrightarrow (\exists y)(P(y)).$$

If $P(x)$ is a propositional function for the domain of discourse U, then the proposition

$$(\forall x)(P(x))$$

asserts that $P(x)$ is true no matter what value (in U) is assigned to x; hence, its denial

$$\sim(\forall x)(P(x))$$

is equivalent to the statement that there exists at least one value of x (in U) for which $P(x)$ is false. Therefore, we have the important rule

$$\sim(\forall x)(P(x)) \Leftrightarrow (\exists x)(\sim P(x)).$$

Similarly, the proposition

$$(\exists x)(P(x))$$

asserts that $P(x)$ is true for at least one value of x (in U); hence, its denial

$$\sim(\exists x)(P(x))$$

is equivalent to the statement that $P(x)$ is false for all values of x (in U). Therefore, we also have the rule

$$\sim(\exists x)(P(x)) \Leftrightarrow (\forall x)(\sim P(x)).$$

These two rules for forming the denial of a quantified statement can be summarized as follows:

> *The denial symbol $\sim$ can be "pushed through" a quantifier at the expense of changing the quantifier from universal to existential and vice versa.*

A person untrained in logic who is asked to deny the proposition *all men are mortal* might respond with *all men are not mortal*. If interpreted literally, the last statement asserts that *all men are immortal*, which is not the denial of the original proposition.

2.7 Example   Write the denial of the proposition *all men are mortal.*

SOLUTION   Let U, the domain of discourse, be the set of all men, and let $M(x)$ be the propositional function asserting that x is mortal. Thus, the proposition *all men are mortal* is expressed by the quantified expression

$$(\forall x)(M(x)),$$

and its denial

$$\sim(\forall x)(M(x))$$

is equivalent to the proposition

$$(\exists x)(\sim M(x)),$$

which translates into the statement *there exists at least one man who is immortal.*   □

The formal study of propositional functions and quantifiers is called **predicate calculus** or **functional calculus**. Predicate (or functional) calculus has roughly the same relationship to propositional calculus as the theory of numerical functions has to ordinary algebra. In multivariate calculus, we consider functions $f(x, y)$, $g(x, y, z)$, and so on, which have more than one independent variable. Likewise, in predicate calculus, we have to deal with propositional functions $P(x, y)$, $Q(x, y, z)$, and so on, which involve more than one object variable. Here are some examples of such propositional functions of more than one variable:

Domain of discourse	Number of variables	Propositional function
1. $U =$ all living humans	2	x is older than y
2. $U =$ all real numbers	2	$x^2 + y^2 = 1$
3. $U =$ all real numbers	2	$x^2 + y^2 \le 1$
4. $U =$ all integers	2	x is an exact divisor of y
5. $U =$ all real numbers	3	$x - \sin y > z$
6. $U =$ all points in the plane	3	x is on the line segment between y and z
7. $U =$ all real numbers	3	$(0 < |x - 3| < \varepsilon) \Rightarrow (|x^2 - 9| < \delta)$
8. $U =$ all real numbers	n	$x_1 > x_2 > x_3 > \cdots > x_n$

If a quantifier is applied to a propositional function of several variables, the result is a propositional function of one fewer variable.

2.8 Example   Let the domain of discourse U consist of all real numbers, and consider the propositional function $x < y$ of the two variables x and y. Quantify this expression:

(a) Universally with respect to y
(b) Existentially with respect to x

Interpret the results.

SOLUTION   **(a)** If we quantify $x < y$ universally with respect to y, we obtain the expression

$$(\forall y)(x < y),$$

which is a propositional function of the single variable x (since y has become an apparent variable). The propositional function $(\forall y)(x < y)$ states that x is less than every number in U.

(b) If we quantify $x < y$ existentially with respect to x, we obtain the expression

$$(\exists x)(x < y),$$

which is a propositional function of the single variable y (since x has become an apparent variable). The propositional function $(\exists x)(x < y)$ states that there exists at least one number that is less than y. □

After a propositional function of several variables has been quantified with respect to one of these variables, the resulting expression can then be quantified with respect to any one of the remaining free variables.

2.9 Example Consider the propositional functions $(\forall y)(x < y)$ and $(\exists x)(x < y)$ obtained in Example 2.8.

(a) Quantify $(\forall y)(x < y)$ existentially with respect to x.
(b) Quantify $(\exists x)(x < y)$ universally with respect to y.

Interpret the results.

SOLUTION **(a)** Quantifying $(\forall y)(x < y)$ existentially with respect to x, we obtain

$$(\exists x)(\forall y)(x < y),$$

which states that there exists a real number x that is less than every real number. Thus, $(\exists x)(\forall y)(x < y)$ is a false statement.
(b) Quantifying $(\exists x)(x < y)$ universally with respect to y, we obtain

$$(\forall y)(\exists x)(x < y),$$

which states that for every real number y there is at least one smaller real number x. Thus, $(\forall y)(\exists x)(x < y)$ is a true statement. □

2.10 Example Rewrite the denial of the proposition

$$(\exists x)(\forall y)(x < y)$$

in such a way that no denial connective appears explicitly, and interpret the result.

SOLUTION Starting with

$$\sim(\exists x)(\forall y)(x < y)$$

we "push the denial connective $\sim$ through the quantifiers," changing existential to universal and universal to existential, to obtain the equivalent proposition

$$(\forall x)(\exists y)(\sim(x < y)).$$

The statement $\sim(x < y)$ says that the real number x is not less than the real number y; hence, it is equivalent to the statement that $x \geq y$. Therefore, the denial of the proposition

$$(\exists x)(\forall y)(x < y)$$

is equivalent to the proposition

$$(\forall x)(\exists y)(x \geq y).$$

In words: To deny that there is a real number x that is smaller than every real number y is to assert that, for every real number x, there is a real number y that is less than or equal to x.   □

Example 2.9 shows that you must pay attention to the *order* in which quantifiers are written! Although the propositions

$$(\exists x)(\forall y)(x < y)$$

and

$$(\forall y)(\exists x)(x < y)$$

differ only in the order in which the quantifiers appear, they are *not equivalent*.

More generally, suppose that $P(x, y)$ is a propositional function of two variables, and consider the two propositions $(\exists x)(\forall y)(P(x, y))$ and $(\forall y)(\exists x)(P(x, y))$, which differ only in the order in which the quantifiers are written. On the one hand:

> The proposition $(\exists x)(\forall y)(P(x, y))$ states that there exists a fixed x in U such that $P(x, y)$ is true for every choice of y in U.

On the other hand:

> The proposition $(\forall y)(\exists x)(P(x, y))$ states that for every choice of y in U, there exists some x (possibly depending on the choice of y), such that $P(x, y)$ is true.

Consequently, we have the rule

> $$(\exists x)(\forall y)(P(x, y)) \Rightarrow (\forall y)(\exists x)(P(x, y))$$

but, in general, the converse implication is false.

It turns out that:

> Two adjacent quantifiers of the same type (universal or existential) can always be transposed.

Thus, we have the rules

$$(\forall x)(\forall y)(P(x, y)) \Leftrightarrow (\forall y)(\forall x)(P(x, y))$$

and

$$(\exists x)(\exists y)(P(x, y)) \Leftrightarrow (\exists y)(\exists x)(P(x, y))$$

For instance, if U consists of the real numbers, the proposition

$$(\forall x)(\forall y)((x + y)^2 = x^2 + 2xy + y^2)$$

asserts that the equation

$$(x + y)^2 = x^2 + 2xy + y^2$$

is true for all choices of the real numbers x and y, and this is exactly the same as the assertion made by the proposition

$$(\forall y)(\forall x)((x + y)^2 = x^2 + 2xy + y^2).$$

Similarly, the proposition

$$(\exists x)(\exists y)(x < y)$$

asserts the existence of a pair of real numbers, one of which is less than the other, and this is exactly the same as the assertion made by the proposition

$$(\exists y)(\exists x)(x < y).$$

It often happens that some, but not all, of the objects in the domain of discourse U have a particularly interesting or desirable property. In this case, it may be convenient to reserve certain symbols to represent objects in U that have this property. For instance, in studying differential and integral calculus, our domain of discourse U is likely to be the set of all real numbers. Here the property of being a *positive* number is often of significance, and the Greek letters ε (epsilon) and δ (delta) are often understood to represent only positive numbers. If this agreement is made, then the symbol $\forall \varepsilon$ would correspond to the phrase *for every positive ε*, and the symbol $\exists \delta$ would correspond to the phrase *there exists a positive δ*.

2.11 Example   Suppose that $f(x)$ is a real-valued function of the real variable x. Write out, in the symbolism of predicate calculus, the definition of

$$\lim_{x \to a} f(x) = L.$$

SOLUTION   Let us use the convention that ε and δ stand for positive real numbers. By definition, the statement

$$\lim_{x \to a} f(x) = L$$

means that, for every positive number ε, there exists a positive number δ such that, if $0 < |x - a| < \delta$, then $|f(x) - L| < \varepsilon$. Thus, by definition,

$$\lim_{x \to a} f(x) = L \Leftrightarrow (\forall \varepsilon)(\exists \delta)((0 < |x - a| < \delta) \Rightarrow (|f(x) - L| < \varepsilon)) \qquad \Box$$

Predicate calculus is an important branch of symbolic logic and can be developed with great rigor and precision. Here we have given only a brief and informal introduction to this subject. Every serious student of mathematics should, at some time, work through the formal theory of the predicate calculus as it is found in any standard textbook of symbolic logic. For future reference, we present the following list of some of the basic theorems of the predicate calculus. In this list, x and y denote object variables ranging over a fixed universe of discourse U; $P(x)$, $Q(x)$, $P(x, y)$, and $Q(x, y)$ denote propositional functions; and R denotes a proposition or a propositional function that does not contain x as a free variable.

1. $(\forall x)(P(x)) \Rightarrow P(y)$

2. $P(y) \Rightarrow (\exists x)(P(x))$

3. $(\forall x)(P(x)) \Rightarrow (\exists x)(P(x))$

4. $(\forall x)(P(x)) \Leftrightarrow (\forall y)(P(y))$

5. $(\exists x)(P(x)) \Leftrightarrow (\exists y)(P(y))$

6. $\sim(\forall x)(P(x)) \Leftrightarrow (\exists x)(\sim P(x))$

7. $\sim(\exists x)(P(x)) \Leftrightarrow (\forall x)(\sim P(x))$

8. $(\exists x)(\forall y)(P(x, y)) \Rightarrow (\forall y)(\exists x)(P(x, y))$

9. $(\forall x)(\forall y)(P(x, y)) \Leftrightarrow (\forall y)(\forall x)(P(x, y))$

10. $(\exists x)(\exists y)(P(x, y)) \Leftrightarrow (\exists y)(\exists x)(P(x, y))$

11. $(\forall x)(P(x) \wedge Q(x)) \Leftrightarrow [(\forall x)(P(x)) \wedge (\forall x)(Q(x))]$

12. $(\exists x)(P(x) \vee Q(x)) \Leftrightarrow [(\exists x)(P(x)) \vee (\exists x)(Q(x))]$

13. $[(\forall x)(P(x)) \vee (\forall x)(Q(x))] \Rightarrow (\forall x)(P(x) \vee Q(x))$

14. $(\exists x)(P(x) \wedge Q(x)) \Rightarrow [(\exists x)(P(x)) \wedge (\exists x)(Q(x))]$

15. $(\forall x)(R \wedge Q(x)) \Leftrightarrow R \wedge (\forall x)(Q(x))$

16. $(\forall x)(R \vee Q(x)) \Leftrightarrow R \vee (\forall x)(Q(x))$

17. $(\exists x)(R \wedge Q(x)) \Leftrightarrow R \wedge (\exists x)(Q(x))$

18. $(\exists x)(R \vee Q(x)) \Leftrightarrow R \vee (\exists x)(Q(x))$

We have stated these theorems without proof, but we do ask you to translate their statements into English and convince yourself of their reasonableness.

2.12 Example   (a) Give a convincing argument for theorem number 14 in the list above.
(b) Show by means of a counterexample that the converse of the implication in theorem number 14 is not a theorem of the predicate calculus.

SOLUTION   (a) The hypothesis $(\exists x)(P(x) \wedge Q(x))$ of theorem number 14 says that there is at least one object x in U such that both $P(x)$ and $Q(x)$ are true. If this is so, then there is an object in U, namely this same x, such that $P(x)$ is true; that is, $(\exists x)(P(x))$. Likewise, $(\exists x)(Q(x))$ follows from the hypotheses; hence, $(\exists x)(P(x)) \wedge (\exists x)(Q(x))$ is a consequence of the hypothesis.
(b) The converse of the implication in theorem number 14 would be the statement

$$[(\exists x)(P(x)) \wedge (\exists x)(Q(x))] \Rightarrow (\exists x)(P(x) \wedge Q(x)). \tag{$*$}$$

Let U, the domain of discourse, consist of all positive integers. Let $P(x)$ be the statement x *is even*, and let $Q(x)$ be the statement x *is odd*. Since there exists an even positive integer, for instance 2, the statement $(\exists x)(P(x))$ is true. Since there exists an odd positive integer, for instance 1, the statement $(\exists x)(Q(x))$ is also true. Thus, the hypothesis $(\exists x)(P(x)) \wedge (\exists x)(Q(x))$ of $(*)$ is true. However, the conclusion $(\exists x)(P(x) \wedge Q(x))$ of $(*)$ is false because it says that there exists a positive integer x that is both even and odd.   □

The propositional function

$$x = y$$

of the two variables x and y has a special role to play in the predicate calculus. This propositional function, called **equality** or **logical identity**, asserts that x and y are *identical*. In symbolic logic, the idea of logical identity is understood as follows:

---

*Two objects a and b in the universe of discourse U are said to be **equal**, and we write*

$$a = b,$$

provided that, for every propositional function $P(x)$,

$$P(a) \Leftrightarrow P(b).$$

In other words, *two objects are equal if and only if anything that can be said about the one object can be said about the other and vice versa.* If $\sim(a = b)$, then we say that a and b are **different** or **distinct** objects in U, and we write

$a \neq b$.

In elementary geometry, two line segments are sometimes said to be "equal" if they have the same length; however, according to the discussion above, this is a misuse of the idea of equality unless the line segments happen to be identical. Even in more advanced mathematics, an occasional abuse of the notion of equality is sometimes tolerated in the interest of avoiding more awkward notation. There is nothing wrong with an occasional abuse of notation, *provided that those who indulge in it know exactly what they are doing and can, upon demand, reformulate their statements with precision!*

The notion of logical identity can be used to help formalize the important idea of *unique existence*. If $P(x)$ is a propositional function, then the expression

$(\exists! x)(P(x))$

is understood to be the proposition asserting that there exists *one and only one* object x in U for which $P(x)$ is true. This symbolism is read in English as

there exists a unique x such that $P(x)$.

The following definition formalizes this idea.

2.13 DEFINITION   **UNIQUE EXISTENCE**   The proposition $(\exists! x)(P(x))$ is true if and only if the following two conditions hold:

(i) $(\exists x)(P(x))$
(ii) $(\forall x)(\forall y)(P(x) \wedge P(y) \Rightarrow x = y)$

According to this definition, if you want to show that

$(\exists! x)(P(x))$

is true, you must do two things: First, you must show that there exists an x in U such that $P(x)$ is true. Second, you must show that there cannot be two different things x and y in U such that both $P(x)$ and $P(y)$ are true.

2.14 Example   Let U be the set of all real numbers. Show that there exists a unique positive real number x such that $x^2 = 4$.

SOLUTION First, because $2^2 = 4$, there exists an x in U such that x is positive and $x^2 = 4$. Second, suppose that x and y are positive numbers such that $x^2 = 4$ and $y^2 = 4$. Then

$$x^2 - y^2 = 0$$

and therefore

$$(x - y)(x + y) = 0.$$

Since x and y are positive, $x + y$ is also positive; hence, $x + y \neq 0$, and it follows that

$$x - y = 0.$$

Therefore, $x = y$. ☐

PROBLEM SET 1.2

In Problems 1–4, let the domain of discourse U consist of all real numbers. Determine the truth value of each proposition.

1. $P(x)$ is the statement $x < 5$.
(a) $P(-1)$ (b) $P(6)$ (c) $P(5)$

2. $Q(x)$ is the statement $x^2 - x - 6 = 0$.
(a) $Q(-3)$ (b) $Q(3)$ (c) $Q(2)$

3. $R(x)$ is the statement x *is a nonnegative integer.*
(a) $R(2)$ (b) $R(0)$ (c) $R(\pi)$

4. $S(x)$ is the statement $|2x + 7| \geq 11$.
(a) $S(-9)$ (b) $S(3)$ (c) $S(0)$

In Problems 5 and 6, let the domain of discourse U be the set of all real numbers. Determine the truth value of each quantified proposition.

5. (a) $(\forall x)((x + 2)^2 = x^2 + 4x + 4)$ (b) $(\forall y)(|7 + 2y| \geq 7)$
(c) $(\exists x)(2x^2 - x - 1 = 0)$ (d) $(\exists y)(2y^2 + 3y + 2 = 0)$

6. (a) $(\forall \theta)(\cos^2 \theta + \sin^2 \theta = 1)$ (b) $(\forall z)(5 \leq 4z + 1 \leq 17)$
(c) $(\exists w)(18w^2 + 61w - 7 = 0)$ (d) $(\exists t)(\cos t \geq 1 + \sin t)$

In Problems 7–16, (a) translate each English statement into a symbolic proposition with quantifiers by using the indicated domain of discourse U and introducing suitable propositional functions: $T(x)$ for x is a teacher, $S(x)$ for x is a sadist, and so on, and (b) write out, in English, the denial of each statement.

7. All teachers are sadists. $U = $ all people.

8. No teachers are sadists. $U = $ all people.

9. All teachers are sadists. $U = $ all teachers.

10. Some teachers are sadists and some are not. U = all teachers.

11. Not all lawyers are judges. U = all people.

12. Some actors are egoists. U = all people.

13. Not all lawyers are judges. U = all lawyers.

14. No pig has wings. U = all animals.

15. There is a prime number that is exactly divisible by 3. U = all positive integers.

16. All prime numbers are greater than 1. U = all positive integers.

In Problems 17–20, let the domain of discourse U be the real numbers. Quantify each expression: (a) with $(\forall y)$, (b) with $(\exists x)$, (c) with $(\exists x)(\forall y)$, and (d) with $(\forall y)(\exists x)$. In each case, interpret the result.

17. $x > y$ 18. $x^2 + y^2 \leq 1$

19. $x + y = 0$ 20. $x^2 + y^2 = 4$

21. Let the domain of discourse U consist of all integers, $0, \pm 1, \pm 2, \pm 3, \ldots,$ and define the propositional function $D(x, y)$ of two variables by $D(x, y) \Leftrightarrow (\exists z)(zx = y)$. Read $D(x, y)$ as x *divides* y. Find the truth value of each of the following:

   (a) $D(3, -12)$ (b) $D(3, 0)$ (c) $D(0, 3)$
   (d) $D(0, 0)$ (e) $(\exists x)(D(0, x))$ (f) $(\forall x)(D(x, x))$
   (g) $(\exists x)(\forall y)(D(x, y))$ (h) $(\forall y)(\exists x)(D(x, y))$ (i) $(\exists y)(\forall x)(D(x, y))$
   (j) $(\forall x)(\exists y)(D(x, y))$ (k) $(\forall x)(\forall y)(D(x, y))$ (l) $(\exists x)(\exists y)(D(x, y))$

22. Let the domain of discourse U consist of all possible *events* in space-time, and let $C(x, y)$ be interpreted to mean that the event x *causes* the event y. Discuss the meanings of the following assertions:

   (a) $(\exists x)(\forall y)(C(x, y))$ (b) $(\forall x)(\forall y)(C(x, y))$ (c) $(\forall y)(\exists x)(C(x, y))$
   (d) $(\exists x)(\exists y)(C(x, y))$ (e) $(\exists x)(C(x, x))$

   If we interpret (a)–(e) as philosophical doctrines, which doctrine is strongest in the sense that it implies all the others? Which doctrines would a classical physicist probably subscribe to? Which doctrines would you describe as being mystical?

In Problems 23 and 24, let the domain of discourse U be the set of all real numbers, and let n denote an integer. Write the denial of each proposition in such a way that no denial connective appears explicitly and interpret the result.

23. (a) $(\forall x)(\exists n)(n \leq x < n + 1)$ (b) $(\exists x)(\forall n)(x < n + 1 \Rightarrow x < n)$

24. (a) $(\forall x)(\exists n)(0 \leq n \wedge x < n/10)$ (b) $(\forall x)(\forall n)(x \leq n \vee n < x)$

25. Simplify each expression by writing an equivalent expression that does not involve the denial connective:

   (a) $\sim(\exists x)(\sim P(x))$ (b) $\sim(\forall x)(\sim P(x))$

26. Explain why it is possible to develop predicate calculus using only the existential quantifier, that is, without using the universal quantifier at all. [Hint: See Problem 25.]

27. Let the universe of discourse U be all real numbers, and use the convention that ε and δ stand for positive real numbers. Suppose that $f(x)$ is a real-valued function of the real variable x.
    (a) If a is a real number, write out, in the symbolism of predicate calculus, the definition of the statement $f(x)$ *is continuous at* $x = a$.
    (b) Write out, in the symbolism of predicate calculus, but without using the denial connective, a statement equivalent to the condition that $f(x)$ *is not continuous at* $x = a$.

28. (a) Continuing with the notation of Problem 27, write out, in the symbolism of predicate calculus, the condition for $f(x)$ to be *continuous*, that is, continuous at every real number.
    (b) By definition, $f(x)$ is said to be *uniformly continuous* if and only if

    $$(\forall\varepsilon)(\exists\delta)(\forall x)(\forall y)(|x - y| < \delta \Rightarrow |f(x) - f(y)| < \varepsilon).$$

    Show that, if $f(x)$ is uniformly continuous, then it is continuous.

29. Let the universe of discourse U be all real numbers, and let $f(x)$ be a real-valued function of the real variable x. Write out, in the symbolism of predicate calculus, the condition that $f(x)$ is a *strictly increasing function of* x.

30. By means of a counterexample, defeat the following conjecture: A continuous function is uniformly continuous. [See Problem 28.]

*In Problems 31—37, we refer to the theorems of propositional calculus numbered 1–18 in the list on **page 28**.*

31. Let U be the set of all people. Replace the propositional functions $P(x)$ and $Q(x)$ in Theorems 1–18 by particular propositional functions such as *x is a politician*, *x is honest*, and so on, and translate the resulting statements into English.

32. Some of the Theorems 1–18 are implications rather than equivalences. For those that are implications, show by means of counterexamples that the converse implications are not theorems. [This has already been done in Part (b) of Example 2.12 for Theorem 14.]

33. Give a convincing argument for Theorem 11.

34. Give a convincing argument for Theorem 12.

35. Give a convincing argument for Theorem 13.

36. Using Theorems 6 and 18, give an argument to show that, if R denotes a proposition or a propositional function that does not contain x as a free variable, then

    $$(\exists x)(Q(x) \Rightarrow R) \Leftrightarrow ((\forall x)(Q(x)) \Rightarrow R)$$

is a theorem of predicate calculus. [Hint: Use the fact that $P \Rightarrow Q$ is equivalent to $(\sim P) \vee Q$.]

37. Give convincing arguments for Theorems 15, 16, 17, and 18.

38. Give an argument to show that, if R denotes a proposition or a propositional function that does not contain x as a free variable, then

$$(\forall x)(Q(x) \Rightarrow R) \Leftrightarrow ((\exists x)(Q(x)) \Rightarrow R)$$

is a theorem of predicate calculus. [Hint: See Problem 36.]

39. Give several examples of instances in which the notion of *equal* is not used in strict accord with the notion of *logical identity*.

40. The *principle of substitution of equals for equals* is stated in 3.10 on page 38.
(a) Explain why this principle is reasonable in view of the idea that *equal* means *logical identity*.
(b) Does this principle work for the other uses of the word *equal* that you cite in Problem 39?

41. One of the important principles governing the use of the notion of equality is that *things equal to the same thing are equal to each other*.
(a) Express this principle in symbolic form with the aid of quantifiers.
(b) Explain why this principle is reasonable in view of the idea that *equal* means *logical identity*.

42. Suppose that an experimental physicist adopts the following operational definition of *equal*: Two physical objects are equal if and only if every measurement of the one object yields the same result (within the limits of accuracy of the measuring instruments) as the corresponding measurement of the other object. Explain why, in this case, the principle that *things equal to the same thing are equal to each other* may fail.

43. Show that there exists a unique real number x such that $x^3 = 8$.

44. Show that there exists a unique even positive integer x such that x is a prime number.

45. Show that the proposition $(\forall x)(P(x)) \Rightarrow (\exists y)(P(y) \wedge x \neq y)$ is equivalent to the denial of the proposition $(\exists! x)(P(x))$.

46. A mathematician, arriving in a certain city to attend a convention, is amused by a sign stating that *this is America's most unique city*. Why does this amuse her?

In Problems 47–49, suppose that the universe of discourse U consists only of the numbers 1 and 2.

47. (a) Explain why the proposition $(\forall x)(P(x))$ is equivalent to $P(1) \wedge P(2)$.
(b) Explain why the proposition $(\exists x)(P(x))$ is equivalent to $P(1) \vee P(2)$.

48. Reasoning as in Problem 47, translate the proposition $(\exists x)(\forall y)(P(x, y))$ into a formula of propositional calculus.

49. Using the results of Problem 47, show that Theorems 6 and 7 in the list on page 28 are consequences of De Morgan's laws.

50. Show that the following is a tautology:

$$((P \wedge Q) \Rightarrow R) \Leftrightarrow ((P \Rightarrow R) \vee (Q \Rightarrow R))$$

Let U, our domain of discourse, consist of all straight lines in Euclidean three-dimensional space. In the tautology above, let P be the statement *x and y lie in the same plane*, let Q be the statement *x and y do not meet*, and let R be the statement *x and y are parallel*. Evidently, the left side of the tautology is true, since two straight lines that lie in the same plane and do not meet are indeed parallel. The statement $P \Rightarrow R$ says that lines that lie in the same plane are parallel, and this is false. The statement $Q \Rightarrow R$ says that lines that do not meet are parallel, and this also is false. Therefore, the right side of the tautology is false. What is the trouble here?

3

PROOFS AND RULES OF PROCEDURE

In its most general sense, the word *proof* means *an argument to convince a person (or persons) that some proposition is true.* Even an argument such as "you'd better believe it or I'll punch you in the nose" is a proof in this most general sense. (This barbaric argument is called the *argumentum baculinum.*) Obviously, we need a more restricted notion of proof that is better suited to the requirements of mathematicians.

It is not easy to give a definition of *mathematical proof* that will be acceptable to all mathematicians—in fact, it is probably impossible. Currently, for instance, there are even some doubts about the validity of certain proofs that involve the extensive use of high-speed computers. Fortunately, for our purposes in this textbook, we can settle for an informal working definition of a mathematical proof.

We begin by introducing some **rules of procedure**, or **rules of inference**, which state conditions under which we agree to say that a proposition is **justified**. The word *justified* should be interpreted as meaning something like "true in context." The most important rule of procedure, called **modus ponens** (Latin for *method of asserting*) is as follows:

3.1 RULE OF PROCEDURE

> **MODUS PONENS** If P is a justified proposition and if $P \Rightarrow Q$ is a justified proposition, then we can infer that Q is a justified proposition.

The rationale for modus ponens is simple enough. If you examine the truth table

P	Q	$P \Rightarrow Q$
1	1	1
0	1	1
1	0	0
0	0	1

for $P \Rightarrow Q$, you see that it is true only for the first, second, and fourth truth combinations. Therefore, if $P \Rightarrow Q$ is true, one of these three truth combinations must hold. But, if P is true, only the first truth combination can hold; hence, Q must be true.

3.2 Example   Consider the following argument: *Edna has good sense. If Edna has good sense, then she will not interrupt Professor Twit's lecture. Therefore, Edna will not interrupt the lecture.* Why is this a valid argument?

SOLUTION   Let P denote the proposition

   Edna has good sense,

and let Q denote the proposition

   she will not interrupt Professor Twit's lecture.

We are given both P and $P \Rightarrow Q$; hence, by modus ponens, we can infer Q. ☐

3.3 DEFINITION

> **MATHEMATICAL PROOF**
>
> (i) A **mathematical proof** is a finite sequence of justified propositions. Each such justified proposition is called a *step* of the proof.
> (ii) A **formal mathematical proof** is a mathematical proof in which each step is accompanied by the rule or rules of procedure that justify it.
> (iii) By a **mathematical proof of a proposition** P, we mean a mathematical proof whose last step is the proposition P.

A proposition P that can be shown to have a mathematical proof is called a **theorem**. To **prove** a theorem means to exhibit its proof. An **abbreviation** or **outline** of a mathematical proof is an indication of some of its more important or less obvious steps. To prove a theorem **informally** means to exhibit an abbreviation or outline of its proof.

Working mathematicians usually give informal proofs of their theorems, replacing omitted portions of their arguments with words such as

it can easily be seen that . . .

or

it is clear that . . . .

Also, they often leave out justifications for steps when they feel that these are "obvious." If the person or persons to whom the proof is being shown do not accept or understand the argument, the mathematician is obliged to fill in the gaps, that is, restore some of the omitted steps and justifications. The process of filling in the gaps usually continues until a consensus is reached that the theorem can be proved formally.

A **lemma** is a theorem that we state and prove because we intend to use it to help us in the proof of a subsequent (and, perhaps, more important) theorem. Thus, a lemma is a helper-theorem. A **corollary** is a theorem whose proof follows easily if we make use of a previously proved theorem. That previously proved theorems can be used in the proofs of theorems is guaranteed by the following rule of procedure.

3.4 RULE OF PROCEDURE

PREVIOUSLY PROVED THEOREMS   A theorem that has already been proved may be inserted in the proof of a subsequent theorem as a justified step.

Because mathematical theorems often have the form $P \Rightarrow Q$, the following two rules of procedure are especially useful.

3.5 RULE OF PROCEDURE

ASSERTION OF HYPOTHESIS   In the proof of a theorem of the form $P \Rightarrow Q$, the hypothesis P may be written down as a justified step.

The rationale for 3.5 is easy to supply. Recall that an implication $P \Rightarrow Q$ is automatically true if its hypothesis P is false. Therefore, if we want to prove that $P \Rightarrow Q$ is true, we need consider only the case in which P is true.

3.6 RULE OF PROCEDURE

JUSTIFICATION OF CONCLUSION   If Q is a justified proposition, then we can infer that $P \Rightarrow Q$ is a justified proposition.

Again, the rationale for 3.6 is simple. Just recall that an implication $P \Rightarrow Q$ is automatically true if its conclusion Q is true.

As we mentioned in Section 1, students of mathematics are often troubled by some of the apparently peculiar features of the implication connective $\Rightarrow$ as defined by its truth table. Note, however, that 3.1, 3.5, and 3.6 are the only rules of procedure pertaining to the use of this connective; and, as we have seen, they can be substantiated on the basis of this truth table. Thus, although the connective $\Rightarrow$ may not conform in every respect to our usual idea of implication, it works perfectly for the creation of mathematical proofs.

We now give several more rules of procedure and introduce a few more concepts relating to the idea of a mathematical proof.

3.7 RULE OF PROCEDURE

> **REPLACEMENT INSTANCES OF TAUTOLOGIES**   If P is a propositional variable occurring in a tautology $\mathcal{T}$, and if $\mathcal{R}$ is the proposition that results when P is replaced in its every occurrence in $\mathcal{T}$ by a particular proposition, then the proposition $\mathcal{R}$ is justified.

In 3.7, the proposition $\mathcal{R}$ is called a **replacement instance** of the tautology $\mathcal{T}$.

3.8 Example   Justify the proposition $(1 = 0) \vee (1 \neq 0)$.

SOLUTION   In the tautology $P \vee (\sim P)$, replace P by the proposition $1 = 0$.   □

3.9 RULE OF PROCEDURE

> **IDENTITY**   If a is any object in the domain of discourse U, then $a = a$ is a justified proposition.

3.10 RULE OF PROCEDURE

> **SUBSTITUTION OF EQUALS**   Let P be a justified proposition about a certain object a in the domain of discourse U. Suppose that $a = b$ is also a justified proposition. Then, the proposition Q that results when b is substituted for a in any or all of its occurrences in P is also a justified proposition.

3.11 Example   If a and b are objects in the domain of discourse U, prove that $a = b \Rightarrow b = a$.

SOLUTION   The theorem to be proved has the form $P \Rightarrow Q$, where P is the proposition $a = b$ and Q is the proposition $b = a$. By 3.5 (assertion of hypotheses), we can assume that the hypothesis $a = b$ is justified. By 3.9, $a = a$ is also justified. By 3.10, we can substitute b for a in any or all of its occurrences in the proposition $a = a$. We choose to substitute b for a only on the left of the equal sign, and thus we obtain $b = a$ as a justified statement. By 3.6 (justification of conclusion), we can infer that

$$a = b \Rightarrow b = a$$

is justified, and the theorem is proved.   □

At this point, you may be wondering, "Why all the fuss? It's obvious that $a = b$ implies $b = a$, so why do we bother to prove it?" The answer is that we are testing our theorem-proving tools. If these tools are incapable of proving simple facts, they can hardly be trusted for more complicated tasks. In testing a new jigsaw, you would begin by making rather simple cuts, not by doing elaborate scrollwork.

Among the more important theorem-proving tools are *axioms*, or *postulates*, and *definitions*. In mathematics, the word **axiom** is usually taken to mean a proposition whose truth we *assume* for the purpose of studying its consequences. The theorems that follow from the assumption of a set of axioms form what is called a **mathematical theory**. There is nothing sacred about axioms; it is not even necessary to believe in them. They are posited only as starting points for the development of theories. The only question is whether a theory is of interest to the mathematical community.

A **definition** in mathematics is usually a notational device in which a certain meaningful cluster of symbols is, for simplicity, represented by a simpler cluster of symbols or by a single symbol. For instance, in calculus, we define

$$e = \lim_{n \to \infty} \left(1 + \frac{1}{n} \right)^n.$$

Here the complicated expression on the right, presumably containing previously defined terms, is set equal to e by definition. Superficially, definitions are nothing but devices of convenience: It is much easier to write e than to write the more complicated expression for which it stands. In fact, however, definitions play a decisive role in the creation of a mathematical theory because they focus our attention on those particular clusters of symbols that the creator of the theory regards as being particularly significant. In other words, definitions establish the fundamental *concepts* with which the theory deals.

3.12 RULE OF PROCEDURE	**AXIOMS AND DEFINITIONS** Any previously introduced axiom or definition may be considered to be a justified proposition.

Occasionally, the word **postulate** is used in mathematics as a synonym for the word *axiom*. (This does not coincide with older usage in which a technical distinction was made between axioms and postulates, as, for example, in Euclid's geometry.) It should also be mentioned that the individual propositions that together make up a definition are often spoken of as the *axioms* or the *postulates* for the concept being defined. (For instance, one speaks of the *postulates for a group* or the *axioms for a topological space*.)

In mathematics, one of the most useful (and also one of the most controversial) rules of procedure is the rule permitting the formation of **indirect proofs**, or **proofs by contradiction**.

3.13 RULE OF PROCEDURE

> **INDIRECT PROOF, OR PROOF BY CONTRADICTION**   For a proposition P, if the assumption that the denial $\sim P$ is justified leads to a contradiction, then we can infer that P is justified.

Tautology 17 on page 10,

$$((\sim P) \Rightarrow (Q \wedge (\sim Q))) \Rightarrow P,$$

provides the rationale for proofs by contradiction using 3.13.

3.14 Example   Prove that there is no smallest positive real number.

SOLUTION   We make a proof by contradiction. Let P denote the proposition

there is no smallest positive real number.

Then $\sim P$ is the proposition

there is a smallest positive real number.

Assume $\sim P$; that is, assume there is a smallest positive real number. Call this number s. Let $t = s/2$, noting that $t < s$ and that t is positive. But, because s is the smallest positive real number, $t \not< s$. Thus, $\sim P$ leads to the contradiction $(t < s) \wedge (t \not< s)$; hence, P is true by rule of procedure 3.13.   □

In practice, the proof by contradiction given in Example 3.14 might be abbreviated as follows:

Theorem   *There is no smallest positive real number.*

PROOF   Suppose there were such a number. Call it s. Then $s/2 < s$, contradicting the supposition that s is the smallest positive real number.   □

The following two rules of procedure pertain to the use of existential and universal quantifiers.

3.15 RULE OF PROCEDURE

> **EXISTENTIAL QUANTIFICATION**   Let $P(x)$ be a propositional function. If b is an object in the domain of discourse U, and if $P(b)$ is justified, then we may infer that $(\exists x)(P(x))$ is justified. Conversely, if $(\exists x)(P(x))$ is justified, then we may infer that, for some suitable choice of an object c in U, $P(c)$ is justified.

3.16 RULE OF PROCEDURE

> **UNIVERSAL QUANTIFICATION**   Let $P(x)$ be a propositional function. If $P(x)$ is justified, where x represents an arbitrary object in the domain of discourse U, then we may infer that $(\forall x)(P(x))$ is justified. Conversely, if $(\forall x)(P(x))$ is justified and if b is an object in U, we may infer that $P(b)$ is justified.

Although we have given no formal definition of the word *justified*, it should be plain that the rules of procedure are actually agreements about the ways in which we intend to use this word. In this sense, the rules of procedure establish an operational definition of the word *justified*. We make no claim for the completeness of the rules of procedure given above; however, they are sufficiently comprehensive to provide a reasonably secure foundation for most of the proofs in this textbook.

The ideas discussed above have given rise to a general notion of a **formal mathematical system**. Such a system begins with a list of *primitive*, or *undefined*, *symbols*. Ordinarily, some attempt is made to indicate just what these symbols might stand for, but this indication is not to be regarded as part of the formal system itself. Next comes a list of *axioms*, or *postulates*, that govern the undefined symbols. These axioms are out-and-out assumptions about relationships among the undefined symbols, and as such, they constitute a contextual or operational definition of these symbols. *Rules of procedure* are then set forth, which determine the specific conditions under which propositions, expressed in terms of the primitive symbols, are considered justified. These rules of procedure enable the person who is constructing the formal theory to prove the *theorems* that form the body of the formal mathematical system. From time to time during the construction of the formal system, it may be convenient to introduce new symbols by *definitions* that fix their meanings in terms of the primitive symbols and previously defined symbols.

The concept of a formal mathematical system has so fascinated logicians and mathematicians that they have created a new discipline called

metamathematics, the study of formal mathematical systems *in general.* One of the most intriguing metamathematical results is a theorem, published by Kurt Gödel in 1931, which states that within a formal mathematical system sufficiently rich to enable the arithmetic of whole numbers, there are propositions that are *undecidable* in the sense that, by working within the system, *they can neither be proved nor disproved.* This astonishing theorem put an end to nearly a century of efforts to create a formal axiomatic basis for all of mathematics. Although Gödel's proof is highly technical, a very readable popular account of this and related matters can be found in Douglas R. Hofstadter, *Gödel, Escher, Bach: An Eternal Golden Braid* (New York: Basic Books), 1979.

We close this section with three remarks:

First: There is nothing inviolable about the rules of procedure given above. Although these rules, or close counterparts, are used (implicitly or explicitly) by most contemporary mathematicians, there are certain people (such as the so-called intuitionists) who find that they cannot subscribe to these particular rules and who do indeed replace them by substantially different rules.

Second: In practice, mathematicians rarely work explicitly within the confines of a formal mathematical system. Most mathematical proofs are given somewhat informally, and the use of rules of procedure is often implicit or tacit. However, when necessary, most practicing mathematicians can reformulate their work in the context of some suitable formal mathematical system.

Third: There is nothing magic about a mathematical theory: No more can be squeezed out of it than is implicit in its axioms and rules of procedure. In a sense, a mathematical theorem is nothing but a (perhaps) inobvious tautology. A person who does not understand and appreciate this fact is subject to being imposed on by charlatans who claim to be able to prove outlandish things "mathematically."

PROBLEM SET 1.3

1. By using the truth table for $P \Rightarrow Q$, explain why, if Q is a justified proposition and if $P \Rightarrow Q$ is a justified proposition, then (in general) we cannot infer that P is a justified proposition.

2. Give a specific example showing that $P \Rightarrow Q$ may be true and Q may be true, but P may be false.

3. If $P \Rightarrow Q$ is justified and $\sim Q$ is justified, explain how $\sim P$ can be justified. [Hint: Use the law of contraposition (tautology number 12 on page 10) and modus ponens.]

4. If $P \vee Q$ is justified and $\sim P$ is justified, explain how Q can be justified. [Hint: Show that $(P \vee Q) \Rightarrow ((\sim P) \Rightarrow Q)$ is a tautology, and then use modus ponens.]

In Problems 5–10, indicate which arguments are valid, which are invalid, and say why.

5. If Joe embezzled the union funds, then he is guilty of a felony. Joe did embezzle the union funds. Therefore, he is guilty of a felony.

6. If Maria can solve the problem, then she has promise as a mathematician. Maria cannot solve the problem. Therefore, she has no promise as a mathematician.

7. If Gwen is a brain surgeon, then she is highly trained. Gwen is highly trained. Therefore, she is a brain surgeon.

8. Carlos is either a genius or he is crazy. Carlos is not crazy. Therefore, he is a genius.

9. If Rodney cheated on the exam, then he passed his calculus class. Rodney passed his calculus class. Therefore, he cheated on the exam.

10. If $f(x)$ is a differentiable function, then it is continuous. The function $f(x)$ is not differentiable. Therefore, it is not continuous.

11. Professor Grumbles wants to prove that if a series $\sum_{j=1}^{\infty} a_j$ converges, then $\lim_{n \to \infty} a_n = 0$. He begins his proof by saying, "Suppose that the series $\sum_{j=1}^{\infty} a_j$ converges." Is this step justified? Why or why not?

12. Professor Twit wants to prove that if a series $\sum_{j=1}^{\infty} a_j$ converges, then $\lim_{n \to \infty} a_n = 0$. He begins his proof by saying, "Suppose that $\lim_{n \to \infty} a_n = 0$." Is this step justified? Why or why not?

13. Professor Keen wants to prove that, if a series $\sum_{j=1}^{\infty} a_j$ converges, then $\lim_{n \to \infty} a_n = 0$. After an argument in which each step is justified, she is able to justify the statement $\lim_{n \to \infty} a_n = 0$. Has she proved the theorem? Why or why not?

14. Archibald knows that if a series $\sum_{j=1}^{\infty} a_j$ converges, then $\lim_{n \to \infty} a_n = 0$. On an exam, he is asked whether the harmonic series $\sum_{j=1}^{\infty} (1/j)$ converges. He argues that, since $\lim_{n \to \infty} (1/n) = 0$, the harmonic series must converge. Criticize Archibald's reasoning.

In Problems 15–18, justify each proposition by showing that it is a replacement instance of a tautology.

15. $\cos 0 = 1 \Leftrightarrow \sim(\cos 0 \neq 1)$

16. Boris is either a spy, or he is not a spy.

17. $((\cos 0 = 1) \wedge ((\cos 0 = 1) \Rightarrow \sin 0 = 0)) \Rightarrow \sin 0 = 0$

18. $(((\sqrt{4} = 2) \vee (\sqrt{4} \neq 2)) \Rightarrow \sqrt{1} = 2) \Rightarrow (\sqrt{1} = 2)$

19. Let x and y denote real numbers. Show that the proposition

$$(x = 1) \wedge ((x = 1) \Rightarrow (y = 2)) \wedge (y \neq 2)$$

is equivalent to a contradiction.

20. Give a rationale for the Rule of Procedure 3.7.

21. If P and Q are propositions, and if $P \wedge Q$ is a justified step in a formal mathematical proof, explain how P can be obtained as a justified step in this proof. [Hint: Use tautology 18 in the list on page 10.]

22. Give a rationale for the Rule of Procedure 3.10.

23. If a, b, and c are objects in the domain of discourse U, prove that

$$((a = b) \wedge (b = c)) \Rightarrow a = c.$$

24. Although the Rule of Procedure 3.10, Substitution of Equals, seems reasonable and works in connection with the usual propositions encountered in mathematics, its unrestricted use for propositions of ordinary discourse can lead to difficulties. For instance, suppose that the proposition

Superman is so called because of his superhuman abilities

is justified. Suppose, also, that the proposition

Superman = Clark Kent

is justified. Using rule 3.10, we infer that

Clark Kent is so called because of his superhuman abilities,

which is absurd. Can you resolve this paradox? Why do such difficulties rarely arise in mathematical proofs?

25. In ordinary conversation, the word *axiom* means a self-evident or universally recognized truth. Contrast this with the *mathematical* use of the word.

26. (a) Explain the sense in which definitions in a dictionary are essentially *circular*.
(b) Explain how circular definitions are avoided in a formal mathematical theory.

27. Prove by contradiction (that is, make an indirect proof) that there is no largest positive integer.

28. Let the domain of discourse U consist of the integers.
(a) Prove that, if n is odd, then n^2 is odd. [Hint: If n is odd, it can be written in the form $n = 2k + 1$ for some integer k.]
(b) Using the result in Part (a), make an indirect proof to show that if n^2 is even, then n is even. [Hint: Use the fact that n is even if and only if it is not odd.]

29. Professor Keen wants to prove a theorem of the form $P \Rightarrow Q$. She elects to use an indirect proof. As her first step, she says, "Assume that P is true, but that Q is false." Explain.

30. Let the domain of discourse U consist of the real numbers. Take it as a previously proved theorem that

$$(0 \leq x \leq y) \Rightarrow x^2 \leq y^2.$$

Using this fact, make an indirect proof showing that

$$(0 < x < y) \Rightarrow (\sqrt{x} < \sqrt{y}).$$

31. Prove by contradiction that if a straight line L intersects a circle with center O at a point P, and if L is perpendicular to the radius $\overline{OP}$ (Figure 1-1), then L does not intersect the circle at another point $Q \neq P$.

32. If P and Q are propositions, denote by the symbolism $P \dashv Q$ the statement that P implies Q in the usual intuitive sense, namely, that Q can be inferred or deduced from P.
    (a) Does it appear to you that

$$(P \dashv Q) \dashv (P \Rightarrow Q)$$

should be a theorem? Why or why not?
    (b) Does it appear to you that

$$(P \Rightarrow Q) \dashv (P \dashv Q)$$

should be a theorem? Why or why not?

33. Find some examples of indirect proofs in your calculus textbook.

34. A person claims to have a mathematical proof that the world is flat. What are some of the pertinent questions that you might ask in connection with the alleged proof?

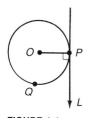

FIGURE 1-1

4
SETS

The idea of a *set* permeates all of modern mathematics. Between the years 1873 and 1895, the basic elements of set theory were created by the German mathematician Georg Cantor (1845–1918), who became aware of the need for such a theory during his research on trigonometric (Fourier) series. Although Cantor's revolutionary ideas were initially scorned by other mathematicians, set theory soon came to be recognized as a legitimate branch of mathematics. Then, in 1902, the English philosopher Bertrand Russell (1872–1970) discovered that the methods used by Cantor and his followers in reasoning about sets lead to a contradiction—the *Russell paradox*. Using a so-called theory of types, Russell and Alfred North

Whitehead (1861–1947) attempted to recast set theory to avoid this paradox in their monumental *Principia Mathematica.* Later, other mathematicians and philosophers formulated several competing brands of set theory, each designed to avoid the paradoxes inherent in Cantor's original work. In this textbook, we ignore these difficulties and treat set theory intuitively and more or less informally. After you have learned the basics of this "naive" set theory, you will be better prepared to study some brand of formal axiomatic set theory, if you are so inclined.

Mathematicians use the word **set** to mean much the same thing that it means in ordinary English—a collection of things. The things comprising a set are understood to be objects from a fixed domain of discourse U. These objects may be physical things (trees, books, molecules, and so on), or they may be nonphysical entities (numbers, geometric points, continuous functions, and so on). In one of the influential early books on set theory, the German mathematician Felix Hausdorff wrote, "eine Menge ist eine Vielheit, als Einheit gedacht," which translates roughly as, "a set is a multiplicity thought of as a unity." The significant words here are *thought of*. Thus, in mathematics, a set should be regarded as a *concept* or an *idea in our minds* and not as a physical collection of objects in space and time. For instance, we may have a bag of marbles, and we may speak of the set of marbles in this bag, but this set is an idea in our minds. It can be spoken about but not exhibited, and it must not be confused with the bag of marbles itself.

Just as we use symbols (numerals) to denote numbers, we can use symbols to denote sets. And just as we use symbols in algebra to denote arbitrary numbers, we shall use symbols to denote arbitrary sets. In this book, we usually use capital letters—such as M or N—to represent arbitrary sets. If M is a set and b is an object in our domain of discourse U, then the symbolism

$$b \in M$$

stands for the proposition stating that the object b **belongs to**, or **is a member of**, the set M. If $b \in M$, we also say that b is an **element** of the set M. The symbol $\in$ is called the **set membership symbol**. Similarly, the symbolism

$$b \notin M$$

is understood to stand for the proposition stating that the object b does not belong to the set M. Therefore, by definition,

$$b \notin M \Leftrightarrow \sim (b \in M).$$

In mathematics, it is always understood that *a set is completely determined by its members.* Thus, a club or syndicate is not an example of a set in the mathematical sense. There is more to a club or syndicate than the mere totality of its members thought of as a unity; after all, even though these members are banded together for some purpose, it may not be possible to determine this purpose from the membership list alone. The requirement that a set is completely determined by its members is made precise by the following axiom.

4.1 AXIOM

AXIOM OF EXTENT   Let M and N be sets. Then

$$M = N \Leftrightarrow (\forall x)(x \in M \Leftrightarrow x \in N).$$

In words, the axiom of extent says that *two sets are equal if and only if they have the same members.*

Because a set is completely determined by its members or elements, it can be specified simply by listing these elements. There is special notation for this. We write

$$M = \{a, b, c, \ldots, k\}$$

to indicate that *M is the set whose elements are a, b, c, . . . , k, and no others.* When a set is specified in this way, we say that it is described **explicitly.**

4.2 Example   Rewrite the set E of all even integers between 1 and 9 in explicit form.

SOLUTION $E = \{2, 4, 6, 8\}.$ □

If M is a set and b is an object in the domain of discourse U, then b either belongs to M or it does not. If b does belong to M, then (as a consequence of the axiom of extent) the number of ways in which it qualifies for membership in M is of no significance. *We can ask only whether or not a given object belongs to a set; it makes no sense to ask how many "times" it belongs to the set.* For instance, if A is the set consisting of all people who hold a doctoral degree in astrophysics or who have orbited the earth, then Sally Ride is a member of A. But we do not say that she belongs to A twice, even though she qualifies for membership in A in two different ways.

4.3 Example   Let $B = \{1, 3, 5, 7\}$ and let $C = \{2, 3, 7, 8\}$. Give an explicit description of the set D consisting of all numbers that belong to B or to C (or to both).

SOLUTION $D = \{1, 2, 3, 5, 7, 8\}.$ □

Note that, in spite of the fact that 3 and 7 qualify in two different ways for membership in D, *we do not write $D = \{1, 2, 3, 3, 5, 7, 7, 8\}$.*

Another consequence of the axiom of extent is that *the elements of a set M should not be regarded as belonging to M in any particular order.* For instance,

$$\{1, 2, 3, 4, 5, 6\} = \{3, 6, 1, 5, 4, 2\},$$

because the set on the left and the set on the right have exactly the same members. It is an accident of our system of notation that when we write the elements of a set explicitly, we must put them down in some particular order; however, this order is not to be regarded as part of the structure of the set.

A set that has exactly one member is called a **singleton set**. If b is an object in the domain of discourse U, then the set $\{b\}$, whose only member is b, is called **singleton** b. *You must be careful to distinguish between the object b and the singleton set* $\{b\}$. The set $\{b\}$ is an idea or concept that certainly concerns the object b, but it is not the *same* as the object b. For instance, Detroit is a city, but $\{Detroit\}$ is a set and not a city.

Because a set is a concept, there is no reason why we cannot form the concept of a set with no members whatsoever. By the axiom of extent, there can be only one such set. We call it the **null set**, or the **empty set**, and we symbolize it by $\varnothing$. Although it is not usually done, we could write the description of the null set in explicit form as

$$\varnothing = \{\ \}.$$

Do not make the mistake of supposing that the null set is nothing—after all, it is a set.

4.4 Example   Is it true that $\{\varnothing\} = \varnothing$?

SOLUTION   No! The set $\{\varnothing\}$ is not the null set because it does have a member: $\varnothing \in \{\varnothing\}$. In fact, $\{\varnothing\}$ is the singleton set whose only member is the null set.   □

Until now, we have only one way to specify sets—namely, by means of explicit description. Of course, explicit description is restricted to sets that have a finite number of elements, and even then it is restricted in practice to small sets. But there is a second way to specify sets that overcomes this limitation. If $P(x)$ is a propositional function, and M is a set, then by the symbolism

$$\{x \in M \,|\, P(x)\}$$

we mean

the set of all objects x in M such that P(x) is true.

Thus, if

$$N = \{x \in M \,|\, P(x)\},$$

then, for all x in U,

$$x \in N \Leftrightarrow (x \in M) \wedge P(x).$$

The following basic axiom of set theory provides for the existence of such a set N.

4.5 AXIOM

AXIOM OF SPECIFICATION If M is a set and if $P(x)$ is a propositional function, then there exists a set N such that

$$(\forall x)(x \in N \Leftrightarrow (x \in M) \wedge P(x)).$$

Notice that the axiom of extent guarantees the uniqueness of the set N in Axiom 4.5. The symbolism $\{x \in M \,|\, P(x)\}$ is sometimes called **set builder notation** for this unique set N.

4.6 Example If M is the set of all politicians, what are the members of the set $H = \{x \in M \,|\, x \text{ is honest}\}$?

SOLUTION H is the set of all honest politicians. (Some cynics believe that $H = \varnothing$.)

□

The symbolism $\{x \in M \,|\, P(x)\}$ provides another example of an apparent variable. Indeed, the set $\{x \in M \,|\, P(x)\}$ does not depend on x in any way; so we have, for instance,

$$\{x \in M \,|\, P(x)\} = \{y \in M \,|\, P(y)\}.$$

When a set is specified in the form $\{x \in M \,|\, P(x)\}$, we say that it is described **implicitly**. For instance, if $\mathbb{R}$ denotes the set of all real numbers, then the implicitly described set

$$\{x \in \mathbb{R} \,|\, x^2 - 5x + 6 = 0\}$$

is the same as the explicitly described set $\{2, 3\}$. In elementary algebra, when you were asked to solve an equation, you were really being asked to convert the description of a certain set from implicit to explicit form.

We now come to a vexing question: Is the universe of discourse U a set? Actually, the unrestricted assumption that U is a set can lead to logical difficulties. Here's how: Suppose that U consists of all concepts that can be formulated by the human mind, and suppose that U is a set. Now, let

$$N = \{x \in U \,|\, x \text{ is a set and } x \notin x\}. \tag{$*$}$$

Because N is a set, it is a concept; so it follows that

$$N \in U.$$

Therefore, by (∗),

$$N \in N \Leftrightarrow N \notin N,$$

a self-contradictory conclusion. This is one form of Russell's paradox. The paradox can be avoided by dropping the assumption that U is a set; then the notation in (∗) becomes meaningless, and we cannot even form the set N. Conclusion: *Certain domains of discourse cannot be considered to be sets. In the remainder of this textbook, we deal only with domains of discourse that can be regarded as being sets.* We leave the consideration of more esoteric domains of discourse (and other so-called *proper classes*) to the more formal treatises on axiomatic set theory.

Let $P(x)$ be a propositional function. Because we are assuming that U is a set, we can form the set

$$\{x \in U \mid P(x)\}.$$

This set is often abbreviated as

$$\{x \mid P(x)\}$$

and is read, in English, as *the set of all x such that P(x)*, it being understood that x represents an object in the domain of discourse U.

4.7 Example   Let U denote the set of all real numbers. Rewrite the implicitly specified set $\{x \mid x^2 = 3\}$ in explicit form.

SOLUTION $$\{x \mid x^2 = 3\} = \{\sqrt{3}, -\sqrt{3}\} \qquad \square$$

If M and N are sets and if every element of M is an element of N, then we say that M is **contained** in N or that M is a **subset** of N. We write

$$M \subseteq N$$

to signify that M is contained in N. The following definition makes the idea of containment more formal.

4.8 DEFINITION

SET CONTAINMENT   Let M and N be sets. Then, by definition,

$$M \subseteq N \Leftrightarrow (\forall x)(x \in M \Rightarrow x \in N)$$

and

$$M \nsubseteq N \Leftrightarrow \sim(M \subseteq N).$$

To prove that $M \subseteq N$, you must show that every member of M is also a member of N. To prove that $M \nsubseteq N$, you must show that there is at least one member of M that is not a member of N.

4.9 Example   In each case, decide whether or not $M \subseteq N$:

(a) M is the set of all mammals. N is the set of all vertebrates.
(b) M is the set of all prime numbers. N is the set of all odd integers.

SOLUTION   (a) $M \subseteq N$ is true because, indeed, every mammal is a vertebrate.
(b) $M \nsubseteq N$, because $2 \in M$ but $2 \notin N$.   □

Note that $M \subseteq N$ is a proposition about the sets M and N; hence, it is either true or false, depending on M and N. Be careful: $M \subseteq N$ is not a set! Also, note that *every set M is a subset of itself*; that is,

$$M \subseteq M$$

is always a true statement. Furthermore, since all objects under consideration belong to the universe of discourse U, then *every set M is a subset of U*; that is,

$$M \subseteq U$$

is always a true statement.

4.10 Example   Prove that *the null set is a subset of every set.*

SOLUTION   Let M be a set. We must prove that $\varnothing \subseteq M$. According to Definition 4.8, we must prove that, for every x in U,

$$x \in \varnothing \Rightarrow x \in M.$$

But this implication is automatically true, because its hypothesis $x \in \varnothing$ is false.   □

If M is any set, then both $\varnothing$ and M itself are subsets of M. We call $\varnothing$ and M the **trivial** subsets of M; all other subsets of M (if there are any) are called **nontrivial** subsets of M. If $N \subseteq M$ and $N \neq M$, then we say that N is a **proper** subset of M. Some authors indicate that N is a proper subset of M by writing

$$N \subset M.$$

If A, B, and C are sets, we write

$$A \subseteq B \subseteq C$$

to mean that $A \subseteq B$ and $B \subseteq C$. In particular, we have

$$\varnothing \subseteq M \subseteq U$$

for any set M. One of the basic properties of set inclusion is the transitivity property:

$$(A \subseteq B \wedge B \subseteq C) \Rightarrow A \subseteq C.$$

We leave the proof of this property as an exercise (Problem 15).

People who work with set theory often use diagrams to illustrate various relations among sets. These are called **Venn diagrams** in honor of the British logician John Venn (1834–1923) who devised them in 1880. First, we draw a rectangle and think of the points that lie within this rectangle as representing the objects belonging to the domain of discourse U. If we draw a circle (or any other simple closed curve) within this rectangle, then we can think of the points enclosed by the circle (or curve) as representing the elements of a set M. Two or more different sets are represented by two or more different circles. For instance, the fact that $M \subseteq N$ can be indicated as in the Venn diagram in Figure 1-2.

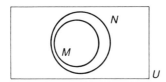

FIGURE 1-2

The following theorem expresses an important relationship between set containment and set equality.

4.11 THEOREM

> **EQUALITY OF SETS**   Let M and N be sets. Then
>
> $(M \subseteq N \wedge N \subseteq M) \Rightarrow M = N.$

PROOF   Assume the hypothesis, so that $M \subseteq N$ and $N \subseteq M$. Since $M \subseteq N$, every element of M is an element of N. Likewise, since $N \subseteq M$, every element of N is an element of M. Therefore, N and M have the same elements, and it follows from the axiom of extent that $M = N$.   □

In words, Theorem 4.11 says that *two sets are equal if each is contained in the other.* Thus, one way to prove that two sets are equal is to prove separately that each set is a subset of the other.

Often we have to consider more sets than can be represented by the twenty-six letters of the alphabet. One way to resolve this difficulty is to use the same letter with different subscripts to represent various sets, for instance,

$M_1, M_2, M_3, \ldots.$

A subscript used in this way is called an **index**, and we refer to

$M_1, M_2, M_3, \ldots$

as an **indexed family of sets**. More generally, if I is any nonempty set and if, for each $i \in I$ we have a corresponding set M_i, then we denote this indexed family of sets by

$$(M_i)_{i \in I}.$$

For such an indexed family of sets, *there is no automatic assumption that sets with different indices are different*; that is, it is quite possible to have

$$M_i = M_j$$

even though $i \neq j$. We consider indexed families of sets in more detail in Section 1.6.

Sometimes we have to consider sets whose elements are themselves sets; that is, we have to deal with *sets of sets*. It is often convenient to use capital script letters to denote sets of sets, for instance,

$$\mathscr{E} = \{A, B, C, D\}$$

would specify that $\mathscr{E}$ is the set whose elements are the sets A, B, C, and D. Notice that an indexed family of sets

$$(M_i)_{i \in I}$$

gives rise to a set of sets

$$\mathscr{E} = \{M_i | i \in I\}.$$

Another example of a set of sets is provided by the set of all subsets of a given set M.

4.12 DEFINITION **POWER SET OF A SET** If M is a set, then the **power set** of M, in symbols $\mathscr{P}(M)$, is the set of all subsets of M; that is,

$$\mathscr{P}(M) = \{N | N \subseteq M\}.$$

4.13 Example If $M = \{1, 2, 3\}$, find $\mathscr{P}(M)$.

SOLUTION $\mathscr{P}(M) = \{\varnothing, \{1\}, \{2\}, \{3\}, \{1, 2\}, \{1, 3\}, \{2, 3\}, \{1, 2, 3\}\}$. □

PROBLEM SET 1.4

In Problems 1 and 2, rewrite each set in explicit form.

1. (a) The set A of positive integers less than or equal to 8.
    (b) The set B of integers whose squares are greater than 5 and less than 90.

(c) The set C consisting of all numbers that belong to both sets A and B in Parts (a) and (b).

(d) The set D consisting of all numbers that belong to set A or to set B in Parts (a) and (b).

(e) The set E of integers between 2 and 42 that are exactly divisible by 4.

(f) The set F consisting of all integers that are positive and satisfy the inequality $x + 3 < 7$, or that satisfy the equation $x^2 + 1 = 0$.

2. (a) The set P of prime numbers between 35 and 50.

(b) The set K of integers whose cubes lie between -27 and 27, inclusive.

(c) The set R of positive integers whose square root is less than 5.

(d) The set M consisting of numbers that are either odd positive integers less than 10 or are even integers between 5 and 13.

(e) The set N consisting of all integers that satisfy the equation $|x + 1| = 5$ and also satisfy the inequality $|x - 1| < 4$.

(f) The set S of real numbers that are solutions of the equation $12x^2 - 17x + 6$.

In Problems 3–8, let the universe of discourse U be the set of all real numbers. Rewrite each implicitly described set in explicit form.

3. $\{x \mid x^2 = 4\}$

4. $\{x \mid x^2 - x - 2 = 0\}$

5. $\{x \mid x^2 + x + 1 = 0\}$

6. $\{x \mid |2x + 7| = 11\}$

7. $\{x \mid (3x - 1)^{1/3} = 2\}$

8. $\{x \mid (|x| < 1) \wedge (x^2 > 4)\}$

9. Which of the following sets are equal to the empty set?

(a) The set of all women who have been President of the United States.

(b) The set of all months of the year whose names in English start with the letter O.

(c) The set of all real numbers whose square is -1.

(d) The set of all signers of the Declaration of Independence who were Presidents of the United States.

10. True or false:

(a) $1 \in \{1\}$ (b) $1 \subseteq \{1\}$ (c) $\{1\} \in \{1\}$ (d) $\{1\} \subseteq \{1\}$

In Problems 11 and 12, decide whether or not $M \subseteq N$.

11. (a) M is the set of U.S. senators, N is the set of U.S. congressmen.

(b) M is the set of all squares, N is the set of all rectangles.

(c) M is the set of all triangles, N is the set of all right triangles.

(d) M is the set of all citizens of the U.S. who are eligible to be President, N is the set of all citizens of the U.S. who are at least 35 years of age.

(e) $M = \{1, 3, 5, 7\}$, $N = \{-1, 1, 3, 5, 7, 9\}$

(f) $M = \{1, 2\}$, $N = \{1, \{2\}\}$

(g) The universe of discourse U is all real numbers, $M = \{x \mid 3x - 2 \leq 4\}$, $N = \{x \mid 8 - 5x \geq 3\}$.

(h) The universe of discourse U is all real numbers,
$$M = \{x \mid x^2 + x - 12 < 0\}, N = \{x \mid |2x + 7| < 11\}.$$

12. (a) $M = \emptyset, N = \{\{\emptyset\}\}$ (b) $M = \emptyset, N = \emptyset$
    (c) $M = \{\emptyset\}, N = \{\emptyset, \{\emptyset\}\}$ (d) $M = \{1\}, N = \{\{1\}\}$
    (e) $M = \{2, 3\}, N = \{1, \{2, 3\}\}$ (f) $M = \{1, 2\}, N = \{1, 2, \{3\}\}$
    (g) $M = \{1, 2, 3\}, N = \{1, 2, \{3\}\}$ (h) $M = \{1, \{2\}\}, N = \{1, 2, \{2\}\}$

13. Let U, the universe of discourse, consist of all real numbers. In each case determine whether the two sets A and B are equal.
    (a) $A = \{x \mid x^2 + x - 2 = 0\}, B = \{1, -2\}$
    (b) $A = \{x \mid 1 \le x \le 3\}, B = \{1, 3\}$
    (c) $A = \{x \mid 3x - 10 \le 5 < x + 3\}, B = \{x \mid 2 < x \le 5\}$
    (d) $A = \{x \mid 2 \sin x = 1\}, B = \{\pi/6, 5\pi/6\}$
    (e) $A = \{x \mid x \text{ is an integer and } |2x - 1| \le 5\}, B = \{-2, -1, 0, 1, 2, 3\}$

14. Let U, the universe of discourse, consist of the twenty-six letters of the alphabet. Explain why the set

    $\{x \mid x \text{ is a letter in the word } \textit{little}\}$

    is equal to the set

    $\{x \mid x \text{ is a letter in the word } \textit{tile}\}.$

15. Prove that, if A, B, and C are sets with $A \subseteq B$ and $B \subseteq C$, then $A \subseteq C$.

16. True or false: If A and B are sets and $A \nsubseteq B$, then $B \subseteq A$. Explain.

17. Draw a Venn diagram illustrating three sets A, B, and C such that $A \subseteq B$, $A \subseteq C$, $B \nsubseteq C$, and $C \nsubseteq B$.

18. Let $M = \{x \mid P(x)\}$ and $N = \{x \mid Q(x)\}$. Suppose that $(\forall x)(P(x) \Rightarrow Q(x))$. Prove that $M \subseteq N$.

19. Let $I = \{1, 2, 3\}$ and consider the family of sets $(M_i)_{i \in I}$ such that $M_1 = \{a, b, d\}$, $M_2 = \{b, c, d\}$, and $M_3 = \{c, d\}$. Find, in explicit form, the set M consisting of all elements that belong to *at least one* of the sets in the family $(M_i)_{i \in I}$.

20. In Problem 19, find, in explicit form, the set D consisting of all elements that belong to all of the sets in the family $(M_i)_{i \in I}$.

In Problems 21–26, find $\mathscr{P}(M)$.

21. $M = \{1\}$ 22. $M = \emptyset$

23. $M = \{1, 2\}$ 24. $M = \{1, 2, 3, 4\}$

25. $M = \{a, b, c\}$ 26. $M = \{1, \{1\}\}$

27. List all of the proper subsets of $\{1, 2, 3\}$.

28. List all of the proper subsets of $\{1, 2, 3, 4\}$.

29. List all of the nontrivial subsets of $\{1, 2, 3\}$.

30. List all of the nontrivial subsets of $\{1, 2, 3, 4\}$.

31. True or false: Every proper subset of a set is a nontrivial subset of the set. Explain.

32. Is it possible to find sets A and B such that $A \neq \varnothing$, $A \subseteq B$, and $A \in B$? If so, give an example; if not, explain why not.

33. If A and B are sets and $A \subseteq B$, prove that $\mathscr{P}(A) \subseteq \mathscr{P}(B)$.

5

THE ALGEBRA OF SETS

In much the same way that propositions are combined by propositional connectives to form new propositions, sets can be combined by suitable operations to form new sets. The result is an **algebra of sets**. If M and N are sets, then the **union** of M and N, written

$$M \cup N,$$

is defined to be *the set of all objects that belong to M or to N or to both M and N*. In other words, $M \cup N$ is the set of all elements that belong to *at least one* of the sets M or N. The **intersection** of M and N, written

$$M \cap N,$$

is defined to be *the set of all objects that belong to both M and N*. The **relative complement** of N in M, written

$$M \setminus N,$$

is defined to be *the set of all objects that belong to M but not to N*. The following definition is more formal:

5.1 DEFINITION   **UNION, INTERSECTION, AND RELATIVE COMPLEMENT**   Let M and N be sets. Then,

   (i) $M \cup N = \{x \mid x \in M \lor x \in N\}$
  (ii) $M \cap N = \{x \mid x \in M \land x \in N\}$
 (iii) $M \setminus N = \{x \mid x \in M \land x \notin N\}$

In words:

$M \cup N$ is read M *union* N

$M \cap N$ is read M *intersection* N

$M \setminus N$ is read M *slash* N

For instance, if M is the set of all politicians and N is the set of all honest people, then $M \cup N$ is the set of all people who are politicians or honest (or both), $M \cap N$ is the set of all honest politicians, and $M \setminus N$ is the set of all dishonest politicians. Some authors write $M \setminus N$ as $M - N$ and refer to it as M *minus* N, but we prefer to use the slash because of the possible confusion of the minus sign with numerical subtraction.

5.2 Example   Let $M = \{1, 3, 6, 9\}$ and $N = \{3, 4, 9\}$. Find:

(a) $M \cup N$ **(b)** $M \cap N$ **(c)** $M \setminus N$

SOLUTION   **(a)** $M \cup N = \{1, 3, 4, 6, 9\}$ **(b)** $M \cap N = \{3, 9\}$
(c) $M \setminus N = \{1, 6\}$ □

In informal mathematical writing, a comma is often used as a substitute for the word *and* or for the connective $\wedge$. For instance, the expression

$$x \in M \wedge x \in N$$

may be written in the equivalent form

$$x \in M, x \in N.$$

Thus, with this notation,

$$M \cap N = \{x \mid x \in M, x \in N\},$$

and likewise,

$$M \setminus N = \{x \mid x \in M, x \notin N\}.$$

With appropriate shading, we indicate the union, intersection, and relative complements of sets M and N in Venn diagrams, as shown in Figures 1-3, 1-4, and 1-5.

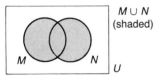

FIGURE 1-3

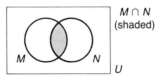

FIGURE 1-4

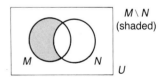

FIGURE 1-5

The set $U \setminus M$ illustrated in Figure 1-6 is called the **absolute complement** of M, or sometimes just the **complement** of M, and is often written as M'. (Other authors use the notation M^c, $\bar{M}$, or $\tilde{M}$ for the complement of M.)

FIGURE 1-6

When you refer to the complement M' of a set M, you must be certain that U, the universe of discourse, is clearly specified or understood. In set algebra, U is usually called the **universal set**. Note that

$$U' = \varnothing \quad \text{and} \quad \varnothing' = U.$$

Also, if you complement a set twice, you obtain the original set; that is,

$$(N')' = N.$$

Because people like to use Venn diagrams to help visualize sets, the elements of a set M are often referred to as the points of M. The word *point*, used in this way, is not necessarily to be taken literally in its usual geometric sense.

Many of the theorems of set algebra state that one set, built up in a certain way from arbitrary sets by means of union, intersection, and relative (or absolute) complementation, is contained in or is equal to another such set. The **pick-a-point process** is one way to prove that a first set is contained in a second set: *Pick an arbitrary point (element) in the first set and argue that it must be in the second set.*

5.3 Example If M and N are sets, prove that $M \cap N \subseteq M$.

SOLUTION We use the pick-a-point process: Let $x \in M \cap N$. Then $x \in M$ and $x \in N$; so, in particular, $x \in M$. This shows that every element x in $M \cap N$ belongs to M, that is, that $M \cap N \subseteq M$. □

A theorem stating that two sets are equal can be proved by using the pick-a-point process twice to show that each set is contained in the other. Sometimes such a proof can be shortened by arguing directly that an arbitrary point belongs to the one set *if and only if* it belongs to the other set. This technique is illustrated in the proof of the following theorem:

5.4 THEOREM **DISTRIBUTIVITY OF $\cap$ OVER $\cup$** If L, M, and N are sets, then

$$L \cap (M \cup N) = (L \cap M) \cup (L \cap N).$$

PROOF Let x be an arbitrary point in the universal set U. Then, by Part (ii) of Definition 5.1,

$$x \in L \cap (M \cup N) \Leftrightarrow x \in L \wedge (x \in M \cup N),$$

and by Part (i) of Definition 5.1,

$$x \in M \cup N \Leftrightarrow x \in M \vee x \in N;$$

hence,

$$x \in L \cap (M \cup N) \Leftrightarrow x \in L \wedge (x \in M \vee x \in N).$$

Similarly,

$$x \in (L \cap M) \cup (L \cap N) \Leftrightarrow (x \in L \wedge x \in M) \vee (x \in L \wedge x \in N).$$

But, by tautology 8 on page 10, with P replaced by $x \in L$, Q replaced by $x \in M$, and R replaced by $x \in N$, we have

$$x \in L \wedge (x \in M \vee x \in N) \Leftrightarrow (x \in L \wedge x \in M) \vee (x \in L \wedge x \in M),$$

and it follows that

$$x \in L \cap (M \cup N) \Leftrightarrow x \in (L \cap M) \cup (L \cap N). \qquad \square$$

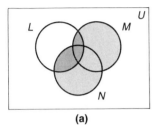

(a)

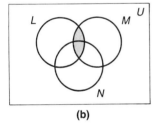

(b)

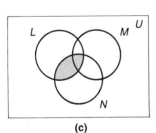

(c)

FIGURE 1-7

The Venn diagrams in Figure 1-7 illustrate the distributive law in Theorem 5.4. Part (a) shows $M \cup N$ shaded and $L \cap (M \cup N)$ with darker shading; Part (b) shows $L \cap M$ shaded, and Part (c) shows $L \cap N$ shaded. Clearly, $L \cap (M \cup N)$ in Part (a) is the union of $L \cap M$ in Part (b) and $L \cap N$ in Part (c).

Notice the analogy between the distributive law

$$L \cap (M \cup N) = (L \cap M) \cup (L \cap N)$$

for sets and the distributive law

$$P \wedge (Q \vee R) \Leftrightarrow (P \wedge Q) \vee (P \wedge R)$$

for propositions, which was, in fact, used in its proof. In a similar way, many of the tautologies of propositional calculus give rise to corresponding theorems for sets, and this creates a far-reaching analogy between propositional calculus and the algebra of sets. There is a further analogy between the distributive law above for sets and the familiar distributive law

$$x(y + z) = xy + xz$$

of ordinary algebra. Such analogies between the laws of combination for sets and the laws of ordinary algebra originally encouraged George Boole to regard sets and set operations as a new kind of algebra. In his honor, the algebra of sets and other similar mathematical systems are now called *Boolean algebras*.

In the following list of some of the basic theorems of set algebra, M, N, and L denote sets and U denotes the universal set.

1. $M = (M')'$	Double complementation
2. $M \cap N = N \cap M$	Commutative law for $\cap$
3. $M \cup N = N \cup M$	Commutative law for $\cup$
4. $(M \cap N) \cap L = M \cap (N \cap L)$	Associative law for $\cap$
5. $(M \cup N) \cup L = M \cup (N \cup L)$	Associative law for $\cup$
6. $M \cup M' = U$	First complementation law
7. $M \cap M' = \varnothing$	Second complementation law

8. $M \cap (N \cup L) = (M \cap N) \cup (M \cap L)$	Distributive law for $\cap$ over $\cup$	
9. $M \cup (N \cap L) = (M \cup N) \cap (M \cup L)$	Distributive law for $\cup$ over $\cap$	
10. $(M \cap N)' = M' \cup N'$	De Morgan's law for $\cap$	
11. $(M \cup N)' = M' \cap N'$	De Morgan's law for $\cup$	
12. $M \setminus N = M \cap N'$	Relative complement law	
13. $M \setminus (N \cap L) = (M \setminus N) \cup (M \setminus L)$	De Morgan's law for relative complements	
14. $M \setminus (N \cup L) = (M \setminus N) \cap (M \setminus L)$	De Morgan's law for relative complements	
15. $(M \cup N) \cap N = N$	First absorption law	
16. $(M \cap N) \cup N = N$	Second absorption law	
17. $M \cap M = M$	Idempotent law for $\cap$	
18. $M \cup M = M$	Idempotent law for $\cup$	

19. $M \cap N \subseteq M$

20. $M \subseteq M \cup N$

21. $\varnothing \subseteq M \subseteq U$

22. $M \subseteq N \Leftrightarrow N' \subseteq M'$

23. $M \subseteq N \Leftrightarrow M \cap N = M$

24. $M \subseteq N \Leftrightarrow M \cup N = N$

25. $M \subseteq N \Leftrightarrow M \cap N' = \varnothing$

26. $M \subseteq N \Leftrightarrow M' \cup N = U$

27. $M = N \Leftrightarrow ((M \subseteq N) \wedge (N \subseteq M))$

28. $M \cup N \subseteq L \Leftrightarrow ((M \subseteq L) \wedge (N \subseteq L))$

29. $L \subseteq M \cap N \Leftrightarrow ((L \subseteq M) \wedge (L \subseteq N))$

All of the theorems in the list above can be proved by the pick-a-point process.

5.5 Example   Prove theorem number 26 in the list above,

$$M \subseteq N \Leftrightarrow M' \cup N = U.$$

SOLUTION   We prove that the two statements $M \subseteq N$ and $M' \cup N = U$ are equivalent by proving that each implies the other. We begin by proving that

$$M \subseteq N \Rightarrow M' \cup N = U.$$

Assume that

$$M \subseteq N.$$

On the basis of this assumption, we must show that

$$M' \cup N = U.$$

Because every set is contained in U, we automatically have

$$M' \cup N \subseteq U,$$

so it remains to argue that

$$U \subseteq M' \cup N.$$

Thus, let $x \in U$. We must show $x \in M' \cup N$. If $x \in M'$, then $x \in M' \cup N$, so we only need to consider the case in which $x \notin M'$; that is, we can suppose that $x \in M$. But then, by the assumption that $M \subseteq N$, we have $x \in N$, and it follows that $x \in M' \cup N$. This completes the proof that

$$M \subseteq N \Rightarrow M' \cup N = U.$$

Now, to prove the converse implication,

$$M' \cup N = U \Rightarrow M \subseteq N,$$

we begin by assuming that

$$M' \cup N = U.$$

On the basis of this assumption, we must show that

$$M \subseteq N.$$

To this end, let $x \in M$. Our proof will be complete if we can argue that $x \in N$. Since

$$x \in U \quad \text{and} \quad U = M' \cup N,$$

it follows that

$$x \in M' \cup N;$$

hence, $x \in M'$ or $x \in N$. Because $x \in M$, we cannot have $x \in M'$, and the only alternative is that $x \in N$. $\square$

Although the following theorem is very interesting, its proof is too difficult to include here.

5.6 THEOREM

> **DEDUCTIVE DEVELOPMENT OF SET ALGEBRA**   Every identity of the algebra of sets is a consequence of the commutative laws 2 and 3, the associative laws 4 and 5, the complementation laws 6 and 7, the distributive laws 8 and 9, the relative complementation law 12, the absorption laws 15 and 16, and the containment laws 21 and 23.

Thus, the thirteen laws mentioned in this theorem could be taken as postulates, and the algebra of sets could be derived deductively from them. (Note, however, that the algebra of sets is only a portion of the theory of sets.) By definition, an abstract mathematical system satisfying these thirteen postulates is called a **Boolean algebra**. Actually, more economical systems of postulates for a Boolean algebra have been found, and one can get along with fewer than thirteen postulates. Also, some of the thirteen laws mentioned above are redundant in the sense that they can be derived deductively from the remaining ones. For instance, the second distributive law, 9, can be derived from the remaining twelve laws.

The following example illustrates how an identity of Boolean algebra can be derived deductively from the thirteen laws mentioned in Theorem 5.6 (as opposed to a pick-a-point argument).

5.7 Example   Give a deductive proof of the idempotent law 18, for union: $M \cup M = M$.

SOLUTION   The first absorption law, 15, holds for all sets M and N, so we can replace the symbol N in 15 by the symbol M to obtain

$$(M \cup M) \cap M = M. \tag{1}$$

Next, we replace the symbol M in the second absorption law, 16, by $M \cup M$ to obtain

$$((M \cup M) \cap N) \cup N = N. \tag{2}$$

Now, replacing N by M in (2), we have

$$((M \cup M) \cap M) \cup M = M. \tag{3}$$

Finally, substituting (1) into (3), we have

$$M \cup M = M. \tag{4}$$

$\square$

It is often possible to count the elements of a set M. The result of such a counting process is a nonnegative integer m called the **number** of elements in M. Counting the elements of a set M in two different ways always results in the same number m (unless a mistake in counting has been made). We introduce the notation $\#M$ for the number of elements in M. Thus, the equation

$$\#M = m$$

means that the result of counting the elements of the set M is the nonnegative integer m. For instance,

$$\#\{3, 5, 7, 9\} = 4.$$

Note that

$$\#\varnothing = 0.$$

In set theory, $\#M$ is called the **cardinal number** of the set M. Other authors use different notation for the cardinal number of a set; one popular alternative is $|M|$.

5.8 Example   If U is the set of all real numbers, find

$$\# \{x \,|\, x^2 - 4x + 3 = 0\}.$$

SOLUTION   Because

$$\{x \,|\, x^2 - 4x + 3 = 0\} = \{1, 3\},$$

we have

$$\# \{x \,|\, x^2 - 4x + 3 = 0\} = 2. \qquad \square$$

At this point in our development, it is not possible to make a rigorous formal definition of $\# M$; however, since we all know how to count, for the time being we are going to make free use of the idea of the number of elements in a set M, at least when M is a finite set. (Later, in Section 4.6, we give a more formal development of the idea of the cardinal number of a set.) For our present purposes, a set M is finite if the process of counting its elements eventually terminates and results in a nonnegative integer m. We assume that finite sets have the following properties:

1. If M is a finite set, then $\# M$ is a nonnegative integer.
2. If $\# M$ is a nonnegative integer, then M is a finite set.
3. A subset of a finite set is a finite set.
4. If $M = \{a\}$ is a singleton set, then $\# M = 1$.
5. If M and N are finite sets, then $M \cup N$ is a finite set.

Of course, we cannot give rigorous proofs of theorems involving the concept of the number of elements in a set until we have a formal definition of $\# M$. In the meantime, we propose to settle for informal intuitive proofs of such theorems based on our usual understanding of the counting process.

5.9 THEOREM

> **THE NUMBER OF ELEMENTS IN A UNION OF TWO SETS**   Let M and N be finite sets. Then,
>
> $$\#(M \cup N) = \# M + \# N - \#(M \cap N).$$

INFORMAL PROOF   If you count the elements of M and N separately, you will have counted the elements in $M \cap N$ twice: once when you count the elements of M and again when you count the elements of N. Thus, you can obtain the number of elements in the union of M and N by adding the number of elements in M to the number of elements in N and then compensating for the elements counted twice by subtracting the number of elements in $M \cap N$. $\qquad \square$

5.10 DEFINITION

DISJOINT SETS Two sets M and N are said to be **disjoint** if $M \cap N = \varnothing.$[†] The union of two disjoint sets M and N is sometimes written as

$$M \cup N$$

and called the **disjoint union** of M and N.

5.11 THEOREM

CARDINAL NUMBER OF A DISJOINT UNION If M and N are disjoint finite sets, then

$$\#(M \cup N) = \#M + \#N.$$

PROOF Since M and N are disjoint,

$$\#(M \cap N) = \#\varnothing = 0.$$

Therefore, by Theorem 5.9,

$$\#(M \cup N) = \#M + \#N - \#(M \cap N)$$
$$= \#M + \#N - 0$$
$$= \#M + \#N.$$ □

5.12 THEOREM

THE NUMBER OF ELEMENTS IN A RELATIVE COMPLEMENT
If M and N are sets and M is finite, then

$$\#(M \setminus N) = \#M - \#(M \cap N).$$

PROOF Note that $M \setminus N$ and $M \cap N$ are disjoint sets (see Problem 15a). In what follows, we use the laws of set algebra on pages 59–60. By the relative complementation law 12, we have

$$(M \cap N) \cup (M \setminus N) = (M \cap N) \cup (M \cap N'), \tag{5}$$

and by the distributive law for $\cap$ over $\cup$, 8, we have

$$(M \cap N) \cup (M \cap N') = M \cap (N \cup N'). \tag{6}$$

Combining Equations (5) and (6), we find that

$$(M \cap N) \cup (M \setminus N) = M \cap (N \cup N'). \tag{7}$$

[†] In definitions, people often write *if* when they really mean *if and only if*. Although this dubious practice would never be followed in writing very formal mathematics, it does not seem to cause any real confusion in more informal situations.

By the first complementation law 6, $N \cup N' = U$; so we can rewrite (7) as

$$(M \cap N) \cup (M \setminus N) = M \cap U. \tag{8}$$

The containment law 21 implies that $M \subseteq U$; hence, by the containment law 23, $M \cap U = M$. Therefore, (8) can be rewritten as

$$(M \cap N) \cup (M \setminus N) = M. \tag{9}$$

By (9) and Theorem 5.11, we have

$$\#(M \cap N) + \#(M \setminus N) = \#M, \tag{10}$$

from which it follows that

$$\#(M \setminus N) = \#M - \#(M \cap N). \tag{11}$$

□

5.13 COROLLARY

If M is a finite set and $N \subseteq M$, then

$$\#(M \setminus N) = \#M - \#N.$$

The proof of Corollary 5.13 follows easily from Theorem 5.12, and we leave it for you as an exercise (Problem 27).

5.14 THEOREM

THE NUMBER OF SUBSETS OF A SET If M is a finite set and $\#M = m$, then there are exactly 2^m different subsets of M. Therefore,

$$\#\mathscr{P}(M) = 2^{\#M}.$$

INFORMAL PROOF The case in which $M = \varnothing$ is addressed in Problem 28. Thus, suppose $M \neq \varnothing$; say,

$$M = \{a_1, a_2, a_3, \ldots, a_m\}.$$

Let N denote an arbitrary subset of M, and for each

$$i = 1, 2, 3, \ldots, m,$$

let P_i represent the proposition stating that a_i is an element of N, so that

$$P_i \Leftrightarrow a_i \in N.$$

By the axiom of extent, N is determined by its elements; hence, N is determined by the truth values of the propositions

$$P_1, P_2, P_3, \ldots, P_m.$$

In this way, the different subsets N of M correspond to the different truth combinations for these m propositions. In Section 1.1 we showed that there are exactly 2^m such truth combinations, and it follows that there are exactly 2^m different subsets of M. □

5.15 Example   In the ABO-system, human blood is grouped according to which of the three antigens A, B, and Rh it contains. For instance, type O Rh+ blood contains the Rh antigen but not the A or the B antigen; and type AB Rh− blood contains both the A and the B antigen but not the Rh antigen. How many different blood types are possible in the ABO-system?

SOLUTION   Let $M = \{A, B, Rh\}$. Corresponding to each subset N of M is a blood type containing all the antigens in N and none of the antigens in $M \setminus N$. By Theorem 5.14, there are $2^3 = 8$ such subsets; hence, in the ABO-system there are eight different blood types. □

PROBLEM SET 1.5

1. Let $M = \{1, 2, 3, 4\}$, $N = \{3, 4, 5\}$, $Q = \{3, 4, 6, 7\}$, and $P = \{6, 8, 9\}$. Find each set (in explicit form).
    (a) $M \cup N$ (b) $M \cap N$ (c) $M \cap Q$
    (d) $Q \cup P$ (e) $M \setminus N$ (f) $Q \setminus P$
    (g) $M \setminus Q$ (h) $Q \setminus N$ (i) $M \cup (Q \cap P)$
    (j) $M \cap (Q \cup P)$ (k) $(N \cup P) \cap Q$ (l) $(Q \cap P) \cup N$
    (m) $(M \cup N) \setminus (Q \cap P)$ (n) $(M \cap N) \cup (M \setminus P)$

2. In Problem 1, which pairs, if any, of the sets M, N, Q, and P are disjoint?

3. If M, N, and L denote three arbitrary sets, draw Venn diagrams and use shading to illustrate the following:
    (a) $M \setminus (N \cup L)$ (b) $(M \cap N) \setminus L$ (c) $M \cap N \cap L$
    (d) $M' \cap N' \cap L'$ (e) $(M \setminus N) \cup (N \setminus M)$

4. If U is the universal set and M is an arbitrary set, indicate which of the following are necessarily true and which could be false.
    (a) $M \cap \emptyset = \emptyset$ (b) $\emptyset \cup U = U$ (c) $M \cap U = U$
    (d) $M \setminus M = \emptyset$ (e) $M \setminus U = M$ (f) $U \setminus M = \emptyset$
    (g) $M \cup U = U$ (h) $\emptyset \setminus M = M'$ (i) $M \setminus \emptyset = M$

5. Let $U = \{2, 3, 4, 5, 6\}$, $M = \{2, 3, 4\}$, and $N = \{2, 4, 5, 6\}$. Find each set (in explicit form).
    (a) M' (b) N' (c) $M' \cup N'$
    (d) $M' \cap N'$ (e) $(M \cap N)'$ (f) $M' \cap M$
    (g) $N' \cup N$ (h) $M \setminus N$ (i) $N \setminus M$

6.   If M and N are sets, prove that each of the following conditions is equiva-
     lent to the condition that $M \subseteq N$:
     (a) $M \cap N' = \emptyset$ (b) $M \cup N = N$ (c) $M \cap N = M$
     (d) $N' \subseteq M'$ (e) $M' \cap N' = N'$ (f) $M' \cup N' = M'$

7.   Prove the distributive law for $\cup$ over $\cap$.

8.   Use the pick-a-point method to prove the relative complement law,
     $M \setminus N = M \cap N'$.

9.   Prove the two De Morgan laws:
     (a) $(M \cap N)' = M' \cup N'$ (b) $(M \cup N)' = M' \cap N'$

10. Let U be the universal set, and let M and N be arbitrary sets. What con-
     ditions must be imposed on M and N to make each of the following true?
     (a) $M \cup N = \emptyset$ (b) $M \cap N = U$ (c) $M \cup U = M$
     (d) $M \cup \emptyset = \emptyset$ (e) $M \cap U = U$ (f) $M \cap U = M$

11. Illustrate the two De Morgan laws in Problem 9 by Venn diagrams.

12. Prove the two De Morgan laws for relative complements (13 and 14 in
     the list on page 60) and illustrate these laws with Venn diagrams.

13. Prove the two absorption laws (15 and 16 on page 60) by the pick-a-point
     method.

14. Give a deductive proof, patterned after the proof of Example 5.7, for the
     idempotent law for intersection: $M \cap M = M$.

15. In the proof of Theorem 5.12, a deductive proof is given of the identity
     $(M \cap N) \cup (M \setminus N) = M$.
     (a) Verify that the sets $M \cap N$ and $M \setminus N$ are disjoint.
     (b) Give a pick-a-point proof of this identity.
     (c) Illustrate this identity with a Venn diagram.

16. Prove that the second distributive law 9 (in the list on page 60) can be
     derived deductively from the remaining twelve laws mentioned in Theo-
     rem 5.6. (You may not have to use all twelve of the remaining laws to do
     this.)

17. True or false: $M \cup L \subseteq N \cup L \Rightarrow M \subseteq N$. If true, prove it; if false, give a
     counterexample.

18. If M, N, L, and P are sets with $M \subseteq N$ and $L \subseteq P$, prove that $L \setminus N \subseteq$
     $P \setminus M$.

     *In Problems 19–22, the notation $M \, \Delta \, N$ is used for the **symmetric differ-
     ence** of the two sets M and N, defined by*

     $$M \, \Delta \, N = (M \setminus N) \cup (N \setminus M).$$

19. If $M = \{1, 2, 3\}$ and $N = \{2, 3, 4, 5\}$, find each set (in explicit form).
     (a) $M \, \Delta \, N$ (b) $N \, \Delta \, M$ (c) $M \, \Delta \, M$ (d) $M \, \Delta \, \emptyset$

20. Draw Venn diagrams for each of the following:
(a) $M \Delta N$ (b) $M \Delta (N \Delta L)$ (c) $(M \Delta N) \Delta L$

21. Prove that $M \Delta N = (M \cup N) \setminus (M \cap N)$.

22. If U is the universal set, and M and N are arbitrary sets, prove that symmetric difference has the following properties:
(a) $M \Delta N = N \Delta M$ (b) $M \Delta M = \varnothing$
(c) $M \Delta M' = U$ (d) $M \Delta \varnothing = M$
(e) $M \Delta U = M'$ (f) $M \Delta (N \Delta L) = (M \Delta N) \Delta L$
(g) $M \cap (N \Delta L) = (M \cap N) \Delta (M \cap L)$

23. Let the universal set U consist of all real numbers. Find:
(a) $\#\{x \mid 2x^2 - x - 1 = 0\}$ (b) $\#\{x \mid 2x^2 + x + 1 = 0\}$
(c) $\#\{x \mid 9x^2 - 12x + 4 = 0\}$ (d) $\#\{x \mid x^3 = 8\}$

24. Let M be a finite set with $\#M = m$. Prove that there are exactly $(3^m + 1)/2$ different pairs of disjoint subsets of M.

25. A farmer has 41 pigs. Every fat pig is greedy; 20 of the pigs are both fat and healthy; 8 of the pigs are greedy but not healthy; and 30 of the pigs are greedy. Also, 1 pig is neither healthy nor greedy; and 5 pigs are greedy but not fat.
(a) How many pigs are healthy but not greedy?
(b) How many pigs are greedy but neither fat nor healthy?
(c) How many pigs are healthy and greedy but not fat? [Hint: Draw a Venn diagram with U = all 41 pigs, G = all greedy pigs, F = all fat pigs, and H = all healthy pigs. Label sets in the diagram with their known cardinal numbers, and deduce from this the remaining cardinal numbers.]

26. Let M be a finite set, let $N \subseteq M$, and let $d = \#(M \setminus N)$. Prove that $\#\{L \mid N \subseteq L \subseteq M\} = 2^d$.

27. Prove Corollary 5.13.

28. The proof of Theorem 5.14 does not treat the case in which $M = \varnothing$.
(a) What is $\mathscr{P}(\varnothing)$?
(b) Show that Theorem 5.14 is true even if $M = \varnothing$.

29. As in Problems 19–22, define the symmetric difference of sets M and N by $M \Delta N = (M \setminus N) \cup (N \setminus M)$. If M and N are finite sets, prove that $M \Delta N$ is a finite set and that $\#(M \Delta N) = \#M + \#N - 2\#(M \cap N)$.

30. In the theory of permutations and combinations, the symbol C_r^n, called the *number of combinations of n things taken r at a time*, is defined for non-negative integers n and r, with $r \leq n$, by

$$C_r^n = \frac{n!}{(n - r)!r!},$$

where $n! = 1 \cdot 2 \cdot 3 \cdots (n - 1)n$ for $n > 0$, and $0! = 1$. Show that if M is a set containing n elements, then the number of different subsets of M containing exactly r elements is C_r^n.

31. Let M be a finite set with $\#M = m$.
(a) How many proper subsets does M have?
(b) How many nontrivial subsets does M have?

32. By combining Theorem 5.14 and Problem 30, show that if n is a nonnegative integer, then

$$C_0^n + C_1^n + C_2^n + \cdots + C_{n-1}^n + C_n^n = 2^n.$$

33. In drawing a Venn diagram to illustrate a theorem of set algebra, you must be careful that the circles or other figures representing the sets are "in a general position." For instance, if only two sets are involved, and if the theorem does not contain a hypothesis that the two sets are disjoint, you must make certain that the circles representing these sets overlap. Try to give a careful definition of the phrase "in a general position" as it would apply to a Venn diagram involving three sets.

34. In set theory, it is possible to prove the following theorem: *A statement asserting the equality of two sets or a containment of one set in another, where the two sets are formed from sets $M_1, M_2, M_3, \ldots, M_n$ using the operations of union, intersection, and relative or absolute complementation, can be tested by replacing the sets $M_1, M_2, M_3, \ldots, M_n$ by the null set $\varnothing$ and the universal set U in every possible way. (This requires 2^n separate tests.) If the statement is true in each of these 2^n special cases, then it is a theorem of set algebra.* Using this theorem as a basis, describe a tabular procedure, similar to the truth-table procedure for checking tautologies, that tests for theorems in set algebra.

35. Suppose that the square array (matrix) of sets

$$\begin{bmatrix} H & L \\ M & N \end{bmatrix}$$

has the following properties: (i) Sets appearing in the same row are disjoint; that is, $H \cap L = M \cap N = \varnothing$. (ii) The union of the sets appearing in any column is a fixed set G; that is, $H \cup M = L \cup N = G$. Prove that the union of the sets appearing in any row must be the fixed set G; that is, prove that $H \cup L = M \cup N = G$.

36. Use the tabular test suggested in Problem 34 to show that

$$M_1 \setminus (M_2 \cup M_3 \cup M_4) = (M_1 \setminus M_2) \cap (M_1 \setminus M_3) \cap (M_1 \setminus M_4)$$

is a theorem of set algebra.

37. Generalize the theorem of Problem 35 to an arbitrary n-by-n square array (matrix) of sets.

6

GENERALIZED UNIONS AND INTERSECTIONS

This section discusses generalizations of the notions of union and intersection to indexed families of sets and to sets of sets. Before we make the appropriate definitions, it is necessary to point out that there is an important (but somewhat subtle) distinction between an indexed family of sets and a set of sets.

For an indexed family of sets

$$(M_i)_{i \in I},$$

we have a definite correspondence that associates a set M_i with each element i in the index set I, whereas for a set of sets

$$\mathscr{E} = \{A, B, C, \ldots\}$$

no such correspondence is automatically provided.

For instance, suppose that

$$I = \{1, 2, 3\}$$

and consider the indexed family of sets

$$(M_i)_{i \in I}$$

given by

$$M_1 = \{\text{Olga, Carlos}\},$$
$$M_2 = \{\text{Rose, Pedro}\},$$
$$M_3 = \{\text{Anna, Boris}\}.$$

Here we have a definite correspondence that associates

the set $\{\text{Olga, Carlos}\}$ with the index $1 \in I$,

the set $\{\text{Rose, Pedro}\}$ with the index $2 \in I$, and

the set $\{\text{Anna, Boris}\}$ with the index $3 \in I$.

However, no such correspondence can be discerned merely from the set of sets $\mathscr{E}$ given by

$$\mathscr{E} = \{\{\text{Olga, Carlos}\}, \{\text{Rose, Pedro}\}, \{\text{Anna, Boris}\}\}.$$

(Recall that no particular order is associated with the elements of a set.)

Thus, although an indexed family of sets

$$(M_i)_{i \in I}$$

gives rise to a set of sets

$$\mathscr{E} = \{M_i \mid i \in I\},$$

it does not automatically work the other way around. However, given a set of sets $\mathscr{E}$, it is possible to form a corresponding indexed family of sets

$(M_i)_{i \in I}$

by choosing an appropriate index set I, setting up a suitable correspondence between the elements $i \in I$ and the sets in $\mathscr{E}$, and taking M_i to be the corresponding set in $\mathscr{E}$ for each $i \in I$. The process of choosing such an I and setting up such a correspondence is called **indexing** the sets in $\mathscr{E}$. Note, however, that there are many different ways in which a given set of sets $\mathscr{E}$ can be indexed—*indexing is not a unique process!*

6.1 Example   Index the set of sets $\mathscr{E}$ given by

$\mathscr{E} = \{\{red, yellow\}, \{green, blue, purple\}, \{pink, green\}\}.$

SOLUTION   We use $I = \{p, q, r\}$ as our indexing set and define the indexed family of sets

$(M_i)_{i \in I}$

by taking

$M_p = \{red, yellow\},$
$M_q = \{green, blue, purple\},$
$M_r = \{pink, green\}.$ □

It is often possible to ignore the distinction between an indexed family of sets and a set of sets without running into mathematical or logical difficulties, and people often do this in the interest of writing or speaking more concisely. We indulge in this practice when it is convenient and harmless. But, *be careful*, in some situations serious logical trouble results from confusing an indexed family of sets with a set of sets.

Now we are ready to give the generalized definitions of union and intersections, beginning with the generalized union.

6.2 DEFINITION

> **UNION OF A SET OF SETS**   Let $\mathscr{E}$ be a set of sets. We define the **union** of the sets in $\mathscr{E}$, denoted $\bigcup \mathscr{E}$, by
>
> $$\bigcup \mathscr{E} = \{x \mid (\exists M)(M \in \mathscr{E} \wedge x \in M)\}.$$

In words, *the union $\bigcup \mathscr{E}$ of the set of sets $\mathscr{E}$ is the set consisting of all elements that belong to at least one of the sets M in $\mathscr{E}$.* Alternative notation for $\bigcup \mathscr{E}$ is

$\bigcup_{M \in \mathscr{E}} M$ or $\bigcup_{M \in \mathscr{E}} M.$

6.3 Example Let $\mathscr{E} = \{A, B, C\}$ where $A = \{2, 4, 6\}$, $B = \{1, 2, 5\}$, and $C = \{2, 6, 8\}$. Find $\bigcup \mathscr{E}$.

SOLUTION $\bigcup \mathscr{E}$ consists of all elements that belong to at least one of the sets A, B, or C. Therefore,

$$\bigcup \mathscr{E} = \{1, 2, 4, 5, 6, 8\}.$$ □

If $\mathscr{E}$ is a finite set of sets, say,

$$\mathscr{E} = \{M_1, M_2, M_3, \ldots, M_n\},$$

then we often write $\bigcup \mathscr{E}$ in the alternative form

$$\bigcup \mathscr{E} = M_1 \cup M_2 \cup M_3 \cup \cdots \cup M_n.$$

For instance, in Example 6.3,

$$\bigcup \mathscr{E} = A \cup B \cup C = \{1, 2, 4, 5, 6, 8\}.$$

Also, if $(M_i)_{i \in I}$ is an indexed family of sets, we define

$$\bigcup_{i \in I} M_i = \bigcup_{i \in I} M_i = \bigcup \{M_i | i \in I\},$$

so that

$$\bigcup_{i \in I} M_i = \{x | (\exists i)(i \in I \wedge x \in M_i)\}.$$

In words, *the union $\bigcup_{i \in I} M_i$ of an indexed family of sets is the set consisting of all elements that belong to at least one of the sets M_i, for some $i \in I$.*

6.4 Example Find the union $\bigcup_{i \in I} M_i$ for the indexed family of sets $(M_i)_{i \in I}$ in Example 6.1.

SOLUTION In Example 6.1, we have $I = \{p, q, r\}$, $M_p = \{\text{red, yellow}\}$, $M_q = \{\text{green, blue, purple}\}$, and $M_r = \{\text{pink, green}\}$. Therefore,

$$\bigcup_{i \in I} M_i = M_p \cup M_q \cup M_r$$
$$= \{\text{red, yellow, green, blue, purple, pink}\}.$$ □

The integers from 1 to n inclusive, or the set of all positive integers, is often used as the index set I for a family of sets $(M_i)_{i \in I}$. If $I = \{1, 2, 3, \ldots, n\}$, then the union of the family of sets $(M_i)_{i \in I}$ can be written in the alternative form

$$\bigcup_{i=1}^{n} M_i \quad \text{or} \quad \bigcup_{i=1}^{n} M_i.$$

Likewise, if $I = \{1, 2, 3, \ldots\}$, the notation

$$\bigcup_{i=1}^{\infty} M_i \quad \text{or} \quad \bigcup_{i=1}^{\infty} M_i$$

may be used for the union of the family of sets $(M_i)_{i \in I}$.

6.5 Example   Let the universal set U consist of all real numbers, and for each positive integer i, let

$$M_i = \{x \mid i - 1 \leq x \leq i\}.$$

Thus, with the usual interval notation used in calculus, M_i is the closed interval $[i - 1, i]$. Find $\bigcup_{i=1}^{\infty} M_i$.

SOLUTION   Since each nonnegative real number belongs to at least one of the intervals $M_i = [i - 1, i]$, we have

$$\bigcup_{i=1}^{\infty} M_i = \{x \mid 0 \leq x\} = [0, \infty).$$ □

6.6 THEOREM

> **GENERALIZED DISTRIBUTIVE LAW FOR ∩ OVER ∪**   Let N be a set and let $(M_i)_{i \in I}$ be an indexed family of sets. Then,
>
> $$N \cap \left(\bigcup_{i \in I} M_i \right) = \bigcup_{i \in I} (N \cap M_i).$$

PROOF   We use the pick-a-point procedure. First, suppose that x is a point in the set on the left side of the equation. Then,

$$x \in N \quad \text{and} \quad x \in \bigcup_{i \in I} M_i.$$

Because

$$x \in \bigcup_{i \in I} M_i,$$

it follows that there exists an index $j \in I$ such that

$$x \in M_j.$$

Therefore,

$$x \in N \quad \text{and} \quad x \in M_j;$$

that is,

$$x \in N \cap M_j.$$

Consequently,

$$x \in \bigcup_{i \in I} (N \cap M_i);$$

so x belongs to the set on the right side of the equation.

Now, suppose that x is a point in the set on the right side of the equation. Then there exists an index $j \in I$ such that

$$x \in N \cap M_j;$$

that is,

$$x \in N \quad \text{and} \quad x \in M_j.$$

Because $x \in M_j$, it follows that

$$x \in \bigcup_{i \in I} M_i.$$

Therefore,

$$x \in N \quad \text{and} \quad x \in \bigcup_{i \in I} M_i,$$

and it follows that x belongs to the set on the left side of the equation.   ☐

The generalized intersection is defined as follows:

6.7 DEFINITION

INTERSECTION OF A SET OF SETS   Let $\mathscr{E}$ be a set of sets. We define the **intersection** of the sets in $\mathscr{E}$, denoted $\bigcap \mathscr{E}$, by

$$\bigcap \mathscr{E} = \{x \,|\, (\forall M)(M \in \mathscr{E} \Rightarrow x \in M)\}.$$

In words, *the intersection $\bigcap \mathscr{E}$ of the set of sets $\mathscr{E}$ is the set consisting of all elements that belong to every one of the sets M in $\mathscr{E}$.* Alternative notation for $\bigcap \mathscr{E}$ is

$$\bigcap_{M \in \mathscr{E}} M \quad \text{or} \quad \bigcap_{M \in \mathscr{E}} M.$$

6.8 Example   Let $\mathscr{E} = \{A, B, C\}$ where $A = \{2, 4, 6\}$, $B = \{1, 2, 5\}$, and $C = \{2, 6, 8\}$ as in Example 6.3. Find $\bigcap \mathscr{E}$.

SOLUTION $\bigcap \mathscr{E}$ consists of all elements that belong to *all three* of the sets A, B, and C. Therefore,

$$\bigcap \mathscr{E} = \{2\}.$$ ☐

If $\mathscr{E}$ is a finite set of sets, say,

$$\mathscr{E} = \{M_1, M_2, M_3, \ldots, M_n\},$$

then we often write $\bigcap \mathscr{E}$ in the alternative form,

$$\bigcap \mathscr{E} = M_1 \cap M_2 \cap M_3 \cap \cdots \cap M_n.$$

For instance, in Example 6.8,

$$\bigcap \mathscr{E} = A \cap B \cap C = \{2\}.$$

Also, if $(M_i)_{i \in I}$ is an indexed family of sets, we define

$$\bigcap_{i \in I} M_i = \bigcap_{i \in I} M_i = \bigcap \{M_i | i \in I\},$$

so that

$$\bigcap_{i \in I} M_i = \{x | (\forall i)(i \in I \Rightarrow x \in M_i\}.$$

In words, *the intersection $\bigcap_{i \in I} M_i$ of an indexed family of sets is the set consisting of all elements that belong to every one of the sets M_i, for $i \in I$.*

6.9 Example   As in Example 6.1, let $I = \{p, q, r\}$ with

$M_p = \{\text{red, yellow}\},$
$M_q = \{\text{green, blue, purple}\},$
$M_r = \{\text{pink, green}\}.$

Find $\bigcap_{i \in I} M_i$.

SOLUTION   There are no elements that belong to all three of the sets M_p, M_q, and M_r; hence,

$$\bigcap_{i \in I} M_i = \varnothing. \qquad \square$$

6.10 DEFINITION

PAIRWISE DISJOINT SETS OF SETS   A set of sets $\mathscr{E}$ is said to be **pairwise disjoint** if, for all sets M and N in $\mathscr{E}$,

$$M \neq N \Rightarrow M \cap N = \varnothing.$$

Likewise, an indexed family of sets $(M_i)_{i \in I}$ is said to be pairwise disjoint if, for every pair of indices i, j in I,

$$i \neq j \Rightarrow M_i \cap M_j = \varnothing.$$

Notice that *the intersection of a pairwise disjoint set (or indexed family) of two or more sets is empty* (Problem 23). However, consider the following:

6.11 Example   Is the indexed family of sets $(M_i)_{i \in I}$ in Example 6.9 pairwise disjoint?

SOLUTION   No! Although the intersection of this family of sets is empty, it is not pairwise disjoint because

$$M_q \cap M_r = \{\text{green}\} \neq \varnothing. \qquad \square$$

We leave the proof of the following generalized distributive law as an exercise for you (Problem 18).

6.12 THEOREM

> **GENERALIZED DISTRIBUTIVE LAW FOR $\cup$ OVER $\cap$**   Let N be a set and let $(M_i)_{i \in I}$ be an indexed family of sets. Then,
>
> $$N \cup \left(\bigcap_{i \in I} M_i \right) = \bigcap_{i \in I} (N \cup M_i).$$

The De Morgan laws also generalize (Problem 24).

6.13 THEOREM

> **GENERALIZED DE MORGAN LAWS**   Let $(M_i)_{i \in I}$ be an indexed family of sets. Then,
>
> (i) $(\bigcup_{i \in I} M_i)' = \bigcap_{i \in I} M_i'.$
> (ii) $(\bigcap_{i \in I} M_i)' = \bigcup_{i \in I} M_i'.$

If $I = \{1, 2, 3, \ldots, n\}$, then the intersection of the family of sets $(M_i)_{i \in I}$ is often written in the alternative form

$$\bigcap_{i=1}^{n} M_i \quad \text{or} \quad \bigcap_{i=1}^{n} M_i.$$

Likewise, if $I = \{1, 2, 3, \ldots\}$, the notation

$$\bigcap_{i=1}^{\infty} M_i \quad \text{or} \quad \bigcap_{i=1}^{\infty} M_i$$

may be used for the intersection of the family of sets $(M_i)_{i \in I}$.

6.14 Example   Let the universal set U consist of all real numbers, and, for each positive integer i, let

$$M_i = \left\{ x \,\middle|\, 1 - \frac{1}{i} < x < 1 + \frac{1}{i} \right\}.$$

Thus, with the usual interval notation used in calculus, M_i is the open interval $(1 - (1/i), 1 + (1/i))$. Find $\bigcap_{i=1}^{\infty} M_i$.

SOLUTION   A real number x belongs to M_i if and only if $|x - 1| < 1/i$. Since $1/i$ becomes arbitrarily small as i gets larger and larger, x belongs to *every* M_i if and only if $x = 1$. Therefore,

$$\bigcap_{i=1}^{\infty} M_i = \{1\}.$$

□

The empty set $\varnothing$ can be regarded as a set of sets—the empty set of sets. As such, the statement $M \in \varnothing$ is false for *every* set M. Of course, we have

$$\bigcup \varnothing = \varnothing;$$

in words, *the union of the empty set of sets is the empty set* (Problem 27). For the intersection of the empty set of sets, there is a bit of a surprise: *The intersection of the empty set of sets is the universal set U; in symbols,*

$$\bigcap \varnothing = U.$$

6.15 Example   Prove that $\bigcap \varnothing = U$.

SOLUTION   Since every set is a subset of U, we have only to prove that $U \subseteq \bigcap \varnothing$. We use the pick-a-point process. Let $x \in U$. To prove that $x \in \bigcap \varnothing$, we must prove that, for every set M,

$$M \in \varnothing \Rightarrow x \in M.$$

But, the hypothesis $M \in \varnothing$ of this implication is *false*; hence, the implication is *true*.   □

PROBLEM SET 1.6

In Problems 1–6, index the set of sets $\mathscr{E}$.

1. $\mathscr{E} = \{\{\text{Alabama, Alaska}\}, \{\text{Michigan, Ohio}\}, \{\text{Maine, Texas}\}\}$

2. $\mathscr{E} = \{\{\text{Washington, Roosevelt}\}, \{\text{Kennedy, Carter, Reagan}\}\}$

3. $\mathscr{E} = \{\{1, 4, 5, 9\}, \{1, 3, 4, 5\}, \{4, 5, 7, 8\}, \{1, 4, 7, 9\}, \{4, 5, 8, 9\}\}$

4. $\mathscr{E} = \mathscr{P}(\{a, b, c\})$

5. $\mathscr{E} = $ the set of all sets of the form

$$\{x \mid x \text{ is an integer multiple of } n\},$$

where n is a positive integer.

6. $\mathscr{E} = \{\{-3, n\} \mid n \text{ is a positive integer}\}$

7.   Let $I = \{1, 2, 3, 4\}$, and for $i \in I$ let $M_i = \{i, i + 2, 3i\}$. Find each set in explicit form:
     (a) M_1 (b) M_2 (c) M_4 (d) $M_1 \cup M_3$
     (e) $M_1 \cap M_3$ (f) $\bigcup_{i=1}^{4} M_i$ (g) $\bigcap_{i=1}^{4} M_i$

8.   Let $(M_i)_{i \in I}$ be the indexed family of sets in Problem 7 and let $\mathscr{E} = \{M_i \mid i \in I\}$. Reindex the set of sets $\mathscr{E}$ using $J = \{p, q, r, s\}$ as the new indexing set.

In Problems 9–12, let $\mathscr{E} = \{A, B, C\}$. Find (a) $\bigcup \mathscr{E}$ and (b) $\bigcap \mathscr{E}$ in explicit form.

9. $A = \{1, 2, 3\}, B = \{2, 3, 5\}, C = \{2, 4, 7\}$

10. $A = \{\text{tall, short, fat}\}, B = \{\text{slim, tall, short}\}, C = \{\text{tall}\}$

11. $A = \{a, b, c, d\}, B = \{b, c, d, e\}, C = \{c, d, e, f\}$

12. $A = \{-1, 0, 1, 3\}, B = \{-1, 0, 2, 4\}, C = \{2, 3, 5, 7\}$

In Problems 13–16, find (a) $\bigcup_{i \in I} M_i$ and (b) $\bigcap_{i \in I} M_i$ in explicit form.

13. $I = \{1, 2, 3\}$
$M_1 = \{red, yellow, green, blue\}$
$M_2 = \{yellow, green, blue, violet\}$
$M_3 = \{green, blue, pink, orange\}$

14. $I = \{a, b, c, d\}$
$M_a = \{1, 2, 3\}$
$M_b = \{2, 4, 5\}$
$M_c = \{2, 5, 6\}$
$M_d = \{2, 7, 8\}$

15. $I = \{1, 2, 3, 4, 5\}$
$M_1 = \{right, left\}$
$M_2 = \{right, left, up\}$
$M_3 = \{right, left, up, down\}$
$M_4 = \{right, left, up, down, front\}$
$M_5 = \{right, left, up, down, front, back\}$

16. $I = \{dog, cat, lion, tiger, canary\}$
$M_{dog} = \{domestic, quadruped, Canidae\}$
$M_{cat} = \{domestic, quadruped, Felidae\}$
$M_{lion} = \{wild, quadruped, Panthera\}$
$M_{tiger} = \{wild, quadruped, Panthera\}$
$M_{canary} = \{domestic, biped, Serinus\}$

17. (a) Which of the sets of sets in Problems 1–6 are pairwise disjoint?
(b) Which of the indexed families of sets in Problems 13–16 are pairwise disjoint?

18. Prove Theorem 6.12.

In Problems 19–22, we use the usual notation for intervals of real numbers: (a, b) for the open interval $\{x \mid a < x < b\}$, $(a, b]$ for the half-open interval $\{x \mid a < x \leq b\}$, $[a, b]$ for the closed interval $\{x \mid a \leq x \leq b\}$, $[a, \infty)$ for the unbounded interval $\{x \mid a \leq x\}$, and so on. The answer to each question should be an interval, a singleton set, or the empty set.

19. Find: (a) $\bigcup_{i=1}^{\infty} [i - 1, i)$　　(b) $\bigcap_{i=1}^{\infty} [i - 1, i)$

20. Find: (a) $\bigcup_{n=1}^{\infty} (-n, n)$　　(b) $\bigcap_{n=1}^{\infty} (-n, n)$

21. Find: (a) $\bigcup_{j=1}^{\infty} [0, 1/j]$　　(b) $\bigcap_{j=1}^{\infty} [0, 1/j]$

22. Find: (a) $\bigcup_{k=1}^{\infty} (1/k, \infty)$　　(b) $\bigcap_{k=1}^{\infty} (1/k, \infty)$

23. Suppose that I contains at least two distinct elements and that the indexed family of sets $(M_i)_{i \in I}$ is pairwise disjoint. Prove that $\bigcap_{i \in I} M_i = \varnothing$.

24. Prove Theorem 6.13.

25. If $\mathscr{E}$ is a set of sets and $\varnothing \in \mathscr{E}$, prove that $\bigcap \mathscr{E} = \varnothing$.

26. If $(M_i)_{i \in I}$ and $(N_j)_{j \in J}$ are two indexed families of sets, express the set $(\bigcup_{i \in I} M_i) \cap (\bigcup_{j \in J} N_j)$ as a union of intersections. [Use Theorem 6.6.]

27. If $\varnothing$ is regarded as a set of sets, prove that $\bigcup \varnothing = \varnothing$.

28. If $(M_i)_{i \in I}$ and $(N_j)_{j \in J}$ are two indexed families of sets, express the set $(\bigcap_{i \in I} M_i) \cup (\bigcap_{j \in J} N_j)$ as an intersection of unions. [Hint: Use Theorem 6.12.]

29. Suppose $(M_i)_{i \in I}$ is an indexed family of sets and that A and B are sets such that $A \subseteq M_i \subseteq B$ for every $i \in I$.
(a) Prove that $A \subseteq \bigcap_{i \in I} M_i$.   (b) Prove that $\bigcup_{i \in I} M_i \subseteq B$.

30. If $\mathscr{E}$ is a set of sets, is it always true that $\bigcap \mathscr{E} \subseteq \bigcup \mathscr{E}$? Why or why not?

31. Let X be a nonempty set. Show that $\mathscr{E}$ is a set of subsets of X if and only if $\mathscr{E} \in \mathscr{P}(\mathscr{P}(X))$.

32. If X is a nonempty set, then a set $\mathscr{C}$ of subsets of X is called a **closure system** for X if it has the following properties: $X \in \mathscr{C}$ and $\varnothing \neq \mathscr{E} \subseteq \mathscr{C} \Rightarrow \bigcap \mathscr{E} \in \mathscr{C}$. Let $\mathscr{C}$ be a closure system for X. If $M \subseteq X$, define the **closure** of M, in symbols $\bar{M}$, by $\bar{M} = \bigcap \{ C \in \mathscr{C} \mid M \subseteq C \}$. If $M, N \subseteq X$, prove:
(a) $M \subseteq \bar{M}$ (b) $M \subseteq N \Rightarrow \bar{M} \subseteq \bar{N}$ (c) $N = \bar{M} \Rightarrow N = \bar{N}$

HISTORICAL NOTES

Analysis is the branch of mathematics that grows out of calculus. So, in some sense, analysis began with Gottfried Wilhelm von Leibniz (1646–1716) in Germany and Sir Isaac Newton (1642–1727) in England. But, as is usually true in history, the situation is much more complicated than that. Leibniz is credited with having published the first work on differential calculus, "Nova methodus pro maximis et minimis," in the *Acta Eruditorum* of Leipzig in 1684. Newton's first publication on calculus, which he called the theory of fluxions, came 3 years later in his masterpiece, *Philosophiae Naturalis Principia Mathematica*. So, for having published first, Leibniz technically has priority, but it is generally conceded that Newton really developed these ideas as early as 1666 in his famous *annus mirabilis*, that miraculous year in which he made so many of his best discoveries: calculus, the generalized binomial theorem, the law of gravitation, and his work on the nature of color. The question of priority was not initially an issue between Newton and Leibniz. But generous recognition of Leibniz's work in the first edition of the *Principia* turned to accusation by the time of the third edition.

In fact, calculus was being developed prior to Newton and Leibniz. René Descartes (1596–1650), the French philosopher–mathematician, and Pierre de Fermat (1601–1665), that greatest of all mathematical amateurs (by profession he was a lawyer and judge), had both come close to discovering the derivative in their investigations of tangent lines. And Newton's own teacher at Cambridge, Isaac Barrow (1630–1677), had anticipated Newton's work and had a method for calculating tangent lines. So one might say that by the time Newton and Leibniz came along, the time was right for calculus. The ideas were definitely "in the air." One important thing to remember, however, is that astronomical investigations and questions in mechanics and other areas of physics were raising problems that would require calculus for solution. These were problems of velocity, and hence of tangent lines, along with related questions.

Some of the other great problems of calculus, those solved by integral calculus, had been solved by Archimedes (c. 287–212 B.C.), as it turns out, though mathematicians in the seventeenth century were unaware of this. It was not until 1906 that a Danish scholar, J. L. Heiberg, learned of a 185-leaf mathematical manuscript on sheepskin (palimpsests) in Constantinople, which, though it seemed to deal with liturgy of the Eastern Orthodox Church, in fact yielded a long-lost work of Archimedes. It was not uncommon at that time that if one had a book (a set of sheepskins) and one grew bored with the content, one just rubbed out the writing and wrote one's own book on top of it. This is what had happened to this work, now called *The Method* of Archimedes. Fortunately, Heiberg could use modern photographic techniques that allowed him to read most of the writing below the top level. The manuscript included the solutions of problems of finding areas, volumes, and centers of gravity. The method of solution, though strongly physically motivated and certainly not rigorous by modern standards, still shows a remarkable similarity to the method used today in elementary calculus courses in solving these problems by adding up slices of a solid. It is interesting to speculate on the effect on later science and technology had this method survived the tenth century, the conjectured date of the manuscript.

The relationship between differential and integral calculus was not understood, however, until the time of Newton and Leibniz, so one cannot deny their genius in putting together a workable system for solving many problems. But calculus at that time was strongly tied to geometrical or physical origins. To a great extent it worked, and for those interested only in the solution of physical problems, this was sufficient. But there were those who were not happy with the way calculus was being done. Foremost among these was a philosopher, George Berkeley (1685–1753), bishop of Cloyne, who wrote in 1734 a devastating attack on calculus in a book called *The Analyst*. There were many, including Berkeley, who were uneasy about the idea of an "infinitesimal," something that was treated in many contexts like zero but was not zero. Berkeley objected to taking nonzero increments, then letting them be zero, to arrive at the derivative, which, he conceded, yielded correct results. As he put it: "By virtue of a twofold mistake you arrive, though not at science, yet at the truth." On Newton's fluxions he expressed himself in memorable language: "And what are these fluxions? The velocities of evanescent increments. And what are these same evanescent increments? They are neither finite quantities, nor quantities infinitely small, nor yet nothing. May we not call them ghosts of departed quantities?" His words may be more poetic than scientific, but they took their toll.

Some of the prominent British mathematicians following Newton, notably Colin Maclaurin (1698–1746), devoted their energies to defending Newton's fluxions. British mathematics stagnated, and analysis in Britain was not revived until the nineteenth century. Objections to calculus on the European continent, however, mainly from Count Ehrenfried Walter von Tschirnhaus (1651–1708), were not very effective, and analysis thrived. It was one of the glorious periods in the history of mathematics, with stunning achievements by the greatest member of that famous Flemish–Swiss mathematical family, Jacques Bernoulli (1654–1705), by his brother, Jean Bernoulli (1667–1748), and by Jean's even more famous student, Leonhard Euler (1707–1783). Euler worked mainly in the courts of St. Petersburg and Berlin and was probably the finest mathematician in analysis in his day, though in this he had a rival in Joseph Louis Lagrange (1736–1813). Their work was different, however, in that Euler was the master of intuitive, algorithmic mathematics and Lagrange concentrated on general theories.

Work on analysis continued with great energy through the eighteenth century and into the nineteenth when, finally, the state of other areas of mathematics permitted a more careful treatment of the fundamental ideas of calculus.

ANNOTATED BIBLIOGRAPHY

Halmos, Paul R. *Naive Set Theory* (New York: Springer-Verlag), 1974.
This is a classic text in set theory, widely used and admired, written by one of the great expositors of mathematics in this century.

Hamilton, A. G. *Logic for Mathematicians* (Cambridge, England: Cambridge University Press), 1978.
A widely used elementary treatment of mathematical logic.

Kline, Morris. *Mathematics: The Loss of Certainty* (New York: Oxford University Press), 1980.
A book for a general audience, by an often controversial author, on the limitations of logic and mathematics in providing sure answers. Sometimes disturbing, but at the same time stimulating.

Lakatos, Imre. *Proofs and Refutations: The Logic of Mathematical Discovery* (Cambridge, England: Cambridge University Press), 1976.
An intellectual tour de force showing the development of a mathematical theorem (Euler's formula for polyhedra) through various false starts, misstatements, and subsequent modifications. Written in the form of a Socratic dialogue.

Newsom, Carroll V. *Mathematical Discourses: The Heart of Mathematical Science* (Englewood Cliffs, N.J.: Prentice-Hall), 1960.

    An interesting account for the general reader of the nature of mathematical discourse and logical argument.

Quine, Willard Van Orman. *Set Theory and Its Logic* (Cambridge, Mass.: Belknap), 1969.

    This is a fairly sophisticated but readable treatment of the subject, written by one of the giants in the field.

Renz, Peter. "Mathematical Proof: What It Is and What It Ought to Be," *Two-Year College Mathematics Journal* **12** (1981), 83–103.

    A somewhat philosophical discourse on what constitutes a proof in mathematics. Interesting examples.

2
RELATIONS AND FUNCTIONS

Contemporary mathematics is written in terms of sets, relations, and functions. It is widely believed that, ultimately, every mathematical notion can be so expressed. As we indicate in this chapter, it is possible to develop the theory of relations and functions on a set-theoretic basis; hence, it is conceivable that all of mathematics can be founded on logic and set theory. Thus, in the authoritative Bourbaki,[†] *Éléments de Mathematique*, we read that:

> Nowadays it is known to be possible, logically speaking, to derive practically the whole of known mathematics from a single source, The Theory of Sets.

The theory of relations and functions is developed on a set-theoretic basis by using the idea of an ordered pair (a, b) of elements of the universe of discourse U. Therefore, to carry out the Bourbaki program of deriving mathematics from a single set-theoretic source it is necessary to define an ordered pair (a, b) purely in terms of sets. The usual definition is

$$(a, b) = \{\{a\}, \{a, b\}\},$$

but we feel that this is a bit too sophisticated for our present purposes. Hence, in what follows, we treat ordered pairs somewhat more informally.

1
ORDERED PAIRS

If a and b are elements of the universe of discourse U, then, since the elements of a set are not supposed to occur in any particular order, we have

$$\{a, b\} = \{b, a\}.$$

[†] A collective pseudonym for a prestigious group of contemporary French mathematicians. See Paul Halmos, "Nicolas Bourbaki," *Scientific American* (May 1957).

For this reason, the set $\{a, b\}$ is sometimes called the *unordered* pair a, b. But suppose we want to form an **ordered** pair consisting of a **first** element a and a **second** element b. The notation

$$(a, b)$$

is used for such an ordered pair. Thus, in contrast to the unordered pair $\{a, b\}$, we understand that for $a \neq b$,

$$(a, b) \neq (b, a).$$

Our ordered pairs will be governed by just two axioms: an *existence* axiom and an *equality* axiom.

1.1 AXIOM

> **EXISTENCE OF ORDERED PAIRS**   If a and b are elements in the universe of discourse U, then the ordered pair (a, b) exists.

Axiom 1.1 is supposed to hold even if $a = b$; so an ordered pair

$$(a, a)$$

consisting of the same first and second elements is perfectly legitimate.

1.2 AXIOM

> **EQUALITY OF ORDERED PAIRS**   If a, b, c, and d are elements of U, then
>
> $$(a, b) = (c, d) \Leftrightarrow a = c \land b = d.$$

In words, Axiom 1.2 says that *two ordered pairs are equal if and only if they have the same first element and the same second element.* As a consequence of Axiom 1.2, we have

$$(a, b) = (b, a) \Leftrightarrow a = b,$$

which you will prove in Problem 4.

In the seventeenth century, the French mathematician and philosopher René Descartes (1596–1650) systematically developed the idea that, by means of a coordinate system, geometry could be translated into algebra and vice versa. In the **Cartesian coordinate system** (named after Descartes), a geometric point P in the plane is assigned x and y coordinates with respect to a fixed pair of coordinate axes. The ordered pair (x, y) of these coordinates provides a numerical "address" for the point P. The resulting correspondence

$$P \leftrightarrow (x, y)$$

between the geometric point P and the ordered pair (x, y) of its coordinates is so compelling that, in some cases, it is convenient to identify P with (x, y) and write

$$P = (x, y).$$

In this way, the set of all points in the plane is identified with the set of all ordered pairs (x, y) of real numbers.

In mathematical writing, the symbol $\mathbb{R}$ is often used to represent the set of all real numbers, and the set of all ordered pairs of real numbers is denoted by

$$\mathbb{R}^2 = \{(x, y) \mid x \in \mathbb{R} \wedge y \in \mathbb{R}\}.$$

Because each $(x, y) \in \mathbb{R}^2$ can be regarded as a point in the plane, the set $\mathbb{R}^2$ is called the **Cartesian plane**.

The idea of a Cartesian plane can be generalized by considering the set of all ordered pairs whose first and second elements belong to specified sets—not necessarily the set $\mathbb{R}$ of real numbers. Thus, if M and N are sets, we define the **Cartesian product** of M and N, in symbols

$$M \times N,$$

to be the set consisting of *all ordered pairs whose first element belongs to M and whose second element belongs to N*. More formally:

1.3 DEFINITION | **CARTESIAN PRODUCT OF SETS**   If M and N are sets, we define

$$M \times N = \{(x, y) \mid x \in M \wedge y \in N\}.$$

1.4 Example   Let $M = \{1, 2, 3\}$, $N = \{a, b\}$, and $S = \{s\}$. Find:

(a) $M \times N$ **(b)** $N \times M$ **(c)** $S \times N$ **(d)** $S \times S$

SOLUTION   **(a)** $M \times N = \{(1, a), (1, b), (2, a), (2, b), (3, a), (3, b)\}$
(b) $N \times M = \{(a, 1), (a, 2), (a, 3), (b, 1), (b, 2), (b, 3)\}$
(c) $S \times N = \{(s, a), (s, b)\}$
(d) $S \times S = \{(s, s)\}$ □

In Parts (a) and (b) of Example 1.4, notice that $M \times N \neq N \times M$.

Just as we use Venn diagrams to illustrate unions, intersections, and complements of sets, we can visualize Cartesian products with suitable diagrams suggested by analogy with the Cartesian plane. Figure 2-1 shows such a diagram illustrating the Cartesian product $M \times N$. In this figure, the points on the lower edge of the rectangle represent the elements of M, the points on the left edge of the rectangle represent the elements of N,

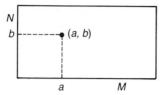

FIGURE 2-1

and the point in the rectangle with "coordinates" $a \in M$ and $b \in N$ represents the ordered pair $(a, b) \in M \times N$. In this way, we can think of the Cartesian product $M \times N$ as the set of all points in the rectangle.

If M and N actually are the sets of points on the lower and side edges of a rectangle R, then we can regard R as being represented by $M \times N$ as in Figure 2-1 and write

$$R = M \times N.$$

In this sense, *the Cartesian product of line segments is a rectangle*. In a similar way, and with a little imagination, you can visualize Cartesian products of simple geometric figures other than line segments.

1.5 Example　Give a geometric interpretation of the Cartesian product of a circle and a line segment.

SOLUTION　We can visualize the Cartesian product $C \times L$ of a circle C and a line segment L as forming the surface S of a right circular cylinder. Think of C as forming the circular base of S and of L as being one of the line segments, perpendicular to the base and lying on the surface S. The location of a point P on S can be specified by "coordinates" $x \in C$ and $y \in L$, determined as follows: x is the point at the foot of the perpendicular dropped from P to C; y is the point where a circle on S, parallel to C and containing P, intersects L (Figure 2-2).　□

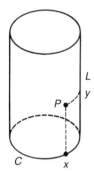

FIGURE 2-2

Example 1.5 should remind you of the "cylindrical coordinates" used in calculus.

1.6 THEOREM

> **THE CARDINAL NUMBER OF A CARTESIAN PRODUCT**　If M and N are finite sets, then $M \times N$ is a finite set and
>
> $$\#(M \times N) = \#M \cdot \#N.$$

INFORMAL PROOF　Let $m = \#M$ and $n = \#N$. In forming an ordered pair $(x, y) \in M \times N$, we have m choices for the first element $x \in M$. For each such choice, we have n choices for the second element $y \in N$; hence, altogether, we have exactly mn such choices, and it follows that there are mn different elements $(x, y) \in M \times N$.　□

1.7 Example   How many different outcomes are possible if a pair of dice are tossed?

SOLUTION   The answer to this question depends on whether we regard, say, a 2 on the first die and a 5 on the second die as being the same as a 5 on the first die and a 2 on the second die. Certainly, a dice player would regard these as equivalent outcomes. However, for mathematical purposes (specifically, for a proper application of the theory of probability to dice tossing), it is necessary to regard them as two distinct outcomes.

The faces of a single die correspond to the elements of the set

$$F = \{1, 2, 3, 4, 5, 6\},$$

so the possible outcomes of a toss of two dice are represented by the Cartesian product $F \times F$. For instance, if the first die falls with the face 2 upward and the second with the face 5 upward, the outcome of the toss would be represented by the ordered pair $(2, 5)$. Because $\#F = 6$, it follows from Theorem 1.6 that

$$\#(F \times F) = \#F \cdot \#F = 6 \cdot 6 = 36;$$

so there are 36 possible outcomes of a toss of the two dice.   □

Theorems concerning the equality or containment of sets built up from combinations of unions, intersections, complements, and Cartesian products can be proved by the pick-a-point procedure.

1.8 Example   If A, B, and C are sets, prove that

$$(A \cup B) \times C \subseteq (A \times C) \cup (B \times C).$$

SOLUTION   We begin by picking an arbitrary point in the set on the left of the inclusion. Such a point has the form (x, y) where

$$x \in (A \cup B) \quad \text{and} \quad y \in C.$$

Since $x \in (A \cup B)$, it follows that

$$x \in A \quad \text{or} \quad x \in B.$$

If $x \in A$, then we have

$$x \in A \quad \text{and} \quad y \in C;$$

hence, in this case,

$$(x, y) \in A \times C.$$

Likewise, if $x \in B$, then we have

$$x \in B \quad \text{and} \quad y \in C,$$

so that

$$(x, y) \in B \times C.$$

This shows that at least one of the two statements,

$$(x, y) \in A \times C \quad \text{or} \quad (x, y) \in B \times C,$$

is true, and therefore (x, y) belongs to the set on the right of the inclusion.

□

In Problem 29, we ask you to show that the inclusion in Example 1.8 can be strengthened to an equality.

PROBLEM SET 2.1

1. If x and y are real numbers and the ordered pairs $(x + 2y, 5)$ and $(3, 7x - 2y)$ are equal, find x and y.

2. True or false: $(1, (2, 3)) = ((1, 2), 3)$. Explain.

3. Find the ordered pairs corresponding to the points $P_1, P_2, P_3,$ and P_4 in the diagram of $\{1, 2, 3, 4\} \times \{1, 2, 3, 4\}$ in Figure 2-3.

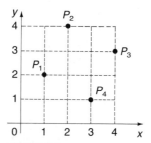

FIGURE 2-3

4. Prove that $(a, b) = (b, a)$ if and only if $a = b$.

In Problems 5–12, let $M = \{2, 4, 5\}$ and $N = \{1, 2\}$. Write out each set explicitly.

5. $M \times N$ 6. $M \times M$

7. $N \times M$ 8. $N \times N$

9. $\varnothing \times M$ 10. $N \times \varnothing$

11. $(M \times N) \cup (N \times M)$ 12. $(M \times N) \cap (N \times M)$

In Problems 13–16, let $M = \{a, b\}$, $N = \{c, d\}$, and $R = \{e, f\}$. Write out each set explicitly.

13. $M \times (N \cup R)$ 14. $M \times (N \cap R)$

15. $(M \times N) \cup (M \times R)$ 16. $(M \times N) \cap (M \times R)$

17. Give a geometric interpretation of the Cartesian product of a triangle and a line segment.

18. Give a geometric interpretation of the Cartesian product of two circles. [Hint: It is not a sphere.]

19. Give a geometric interpretation of the Cartesian product of a circular disk and a line segment.

20. Prove: $A \times B = \varnothing \Rightarrow A = \varnothing \vee B = \varnothing$.

In Problems 21–24, find $\#(M \times N)$.

21. $M = \{1, 2, 3\}$, $N = \{2, 7\}$

22. $M = N =$ the set of major planets in the solar system.

23. $M = \{1, 2, 3\}$, $N =$ the set of letters in the alphabet.

24. $M = N =$ the set of cards in a poker deck (excluding the joker).

25. How many different outcomes are possible if three dice are tossed?

26. If $A \neq \varnothing$ and $B \neq \varnothing$, prove that $A \times B = B \times A \Rightarrow A = B$.

27. If $A \subseteq X$ and $B \subseteq Y$, prove that $A \times B \subseteq X \times Y$.

28. Prove that $(A \cap B) \times C = (A \times C) \cap (B \times C)$.

29. Prove that $(A \cup B) \times C = (A \times C) \cup (B \times C)$.

30. True or false: $(A \setminus B) \times C = (A \times C) \setminus (B \times C)$. If true, prove it; if false, give a counterexample.

31. Prove that $(A \cap B) \times (C \cap D) = (A \times C) \cap (B \times D)$.

32. Generalize Problem 28, replacing $A \cap B$ by an intersection of an indexed family of sets.

33. Illustrate Problem 29 with diagrams.

34. True or false: $(A \cup B) \times (C \cup D) = (A \times C) \cup (B \times D)$. If true, prove it; if false, give a counterexample.

35. If M is a finite set with $\#M = m$, what is $\#(\mathscr{P}(M) \times \mathscr{P}(M))$?

36. If we define $(a, b) = \{\{a\}, \{a, b\}\}$, show that Axiom 1.2 becomes a theorem.

2

RELATIONS

The word *relation* is used in mathematics in much the same way as in English: A certain relation holds between two objects when there is a particular type of connection between them. For instance, in conversation we speak of the relation of parenthood between a mother or father and their children, and in mathematics we speak about the relation of similarity between two triangles.

Suppose we have in mind a definite mathematical relation, say, the relation of similarity between triangles or, perhaps, the relation of exact divisibility between whole numbers. Some triangles are similar to others, and some are not; some whole numbers divide others exactly, and some do not. Now, consider the propositional function

$P(x, y)$

of two variables x and y in U which states that *the relation in question holds for x and y.* For instance, if U is the set of all triangles and the relation under consideration is similarity, then $P(x, y)$ would be the statement

x is similar to y.

Likewise, if U is the set of all whole numbers and the relation in question is exact divisibility, then $P(x, y)$ would be the statement

x divides y.

According to the discussion above, *a mathematical relation can be represented by a propositional function $P(x, y)$ of two variables.* Conversely, *a propositional function $P(x, y)$ of two variables can be regarded as stating that a certain definite relation holds between x and y.* What relation? The relation that holds between x and y when $P(x, y)$ is true. For this reason, logicians refer to a propositional function of two variables as an **intensional binary relation**. The word *intensional* is a technical term used in logic to indicate that all of the qualities or properties inherent in the relation are expressed by the propositional function. The word *binary* simply means that *two* variables are involved.

If $P(x, y)$ is an intensional binary relation (that is, a propositional function of two variables), then the set

$\{(x, y) \mid P(x, y)\},$

consisting of all ordered pairs (x, y) for which $P(x, y)$ is true, is called the **graph** of the relation. Logicians refer to this graph as the binary relation in **extension**.

2.1 Example   If U, the universe of discourse, consists of all the integers, find the graph R of

$(2 < x < 7) \wedge (y = 3x \vee x = 2y).$

SOLUTION   The only possible values of x are 3, 4, 5, and 6. For these values, $y = 3x$ has corresponding values 9, 12, 15, and 18; so the ordered pairs (3, 9), (4, 12), (5, 15), and (6, 18) belong to the graph R. Furthermore, if x is 4 or 6, we have $x = 2y$ when y is 2 or 3, respectively; so the ordered pairs (4, 2) and (6, 3) also belong to R. Since these are the only pairs in R, it follows that

$R = \{(3, 9), (4, 2), (4, 12), (5, 15), (6, 3), (6, 18)\}.$ □

2.2 Example Show that every set R of ordered pairs is the graph of some propositional function $P(x, y)$.

SOLUTION Let R be a set of ordered pairs. Define the propositional function $P(x, y)$ of two variables to be the statement asserting that $(x, y) \in R$. Then the graph of $P(x, y)$ is given by

$$\{(x, y) \mid P(x, y)\} = \{(x, y) \mid (x, y) \in R\}$$
$$= R. \qquad \square$$

An intensional binary relation $P(x, y)$ uniquely determines its graph,

$$R = \{(x, y) \mid P(x, y)\},$$

but it is not clear whether the original statement $P(x, y)$ can be recaptured from the graph R. You might stare at the set

$$R = \{(3, 9), (4, 2), (4, 12), (5, 15), (6, 3), (6, 18)\}$$

obtained in Example 2.1 for a long time without realizing that it is the graph of

$$(2 < x < 7) \wedge (y = 3x \vee x = 2y).$$

Indeed, you might never hit on this fact. This should make it plain that there is a difference between binary relations in *intension* (propositional functions of two variables) and binary relations in *extension* (sets of ordered pairs). However, when speaking or writing informally, mathematicians usually do not bother to distinguish between relations in these two senses.

In more formal mathematical definitions, theorems, and proofs, it is usually convenient to deal with binary relations in extension (sets of ordered pairs); hence, we make the following definition:

2.3 DEFINITION

> **RELATION** A **binary relation** or, in what follows, simply a **relation**, is a set of ordered pairs of elements of the domain of discourse U.

Thus, in all of our definitions and proofs, a relation R is a subset of $U \times U$. If x and y are in U, we say that x *is related to y by R* if and only if

$$(x, y) \in R.$$

The statement $(x, y) \in R$ is often abbreviated by writing

$$xRy.$$

Whether a relation R is expressed in words or specified in mathematical symbols, it is always understood that R is the *set of all ordered pairs that satisfy the condition or conditions expressed by the words or symbols.* For instance, if the domain of discourse U is the set of all real numbers and if the relation R is defined by

$$xRy \Leftrightarrow x = y^2,$$

then,

$$R = \{(x, y) \mid x = y^2\}.$$

2.4 DEFINITION

DOMAIN AND CODOMAIN   Let R be a relation.

(i) The **domain** of R, written dom(R), is the set given by

$$\mathrm{dom}(R) = \{x \mid (\exists y)(xRy)\}.$$

(ii) The **codomain** of R, written codom(R), is the set given by

$$\mathrm{codom}(R) = \{y \mid (\exists x)(xRy)\}.$$

(iii) If dom(R) $\subseteq X$ and codom(R) $\subseteq X$, then we say that R is a relation **on** the set X.

In words, *the domain of R is the set of all first elements of ordered pairs in R, and the codomain of R is the set of all second elements of ordered pairs in R.* Note that R is a relation on the set X if and only if $R \subseteq X \times X$ (Problem 3).

2.5 Example   If $R = \{(\pi, 3.14), (e, 2.71), (\sqrt{2}, 1.41)\}$, find:

(a) dom(R)   (b) codom(R)

SOLUTION   (a) dom(R) $= \{\pi, e, \sqrt{2}\}$
(b) codom(R) $= \{3.14, 2.71, 1.41\}$ □

Other authors use other words for the codomain of a relation. Some people call it the *range*, and others call it the *image* of the relation.

2.6 DEFINITION

CONVERSE, OR INVERSE   If R is a relation, then the **converse**, or **inverse**, of R, written R^{-1}, is the relation defined by

$$R^{-1} = \{(y, x) \mid (x, y) \in R\}.$$

Note that *the converse, or inverse, of R is obtained by reversing all of the ordered pairs in R.* Thus,

$$yR^{-1}x \Leftrightarrow xRy.$$

For instance, let U be the set of all human beings, and let M be the relation defined by xMy if and only if x is the biological mother of y. Then $yM^{-1}x$ holds if and only if x is a female and y is a son or daughter of x.

2.7 Example   If R is the relation of Example 2.5, find R^{-1}.

SOLUTION   Since

$$R = \{(\pi, 3.14), (e, 2.71), (\sqrt{2}, 1.41)\},$$

it follows that

$$R^{-1} = \{(3.14, \pi), (2.71, e), (1.41, \sqrt{2})\}. \qquad \square$$

Note that *the domain of the converse of a relation R is the codomain of R.* This accounts for the terminology *codomain*, which stands for *converse domain*.

2.8 DEFINITION | **DIAGONAL AND UNIVERSAL RELATION**   Let X be a set. Then:

(i) The **diagonal relation** on X, denoted by Δ_X or simply by Δ if X is understood, is the relation defined by

$$\Delta_X = \{(x, x) \mid x \in X\}.$$

(ii) The **universal relation** on X is the relation $X \times X$.

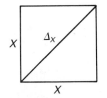

X

Δ_X

X

FIGURE 2-4

Note that $x(\Delta_X)y$ holds if and only if x and y are elements of X and $x = y$. Graphically, the relation Δ_X can be visualized as shown in Figure 2-4, which explains why it is called the *diagonal* relation on X. The universal relation $X \times X$ holds between *any* two elements x and y in X, and so it can be visualized as *all* points in the square shown in the figure.

2.9 DEFINITION | **REFLEXIVE AND TRANSITIVE RELATIONS**   Let R be a relation, and let X be a set.

(i) R is said to be **reflexive on** X if xRx holds for all $x \in X$.

(ii) R is said to be **transitive** if, for all x, y, and z,

$$xRy \wedge yRz \Rightarrow xRz.$$

Note that both Δ_X and $X \times X$ are reflexive on X (Problem 9a). Also, both Δ_X and $X \times X$ are transitive (Problem 9c).

2.10 Example   Let $\mathbb{R}$ denote the set of all real numbers, and let L be the relation on $\mathbb{R}$ defined by $xLy \Leftrightarrow x < y$.

(a) Is L reflexive on $\mathbb{R}$?   **(b)** Is L transitive?

SOLUTION   **(a)** L is *not* reflexive on $\mathbb{R}$; indeed, no real number is less than itself; so xLx never holds.

(b) L is transitive because, if $x < y$ and $y < z$, then it follows that $x < z$.
       □

2.11 DEFINITION   **SYMMETRIC AND ANTISYMMETRIC RELATIONS**   Let R be a relation.

(i) R is said to be **symmetric** if, for every x and every y,

$$xRy \Rightarrow yRx.$$

(ii) R is said to be **antisymmetric** if, for every x and every y,

$$xRy \wedge yRx \Rightarrow x = y.$$

Note that both Δ_X and $X \times X$ are symmetric relations (Problem 9b). Furthermore, Δ_X is antisymmetric (Problem 8a), but $X \times X$ is antisymmetric if and only if X is either a singleton set or the empty set (Problem 8b).

2.12 Example   Let $\mathbb{R}$ denote the set of all real numbers, and let G be the relation on $\mathbb{R}$ defined by $xGy \Leftrightarrow x \geq y$.

(a) Is G symmetric?   **(b)** Is G antisymmetric?

SOLUTION   **(a)** G is *not* symmetric because, in general, $x \geq y$ *does not imply* that $y \geq x$. For instance, $3 \geq 2$, but $2 \not\geq 3$.

(b) G is antisymmetric because, if both $x \geq y$ and $y \geq x$ hold, then it follows that $x = y$.   □

Relations are often used to order, rank, or otherwise organize the elements of a set. Among the relations that are used in this way, perhaps the most general are the preorder relations, defined as follows:

2.13 DEFINITION   **PREORDER**   Let X be a set. By a **preorder** relation on X we mean a relation R on X such that

(i) R is reflexive on X, and
(ii) R is transitive.

Note that both Δ_X and $X \times X$ are preorder relations on X (Problem 13a). Also, the relation G in Example 2.12 is a preorder on the set $\mathbb{R}$ of all real numbers (Problem 13b), but the relation L in Example 2.10 is not (because it is not reflexive on $\mathbb{R}$).

2.14 Example   Let T denote the set of all triangles in the plane, and let R be the relation defined on T by the condition that xRy if and only if the area of triangle y is at least as large as the area of triangle x. Is R a preorder relation on T?

SOLUTION   Because the area of a triangle is at least as large as the area of that same triangle, R is reflexive on T. To see that R is transitive, suppose that

$$xRy \quad \text{and} \quad yRz.$$

Let A_x, A_y, and A_z denote the areas of triangles x, y, and z, respectively. Since xRy and yRz, we have

$$A_x \leq A_y \quad \text{and} \quad A_y \leq A_z,$$

and it follows that

$$A_x \leq A_z,$$

that is,

$$xRz.$$

Therefore, R is both reflexive on T and transitive; so it is a preorder on T. □

2.15 DEFINITION

> **PARTIAL ORDER**   Let X be a set. By a **partial order** relation on X we mean a relation R on X such that
>
> (i) R is reflexive on X,
> (ii) R is antisymmetric, and
> (iii) R is transitive.

If you compare Definitions 2.13 and 2.15, you will see that *a partial order relation is the same thing as a preorder relation that is antisymmetric.* A memory device for partial order relations is RAT (Reflexive, Antisymmetric, and Transitive). Note that the diagonal relation Δ_X is a partial order relation on X; but, unless X is either empty or a singleton set, the universal relation $X \times X$ on X is not (Problem 14).

2.16 Example   Let $\mathbb{R}$ denote the set of all real numbers, and let R be the relation defined on $\mathbb{R}$ by

$$xRy \Leftrightarrow x \leq y.$$

Is R a partial order relation on $\mathbb{R}$?

SOLUTION   Yes. Because

$$x \leq x$$

holds for all $x \in \mathbb{R}$, it follows that x is reflexive on $\mathbb{R}$. Also,

$$x \leq y \wedge y \leq x \Rightarrow x = y,$$

and so R is antisymmetric. Finally,

$$x \leq y \wedge y \leq z \Rightarrow x \leq z,$$

which shows that R is transitive.   □

2.17 Example   Is the relation R in Example 2.14 a partial order relation on the set T of all triangles in the plane?

SOLUTION   No, because in Example 2.14 the relation R is *not* antisymmetric. Indeed, if x and y are triangles in the plane and if both xRy and yRx hold, then x and y are triangles with the same area; but this does not imply that they are the same triangle.   □

The relation $\leq$ in Example 2.16 on the set $\mathbb{R}$ of all real numbers is the prototype for all partial order relations; and for this reason, an arbitrary partial order relation on a set X is frequently denoted by $\leq$ rather than by R. Thus, if $\leq$ is a partial order relation on X, we read the statement $x \leq y$ as x *is less than or equal to* y. This has the advantage of making general partial order relations seem less abstract, but it sometimes has the disadvantage of suggesting *too much*. For instance, if $\leq$ is a partial order relation on a set X and if the statement $x \leq y$ is false, then you might be tempted to conclude that $y \leq x$ must be true (by analogy with the usual order relation for the real numbers). However, as the following example shows, this conclusion is not necessarily correct for partial order relations in general.

2.18 Example   Let X denote the set of all positive integers and define the relation $\leq$ on X by

$$x \leq y \Leftrightarrow y \text{ is an integer multiple of } x.$$

(Thus, in this example, $\leq$ does not have its usual meaning.) Show that it is possible for *both* of the statements $x \leq y$ and $y \leq x$ to be false.

SOLUTION   Let $x = 2$, and let $y = 3$. Then, neither x nor y is an integer multiple of the other. (In Problem 15, we ask you to show that $\leq$, as defined here, is a partial order relation on X.)   □

Example 2.18 should convince you of the need to be careful when dealing with general partial order relations denoted by $\leq$. With this firmly in mind, we make the following definition:

2.19 DEFINITION

PARTIALLY ORDERED SET A **partially ordered set**, or **poset** for short, is a pair $(X, \leq)$ consisting of a nonempty set X and a partial order relation $\leq$ on X.

Thus, $(X, \leq)$ is a poset if and only if $\leq$ is a relation, defined on the non-empty set X and satisfying the following three conditions for all elements x, y, and z in X:

(i) $x \leq x$ (reflexivity)

(ii) $x \leq y$ and $y \leq x \Rightarrow x = y$ (antisymmetry)

(iii) $x \leq y$ and $y \leq z \Rightarrow x \leq z$ (transitivity)

Notice that a partially ordered set, or poset, is more than just a set; it is a set *together with* a specified partial order relation on that set. Nevertheless, people often speak (incorrectly) of "the poset X," it being understood that a partial order relation $\leq$ is specified. In the interest of avoiding wordiness, mathematicians occasionally indulge in such abuses of language.

2.20 DEFINITION

UPPER AND LOWER BOUNDS Let $(X, \leq)$ be a poset, suppose that $M \subseteq X$, and let $a, b \in X$.

(i) We say that a is an **upper bound** for the set M in X if the relation

$$m \leq a$$

holds for every element $m \in M$.

(ii) We say that b is a **lower bound** for the set M in X if the relation

$$b \leq m$$

holds for every element $m \in M$.

2.21 Example In the poset $(\mathbb{R}, \leq)$ of real numbers, ordered in the usual way by $\leq$, find, if possible:

(a) A lower bound for the set P of all positive real numbers

(b) An upper bound for the set P of all positive real numbers

SOLUTION   **(a)** Because $0 \leq p$ holds for every $p \in P$, it follows that 0 is a lower bound for P. Of course, there are many other lower bounds for P; for instance, any negative number is such a lower bound.

(b) There is no real number that is greater than or equal to every positive number; so P has no upper bound in $\mathbb{R}$.   □

    A set, such as P in Example 2.21, which has a lower bound in a poset is said to be **bounded below**. Likewise, a set that has an upper bound is said to be **bounded above**. The set P in Example 2.21 is bounded below, but unbounded above. Notice that 0 is a lower bound for P, but it does not belong to P. In general, an upper or lower bound for a set might or might not belong to the set. This leads to a definition.

2.22 DEFINITION

GREATEST AND LEAST ELEMENTS   Let $(X, \leq)$ be a poset, and let $M \subseteq X$.

  (i) If $a \in M$, and a is an upper bound for M, then we say that a is the **greatest element** of M.

  (ii) If $b \in M$, and b is a lower bound for M, then we say that b is the **least element** of M.

    If M has a least element b or a greatest element a, then this least or greatest element is unique (Problem 37). For instance, in Example 2.21, if M is the set of all positive integers, then 1 is the (unique) least element of M; but M has no upper bound and hence has no greatest element.

    In a poset $(X, \leq)$, we write $x \geq y$ (by definition) to mean $y \leq x$. Statements of the form $x \geq y$ or $y \leq x$ are called **inequalities**. By definition, $x < y$ means that

$$x \leq y \wedge x \neq y.$$

The relation $x < y$, which is called a *strict* inequality, can also be written in the alternative form $y > x$. If x, y, and z are elements of X, we write the *compound* inequality

$$x \leq y \leq z$$

to mean that $x \leq y$ *and* $y \leq z$. This can also be written in the alternative form

$$z \geq y \geq x.$$

Both strict and nonstrict inequalities can be combined in compound inequalities; for instance, $x < y \leq z$.

    If $(X, \leq)$ is a poset and M is a subset of X, then, by the **least upper bound** of M we mean the least element (if it exists) of the set A of all upper bounds of M in X. (Note that A could be empty.) Likewise, by the **greatest lower bound** of M, we mean the greatest element (if it exists) of the set B of all lower bounds of M in X. The least upper and greatest lower bounds of M (if they exist) are written as

LUB(M) and GLB(M),

respectively. The least upper bound of M is also called the **supremum** of M, and the greatest lower bound of M is also called the **infimum** of M. We denote the supremum and infimum of M by

sup(M) and inf(M),

respectively.

2.23 Example   In the poset ($\mathbb{R}$, $\leq$) of Example 2.21, let

$$I = \{x \mid -1 < x \leq 1\}.$$

Find LUB(I) = sup(I) and GLB(I) = inf(I).

SOLUTION   The set of all upper bounds of I is the set

$$A = \{a \mid 1 \leq a\},$$

and the set of all lower bounds of I is the set

$$B = \{b \mid b \leq -1\}.$$

The least element of A is 1, and the greatest element of B is -1; hence,

LUB(I) = sup(I) = 1

and

GLB(I) = inf(I) = -1.   □

In Example 2.23, notice that $-1 = \inf(I)$ does not belong to I, but $1 = \sup(I)$ does belong to I. In general, practically anything can happen: A subset M of a poset X might have a supremum and an infimum, it might have one but not the other, or it might have neither. Even if M has a supremum or an infimum, this supremum or infimum might or might not belong to the set M.

If x and y are elements of a poset X, we say that x and y are **comparable** if at least one of the two conditions $x \leq y$ or $y \leq x$ holds. In the poset ($\mathbb{R}$, $\leq$) of Examples 2.21 and 2.23, any two elements are comparable, but Example 2.18 shows that this is not true for posets in general. This leads us to our final definition in this section:

2.24 DEFINITION   **TOTALLY ORDERED OR LINEARLY ORDERED SET**   A poset $(X, \leq)$ is said to be **totally ordered**, or **linearly ordered**, if, for every pair of elements x, y in X,

$$x \leq y \vee y \leq x.$$

In words, *a totally ordered or linearly ordered set is a poset in which any two elements are comparable.* Thus, the real numbers $\mathbb{R}$, ordered in the usual way by $\leq$, form a totally ordered set; however, the positive integers, ordered by the relation given in Example 2.18, do not. If $(X, \leq)$ is a totally ordered set, we refer to $\leq$ as a **total order relation**.

PROBLEM SET 2.2

In Problems 1 and 2, let U, the universe of discourse, consist of all integers. Find the graph R of each relation.

1. (a) $(1 < x < 4) \wedge (y = x^2)$ (b) $(1 < x < 4) \wedge (x = 5 - y)$
   (c) $(1 < x < 4) \wedge (y = x^2 \vee x = 5 - y)$

2. (a) $(3 < x < 10) \wedge (x = y^2 \vee x = 5 - y)$
   (b) $(3 < x < 10) \wedge (y = \sin(\pi x/2) \vee x = 3y)$

3. Prove that R is a relation on the set X if and only if $R \subseteq X \times X$.

4. If $X = \{2, 3, 4, 5, 6\}$, write the relation

   $$R = \{(x, y) \in X \times X \mid |x - y| \text{ is exactly divisible by 3}\}$$

   explicitly as a set of ordered pairs.

5. For each relation R, find $\text{dom}(R)$, $\text{codom}(R)$, and R^{-1}.
   (a) $R = \{(1, 5), (2, 5), (1, 4), (2, 6), (3, 7), (7, 6)\}$
   (b) $R = \{(c, b), (b, g), (c, e), (b, b), (b, e), (a, e), (a, c)\}$
   (c) $\mathbb{R}$ is the set of all real numbers, and
   $R = \{(x, y) \in \mathbb{R} \times \mathbb{R} \mid 4x^2 + 9y^2 = 36\}$
   (d) $T = \{1, 2, 3, 4, 5\}$ and $R \subseteq T \times T$ is the set of points displayed in Figure 2-5.

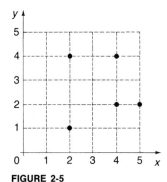

FIGURE 2-5

6. If X is a set, find:
   (a) $\text{dom}(\Delta_X)$ (b) $\text{codom}(\Delta_X)$ (c) $(\Delta_X)^{-1}$
   (d) $\text{dom}(X \times X)$ (e) $\text{codom}(X \times X)$ (f) $(X \times X)^{-1}$

7. Let $R = \{(1, 2), (3, 4), (2, 2), (3, 3), (2, 1)\}$.
   (a) Is R symmetric?   (b) Is R antisymmetric?
   Explain.

8. If X is a set, prove:
   (a) Δ_X is antisymmetric.
   (b) $X \times X$ is antisymmetric if and only if X is either a singleton set or the empty set.

9. If X is a set, prove that both of the relations Δ_X and $X \times X$ are:
   (a) Reflexive   (b) Symmetric   (c) Transitive

10. True or false: If a relation R is both symmetric and antisymmetric, then there exists a set X such that $R = \Delta_X$. If true, prove it; if false, give a counterexample.

11. Is the relation $R = \{(1, 2), (2, 2)\}$ transitive? Explain.

12. Let R be a relation. Prove:
    (a) If R is transitive, so is R^{-1}.   (b) If R is symmetric, so is R^{-1}.

13. (a) If X is a set, prove that both Δ_X and $X \times X$ are preorder relations on X.
    (b) Prove that the relation G in Example 2.12 is a preorder relation on the set $\mathbb{R}$ of all real numbers.

14. If X is a set, prove that Δ_X is a partial order relation on X but that, unless X is either empty or a singleton set, $X \times X$ is not.

15. In Example 2.18, show that the relation $\leq$ is a partial order relation on the set X of all positive integers.

16. Let X and Y be sets and let $R \subseteq X \times Y$. Then, picturing R as in Figure 2-6, explain why dom(R) can be pictured as the vertical projection of R down onto X and codom(R) can be pictured as the horizontal projection of R over onto Y.

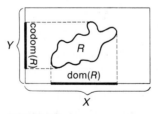

FIGURE 2-6

17. Decide which of the given relations on the indicated set X are (i) reflexive, (ii) symmetric, (iii) antisymmetric, (iv) transitive, (v) preorder relations, (vi) partial order relations, and/or (vii) total order relations.
    (a) X is any set, and xRy means $x \neq y$.
    (b) X is the set of humans, and xRy means that $x = y$ or x is a sibling of y.

(c) X is the set of humans, and xRy means that $x = y$ or x is an ancestor of y.

(d) X is the set of all triangles, and xRy means that x is congruent to y.

(e) X is the set of all circles in the plane, and xRy means that x and y are concentric and the radius of x does not exceed the radius of y.

(f) X is the set of all real numbers, and xRy means $x^3 = y^3$.

(g) X is the set of all integers, n is a fixed positive integer, and xRy means that $x - y$ is exactly divisible by n.

(h) X is the set of all line segments in the plane, and xRy means that x is at least as long as y.

18. Let X be a set, and let R be a relation on X. Picturing R as a subset of $X \times X$, give a pictorial description of each of the following:

(a) R^{-1}

(b) The condition that R is symmetric

(c) The condition that R is reflexive

19. Let X be a set, and let R be a relation on X. Prove:

(a) R is reflexive on X if and only if $\Delta_X \subseteq R$.

(b) R is symmetric if and only if $R^{-1} = R$.

20. Let X be a finite set with $\# X = n$.

(a) Prove that there are exactly 2^{n^2} different relations on X.

(b) Of the 2^{n^2} relations in Part (a), prove that there are exactly $(2^n - 1)^n$ relations whose domain is X.

(c) Of the $(2^n - 1)^n$ relations in Part (b), prove that there are exactly $2^{n^2 - n}$ relations that are reflexive on X.

(d) Of the 2^{n^2} relations in Part (a), prove that there are exactly $2^{(n^2 + n)/2}$ symmetric relations.

(e) Of the $2^{(n^2 + n)/2}$ relations in Part (d), show that there are exactly $2^{(n^2 - n)/2}$ relations that are both symmetric and reflexive on X.

(f) Of the $2^{n^2 - n}$ relations in Part (c), prove that there are exactly $n!$ relations that totally order X.

In Problems 21–32, let X be a set, and let R and S be relations on X. Use the fact that relations are sets, so that you can form unions, intersections, and complements of relations just as you can for any sets.

21. Prove that $R \cup \Delta_X$ is reflexive on X.

22. If R and S are reflexive on X, prove that $R \cap S$ is reflexive on X.

23. If R is reflexive on X, prove that $R \cup S$ is reflexive on X.

24. If R and S are transitive, prove that $R \cap S$ is transitive.

25. Prove:

(a) $(R \cap S)^{-1} = R^{-1} \cap S^{-1}$

(b) $(R \cup S)^{-1} = R^{-1} \cup S^{-1}$

26.   Prove:
    (a) $R \cap R^{-1}$ is a symmetric relation.
    (b) $R \cup R^{-1}$ is a symmetric relation.

27.   If R is symmetric, prove that $(X \times X) \setminus R$ is symmetric.

28.   If R is a partial order relation on X, prove that $R \cap R^{-1} = \Delta_X$.

29.   If R is a partial order relation on X, prove that R^{-1} is also a partial order relation on X.

30.   If R is a partial order relation on X, prove that R is a total order relation on X if and only if $R \cup R^{-1} = X \times X$.

31.   If R and S are partial order relations on X, prove that $R \cap S$ is a partial order relation on X.

32.   True or false: If R and S are partial order relations on X, then $R \cup S$ is a partial order relation on X. If true, prove it; if false, give a counterexample.

33.   Let $\mathbb{R}$ denote the set of all real numbers, and let $\le$ be the usual order relation on $\mathbb{R}$. If $I = \{x \in \mathbb{R} \mid 0 < x \le 1\}$, find:
    (a) LUB(I)   (b) GLB(I)
    (c) The greatest element of I, if it exists.
    (d) The least element of I, if it exists.

34.   Let $\mathbb{Q}$ denote the set of all rational numbers (ratios of integers with non-zero denominators) ordered in the usual way by the relation $\le$. Let $M = \{x \in \mathbb{Q} \mid x^2 < 3\}$. Find, if it exists:
    (a) An upper bound for M.   (b) A lower bound for M.
    (c) sup(M)   (d) inf(M)

35.   Let X be a nonempty set.
    (a) Prove that $(\mathscr{P}(X), \subseteq)$ is a poset.
    (b) If $M, N \in \mathscr{P}(X)$, find $\sup\{M, N\}$ and $\inf\{M, N\}$ in the poset $(\mathscr{P}(X), \subseteq)$.

36.   Prove that a poset X is totally ordered if and only if every nonempty finite subset of X contains a least element.

37.   Let X be a poset, and let $M \subseteq X$.
    (a) If M has a greatest element a, prove that this greatest element is unique.
    (b) If M has a least element b, prove that this least element is unique.

3

FUNCTIONS

In elementary algebra and calculus textbooks, a function is usually defined to be a *rule of correspondence f that assigns to each element x in a set X a unique element f(x) in a set Y.* Although this definition conveys the intuitive idea of a function, it is not sufficiently precise for the purposes of more advanced mathematics. What, exactly, is a rule of correspondence? A more formal definition of a function avoids imprecision but is considerably more abstract:

3.1 DEFINITION

> **FUNCTION**   A **function** is a relation R such that, for every x in dom(R) and every y and z in codom(R),
>
> $$xRy \wedge xRz \Rightarrow y = z.$$

Thus, a *function is a special kind of relation; it is a relation such that no object in the domain is related to two different objects in the codomain.* At first glance, this definition seems to have little to do with the more intuitive definition of a function as a rule of correspondence. In the end, however, you will see that it is merely a logically clean statement of the same basic idea.

3.2 Example   Which of the following sets of ordered pairs are functions?

(a) $R = \{(a, 2), (a, 3), (b, 4), (c, 1)\}$
(b) $S = \{(a, 4), (b, 4), (c, 4), (d, 2)\}$

SOLUTION   **(a)** R is not a function because it contains two ordered pairs $(a, 2)$ and $(a, 3)$ with the same first element but different second elements.
(b) S is a function because no two ordered pairs in S have the same first element and different second elements. □

Although relations in general are usually symbolized by capital letters, relations that are functions are often denoted by small letters. Also, if f is a function, we usually write $(x, y) \in f$ in preference to xfy, reserving the latter notation for the study of general relations. With this notation, Definition 3.1 can be rewritten as follows:

> A *function* is a relation f such that, for every $x \in$ dom(f), there exists a unique $y \in$ codom(f) such that $(x, y) \in f$.

3.3 DEFINITION

> **IMAGE OF AN ELEMENT UNDER A FUNCTION**   If f is a function and x is an element in the domain of f, then the unique element y in the codomain of f such that $(x, y) \in f$ is denoted by $f(x)$ and called the **image of x under** f, or the **value of f at** x.

Thus, if f is a function and $x \in \text{dom}(f)$, we have

$$y = f(x) \Leftrightarrow (x, y) \in f.$$

The expression $f(x)$ is read in English as f *of* x.

3.4 Example   For the function $f = \{(a, 2), (b, 1), (c, 2)\}$ find $f(a)$, $f(b)$, and $f(c)$.

SOLUTION $f(a) = 2$, $f(b) = 1$, and $f(c) = 2$.   □

We leave it as a simple exercise (Problem 2) for you to show that, if f is a function, then

$$\text{codom}(f) = \{y \,|\, (\exists x)(x \in \text{dom}(f) \wedge y = f(x))\},$$

or, with a slight abuse of notation,

$$\text{codom}(f) = \{f(x) \,|\, x \in \text{dom}(f)\}.$$

For functions, the word *range* is commonly used in preference to *codomain*. Let's make this official:

3.5 DEFINITION

RANGE OF A FUNCTION   If f is a function, then the **range** of f, in symbols $range(f)$, is defined by

$$\text{range}(f) = \{f(x) \,|\, x \in \text{dom}(f)\}.$$

Thus, *the range of f is the set of all images under f of elements in the domain of f.*

3.6 Example   Find the domain and range of the function

$$f = \{(a, 2), (b, 1), (c, 2)\}$$

in Example 3.4.

SOLUTION $\text{dom}(f) = \{a, b, c\}$

and

$$\text{range}(f) = \{f(a), f(b), f(c)\} = \{2, 1\}.$$ □

If $\mathbb{R}$ denotes the set of all real numbers, then a function f with $\text{dom}(f) \subseteq \mathbb{R}$ and $\text{range}(f) \subseteq \mathbb{R}$ is called a **real-valued** function of a **real variable**. In your calculus courses you learned to draw the graphs of such functions in a Cartesian coordinate system. The same idea can be applied to provide schematic diagrams (still called graphs) to aid your intuition when dealing with more general functions. Thus, if X and Y are sets and

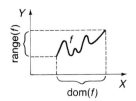

FIGURE 2-7

f is a function such that $\mathrm{dom}(f) \subseteq X$ and $\mathrm{range}(f) \subseteq Y$, you can visualize the function f as the set of ordered pairs

$$(x, f(x))$$

such that $x \in \mathrm{dom}(f)$ (Figure 2-7).

In the figure, notice that the domain of f is the vertical projection of f down onto the X axis, and the range of f is the horizontal projection of f over onto the Y axis.

3.7 DEFINITION

MAPPING NOTATION   The symbolism

$$f:X \rightarrow Y,$$

called **mapping notation**, indicates that

(i) f is a function,
(ii) $\mathrm{dom}(f) = X$, and
(iii) $\mathrm{range}(f) \subseteq Y$.

Note carefully that, in Definition 3.7, we do *not* require that Y is the range of f, only that it *contains* the range of f.

If

$$f:X \rightarrow Y,$$

we say that f is a function, or mapping, **from X into Y**. The alternative notation

$$X \xrightarrow{f} Y$$

is also used to indicate that f is a mapping from X into Y. When mapping notation is used, the statement that

$$y = f(x)$$

is often written in the alternative form

$$x \xmapsto{f} y,$$

or, if f is understood, simply as

$$x \mapsto y.$$

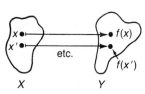

FIGURE 2-8

If $x \mapsto y$, we say that x is **mapped**, or **transformed**, into y. (Notice that a "tail" is placed on the arrow when mapping notation is used in this fashion.)

FIGURE 2-9

The idea of a function as a mapping, or transformation,

$$f:X \rightarrow Y,$$

is nicely depicted by a **transformation diagram** in which arrows are drawn connecting each element $x \in X$ with the element $f(x) \in Y$ into which it is mapped, or transformed, by f (Figure 2-8). In fact, if the sets X and Y are *finite*, a mapping $f: X \rightarrow Y$ can be specified completely by such a diagram (Figure 2-9).

3.8 Example Determine as an explicit set of ordered pairs the function f depicted in the transformation diagram in Figure 2-9.

SOLUTION $\quad f = \{(a, 2), (b, 3), (c, 1), (d, 3)\}.$ $\qquad \square$

An interesting variation of the idea of a mapping, or transformation, is the picture of a function as a machine that takes objects x in the domain X and converts them into objects $f(x)$ in the set Y. The function f can be depicted as a "black box" with some mysterious inner workings that can accept any element x in X as an **input**, process it internally, and eject $f(x)$ as an **output** (Figure 2-10). This machine picture is useful in the study of computing devices.

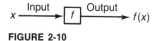

FIGURE 2-10

Functions and variables are closely related. In the old days, when mathematicians spoke of a *variable*, they meant a quantity that is changing, or at least capable of changing. Although it is easy to think about a variable as a quantity subject to change, this idea is somewhat imprecise by modern standards of mathematical rigor. Contemporary mathematicians often regard a variable as a "blank spot" that can be "filled" by any particular object from the set of objects under consideration. From this point of view, an equation such as

$$(x + 1)^2 = x^2 + 2x + 1$$

can be thought of as having the form

$$(\Box + 1)^2 = \Box^2 + 2\Box + 1,$$

where any particular real number can be inserted into the three blank spots marked by the empty boxes. When a number is placed in the boxes, the result is a definite proposition about that number. The advantage of this viewpoint is that the philosophically elusive ideas of time and change are not required; the disadvantage is that variables become rather dull, more or less static entities, rather than dynamic quantities that change even as we speak about them.

We recommend that you adopt the following attitude about variables: When you use variables in a formal mathematical argument, think of them as blank spots that can be filled by particular objects, but, when you are trying to gain an intuitive understanding of some concept involving variables, think of them as quantities that change in time by ranging over the set of objects involved.

If f is a function and

$$y = f(x),$$

we call x the **argument** of f, or the **independent variable**; and, since y depends on x, we often refer to y as the **dependent variable**. If we think of x and y as variable quantities, then, as x changes, y is subject to a corresponding change, and we obtain a compelling dynamic picture of the function f. This dynamic picture is especially useful in calculus, where the derivative dy/dx can be interpreted as the *rate of change of y with respect to x*.

Now let us consider the idea, mentioned at the beginning of this section, that a function is a rule of correspondence. Suppose we have two sets, X and Y, and a specific *rule, formula, correspondence*, or *prescription* that associates with each object x in X a unique object y in Y. This determines a definite function, or mapping,

$$f : X \to Y,$$

where f is the set of all ordered pairs (x, y) such that $x \in X$, and y is the unique object in Y that is associated with x according to the given rule, formula, correspondence, or prescription. If you want to define a particular function, or mapping, in this way, here is what you should do:

Step 1.   Specify the set X that is to be the domain of f.

Step 2.   Specify the set Y that is to contain the range of f.

Step 3.   Be certain that the rule, formula, correspondence, or prescription unambiguously assigns to each object x in X one *and only one* object y in Y.

In carrying out Step 3, you are making sure that the function, or mapping, is **well-defined.**

In less formal mathematical writing, especially in calculus textbooks, Steps 1 and 2 are often ignored. Step 1 is ignored because it is understood that the domain X is the set of all elements in the domain of discourse for which the rule, formula, or prescription is meaningful. Step 2 is ignored because it is understood that Y comprises all elements y that correspond to elements x in X according to the rule, formula, or prescription.

3.9 Example   In a calculus book, we find the words "Consider the function $y = \sqrt{1 - x}$." Translate these words into more precise mathematical language.

SOLUTION Here it is understood that the universe of discourse is the set $\mathbb{R}$ of all real numbers. We carry out the three steps suggested above:

Step 1. Since the domain is not given, we assume that it consists of all real numbers for which the expression $\sqrt{1-x}$ is defined; that is, all real numbers x for which $x \leq 1$, so that $0 \leq 1 - x$. Therefore, the domain of the function is the set

$$X = \{x \mid x \leq 1\},$$

or, using interval notation,

$$X = (-\infty, 1].$$

Step 2. We can take Y to be any set containing all of the numbers $y = \sqrt{1-x}$ as x runs through X. The simplest choice seems to be $Y = \mathbb{R}$ (although any subset of $\mathbb{R}$ containing all nonnegative numbers would do as well).

Step 3. The formula $y = \sqrt{1-x}$ assigns one and only one value of y to each value of x in X. (Recall that the square root symbol is always understood to represent the *principal* square root, that is, the nonnegative square root.) Therefore, the function

$$f : X \rightarrow Y$$

under consideration is given by

$$f = \{(x, y) \mid x \in \mathbb{R} \land y \in \mathbb{R} \land x \leq 1 \land y = \sqrt{1-x}\}. \qquad \square$$

Summarizing our discussion so far, a function, or mapping,

$$f : X \rightarrow Y$$

can be thought of or visualized in (at least) six different ways:

1. The *mathematical definition* (Definition 3.1) as a set of ordered pairs in which no two different pairs have the same first element.
2. The *graphical picture*, either as a graph (Figure 2-7) or as a transformation diagram (Figures 2-8 and 2-9).
3. The *mapping*, or *transformation*, point of view in which f is thought of as transforming each element $x \in X$ into a corresponding element $y \in Y$, denoted by $x \mapsto y$.
4. The *machine* picture as a *processing device* that converts each input object $x \in X$ into a corresponding output object $f(x) \in Y$.
5. The *dynamic view*, as a relation between variable quantities x and y.
6. The *rule, formula, correspondence,* or *prescription* view.

In a formal mathematical proof, only the official definition (Definition 3.1) of a function is acceptable; the other five interpretations are useful for informal proofs and for building your intuition about functions.

Note that there is a technical distinction between a function f and a mapping $f:X \to Y$. A mapping $f:X \to Y$ is more than the function f because it involves the specification of a set Y that contains, but is not necessarily equal to, range(f). In practice, however, mathematicians often disregard this distinction and speak of functions and mappings more or less interchangeably.

Although Definition 3.1 requires that no two different ordered pairs in f have the same *first* element, it is noncommittal about whether two different ordered pairs can have the same *second* element. Thus, as far as this definition is concerned, it is quite possible to have ordered pairs (say)

$$(a, c) \in f \quad \text{and} \quad (b, c) \in f \qquad \text{with } a \neq b.$$

This is the same as having

$$f(a) = f(b) \qquad \text{with } a \neq b.$$

Functions for which this does not happen are the subject of the following definition.

3.10 DEFINITION

> **ONE-TO-ONE, OR INJECTIVE, FUNCTIONS**   A function, or mapping, $f:X \to Y$ is said to be **one-to-one**, or **injective**, if, for all elements a and b in X,
>
> $$f(a) = f(b) \Rightarrow a = b.$$

The condition that f is one-to-one, or injective, can also be expressed by the contrapositive

$$a \neq b \Rightarrow f(a) \neq f(b)$$

of the implication in Definition 3.10. Thus, *a function, or mapping, is one-to-one, or injective, if and only if distinct elements of its domain have distinct images under the function.*

The terminology *one-to-one* is used in most calculus textbooks, but the word *injective* is often used in more advanced mathematical writing. An injective function, or mapping, is sometimes referred to as an **injection**. The so-called **horizontal-line test** determines whether a real-valued function f of a real variable is injective: *f is injective if and only if no straight line parallel to the x axis intersects the graph of f more than once* (Problem 12). You can also tell from its transformation diagram whether a mapping is injective: *f:X → Y is injective if and only if no two arrows start at different elements of X and point to the same element of Y* (Problem 15).

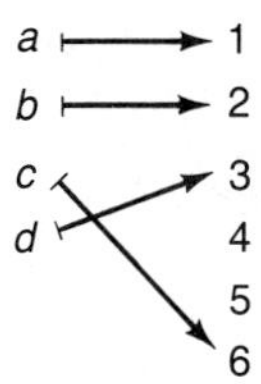

a $\longmapsto$ 1   **3.11 Example**   Is the mapping $f: X \to Y$ whose transformation diagram is shown in Fig-
b $\longmapsto$ 2   ure 2-11 an injection?

c ⟍ 3
d ✕ 4   SOLUTION   Since no two arrows point to the same element in Y, it follows that f is
   5   an injection.   □
   6

FIGURE 2-11   **3.12 Example**   Determine which of the following real-valued functions of a real variable
                                       is an injection:

(a) $f(x) = x^2$
(b) $g(x) = 2x - 5$

SOLUTION   (a) f is not an injection because, for instance,

$$f(-1) = f(1), \quad \text{but} \quad -1 \neq 1.$$

(b) g is an injection because, if $g(a) = g(b)$, then

$$2a - 5 = 2b - 5,$$

so

$$2a = 2b,$$

and therefore

$$a = b.$$ □

In using the mapping notation

$$f: X \to Y$$

you have considerable latitude in your choice of the set Y; the only
requirement is that it *contain* the range of f. The "smallest" possible Y
would be range(f) itself. This leads us to the following definition:

3.13 DEFINITION   **ONTO, OR SURJECTIVE, MAPPING**   A mapping $f: X \to Y$ is said to
be **onto**, or **surjective**, if, for every element $y \in Y$, there exists at least
one element $x \in X$ such that

$$y = f(x).$$

Thus, $f: X \to Y$ is onto, or surjective, if and only if

$$Y = \text{range}(f).$$

A surjective mapping, $f:X \to Y$ is also called a **surjection**. You can tell from its transformation diagram whether a mapping is surjective: $f:X \to Y$ *is surjective if and only if every element in Y has at least one arrow pointing to it.*

3.14 Example   Is the mapping in Figure 2-11 a surjection?

SOLUTION   No, because there are elements in Y—namely, 4 and 5—with no arrows pointing to them.   □

3.15 Example   If $\mathbb{R}$ denotes the set of all real numbers, determine which of the following mappings is a surjection:

(a) $f:\mathbb{R} \to \mathbb{R}$ given by $f(x) = x^2$
(b) $g:\mathbb{R} \to \mathbb{R}$ given by $g(x) = 2x - 5$

SOLUTION   (a) $f:\mathbb{R} \to \mathbb{R}$ is not a surjection because the range of f is the set of all *nonnegative* real numbers, a *proper* subset of $\mathbb{R}$.
(b) $g:\mathbb{R} \to \mathbb{R}$ is a surjection because every real number y can be written in the form

$$y = g(x) = 2x - 5$$

for a suitable choice of x, namely,

$$x = \frac{y + 5}{2}.$$

(This value of x was obtained by solving the equation $y = 2x - 5$ for x in terms of y.)   □

There are mappings that are neither injective nor surjective, mappings that are one but not the other, and mappings that are both injective and surjective. The last possibility suggests the following definition:

3.16 DEFINITION   **ONE-TO-ONE   CORRESPONDENCE,   OR   BIJECTION**   A mapping $f:X \to Y$ is said to be a **one-to-one correspondence**, or a **bijective mapping**, if it is both injective (one-to-one) and surjective (onto).

A bijective mapping is also called a **bijection**.

3.17 Example   Show that the mapping $g:\mathbb{R} \to \mathbb{R}$ defined by

$$g(x) = 2x - 5$$

is a bijection.

SOLUTION By Example 3.12b, $g:\mathbb{R} \to \mathbb{R}$ is an injection, and by Example 3.15b, it is a surjection. Hence, g is a bijection. $\square$

Now, suppose

$$f:X \to Y,$$

so that f is a function, dom$(f) = X$, and range$(f) \subseteq Y$. Then f is a relation (a set of ordered pairs), and

$$f \subseteq X \times Y.$$

It follows that the converse, or inverse, f^{-1} of f is a relation and that

$$f^{-1} \subseteq Y \times X;$$

however, in general, it would be incorrect to write

$$f^{-1}:Y \to X.$$

In the first place, unless $f:X \to Y$ is surjective, so that range$(f) = Y$, *the domain of f^{-1} need not be all of Y.* In the second place, unless $f:X \to Y$ is injective, f^{-1} *need not be a function.* However, we do have the following theorem, whose proof we leave to you as an exercise (Problem 32):

3.18 THEOREM

THE INVERSE OF A BIJECTION Let $f:X \to Y$. Then:

(i) f is injective (one-to-one) if and only if f^{-1} is a function.
(ii) If $f:X \to Y$ is a bijection (one-to-one correspondence), then so is $f^{-1}:Y \to X$.

Some of the concepts that were introduced in earlier sections can be tied in quite nicely with the idea of a function or mapping. For instance, an indexed family $(M_i)_{i \in I}$ of subsets of a set X is just a mapping

$$i \mapsto M_i$$

from the indexing set I into the set $\mathcal{P}(X)$ of all subsets of X. Also, a predicate, or propositional function, $P(x)$ really is a function whose domain is the universe of discourse U and which maps each object $x \in U$ into a corresponding proposition $P(x)$,

$$x \mapsto P(x).$$

Furthermore, the idea of a function, or mapping, can be used to clarify ideas that you may have encountered in your study of algebra or calculus. For instance, a **permutation** of a set X is the same thing as a bijection

$f : X \to X$. Also, a **sequence of real numbers** is just a mapping

$$s : \mathbb{N} \to \mathbb{R}$$

from the set $\mathbb{N} = \{1, 2, 3, \ldots\}$ of positive integers into the set $\mathbb{R}$ of real numbers. If s is such a sequence and $n \in \mathbb{N}$, we usually write $s_n = s(n)$, and refer to s_n as the **nth term** of the sequence.

PROBLEM SET 2.3

1. Determine which sets of ordered pairs are functions. [$\mathbb{R}$ denotes the set of all real numbers.]
   (a) $\{(1, 1), (2, 1), (1, 2)\}$
   (b) $\{(1, 1), (2, 1), (3, 1)\}$
   (c) $\{(x, y) \mid x$ is a triangle in the plane and $y = $ area of $x\}$
   (d) $\{(x, y) \mid x, y \in \mathbb{R}, x^2 = y\}$
   (e) $\{(x, y) \mid x, y \in \mathbb{R}, y^2 = x\}$
   (f) $\{(x, y) \mid x, y \in \mathbb{R}, x = \sin y\}$
   (g) $\{(x, y) \mid x, y \in \mathbb{R}, x \leq y\}$
   (h) $\{(x, y) \mid x, y \in \mathbb{R}, x^2 = y^2\}$

2. If f is a function, prove that $\mathrm{codom}(f) = \{y \mid (\exists x)(x \in \mathrm{dom}(f) \wedge y = f(x))\}$.

3. For the function $f = \{(1, c), (2, d), (3, a), (5, b), (7, c)\}$, find:
   (a) $\mathrm{dom}(f)$　　(b) $\mathrm{range}(f)$　　(c) $f(1)$
   (d) $f(2)$　　　　(e) $f(5)$　　　　(f) $f(7)$

4. In calculus, we often refer to $f(x)$ as a function; however, this is really an abuse of language. Why?

5. Find the domain and range of each function. [$\mathbb{R}$ denotes the set of all real numbers.]
   (a) $f = \{(1, a), (2, a), (4, c), (6, b)\}$
   (b) $g = \{(1, b), (2, b), (3, b), (5, b)\}$
   (c) $h = \{(x, y) \mid x, y \in \mathbb{R}, y = \sin x\}$
   (d) $F = \{(x, y) \mid x, y \in \mathbb{R}, y = x^2 + 4\}$
   (e) $G = \{(x, y) \mid x, y \in \mathbb{R}, y = \sec x\}$
   (f) $H = \{(x, y) \mid x, y \in \mathbb{R}, y = 1/(x + 2)\}$

6. (a) Is the null set a function?
   (b) If Y is a set, does there exist a mapping $f : \varnothing \to Y$? If so, how many such mappings are there?

7. Determine, as an explicit set of ordered pairs, the function f depicted in the transformation diagram in Figure 2-12.

8. A function f (or a mapping $f : X \to Y$) is called a **constant function** (or a **constant mapping**) if $f(a) = f(b)$ holds for every $a, b \in \mathrm{dom}(f)$.

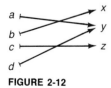

FIGURE 2-12

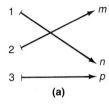

(a)

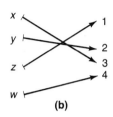

(b)

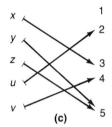

(c)

FIGURE 2-13

(a) Describe the graph of a constant function.
(b) If X is a nonempty set and Y is a finite set with $\#Y = m$, how many different constant mappings $f:X \to Y$ are there?

9. In a calculus textbook, we find the words "Consider the function $y = 1/\sqrt{x} + 2$." Translate these words into more precise mathematical language.

10. In a calculus textbook, we find the words "Consider the function $f(x) = (x + 1)/(x - 1)$." Translate these words into more precise mathematical language.

11. Which of the mappings whose transformation diagrams are shown in Figure 2-13 are injections?

12. If $\mathbb{R}$ denotes the set of all real numbers and f is a function with $f \subseteq \mathbb{R} \times \mathbb{R}$, show that f is injective if and only if no horizontal straight line in the Cartesian plane $\mathbb{R} \times \mathbb{R}$ intersects f more than once.

13. Determine which of the following real-valued functions of a real variable are injections:
(a) $f(x) = x$ (b) $g(x) = x^3 - 1$ (c) $h(x) = \sin x$

14. If f is a function and A is a subset of $\mathrm{dom}(f)$, then the **image** of A under f, written $f(A)$, is defined by

$$f(A) = \{f(x) \mid x \in A\}.$$

Prove that f is an injection if and only if $f(A \cap B) = f(A) \cap f(B)$ holds for every pair of subsets A, B of $\mathrm{dom}(f)$.

15. Show that a mapping $f:X \to Y$ is injective if and only if, in its transformation diagram, no two arrows start at different elements of X and point to the same element of Y.

16. There are eight different mappings from $X = \{1, 2, 3\}$ into $Y = \{a, b\}$.
(a) Draw transformation diagrams for each of these mappings.
(b) How many of these mappings are injective?
(c) How many of these mappings are surjective?
(d) How many of these mappings are bijective?

17. Let $X = \{1, 2, 3, 4, 5\}$, $Y = \{a, b, c, d, e\}$, and

$$f = \{(1, d), (2, c), (3, d), (4, c), (5, d)\}.$$

(a) Show that $f:X \to Y$ is a mapping.
(b) Find range(f).
(c) Is $f:X \to Y$ injective?
(d) Is $f:X \to Y$ surjective?

18. Suppose that f is a function. Does it make any sense to ask whether f is surjective? Explain.

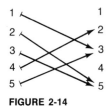

FIGURE 2-14

19.   Let $X = \{1, 2, 3, 4, 5\}$, and let the mapping $f:X \to X$ be defined by the diagram in Figure 2-14. Is f a surjection? Why or why not?

20.   If X is a nonempty set and $M \subseteq X$, we define the **characteristic function** of M, in mapping notation $\chi_M:X \to \{0, 1\}$, by

$$\chi_M(x) = \begin{cases} 1, & \text{if } x \in M \\ 0, & \text{if } x \notin M \end{cases} \quad \text{for all } x \in X.$$

Show that $\chi_M:X \to \{0, 1\}$ is a surjection if and only if M is a nontrivial subset of X.

21.   If $\mathbb{R}$ denotes the set of all real numbers, then each formula below defines a mapping $f:\mathbb{R} \to \mathbb{R}$. In each case, determine whether $f:\mathbb{R} \to \mathbb{R}$ is injective and whether it is surjective.
(a) $f(x) = x$ (b) $f(x) = \sin x$ (c) $f(x) = 3x + 2$
(d) $f(x) = \text{Arctan } x$ (e) $f(x) = 2$ for all $x \in \mathbb{R}$

22.   Let $\mathbb{R}$ denote the set of real numbers, and define $f:\mathbb{R} \to \mathbb{R}$ by

$$f(x) = \begin{cases} 2 - x, & \text{if } x \le 1 \\ 1/x, & \text{if } x > 1 \end{cases} \quad \text{for all } x \in \mathbb{R}.$$

(a) Is $f:\mathbb{R} \to \mathbb{R}$ injective?
(b) Is $f:\mathbb{R} \to \mathbb{R}$ surjective?

23.   If $X = \{1, 2, 3\}$, $Y = \{a, b, c\}$, and $f = \{(1, a), (2, c), (3, b)\}$, show that $f:X \to Y$ is a bijection.

24.   If $\mathbb{R}$ denotes the set of all real numbers, then show that the mapping $T:\mathbb{R} \times \mathbb{R} \to \mathbb{R} \times \mathbb{R}$ defined by $T(x, y) = (x + y, x - y)$ is a bijection.

25.   In the set $\mathbb{R}$ of all real numbers, let $(0, 1)$ be the open interval from 0 to 1, and let $(0, \infty)$ be the unbounded open interval of all positive numbers. Prove that the mapping $f:(0, 1) \to (0, \infty)$ defined by $f(x) = x/(1 - x)$ for all $x \in (0, 1)$ is a bijection.

26.   Let X be a finite set. Explain why a mapping $f:X \to X$ is an injection if and only if it is a bijection. Show by a suitable example that this is not true for an infinite set X.

27.   If $\mathbb{R}$ denotes the set of all real numbers, give examples of mappings $f:\mathbb{R} \to \mathbb{R}$ that are
(a) injective but not surjective;
(b) surjective but not injective;
(c) neither injective nor surjective; and
(d) bijective.

28.   In calculus, the natural logarithm function and the exponential function are inverses of each other. Are the sine and the Arcsine functions inverses of each other? Explain.

29. Each of the formulas given below defines a real-valued function of a real variable. (Assume, as usual, that the domain of each function is the set of all real numbers for which the formula makes sense.) In each case, write a formula for the inverse of the function.
    (a) $f(x) = -7x + 2$ (b) $g(x) = 1/x$
    (c) $h(x) = (2x - 3)/(3x - 2)$ (d) $F(x) = 1 + \sqrt{x}$
    (e) $G(x) = (e^x - e^{-x})/2$

30. Define a real-valued function f of a real variable by

    $$f(x) = \frac{1}{2} \ln \frac{1 + x}{1 - x}, \quad \text{for } |x| < 1.$$

    Prove that f is injective and find a formula for f^{-1}.

31. Two sets X and Y are said to be **equinumerous**, or to have the same **cardinal number**, if there exists a bijection $f : X \to Y$.
    (a) Explain, in your own words, the rationale for this definition.
    (b) Show that the set X of all positive integers and the set Y of all even positive integers are equinumerous.

32. Prove Theorem 3.18.

    In Problems 33–36, let X and Y be finite sets with $\#X = m$ and $\#Y = n$.

33. Prove that there are n^m different mappings $f : X \to Y$.

34. Prove that there are $(n + 1)^m$ different functions g with $g \subseteq X \times Y$. [Hint: Let u be an element that does not belong to Y and let $Y^* = Y \cup \{u\}$. Note that $\#Y^* = n + 1$. If g is a function with $g \subseteq X \times Y$, define $g^* : X \to Y^*$ by $g^*(x) = g(x)$ for $x \in \text{dom}(g)$ and $g^*(x) = u$ for $x \notin \text{dom}(g)$. Show that $g \mapsto g^*$ produces a one-to-one correspondence between the functions $g \subseteq X \times Y$ and the mappings $g^* : X \to Y^*$ and then use the result of Problem 33.]

35. Of the n^m different mappings $f : X \to Y$, prove that, if $m > n$, none of them are injections, whereas if $m \le n$, precisely $n!/(n - m)!$ of them are injections.

36. If $m \ge n > 0$, it can be shown that, of the n^m different mappings $f : X \to Y$, exactly

    $$\sum_{k=1}^{n} (-1)^{n-k} C_k^n \cdot k^m$$

    are surjections, where C_k^n denotes the number of combinations of n things taken k at a time. [See Problem 30 in Problem Set 1.5.] Use this formula to compute the number of surjections $f : X \to Y$ if $X = \{1, 2, 3\}$ and $Y = \{a, b\}$ and compare the result with your answer to Part (c) of Problem 16.

4
_____

MORE ABOUT FUNCTIONS—COMPOSITION AND IMAGES

The idea of a function as a processing device or machine, and the fact that machines can be operated in tandem suggests the concept of *function composition*. In the machine picture, a function f is regarded as a black box that can accept as an input any object x in the domain X, convert it into a new object $f(x)$, and eject $f(x)$ as an output (Figure 2-15).

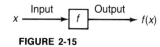

FIGURE 2-15

If f and g are two functions, regarded as machines, we can imagine hooking up the f-machine and the g-machine in tandem, feeding the output of the f-machine directly into the input of the g-machine. If we enclose the resulting assembly in a larger black box called h, we obtain the diagram shown in Figure 2-16.

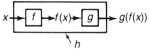

FIGURE 2-16

The composite h-machine can be regarded as a function that transforms x into

$$h(x) = g(f(x)).$$

The *composite function* h is usually written as

$$h = g \circ f,$$

read g *composed with* f. Thus, we have

$$(g \circ f)(x) = g(f(x)).$$

Because of the last equation in the paragraph above, the notation

$$g \circ f$$

for the composition of g and f seems reasonable; however, from the point of view of Figure 2-16, it seems inappropriate because the f-machine is *first* and the g-machine *second* in the diagram. This notational inversion of order is annoying—and can be confusing—but the symbolism $g \circ f$ is entrenched in the mathematical literature. Thus, when translating from diagrams to the symbolism of function composition, you must pay careful attention to the resulting inversion of order.

Let us give a formal definition of function composition:

4.1 DEFINITION	**COMPOSITION OF FUNCTIONS** Let f and g be functions. Then, the relation $g \circ f$ defined by $$g \circ f = \{(x, z) \mid (\exists y)((x, y) \in f \wedge (y, z) \in g)\}$$ is called the **composition** of g and f.

Our first task is to prove that $g \circ f$ is actually a function.

4.2 THEOREM

THE COMPOSITION OF FUNCTIONS IS A FUNCTION If f and g are functions, then $g \circ f$ is a function.

PROOF   We have to show that, if two ordered pairs in the relation $g \circ f$ have the same first element, then they have the same second element. Thus, suppose that

$$(x, z) \in g \circ f \quad \text{and} \quad (x, w) \in g \circ f.$$

We must prove that $z = w$. Because $(x, z) \in g \circ f$, there exists an element y with

$$(x, y) \in f \quad \text{and} \quad (y, z) \in g.$$

Therefore, since f and g are functions, we can write

$$y = f(x) \quad \text{and} \quad z = g(y).$$

Because $(x, w) \in g \circ f$, there exists an element u with

$$(x, u) \in f \quad \text{and} \quad (u, w) \in g.$$

Therefore, since f and g are functions, we can write

$$u = f(x) \quad \text{and} \quad w = g(u).$$

Because $y = f(x)$ and $u = f(x)$, we have $y = u$; hence,

$$w = g(u) = g(y) = z. \qquad \square$$

In the next theorem, we show that Definition 4.1 produces the desired formula $(g \circ f)(x) = g(f(x))$ for the composite function.

4.3 THEOREM

THE IMAGE OF AN ELEMENT UNDER A COMPOSITE FUNCTION   Let f and g be functions, let $x \in \text{dom}(f)$, and suppose that $f(x) \in \text{dom}(g)$. Then,

$$(g \circ f)(x) = g(f(x)).$$

PROOF   Let $x \in \text{dom}(f)$. Then, there exists an element y such that $(x, y) \in f$. Since f is a function, we can write $y = f(x)$. Because

$$y = f(x) \in \text{dom}(g),$$

there exists an element z such that $(y, z) \in g$, that is,

$$z = g(y).$$

Since we have

$$(x, y) \in f \wedge (y, z) \in g,$$

it follows from Definition 4.1 that

$$(x, z) \in g \circ f.$$

Because $g \circ f$ is a function, we can write

$$z = (g \circ f)(x).$$

Therefore,

$$(g \circ f)(x) = z = g(y) = g(f(x)). \qquad \square$$

4.4 Example   Let f and g be the real-valued functions of a real variable defined by

$$f(x) = 2x - 1 \quad \text{and} \quad g(x) = 3x + 5.$$

Find **(a)** $(g \circ f)(x)$ and **(b)** $(f \circ g)(x)$.

SOLUTION   **(a)** $(g \circ f)(x) = g(f(x)) = g(2x - 1) = 3(2x - 1) + 5 = 6x + 2$
(b) $(f \circ g)(x) = f(g(x)) = f(3x + 5) = 2(3x + 5) - 1 = 6x + 9 \qquad \square$

As this example shows, when you use function composition, you must be careful about the order in which you write the functions. In general, $g \circ f$ is not the same as $f \circ g$. In other words, *function composition is not a commutative operation.*

The *chain rule*, one of the most important theorems of differential calculus, is actually a rule for computing the derivative of the composition of two functions. According to the chain rule, *if g is differentiable at x, and if f is differentiable at $g(x)$, then $f \circ g$ is differentiable at x and*

$$(f \circ g)'(x) = f'(g(x))g'(x).$$

In employing the chain rule and in dealing with function composition in general, *you must be careful about the domains and ranges of the functions involved.*

4.5 THEOREM   **DOMAIN AND RANGE OF A COMPOSITION OF FUNCTIONS**   Let f and g be functions. Then:

(i) $\operatorname{dom}(f \circ g) = \{x \mid x \in \operatorname{dom}(g) \land g(x) \in \operatorname{dom}(f)\}$
(ii) $\operatorname{range}(f \circ g) = \{f(y) \mid y \in \operatorname{range}(g)\}$

We leave the proof of Theorem 4.5 to you as an exercise (Problem 16).

4.6 Example   In a calculus textbook, we are asked to consider the function

$$y = \sin^{-1} \sqrt{x} = \text{Arcsin } \sqrt{x}.$$

What is the domain of this function?

SOLUTION   Here we are working with a composition $f \circ g$, where

$$g(x) = \sqrt{x}$$

and

$$f(x) = \sin^{-1} x = \text{Arcsin } x.$$

The domain of g is the unbounded interval $[0, \infty)$ consisting of all non-negative real numbers, and the domain of f is the closed interval $[-1, 1]$ consisting of all real numbers between -1 and 1. By Part (i) of Theorem 4.5, the domain of $f \circ g$ is the set of all numbers x in $[0, \infty)$ such that $g(x) \in [-1, 1]$; that is, all real numbers $x \geq 0$ such that

$$-1 \leq \sqrt{x} \leq 1.$$

Thus, the domain of $f \circ g$ is the closed interval $[0, 1]$. $\square$

Although function composition is not, in general, commutative, it turns out that it is *associative*.

4.7 THEOREM

> **ASSOCIATIVE LAW FOR FUNCTION COMPOSITION**   If f, g, and h are any three functions, then,
>
> $$(f \circ g) \circ h = f \circ (g \circ h).$$

PROOF   We are going to prove that, as sets of ordered pairs,

$$(f \circ g) \circ h = f \circ (g \circ h).$$

We do this by the pick-a-point process. To begin with, suppose

$$(x, z) \in (f \circ g) \circ h,$$

so that, by Definition 4.1, there exists an element y such that

$$(x, y) \in h \quad \text{and} \quad (y, z) \in f \circ g.$$

Because $(y, z) \in f \circ g$, it follows from Definition 4.1 that there exists an element w such that

$$(y, w) \in g \quad \text{and} \quad (w, z) \in f.$$

Now, we have

$$(x, y) \in h, \quad (y, w) \in g, \quad \text{and} \quad (w, z) \in f.$$

Because $(x, y) \in h$ and $(y, w) \in g$, it follows from Definition 4.1 that

$$(x, w) \in g \circ h.$$

Now, we have

$$(x, w) \in g \circ h \quad \text{and} \quad (w, z) \in f;$$

so a final application of Definition 4.1 shows that

$$(x, z) \in f \circ (g \circ h).$$

This proves that

$$(f \circ g) \circ h \subseteq f \circ (g \circ h).$$

The proof of the opposite inclusion is much the same and is left as an exercise (Problem 17).   □

Because of Theorem 4.7, you can write compositions of three or more functions without the aid of parentheses. For instance,

$$f \circ g \circ h$$

is completely unambiguous; it makes no difference whether it is interpreted as $(f \circ g) \circ h$ or as $f \circ (g \circ h)$. In particular, the functions

$$f, f \circ f, f \circ f \circ f, f \circ f \circ f \circ f, \ldots,$$

which are called the **iterates** of f, can be written without parentheses.

4.8 Example   If f is the real-valued function of a real variable defined by the formula

$$f(x) = 2x - x^2,$$

find a formula for the third iterate $f \circ f \circ f$ of f.

SOLUTION
$$\begin{aligned}
(f \circ f)(x) &= f(f(x)) \\
&= f(2x - x^2) \\
&= 2(2x - x^2) - (2x - x^2)^2 \\
&= 4x - 2x^2 - (4x^2 - 4x^3 + x^4) \\
&= -x^4 + 4x^3 - 6x^2 + 4x.
\end{aligned}$$

Therefore,

$$\begin{aligned}
(f \circ f \circ f)(x) &= f((f \circ f)(x)) \\
&= f(-x^4 + 4x^3 - 6x^2 + 4x) \\
&= 2(-x^4 + 4x^3 - 6x^2 + 4x) - (-x^4 + 4x^3 - 6x^2 + 4x)^2 \\
&= -x^8 + 8x^7 - 28x^6 + 56x^5 - 70x^4 + 56x^3 - 28x^2 + 8x.
\end{aligned}$$
□

The mapping notation makes it easier to deal with function composition by automatically taking care of questions involving the domains and ranges of the functions involved. This is accomplished by adopting the following convention: We speak about the composition of the mappings

$$f:X \to Y \quad \text{and} \quad g:W \to Z$$

only when $Y = W$. Then we have

$$f:X \to Y \quad \text{and} \quad g:Y \to Z,$$

in which case,

range(f) $\subseteq Y = $ dom(g),

and it follows that

dom($g \circ f$) $= X$

(Problem 19a) and

range($g \circ f$) $\subseteq Z$

(Problem 19b), so that

$$(g \circ f):X \to Z.$$

This **composition of mappings** is nicely illustrated by the diagram in Figure 2-17. Again, notice the unfortunate order inversion in this diagram: We see f *first* and g *second*, but the composition is $g \circ f$.

FIGURE 2-17

4.9 THEOREM

(i) The composition of two injective mappings is again injective.
(ii) The composition of two surjective mappings is again surjective.
(iii) The composition of two bijective mappings is again bijective.

PROOF

(i) Suppose that $f:X \to Y$ and $g:Y \to Z$ are injections. To prove that $g \circ f:X \to Z$ is an injection, suppose that $a, b \in X$ and that

$$(g \circ f)(a) = (g \circ f)(b).$$

Then

$$g(f(a)) = g(f(b))$$

and, since g is injective, it follows that

$$f(a) = f(b).$$

Since f is injective, the last equation implies that

$$a = b,$$

proving that $g \circ f:X \to Z$ is injective.

(ii) Suppose that $f:X \to Y$ and $g:Y \to Z$ are surjections. To prove that $g \circ f:X \to Z$ is a surjection, suppose that z is an arbitrary element of Z. Then, since $g: Y \to Z$ is surjective, there exists an element $y \in Y$ such that

$$g(y) = z.$$

Since $f: X \to Y$ is surjective, there exists an element $x \in X$ such that

$$f(x) = y.$$

Thus,

$$(g \circ f)(x) = g(f(x)) = g(y) = z,$$

which shows that $g \circ f: X \to Z$ is surjective.

(iii) Part (iii) is a direct consequence of Parts (i) and (ii).   □

4.10 DEFINITION

IMAGE AND INVERSE IMAGE OF A SET   Let f be a function and let M be a set. We define the **image of M under f**, in symbols $f(M)$, and the **inverse image of M under f**, in symbols $f^{-1}(M)$, as follows:

(i) $f(M) = \{y \in \text{range } (f) | (\exists x)(x \in M \cap \text{dom}(f) \wedge y = f(x))\}$.
(ii) $f^{-1}(M) = \{x \in \text{dom}(f) | f(x) \in M\}$.

If $M \subseteq \text{dom}(f)$, then the definition of $f(M)$ is often abbreviated as

$$f(M) = \{f(x) | x \in M\},$$

so that *$f(M)$ is the set of all elements of the form $f(x)$, as x runs through the set M.*

4.11 Example   If f is a function and if M and N are sets, show that

$$f(M \cap N) \subseteq f(M) \cap f(N).$$

SOLUTION   We use the pick-a-point procedure. First, suppose that $y \in f(M \cap N)$. By Part (i) of Definition 4.10 there exists

$$x \in M \cap N \cap \text{dom}(f)$$

such that

$$y = f(x).$$

Because $x \in M \cap N \cap \text{dom}(f)$ and $y = f(x)$, it follows that

$$x \in M \cap \text{dom}(f) \wedge y = f(x);$$

hence, $y \in f(M)$. Likewise,

$$x \in N \cap \text{dom}(f) \wedge y = f(x);$$

hence, $y \in f(N)$. Therefore,

$$y \in f(M) \cap f(N). \qquad \square$$

The following example shows that the inclusion obtained in Example 4.11 cannot, in general, be strengthened to an equality.

4.12 Example   Give an example of a function f and two sets M and N, contained in the domain of f, such that

$$f(M) \cap f(N) \nsubseteq f(M \cap N).$$

SOLUTION   Let $f = \{(1, 1), (2, 1)\}$, $M = \{1\}$, and $N = \{2\}$. Then

$$f(M) = f(\{1\}) = \{f(1)\} = \{1\}$$

and

$$f(N) = f(\{2\}) = \{f(2)\} = \{1\},$$

so that

$$f(M) \cap f(N) = \{1\} \cap \{1\} = \{1\}.$$

However,

$$f(M \cap N) = f(\{1\} \cap \{2\}) = f(\varnothing) = \varnothing$$

(see Problem 29c); hence, in this case,

$$f(M) \cap f(N) \nsubseteq f(M \cap N)$$

because

$$\{1\} \nsubseteq \varnothing. \qquad \square$$

4.13 THEOREM

> **IMAGES, INVERSE IMAGES, AND COMPOSITION**   Let M be a set and let f and g be functions. Then:
>
> (i) $(g \circ f)(M) = g(f(M))$
> (ii) $(g \circ f)^{-1}(M) = f^{-1}(g^{-1}(M))$

PROOF   We prove Part (ii) here and leave the proof of Part (i) as an exercise (Problem 30). Suppose

$$x \in (g \circ f)^{-1}(M).$$

Then, by Part (ii) of Definition 4.10, $x \in \mathrm{dom}(g \circ f)$ and

$$g(f(x)) = (g \circ f)(x) \in M.$$

Consequently,

$$f(x) \in g^{-1}(M),$$

from which it follows that

$$x \in f^{-1}(g^{-1}(M)).$$

This proves that

$$(g \circ f)^{-1}(M) \subseteq f^{-1}(g^{-1}(M)).$$

Conversely, suppose that

$$x \in f^{-1}(g^{-1}(M)).$$

Then $x \in \text{dom}(f)$ and

$$f(x) \in g^{-1}(M).$$

Therefore, $f(x) \in \text{dom}(g)$ and

$$(g \circ f)(x) = g(f(x)) \in M,$$

from which it follows that

$$x \in (g \circ f)^{-1}(M).$$

Hence, we have

$$f^{-1}(g^{-1}(M)) \subseteq (g \circ f)^{-1}(M),$$

and the proof is complete.   ☐

Additional features of images and inverse images of sets under functions are developed in Problems 29–40.

PROBLEM SET 2.4

In Problems 1–10, let f and g be the real-valued functions of a real variable defined by the indicated formulas. In each case, assume that the domains of f and g consist of all real numbers x for which the formula makes sense. Find: (a) $\text{dom}(g \circ f)$, *(b)* $(g \circ f)(x)$ *for* $x \in \text{dom}(g \circ f)$, *(c)* $\text{dom}(f \circ g)$, *and (d)* $(f \circ g)(x)$ *for* $x \in \text{dom}(f \circ g)$.

1. $f(x) = 7x + 2$, $g(x) = -3x$ 2. $f(x) = 2x$, $g(x) = 1/x$

3. $f(x) = 3x$, $g(x) = \sqrt{x + 1}$ 4. $f(x) = 1 - 3x$, $g(x) = (x + 7)^{1/3}$

5. $f(x) = \dfrac{x + 1}{x + 3}$, $g(x) = x^2 + 2$ 6. $f(x) = x^3 + 1$, $g(x) = \sqrt{x - 1}$

7. $f(x) = \sin x$, $g(x) = 3x^2 + 2$ 8. $f(x) = \tan x$, $g(x) = 4x - 3$

9. $f(x) = |5x + 1|, g(x) = 3$ **10.** $f(x) = |x|, g(x) = \dfrac{1}{2x - 3}$

In Problems 11 and 12, find (a) $f \circ g$ and (b) $g \circ f$ as an explicit set of ordered pairs.

11. $f = \{(1, b), (2, c), (3, a)\}, g = \{(a, 1), (b, 2), (c, 2)\}$

12. $f = \{(1, 1), (-1, 5), (2, 1), (-3, 9)\}, g = \{(1, 3), (5, 2), (9, 4), (3, -3)\}$

13. In a calculus textbook, we are asked to consider the real-valued functions of a real variable defined by the indicated formulas. In each case, write the function as a composition $f \circ g$ of real-valued functions f, g of real variables, and find the domain of the function.

(a) $y = \sqrt{\dfrac{x + 1}{x - 1}}$ (b) $y = \cos^{-1} \sqrt{x} = \text{Arccos } \sqrt{x}$

(c) $y = \ln(x^2 - x - 6)$

14. Let f be the real-valued function of a real variable defined by the indicated formula. Find a formula for the fourth iterate $f \circ f \circ f \circ f$ of f.
(a) $f(x) = 2x - 3$ (b) $f(x) = 1 - 4x$

15. Let $a, b, c,$ and d be fixed real numbers and let real-valued functions f and g of real variables be given by $f(x) = ax + b$ and $g(x) = cx + d$ for all real numbers x.
(a) Find necessary and sufficient conditions on $a, b, c,$ and d so that $f \circ g = g \circ f$.
(b) If $a \neq 1, c \neq 1,$ and $f \circ g = g \circ f$, show that there exists a unique real number x such that $f(x) = g(x) = x$.

16. Prove Theorem 4.5.

17. Finish the proof of Theorem 4.7 by proving that $f \circ (g \circ h) \subseteq (f \circ g) \circ h$.

18. Let $\mathbb{R}$ denote the set of all real numbers and suppose that $f : \mathbb{R} \to \mathbb{R}$ and $g : \mathbb{R} \to \mathbb{R}$. If $x \in \mathbb{R}$, show that $(f \circ g)(x)$ can be obtained graphically as follows: Start at the point $(x, 0)$ on the x axis, move vertically to the graph of g, then horizontally to the graph of $y = x$, then vertically to the graph of f, and finally horizontally to the point $(0, y)$ on the y axis (Figure 2-18).

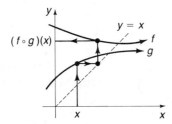

FIGURE 2-18

Conclude that $y = (f \circ g)(x)$. Show how to modify this procedure to find the values $f(x)$, $(f \circ f)(x)$, $(f \circ f \circ f)(x)$, and so on, of the successive iterates of f at x. The set consisting of x and these values is called the **orbit** of x under the iterates of the function f.

19. Let $f: X \to Y$ and $g: Y \to Z$. Prove:
    (a) $\operatorname{dom}(g \circ f) = X$ (b) $\operatorname{range}(g \circ f) \subseteq Z$

20. Give an example of a pair of mappings $f: X \to Y$ and $g: Y \to Z$ such that $(g \circ f): X \to Z$ is bijective but $g: Y \to Z$ is not injective and $f: X \to Y$ is not surjective.

21. Let the mappings $f: X \to Y$ and $g: Y \to Z$ be defined by the diagram in Figure 2-19.
    (a) Draw a diagram for $(g \circ f): X \to Z$.
    (b) Find range $(g \circ f)$.

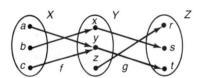

FIGURE 2-19

22. A mapping $g: Y \to X$ is called a **cross section** of the mapping $f: X \to Y$ if $f \circ g = \Delta_Y$.
    (a) If $f: X \to Y$ has a cross section, show that $f: X \to Y$ is surjective.
    (b) If $g: Y \to X$ is a cross section of $f: X \to Y$, show that $g: Y \to X$ is injective.

23. Let $f: X \to Y$ and $g: Y \to Z$.
    (a) If $(g \circ f): X \to Z$ is injective, prove that $f: X \to Y$ is injective.
    (b) If $(g \circ f): X \to Z$ is surjective, prove that $g: Y \to Z$ is surjective.

24. If f and g are functions, prove that $g \circ f = \varnothing$ if and only if the condition range $(f) \cap \operatorname{dom}(g) = \varnothing$ holds. [Note that the empty set, regarded as an empty set of ordered pairs, is a function!]

25. Let $X = \{1, 2, 3\}$ and let $f: X \to X$ be given by the rules $f(1) = 2$, $f(2) = 3$, and $f(3) = 1$. There are 27 different mappings $g: X \to X$. Find all such mappings $g: X \to X$ such that $f \circ g = g \circ f$.

26. Let X be a nonempty set. By a **permutation** of X, we mean a bijection $f: X \to X$. The **active set** of a permutation $f: X \to X$ is defined to be the set $\{x \in X \mid f(x) \neq x\}$.
    (a) Prove that, if two permutations of X have disjoint active sets, then they commute under composition.
    (b) Give an example of two permutations that commute under composition but whose active sets are not disjoint.

27. Let f be the function given by

$$f = \{(1, a), (2, b), (3, a), (4, c), (5, b), (6, d)\}.$$

Find:
(a) $f(\{1, 3\})$ (b) $f(\{2, 5, 6\})$ (c) $f(\{1, 2, 4, 6\})$
(d) $f^{-1}(\{a\})$ (e) $f^{-1}(\{b\})$ (f) $f^{-1}(\{b, c\})$
(g) $f^{-1}(\{c, d\})$

28. Let $f:\mathbb{R} \to \mathbb{R}$ be given by $f(x) = \sin x$. Find:
(a) $f(\{x \mid 0 \leq x \leq \pi/2\})$ (b) $f(\{x \mid 0 \leq x \leq \pi\})$
(c) $f^{-1}(\{0\})$ (d) $f^{-1}(\{-1, 1\})$

29. Let $f:X \to Y$. Show that:
(a) $f(X) = \text{range}(f)$ (b) $f^{-1}(Y) = X$
(c) $f(\varnothing) = \varnothing$ (d) $f^{-1}(\varnothing) = \varnothing$

30. Prove Part (i) of Theorem 4.13.

31. Let f be a function and let M and N be sets. Show that each of the following statements is true:
(a) $f(M) = f(M \cap \text{dom}(f))$ (b) $f^{-1}(N) = f^{-1}(N \cap \text{range}(f))$
(c) $M \subseteq N \Rightarrow f(M) \subseteq f(N)$ (d) $M \subseteq N \Rightarrow f^{-1}(M) \subseteq f^{-1}(N)$
(e) $M \subseteq \text{dom}(f) \Rightarrow M \subseteq f^{-1}(f(M))$
(f) $N \subseteq \text{range}(f) \Rightarrow N = f(f^{-1}(N))$

32. Let $f:X \to Y$.
(a) Give an example to show that it is possible to have $M \subseteq X$ and $M \neq f^{-1}(f(M))$.
(b) Show that $f:X \to Y$ is injective if and only if $M = f^{-1}(f(M))$ holds for all sets $M \subseteq X$.
(c) Give an example to show that it is possible to have $N \subseteq Y$ and $N \neq f(f^{-1}(N))$.
(d) Show that $f:X \to Y$ is surjective if and only if $N = f(f^{-1}(N))$ holds for all sets $N \subseteq Y$.

33. Let f be a function and let M and N be sets. Show that each of the following statements is true:
(a) $f(M \cup N) = f(M) \cup f(N)$ (b) $f^{-1}(M \cup N) = f^{-1}(M) \cup f^{-1}(N)$
(c) $f^{-1}(M \cap N) = f^{-1}(M) \cap f^{-1}(N)$ [Compare with Examples 4.11 and 4.12.]

34. Let f and g be functions.
(a) If $f(M) = g(M)$ holds for every set M, show that $f = g$.
(b) If $f^{-1}(N) = g^{-1}(N)$ holds for every set N, show that $f = g$.

35. Let f be a function and let M and N be sets. Show that each of the following statements is true:
(a) $f(M) \setminus f(N) \subseteq f(M \setminus N)$ (b) $f^{-1}(M) \setminus f^{-1}(N) = f^{-1}(M \setminus N)$

36. Give an example of a function f and two sets M and N with $M \subseteq N \subseteq \text{dom}(f)$ and $f(N \setminus M) \not\subseteq f(N) \setminus f(M)$.

37. Let f be a function and let $(M_i)_{i \in I}$ be an indexed family of sets. Show that each of the following statements is true:

(a) $f(\bigcap_{i \in I} M_i) \subseteq \bigcap_{i \in I} f(M_i)$ (b) $f(\bigcup_{i \in I} M_i) = \bigcup_{i \in I} f(M_i)$

(c) $f^{-1}(\bigcap_{i \in I} M_i) = \bigcap_{i \in I} f^{-1}(M_i)$ (d) $f^{-1}(\bigcup_{i \in I} M_i) = \bigcup_{i \in I} f^{-1}(M_i)$

38. Let $f: X \to Y$ and let $g: Y \to Z$. Define $\phi: (X \times Y) \times Z \to X \times Z$ and $\theta: X \times (Y \times Z) \to (X \times Y) \times Z$ by $\phi((x, y), z) = (x, z)$ and $\theta(x, (y, z)) = ((x, y), z)$ for $x \in X$, $y \in Y$, and $z \in Z$. Prove that

$$g \circ f = \phi((f \times Z) \cap \theta(X \times g)).$$

39. Let $f: X \to Y$ and define $\pi_1: X \times Y \to X$ and $\pi_2: X \times Y \to Y$ by $\pi_1((x, y)) = x$ and $\pi_2((x, y)) = y$ for all $x \in X$ and all $y \in Y$. Let $M \subseteq X$ and let $N \subseteq Y$. Prove:

(a) $f(M) = \pi_2(f \cap \pi_1^{-1}(M))$ (b) $f^{-1}(N) = \pi_1(f \cap \pi_2^{-1}(N))$

40. Let $f: X \to Y$ and suppose that $M \subseteq X$. Show that $f^{-1}(Y \setminus f(X \setminus M)) \subseteq M$.

5

PARTITIONS AND EQUIVALENCE RELATIONS (OPTIONAL)

As we showed in Section 2.2, the elements of a set can be ranked or arranged by means of a preorder, partial order, or total order relation. Another useful procedure for organizing the elements of a set is to classify them according to certain features or properties that some may possess and others may not. This is accomplished by subdividing the set in such a way that elements possessing common features or properties are grouped together. For instance, the set of all real numbers can be classified according to algebraic sign by subdividing it into three parts: the positive numbers, the negative numbers, and zero.

In general, the result of a classification of the elements of a set X is a collection $\mathscr{D}$ of pairwise disjoint subsets of X called **equivalence classes**. Elements of X that possess certain features or properties in common are placed in the same equivalence class; otherwise, they are placed in different equivalence classes. Thus, $\mathscr{D}$ is a set of nonempty subsets of X such that every element of X belongs to one and only one of the sets in $\mathscr{D}$. This suggests the following definition:

5.1 DEFINITION

PARTITION OF A SET   By a **partition** of a set X, we mean a set of sets $\mathscr{D}$ such that

  (i) every set $C \in \mathscr{D}$ is a nonempty subset of X;

  (ii) if $C \in \mathscr{D}$ and $D \in \mathscr{D}$ with $C \neq D$, then $C \cap D = \varnothing$; and

  (iii) if $x \in X$, then there exists $C \in \mathscr{D}$ with $x \in C$.

In other words, *a partition of X is a decomposition, or "chop up," of X into nonempty, pairwise disjoint pieces called equivalence classes.* Condition (ii) in Definition 5.1, that these equivalence classes are pairwise disjoint (so that no two different classes have an element in common), is sometimes expressed by saying that they are **mutually exclusive**. Condition (iii), that every element of X belongs to at least one equivalence class (so that X is the union of these classes), is sometimes expressed by saying that they are **exhaustive**.

A partition $\mathscr{D}$ of X into mutually exclusive and exhaustive equivalence classes can be visualized as shown in Figure 2-20, where the set X is partitioned into 21 equivalence classes,

$$C_1, C_2, C_3, \ldots, C_{21}$$

and

$$\mathscr{D} = \{C_1, C_2, C_3, \ldots, C_{21}\}.$$

As suggested by the appearance of Figure 2-20, the equivalence classes in a partition are sometimes called the *cells* of the partition.

Here are several examples of partitions $\mathscr{D}$:

1. The registered voters in a certain district are classified according to political party; $\mathscr{D} = \{R, D, I\}$, where R is the set of Republicans, D the set of Democrats, and I the set of Independents or members of other parties.
2. The integers are classified according to *parity* (even or odd); $\mathscr{D} = \{E, O\}$, where E is the set of even integers and O the set of odd integers.
3. The points in the Cartesian plane are classified according to the quadrant in which they lie; $\mathscr{D} = \{Q_1, Q_2, Q_3, Q_4, A\}$, where $Q_1, Q_2, Q_3,$ and Q_4 are the points lying in quadrants I, II, III, and IV, respectively, and A is the set of points lying on the coordinate axes.
4. The line segments in three-dimensional space are classified according to their lengths; $\mathscr{D} = \{S_x | x \in \mathbb{R} \wedge x > 0\}$, where $\mathbb{R}$ is the set of real numbers, and for each positive real number x, S_x is the set of all line segments of length x.

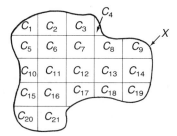

FIGURE 2-20

If $\mathcal{D}$ is a partition of the set X, we can define a relation E on X by specifying that xEy holds if and only if x and y *belong to the same cell of the partition.* For instance, if E is defined in this way for the last example above, then xEy holds if and only if the line segments x and y *have the same length.* This idea is sufficiently important to warrant a more formal definition.

5.2 DEFINITION

> **EQUIVALENCE RELATION DETERMINED BY $\mathcal{D}$**   Let $\mathcal{D}$ be a partition of the set X. Define the relation E on X as follows: If x and y are elements of X, then
>
> $$xEy \Leftrightarrow (\exists C)(C \in \mathcal{D} \wedge x \in C \wedge y \in C).$$
>
> The relation E is called the **equivalence relation determined by the partition $\mathcal{D}$.**

The condition in Definition 5.2, that both x and y belong to C, is often abbreviated by writing

$$x, y \in C$$

rather than

$$x \in C \wedge y \in C.$$

5.3 Example   Consider the partition $\mathcal{D} = \{P, N, Z\}$ of the set $\mathbb{R}$ of all real numbers, where

$$P = \{x \mid x > 0\},$$
$$N = \{x \mid x < 0\},$$
$$Z = \{0\}.$$

Describe in words the equivalence relation E determined by this partition.

SOLUTION xEy holds if and only if either x and y have the same algebraic sign, or they are both zero.   □

5.4 Example   Let $\mathcal{D}$ be a partition of a set X, and let E be the equivalence relation determined by $\mathcal{D}$. Show that the relation E is:

(a) Reflexive on X **(b)** Symmetric   **(c)** Transitive

SOLUTION   **(a)** Let $x \in X$. Then there exists a cell $C \in \mathcal{D}$ such that $x \in C$. It follows that $(\exists C)(C \in \mathcal{D} \wedge x \in C \wedge x \in C)$; hence, xEx holds.

(b) Suppose that xEy, so that there exists a cell $C \in \mathcal{D}$ with $x, y \in C$. Then we have $y, x \in C$, and it follows that yEx.

(c) Suppose that xEy and yEz. Since xEy, there exists a cell $C \in \mathcal{D}$ such that $x, y \in C$. Since yEz, there exists a cell $D \in \mathcal{D}$ such that $y, z \in D$.

Because no two different cells in $\mathscr{D}$ have an element in common, the fact that $y \in C$ and $y \in D$ implies that $C = D$. From $z \in D$ and $D = C$, we conclude that $z \in C$; hence, we have $x, z \in C$, and it follows that xEz. $\square$

Example 5.4 suggests the following definition.

5.5 DEFINITION

EQUIVALENCE RELATION ON A SET Let X be a set. By an **equivalence relation on** X, we mean a relation E that is reflexive on X, symmetric, and transitive.

Example 5.4 shows that the relation E determined by a partition $\mathscr{D}$ of a set X is, in fact, an equivalence relation on X. The diagonal relation Δ_X and the universal relation $X \times X$ are examples of equivalence relations on X (Problem 6).

Here are more examples of equivalence relations E:

1. Define the relation E on the set $\mathbb{R}$ of all real numbers by

$$E = \{(x, y) \mid x^2 = y^2\}.$$

2. Define the relation E on the set $\mathbb{R}$ of all real numbers by

$$E = \{(x, y) \mid \sin x = \sin y\}.$$

3. Define the relation E on the set X of all human beings by xEy if and only if x and y have the same biological father.

4. Let X denote the set of all ordered pairs (m, n) of integers m and n such that $n \neq 0$, and define $(m, n)E(p, q)$ to mean that

$$\frac{m}{n} = \frac{p}{q}.$$

In Problems 9–12, we ask you to show that these relations are equivalence relations on the indicated sets.

We have seen above that *every partition determines an equivalence relation*. Now we are going to show that, vice versa, *every equivalence relation determines a partition*. We begin with the following definition:

5.6 DEFINITION

***E*-EQUIVALENCE CLASSES** Let E be an equivalence relation on the set X. If $x \in X$, we define the ***E*-equivalence class determined by** x, in symbols $[x]_E$, by

$$[x]_E = \{y \in X \mid yEx\}.$$

If the equivalence relation E under consideration is understood, we sometimes write $[x]$ rather than $[x]_E$ and simply refer to it as the *equivalence class determined by* x. Because of the square brackets, $[x]$ is also called *brackets* x.

5.7 Example Let E be the equivalence relation defined on the set $\mathbb{R}$ of all real numbers by

$$E = \{(x, y)\,|\,\sin x = \sin y\}.$$

Find the equivalence class $[0]$ determined by 0.

SOLUTION $$\begin{aligned}[0] = [0]_E &= \{y\,|\,yE0\} \\ &= \{y\,|\,\sin y = \sin 0\} \\ &= \{y\,|\,\sin y = 0\} \\ &= \{y\,|\,y = n\pi \text{ for some integer } n\}.\end{aligned}$$

In other words, the equivalence class $[0]$ is the set of all integer multiples of π. Although it is not in strict accordance with the usual notation for implicitly defined sets, it is permissible to write this in the slightly abbreviated form

$$[0] = \{n\pi\,|\,n \text{ is an integer}\}. \qquad \square$$

5.8 LEMMA

Let E be an equivalence relation on a set X, and let $x, y \in X$. Then:

(i) $y \in [x] \Leftrightarrow yEx.$
(ii) $x \in [x].$
(iii) $[x]$ is a nonempty subset of X.

PROOF (i) By Definition 5.6, $y \in [x] \Leftrightarrow yEx.$
(ii) That $x \in [x]$ is a consequence of Part (i) above and the fact that E is reflexive on X, so that xEx.
(iii) By Part (i) again, if $y \in [x]$, then [by Part (i) of Definition 2.4], $y \in \text{dom}(E)$. Since E is a relation on x, it follows [from Part (iii) of Definition 2.4] that $\text{dom}(E) \subseteq X$; hence, $y \in X$. This shows that $[x]$ is a subset of X. That $[x]$ is nonempty follows from Part (ii) above. $\square$

5.9 THEOREM

PROPERTIES OF [x] Let E be an equivalence relation on a set X, and let $x, y \in X$. Then:

(i) $[x] \cap [y] \neq \varnothing \Leftrightarrow xEy.$
(ii) $[x] \cap [y] \neq \varnothing \Leftrightarrow [x] = [y].$

PROOF   (i) Suppose that $[x] \cap [y] \neq \varnothing$, so there exists $z \in [x] \cap [y]$. Then $z \in [x]$; so by Part (i) of Lemma 5.8, zEx. Likewise, $z \in [y]$; so zEy. Because E is symmetric, it follows from zEx that xEz. Now, from xEz, zEy, and the fact that E is transitive, we have xEy. This shows that

$$[x] \cap [y] \neq \varnothing \Rightarrow xEy.$$

To prove the converse, we begin by assuming that xEy. Then, by Part (i) of Lemma 5.8 again, we have $x \in [y]$. Since $x \in [x]$ by Part (ii) of Lemma 5.8, it follows that $x \in [x] \cap [y]$ and, hence, that $[x] \cap [y] \neq \varnothing$. This proves that

$$xEy \Rightarrow [x] \cap [y] \neq \varnothing,$$

and completes the proof of Part (i).

(ii) Again, we begin by supposing that $[x] \cap [y] \neq \varnothing$. Then, by Part (i) (which has been proved), xEy. Now we must prove that $[x] = [y]$. We do this by the pick-a-point process. Let $z \in [x]$, so that zEx by Part (i) of Lemma 5.8. From zEx, xEy, and the transitivity of E, it follows that zEy; and therefore, by Part (i) of Lemma 5.8 again, $z \in [y]$. We have shown that, if $z \in [x]$, then $z \in [y]$, and it follows that $[x] \subseteq [y]$. The inclusion $[y] \subseteq [x]$ follows from a similar argument, which we leave as an exercise for you (Problem 20). Thus, we have

$$[x] \cap [y] \neq \varnothing \Rightarrow [x] = [y].$$

To prove the converse, we begin by assuming that $[x] = [y]$. By Part (ii) of Lemma 5.8, $x \in [x]$, and it follows that $x \in [y]$. Therefore, $x \in [x] \cap [y]$, and so $[x] \cap [y] \neq \varnothing$. This shows that

$$[x] = [y] \Rightarrow [x] \cap [y] \neq \varnothing,$$

and the proof is complete.   □

If E is an equivalence relation on a set X, then the set of all E-equivalence classes is denoted by

$$X/E.$$

This set of sets is called X **modulo** E, or X **mod** E for short. (The Latin word *modulo* is used in mathematics to mean *with respect to*.) More formally, we make the following definition:

5.10 DEFINITION

> **X MODULO E**   Let E be an equivalence relation on a set X. Then
>
> $$X/E = \{[x]_E \mid x \in X\}.$$

If $[x]$ is understood to mean $[x]_E$, we can simply write

$$X/E = \{[x] \mid x \in X\}.$$

5.11 Example   Let $X = \{$Matilda, Carlos, Emily$\}$, and let E be the relation defined on X by xEy if and only if x and y are of the same sex. Find X/E.

SOLUTION   Here we have

$$[\text{Matilda}] = \{\text{Matilda, Emily}\},$$
$$[\text{Emily}] = \{\text{Matilda, Emily}\},$$

and

$$[\text{Carlos}] = \{\text{Carlos}\}.$$

Therefore,

$$X/E = \{\{\text{Matilda, Emily}\}, \{\text{Carlos}\}\}. \qquad \square$$

Notice that X/E in Example 5.11 is a partition of X. That this is true for any equivalence relation is the content of the next theorem.

5.12 THEOREM

> **THE PARTITION DETERMINED BY E**   If E is an equivalence relation on a set X, then X/E is a partition of X.

PROOF   By Part (iii) of Lemma 5.8, X/E is a set of nonempty subsets of X. By Part (ii) of Theorem 5.9, the sets in X/E are mutually exclusive; and by Part (ii) of Lemma 5.8, they are exhaustive. $\qquad \square$

5.13 Example   Let E be the equivalence relation of Example 5.11, so that $X/E = \{\{\text{Matilda, Emily}\}, \{\text{Carlos}\}\}$. Find the equivalence relation determined by the partition X/E.

SOLUTION   Let F be the equivalence relation determined by the given partition X/E. Then, xFy holds if and only if x and y belong to the same cell of the partition, that is, if and only if x and y are of the same sex. Therefore, we have $F = E$. $\qquad \square$

The following theorem, whose proof we leave to you as an exercise (Problem 21), shows that the phenomenon illustrated in Example 5.13 is general.

5.14 THEOREM

> **RECAPTURING E FROM X/E**   If you start with an equivalence relation E on X and form the partition $\mathscr{D} = X/E$, then the equivalence relation determined by $\mathscr{D}$ is the original equivalence relation E.

5.15 THEOREM

E AS A UNION OF CARTESIAN PRODUCTS Let $\mathscr{D}$ be a partition of X into a finite number of cells,

$$\mathscr{D} = \{C_1, C_2, C_3, \ldots, C_n\}.$$

Then, the equivalence relation E determined by $\mathscr{D}$ is given by

$$E = (C_1 \times C_1) \cup (C_2 \times C_2) \cup (C_3 \times C_3) \cup \cdots \cup (C_n \times C_n).$$

PROOF   By definition, xEy holds if and only if both x and y belong to the same cell, say, C_i of the partition $\mathscr{D}$. But the condition $x, y \in C_i$ is equivalent to $(x, y) \in C_i \times C_i$; hence, E is the union of the sets $C_i \times C_i$ as i runs from 1 to n.   □

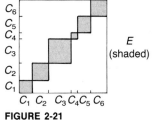

E
(shaded)

FIGURE 2-21

As a matter of fact, Theorem 5.15 is true without any finiteness restriction on the partition $\mathscr{D}$, but then E is the union of an *infinite* set of Cartesian products. (See Section 1.6.) The proof of the more general theorem is similar to the one given above (Problem 22). In view of this theorem, the equivalence relation E determined by a partition $\mathscr{D}$ can be visualized as shown in Figure 2-21 (in which, for purposes of illustration, we have taken $\mathscr{D}$ to consist of six cells).

5.16 Example   Let $X = \{a, b, c, d, e\}$, and consider the partition

$$\mathscr{D} = \{\{a, b\}, \{c, d\}, \{e\}\}$$

of X. Find **(a)** the equivalence relation E on X determined by $\mathscr{D}$ and **(b)** the partition X/E.

SOLUTION   **(a)** By Theorem 5.15,

$$\begin{aligned}
E &= (\{a, b\} \times \{a, b\}) \cup (\{c, d\} \times \{c, d\}) \cup (\{e\} \times \{e\}) \\
&= \{(a, a), (a, b), (b, a), (b, b)\} \cup \{(c, c), (c, d), (d, c), (d, d)\} \cup \{(e, e)\} \\
&= \{(a, a), (a, b), (b, a), (b, b), (c, c), (c, d), (d, c), (d, d), (e, e)\}.
\end{aligned}$$

(b) From Part (a), we have

$$[a] = [b] = \{a, b\},$$
$$[c] = [d] = \{c, d\},$$

and

$$[e] = \{e\}.$$

Hence,

$$X/E = \{\{a, b\}, \{c, d\}, \{e\}\} = \mathscr{D}.$$ □

In Part (b) of Example 5.16, we found that X/E coincides with the original partition $\mathscr{D}$. Again, this phenomenon is general, and we have the following theorem, whose proof we leave as an exercise (Problem 23).

5.17 THEOREM

> **RECAPTURING $\mathscr{D}$ FROM E**   If you start with a partition $\mathscr{D}$ of a set X and form the equivalence relation E determined by $\mathscr{D}$, then the partition X/E coincides with the original partition $\mathscr{D}$.

In view of Theorems 5.14 and 5.17, partitions and equivalence relations are intimately connected; they are "two sides of the same coin." This one-to-one correspondence between partitions and equivalence relations is exploited throughout modern mathematics. Sometimes it is convenient to translate a problem about equivalence relations into a problem about partitions; and, vice versa, it sometimes helps to translate a problem about partitions into a problem about equivalence relations. There is also an interesting connection between preorder relations (Definition 2.13) and equivalence relations.

5.18 THEOREM

> **THE EQUIVALENCE RELATION DETERMINED BY A PREORDER**   Let R be a preorder relation on a set X. Then the relation E defined on X by
>
> $$xEy \Leftrightarrow xRy \land yRx$$
>
> is an equivalence relation on X.

PROOF   We must show that E is reflexive on X, symmetric, and transitive. Because the preorder relation R is reflexive on X, we have xRx for every element x in X, and it follows from the definition of E that xEx holds as well. Thus, E is reflexive on X. Since

$$xRy \land yRx \Leftrightarrow yRx \land xRy,$$

it is clear that E is symmetric. Finally, to prove that E is transitive, suppose that

$$xEy \land yEz.$$

Then, from the definition of E we have

$$xRy \land yRx \land yRz \land zRy.$$

From xRy, yRz, and the fact that the preorder R is transitive, we infer that xRz. Likewise, from zRy and yRx, we infer that zRx. Thus, we have

$$xRz \wedge zRx,$$

and it follows that xEz. □

PROBLEM SET 2.5

In Problems 1–4, determine whether or not each set of sets is a partition of the given set X.

1. $\{C_1, C_2, C_3, C_4\}$, where $C_1 = \{1, 3\}$, $C_2 = \{7, 8, 10\}$, $C_3 = \{2, 5, 6\}$, $C_4 = \{4, 9\}$, and $X = \{1, 2, 3, 4, 5, 6, 7, 8, 9, 10\}$

2. $\{A, B, C\}$, where $A = \{a, c, e\}$, $B = \{b\}$, $C = \{d, g\}$, and $X = \{a, b, c, d, e, f, g\}$

3. $\{\{1, 3, 5\}, \{2, 6, 10\}, \{4, 8, 9\}\}$, where $X = \{1, 2, 3, 4, 5, 6, 7, 8, 9, 10\}$

4. $\{\{a, b, c, d, e, f, g\}\}$, where $X = \{a, b, c, d, e, f, g\}$

5. Decide which of the relations given in Parts (a) through (h) of Problem 17 in Problem Set 2.2 are equivalence relations.

6. If X is a set, show that the diagonal relation Δ_X and the universal relation $X \times X$ are equivalence relations on X.

7. Let $X = \{$Los Angeles, Miami, Dallas, San Francisco, Fort Lauderdale, San Antonio, Daytona Beach, Chicago$\}$. Define a relation E on X by xEy if and only if x and y are in the same state.
   (a) Show that E is an equivalence relation on X.
   (b) Find the equivalence classes in the partition X/E determined by E.

8. In Problem 6, find the partition X/Δ_X and the partition $X/(X \times X)$.

In Problems 9–12, show that the relation E is an equivalence relation on the indicated set.

9. $\mathbb{R}$ is the set of all real numbers, and E is defined on $\mathbb{R}$ by $E = \{(x, y)\,|\,x^2 = y^2\}$.

10. $\mathbb{R}$ is the set of all real numbers, and E is defined on $\mathbb{R}$ by $E = \{(x, y)\,|\,\sin x = \sin y\}$.

11. X is the set of all humans, and E is defined on X by xEy if and only if x and y have the same biological father.

12. X is the set of all ordered pairs (m, n) if integers m and n are such that $n \neq 0$, and $(m, n)E(p, q)$ if and only if $m/n = p/q$.

13. Let E be the equivalence relation defined on the set $\mathbb{R}$ of all real numbers by $E = \{(x, y)\,|\,\cos x = \cos y\}$. Find the equivalence class $[0]_E$.

14. (a) In Problem 9, find $\mathbb{R}/E$.
(b) In Problem 12, find X/E.

15. Describe the partitions X/R determined by those relations in Problem 5 that are equivalence relations.

16. In Problem 10, show that, for $x \in \mathbb{R}$,

$$[x]_E = \{x + 2n\pi \,|\, n \in \mathbb{Z}\} \cup \{(2n + 1)\pi - x \,|\, n \in \mathbb{Z}\},$$

where $\mathbb{Z}$ denotes the set of all integers.
Hint: Use the trigonometric identity

$$\sin x - \sin y = 2 \cos\left(\frac{x + y}{2}\right) \sin\left(\frac{x - y}{2}\right).$$

In Problems 17 and 18, (a) find, as an explicit set of ordered pairs, the equivalence relation E on the set X determined by the partition $\mathscr{D}$, and (b) find the partition of X determined by E.

17. $X = \{a, b, c, d, e\}$ and $\mathscr{D} = \{\{a, d\}, \{b, e\}, \{c\}\}$

18. $X = \{1, 2, 3, 4, 5, 6, 7, 8, 9, 10\}$ and $\mathscr{D} = \{\{2, 5, 8\}, \{3, 6, 9\}, \{1, 4, 7, 10\}\}$

19. Find, as an explicit set of ordered pairs, the equivalence relation on X determined by the partition found in Problem 7.

20. Complete the proof of Part (ii) of Theorem 5.9 by proving that $[x] \cap [y] \neq \varnothing \Rightarrow [y] \subseteq [x]$.

21. Prove Theorem 5.14.

22. State and prove a general version of Theorem 5.15 in which no assumption is made that $\#\mathscr{D}$ is finite.

23. Prove Theorem 5.17.

24. True or false: Every equivalence relation is a preorder relation. If true, prove it; if false, give a counterexample.

25. Prove that, if R is a symmetric and transitive relation and if $X = \text{dom}(R)$, then R is an equivalence relation on X. Does this mean that the reflexivity condition is not needed in Definition 5.5?

26. Let R be a reflexive and transitive relation on X. Prove that $E = R \cap R^{-1}$ is an equivalence relation on X.

27. Suppose that R is both a partial order relation on X and an equivalence relation on X. Prove that $R = \Delta_X$.

28. Let R be a symmetric relation that is reflexive on the set X. Define a relation E on X as follows: xEy holds if and only if there exists a finite sequence of elements $x_1, x_2, \ldots, x_n$ in X such that $x_j R x_{j+1}$ holds for $j = 1, 2, \ldots, n - 1$. Prove that E is an equivalence relation on X.

29. Let E and F be equivalence relations on the set X. Since E and F are sets (of ordered pairs), we can form the intersection $G = E \cap F$.
(a) Prove that G is an equivalence relation on X.
(b) If $x \in X$, find $[x]_G$ in terms of $[x]_E$ and $[x]_F$.

30. Let E and F be equivalence relations on the set X. Give necessary and sufficient conditions on the partitions X/E and X/F so that (as sets of ordered pairs) $E \subseteq F$.

31. Let X be a nonempty set.
(a) Prove that $\{X\}$ is a partition of X.
(b) Find the equivalence relation determined by the partition $\{X\}$ of X.

32. If X is a finite set and E is an equivalence relation on X, prove that $\#E$ is a sum of squares of positive integers.

In Problems 33–36, let R be a preorder relation on the set X, and let E be the corresponding equivalence relation as in Theorem 5.18. If $x \in X$, denote the equivalence class $[x]_E$ by $[x]$, and let $\mathscr{D} = X/E$.

33. If $a, b, c, d \in X$ with $[a] = [c]$ and $[b] = [d]$, show that $aRb \Rightarrow cRd$.

34. We propose to define a relation $\leq$ on $\mathscr{D}$ by $[a] \leq [b]$ if and only if aRb. However, there could be some difficulty with this definition because $[a]$ does not determine the element $a \in X$ uniquely. Using the result of Problem 33, show that there is no difficulty with the definition.

35. If R is a partial order relation on X, show that $E = \Delta_X$.

36. Show that the relation $\leq$ defined in Problem 34 is a partial order relation on $\mathscr{D}$.

37. Let X be a set, and let $\mathscr{E}$ be the set of all equivalence relations on X. If $E, F \in \mathscr{E}$, define $E \leq F$ to mean that (as sets of ordered pairs) $E \subseteq F$. Prove that $\leq$ is a partial order on the set $\mathscr{E}$.

38. In Problem 37, if X has at least three members, prove that $\leq$ is not a total order on $\mathscr{E}$.

39. Let X be a finite nonempty set, let m be a positive integer, and let $\mathscr{D}$ be a partition of X such that, for every $C \in \mathscr{D}$, $\#C = m$.
(a) Prove that m divides $\#X$ exactly.
(b) Prove that $\#\mathscr{D}$ divides $\#X$ exactly.

40. If $\mathbb{R}$ denotes the set of all real numbers and $f(x)$ is a real-valued function defined on all of $\mathbb{R}$, show that the relation

$$E = \{(x, y) \mid f(x) = f(y)\}$$

is an equivalence relation on $\mathbb{R}$.

41. If X is a set with $\#X = n$, let B_n denote the number of different equivalence relations on X. Note that B_n is also the number of different partitions of X. For convenience, put $B_0 = 1$. By listing all possibilities, show that $B_2 = 2$ and that $B_3 = 5$. It can be shown that

$$B_{n+1} = \sum_{k=0}^{n} C_k^n \cdot B_k,$$

where C_k^n denotes the number of combinations of n things taken k at a time. [See Problem 30 in Problem Set 1.5.] Using this formula, compute B_4, B_5, B_6, B_7, and B_8.

42. Let X be a set with $\#X = 8$. A relation on X is selected at random. What is the probability that it is an equivalence relation? [Hint: See Problem 41.]

43. With the notation of Problem 41, explain why B_n is the number of different rhyme schemes for a poem with n lines.

44. It can be shown that the number B_n defined in Problem 41 can be obtained as the sum of an infinite series,

$$B_n = \frac{1}{e} \sum_{k=1}^{\infty} \frac{k^n}{k!},$$

where $e = 2.718281828459\ldots$ is the base of the natural logarithm. Try this out, using a calculator or a computer and forming the partial sums of the first 20 or 30 terms of the series, for $n = 2, 3, 4, 5, 6, 7,$ and 8.

HISTORICAL NOTES

Leibniz will be remembered not only for his independent discovery of calculus but also for his remarkable ideas on logic. In his first paper on combinatorial problems (1666), he outlined a formal symbolic logic. He proposed the introduction of a set of symbols for universal human thoughts so that these symbols could be manipulated in formal ways to produce new ideas. Truth or falsity could be determined by the correctness of calculations within the system, and the proof of theorems could be reduced to a fairly mechanical process. Like many of his other writings, this work displayed the same sort of optimism that brought down on Leibniz the ridicule of Voltaire in *Candide*. It was simply much too ambitious a plan. But the idea came to be used later in more modest ways.

As discussed in Chapter 1, the idea of a set as a mathematical object was not really formulated until the work of George Boole (1815–1864), who developed an algebra of sets in order to examine problems in logic and probability. In 1854 he published his ideas in his famous book that bears the unwieldly title, *An Investigation of the Laws of Thought on Which Are Founded the Mathematical Theories of Logic and Probabilities*.

Georg Cantor (1845–1918) then raised deep and subtle questions about infinite sets and their cardinality. Cantor's ideas led to the theory of sets becoming a branch of mathematics with far-reaching consequences. The subject was controversial during Cantor's lifetime, contributing, possibly, to his bouts with mental illness during the last 33 years of his life. By 1900, however, David Hilbert (1862–1943) could remark that, "From the paradise created for us by Cantor, no one will drive us forth." It is curious to note that though it seems that Cantor followed Boole in developing his ideas, Cantor in fact never accepted Boole's work.

The paradoxes in Cantor's system—and there were some—were investigated by Ernst Zermelo (1871–1953), who tried to axiomatize the system. Gottlob Frege (1848–1925) also developed these ideas to derive number systems from set theoretic foundations. The development of arithmetic entirely from a precise set of axioms was carried further by Bertrand Russell (1872–1970) and Alfred North Whitehead (1861–1947) in their famous work, *Principia Mathematica*. But this difficult, if not impossible, to read work has probably had more influence in logic than in mathematics. In 1931, Kurt Gödel (1906–1978) showed that in a system such as Russell and Whitehead's there will be propositions that can be formulated within the system that will be undecidable within the axioms of the system, that is, there will be propositions that can neither be proved nor disproved. This was a startling discovery and shook the mathematical world.

Let us return, however, to earlier times and some of the ideas introduced in this chapter. A modern view of the Cartesian plane, as a set of points, could not have been part of the original formulation. Descartes, for whom Cartesian coordinates, the Cartesian plane, and the Cartesian product are named, would probably be surprised by the definitions of these terms. Descartes is credited with the invention of analytic geometry, but his contribution is widely misunderstood. Although in some sense he assigned coordinates to points in the plane, he did not choose coordinate axes, and the axes developed in his constructions are rarely, if ever, perpendicular. And there was no preassigned origin. The coordinates were fixed by measuring off distances on two lines that occurred otherwise in the construction. Nevertheless, this was enough to earn for this very ingenious philosopher a firm place in the history of mathematics.

Relations were first studied by Augustus De Morgan (1806–1871), with further work by Charles S. Peirce (1839–1914) (pronounced "purse"), an admirer of Boole, and by Ernst Schroeder (1841–1902). The idea was not taken too seriously until the work of Russell and Whitehead, who considered relations (and functions) to be sets of ordered pairs. The idea of a function goes back much further, but it was not until the development of analytic geometry that a geometric curve was identified with a function. In 1755, Euler gave the modern definition of function, though for numerical functions only. Peter Gustav Lejeune Dirichlet (1805–1859) and Nikolai

Lobachevski (1793–1856) gave the same definition with the assumption of continuity; then Hermann Hankel (1839–1873) gave the modern definition without any additional restrictions.

The formal idea of an equivalence relation probably arose in 1801 when Karl Friedrich Gauss (1777–1855) published his *Disquisitiones Arithmeticae* and introduced the idea of numerical congruence. Of course, the idea can really be traced back to the Greeks, with their interest in the congruence and similarity of triangles.

ANNOTATED BIBLIOGRAPHY

Hillman, A. P., et al. *Discrete and Combinatorial Mathematics* (San Francisco: Dellen), 1987.

A recent, fairly comprehensive text on many topics from discrete mathematics, including relations, functions, and binary operations.

Pfleeger, Shari Lawrence, and David W. Straight. *Introduction to Discrete Structures* (New York: Wiley), 1985.

A detailed treatment of sets, relations, functions, Boolean algebras, lattices, and other topics in discrete mathematics.

Ross, Kenneth A., and Charles R. B. Wright. *Discrete Mathematics* (Englewood Cliffs, N.J.: Prentice-Hall), 1985.

A fairly detailed treatment of relations and functions, along with related material, all with applications in computer science in mind. A text by well-respected mathematicians.

3

THE REAL NUMBER SYSTEM

In the previous chapters, we have often spoken about the real number system $\mathbb{R}$, and we have given several examples and problems involving real numbers; however, we have never provided a formal definition of $\mathbb{R}$. Instead, we have assumed an intuitive understanding of the real numbers and a working knowledge of their properties.

A formal definition of the real number system that is acceptable by modern standards of mathematical rigor must be based on (axiomatic) set theory. Starting from such a basis, it is possible to build up a succession of number systems, each including its predecessor, culminating with $\mathbb{R}$. This process, which is known as *constructing the real numbers*, begins by building up the system

$$\mathbb{N} = \{1, 2, 3, \ldots\}$$

of *natural numbers* using nothing but the machinery of set theory. Next, the system of *integers*

$$\mathbb{Z} = \{0, \pm 1, \pm 2, \pm 3, \ldots\}$$

is constructed from the natural numbers. With $\mathbb{Z}$ at hand, it is possible to form the system

$$\mathbb{Q} = \left\{ \frac{m}{n} \,\middle|\, m, n \in \mathbb{Z} \wedge n \neq 0 \right\}$$

of *rational numbers*. Finally, by applying a suitable *completion process* to $\mathbb{Q}$, the system $\mathbb{R}$ of real numbers is obtained. The succession of number systems does not stop here but proceeds to the *complex numbers*

$$\mathbb{C} = \{x + yi \,|\, x, y \in \mathbb{R}\}$$

and beyond.

The process of building up the succession of number systems

$$\mathbb{N} \subseteq \mathbb{Z} \subseteq \mathbb{Q} \subseteq \mathbb{R} \subseteq \mathbb{C} \subseteq \cdots$$

from below, starting with $\mathbb{N}$, is somewhat involved (and even a bit tedious!).

For our purposes, it is simpler (and perhaps more instructive) to study the characteristic *properties* of the real number system $\mathbb{R}$, leaving the proof of its existence to more advanced courses. Thus, in Sections 1 and 2, we discuss the *algebraic* structure of $\mathbb{R}$; in Section 3, we look at the structure of $\mathbb{R}$ as an *ordered* set; and in Sections 4 and 5, we extract from $\mathbb{R}$ the sub-systems $\mathbb{N}$, $\mathbb{Z}$, and $\mathbb{Q}$ and study their properties.

The algebraic ideas introduced in Sections 1 and 2 are of considerable interest in their own right, so they are treated at a somewhat more general level than is strictly necessary for elucidating the properties of the real number system $\mathbb{R}$.

1

BINARY OPERATIONS AND GROUPS

A *binary operation*, or *binary composition*, on a set S is a way of combining or composing pairs of elements x and y in S to obtain a third element $x * y$ in S, in symbols,

$(x, y) \mapsto x * y.$

For instance, take S to be the set $\mathbb{R}$ of all real numbers and let $*$ be addition:

$(x, y) \mapsto x + y;$

or, take S to be $\mathbb{R}$ and let $*$ be multiplication:

$(x, y) \mapsto x \cdot y;$

or, take S to be all positive real numbers and let $*$ be division:

$(x, y) \mapsto x \div y.$

More formally, we have the following definition:

1.1 DEFINITION   **BINARY OPERATION, OR COMPOSITION**   If S is a nonempty set, then by a **binary operation**, or **binary composition**, on S, we mean a mapping

$*: S \times S \to S.$

If $*: S \times S \to S$ is a binary operation on S, we usually write

$x * y$

rather than

$*((x, y))$

for the image under $*$ of the ordered pair $(x, y) \in S \times S$. The abbreviation $x * y$ for $*((x, y))$, which is called **infix notation**, greatly simplifies calculations with binary operations. Imagine how complicated ordinary arithmetic would look if we wrote

$$+((9, 16)) = 25$$

rather than using the usual infix notation

$$9 + 16 = 25.$$

However, there are situations in more advanced mathematics where it is useful to regard $+$ as what it really is, a mapping

$$+ : \mathbb{R} \times \mathbb{R} \to \mathbb{R}.$$

If $*$ is a binary operation on a set $S = \{a, b, c, \ldots\}$ with a finite number of elements, $\#S = n$, then we can give a complete description of $*$ by writing its **operation table**. To make such a table, we label n horizontal rows and n vertical columns by the symbols $a, b, c, \ldots$, and we fill in the table by placing the value of $x * y$ at the intersection of the row labeled by x and the column labeled by y. The resulting operation table will have the following form:

$*$	a	b	c	$\cdots$
a	$a * a$	$a * b$	$a * c$	$\cdots$
b	$b * a$	$b * b$	$b * c$	$\cdots$
c	$c * a$	$c * b$	$c * c$	$\cdots$
$\vdots$	$\vdots$	$\vdots$	$\vdots$	$\cdots$

1.2 Example   Let $S = \{a, b, c, e\}$ and consider the binary operation $* : S \times S \to S$ defined by the table

$*$	a	b	c	e
a	e	c	b	a
b	c	e	a	b
c	b	a	e	c
e	a	b	c	e

Find: **(a)** $a * e$ **(b)** $a * b$ **(c)** $b * a$ **(d)** $a * (b * c)$

SOLUTION   From the table:

(a) $a * e = a$ **(b)** $a * b = c$
(c) $b * a = c$ **(d)** $a * (b * c) = a * a = e$ $\square$

1.3 DEFINITION   **ALGEBRAIC SYSTEM WITH ONE BINARY OPERATION**   An **algebraic system with a single binary operation** is an ordered pair $(S, *)$ whose first element is a nonempty set S and whose second element is a binary operation $*$ on S.

It is customary to speak of an *algebraic system S with a single binary operation* ∗, although, technically, this is an abuse of language. After all, *S* is just a nonempty *set*, the *algebraic system* is really the ordered pair (*S*, ∗). However, provided there is a clear understanding of what is really meant, no harm seems to come from this custom.

Later we are going to study algebraic systems (for example, the real number system ℝ) with more than one binary operation (for example, addition and multiplication); however, in the present section we confine our attention to algebraic systems *S* with only one binary operation ∗. Therefore, in this section, when we speak of an algebraic system, we always mean an algebraic system with one binary operation.

1.4 DEFINITION

> **ASSOCIATIVE BINARY OPERATION**   A binary operation $*:S \times S \to S$ is said to be **associative** if
>
> $$x * (y * z) = (x * y) * z$$
>
> holds for every x, y, and z in S.

If ∗ is associative, we say that (*S*, ∗) is an **associative** algebraic system. For instance, if ℝ denotes the set of real numbers and + is the usual operation of addition on ℝ, then (ℝ, +) is an associative algebraic system. Likewise, if · denotes the usual operation of multiplication on ℝ, then (ℝ, ·) is associative.

1.5 Example   If − denotes the usual operation of subtraction on the set ℝ of real numbers, show that (ℝ, −) is *not* an associative system.

SOLUTION   We must find real numbers x, y, and z such that

$$x - (y - z) \neq (x - y) - z.$$

Let's investigate the question of exactly when

$$x - (y - z) = (x - y) - z.$$

The latter equation is equivalent to

$$x - y + z = x - y - z,$$

or,

$$z = -z.$$

The only real number z for which $z = -z$ is $z = 0$; hence, unless $z = 0$, we have

$$x - (y - z) \neq (x - y) - z.$$

In particular, for (say) $x = y = z = 1$, we have

$$1 - (1 - 1) \neq (1 - 1) - 1. \qquad \square$$

If S is an algebraic system and $\#S = n$ is finite, then we can write n^3 different equations of the form

$$x * (y * z) = (x * y) * z$$

with x, y, and z in S. Therefore, to check associativity directly from the definition, we would have to verify n^3 different equations. Needless to say, if n is larger than 2 or 3, this could require considerable effort. For instance, the algebraic system in Example 1.2 is associative, but to prove this by direct computation would require verifying $4^3 = 64$ different equations.

1.6 Example For the algebraic system in Example 1.2, verify by direct calculation that

$$x * (y * z) = (x * y) * z$$

for the following cases:

(a) $x = a, \quad y = b, \quad z = c$ **(b)** $x = b, \quad y = c, \quad z = c$

SOLUTION **(a)** From the operation table in Example 1.2:
$a * (b * c) = a * a = e$ and $(a * b) * c = c * c = e$
(b) $b * (c * c) = b * e = b$ and $(b * c) * c = a * c = b$ $\qquad \square$

To complete the verification (by direct calculation) that the algebraic system in Example 1.2 is associative would require checking 62 additional cases. However, by bringing some theory to bear, it is often possible to shorten the labor of checking whether an algebraic system is associative (Problem 7). Also, it is not difficult to write a computer program to check the associativity of algebraic systems.

1.7 DEFINITION

> **COMMUTATIVE BINARY OPERATION** A binary operation $*: S \times S \rightarrow S$ is said to be **commutative** if
>
> $$x * y = y * x$$
>
> holds for every x and y in S.

If $*$ is commutative, we say that $(S, *)$ is a **commutative** algebraic system. For instance, if $\mathbb{R}$ denotes the set of real numbers and $+$ is the usual operation of addition on $\mathbb{R}$, then $(\mathbb{R}, +)$ is a commutative algebraic system. Likewise, if $\cdot$ denotes the usual operation of multiplication on $\mathbb{R}$, then $(\mathbb{R}, \cdot)$

is commutative. It is not difficult to verify that an algebraic system is commutative by examining its operation table—you just have to check that the table is symmetric about the diagonal going from the upper left corner to the lower right corner. For instance, the algebraic system in Example 1.2 is commutative.

1.8 Example   If $S = \{e, a, b, c, d, f\}$, show that the algebraic system $(S, *)$ defined by the following table is not commutative.

*	e	a	b	c	d	f
e	e	a	b	c	d	f
a	a	b	e	d	f	c
b	b	e	a	f	c	d
c	c	f	d	e	b	a
d	d	c	f	a	e	b
f	f	d	c	b	a	e

SOLUTION   From the table we have, for instance, $a * c = d$, but $c * a = f$, so $a * c \neq c * a$.   □

1.9 DEFINITION

> **NEUTRAL ELEMENT**   An element $e \in S$ is said to be a **neutral element** for the algebraic system $(S, *)$ provided that, for every $x \in S$,
>
> $$e * x = x \quad \text{and} \quad x * e = x.$$

A neutral element is also called a **unity element**, a **unit element**, or an **identity element**. For instance, if $\mathbb{R}$ denotes the set of real numbers, $+$ is the usual operation of addition on $\mathbb{R}$, and $\cdot$ is the usual operation of multiplication on $\mathbb{R}$, then 0 is a neutral element for the algebraic system $(\mathbb{R}, +)$ and 1 is a neutral element for the algebraic system $(\mathbb{R}, \cdot)$. In Example 1.2, and also in Example 1.8, the element denoted by e is a neutral element.

According to the following theorem, if an algebraic system has a neutral element, it has exactly one such, and therefore we are entitled to speak of *the* neutral element for the system.

1.10 THEOREM

> **UNIQUENESS OF THE NEUTRAL ELEMENT**   If an algebraic system $(S, *)$ has a neutral element, then this neutral element is unique.

PROOF   Suppose that e and f are two (possibly different) neutral elements for S. Then, since e is a neutral element for S, and since $f \in S$, we have

$$e * f = f.$$

But, since f is a neutral element for S, and since $e \in S$, we also have

$e * f = e$.

Therefore, $e = f$. □

1.11 DEFINITION

> **SEMIGROUP, MONOID** An associative algebraic system $(S, *)$ is called a **semigroup**. A semigroup with a neutral element is called a **monoid**.

The algebraic systems $(\mathbb{R}, +)$ and $(\mathbb{R}, \cdot)$ are monoids and so are the algebraic systems given by the tables in Examples 1.2 and 1.8.

If S is a semigroup, then the identity

$x * (y * z) = (x * y) * z$

holds for all choices of x, y, and z in S; that is, *the result of a composition of three elements of a semigroup is independent of the way in which they are grouped*. This leads us to define

$x * y * z = (x * y) * z,$

or, what is the same thing,

$x * y * z = x * (y * z).$

Similarly, if x, y, z, and w are elements of S, we define

$x * y * z * w = (x * y * z) * w.$

There are five different ways in which the four elements x, y, z, and w can be grouped under the binary composition $*$ (Problem 13); for instance,

$(x * y) * (z * w)$ and $(x * (y * z)) * w$

are two of the ways. It is not difficult to show (Problem 15) that all five of these different groupings produce the same result, namely, $x * y * z * w$. If $x_1, x_2, x_3, \ldots, x_n$ are elements of S, then, proceeding by induction, we define the **n-fold composition** $x_1 * x_2 * x_3 * \cdots * x_n$ by

$x_1 * x_2 * x_3 * \cdots * x_n = (x_1 * x_2 * x_3 * \cdots * x_{n-1}) * x_n.$

By using the associative law over and over again, it can be seen that *in a semigroup S, the result of a composition of any finite number of elements is independent of the way in which they are grouped*. Note, however, that unless S is commutative, the result of a composition

$x_1 * x_2 * x_3 * \cdots * x_n$

might very well depend on the *order* in which the elements appear.

1.12 Example Let S be a semigroup, and let x, y, and z be elements in S. If S is commutative, show that

(a) $x * y * z = y * x * z$
(b) $x * y * z = z * y * x$

SOLUTION **(a)** $x * y * z = (x * y) * z = (y * x) * z = y * x * z$
(b) $x * y * z = (x * y) * z = z * (x * y) = z * (y * x) = (z * y) * x$
$$= z * y * x \qquad \square$$

By arguing as in Example 1.12, and using the associative and commutative laws over and over again, it can be seen that, *in a commutative semigroup, the value of an n-fold composition* $x_1 * x_2 * x_3 * \cdots * x_n$ *is unaffected by the rearrangement of the order in which the elements appear.*

1.13 DEFINITION

LEFT- AND RIGHT-INVERSE   Let $(S, *)$ be a monoid with neutral element e, and let $x \in S$.

(i) An element $a \in S$ is called a **left-inverse** for x if

$$a * x = e.$$

(ii) An element $b \in S$ is called a **right-inverse** for x if

$$x * b = e.$$

(iii) An element $c \in S$ is called an **inverse** for x if

$$c * x = e = x * c.$$

Thus, an inverse for x is an element that is both a right-inverse and a left-inverse for x. Because

$$e * e = e,$$

it follows that *the neutral element e is its own inverse.*

1.14 DEFINITION

EQUALITY OF LEFT- AND RIGHT-INVERSES   Let $(S, *)$ be a monoid, and let $x \in S$. If $a, b \in S$, a is a left-inverse for x, and b is a right-inverse for x, then $a = b$.

PROOF If e is the neutral element of S, then we have

$$a * x = e \qquad \text{and} \qquad x * b = e.$$

Therefore,

$$a = a * e = a * (x * b) = (a * x) * b = e * b = b. \qquad \square$$

1.15 COROLLARY

> **UNIQUENESS OF INVERSES** Let $(S, *)$ be a monoid, and let $x \in S$. If x has an inverse in S, then this inverse is unique.

PROOF Suppose that both a and b are inverses of x in S. Then, in particular, a is a left-inverse for x in S and b is a right-inverse for x in S. Consequently, by Theorem 1.14, $a = b$. □

1.16 DEFINITION

> **INVERTIBLE ELEMENT OF A MONOID** Let $(S, *)$ be a monoid, and let $x \in S$. If x has an inverse in S, we say that x is an **invertible** element of S.

Rather than saying that x is an invertible element of S, we often just say that x is **invertible in** S. If x is invertible in S, then, by Corollary 1.15, it has a unique inverse in S.

1.17 DEFINITION

> **NOTATION FOR THE INVERSE OF AN ELEMENT** If $(S, *)$ is a monoid and x is an invertible element of S, we denote the unique inverse of x in S by the notation x^-.

If $(S, *)$ is a monoid and x is invertible in S, the element x^- is called *x-inverse*. If e is the neutral element of $(S, *)$, we have

$$(x^-) * x = x * (x^-) = e.$$

As we remarked above, the neutral element e is its own inverse, so that

$$e^- = e.$$

In the algebraic system $(\mathbb{R}, +)$, we have

$$(-x) + x = x + (-x) = 0$$

for every element x; hence, every element x is invertible and its inverse is given by

$$x^- = -x.$$

In the algebraic system $(\mathbb{R}, \cdot)$, we have

$$x^{-1} \cdot x = x \cdot x^{-1} = 1$$

for every *nonzero* element x; hence, every element $x \neq 0$ is invertible and its inverse is given by

$$x^- = x^{-1} = \frac{1}{x}.$$

The following theorem establishes two of the most important properties of inverses.

1.18 THEOREM

PROPERTIES OF INVERSES   Let $(S, *)$ be a monoid, and let $x, y \in S$. Then:

(i) If x is invertible in S, so is x^-, and

$$(x^-)^- = x.$$

(ii) If x and y are invertible in S, so is $x * y$, and

$$(x * y)^- = (y^-) * (x^-).$$

PROOF   (i) If x is invertible, then we have

$$(x^-) * x = x * (x^-) = e,$$

where e is the neutral element of S. This shows that x^- is invertible and that its inverse is x, that is,

$$(x^-)^- = x.$$

(ii) Suppose that x and y are invertible in S. To show that $x * y$ is invertible in S, we must find an element $z \in S$ such that

$$(x * y) * z = z * (x * y) = e,$$

and, if such an element z can be found, then $z = (x * y)^-$. Thus, let $z = (y^-) * (x^-)$. Then, since $*$ is associative,

$$(x * y) * z = x * y * z = x * y * (y^-) * (x^-)$$
$$= x * [y * (y^-)] * (x^-) = x * e * (x^-)$$
$$= x * (x^-) = e.$$

A similar calculation (Problem 23) shows that $z * (x * y) = e$, and completes the proof.   □

1.19 Example   Interpret Theorem 1.18 for the monoid $(\mathbb{R}, \cdot)$ of real numbers under the operation of multiplication.

SOLUTION   The neutral element is the real number 1 and $x \in \mathbb{R}$ is invertible if and only if $x \neq 0$, in which case $x^- = x^{-1} = 1/x$. Thus, in this example, Part (i) of Theorem 1.18 is just the familiar fact that

$$x \neq 0 \Rightarrow \frac{1}{x} \neq 0 \quad \text{and} \quad \frac{1}{1/x} = x.$$

Part (ii) is the equally familiar fact that

$$x \neq 0 \wedge y \neq 0 \Rightarrow xy \neq 0 \wedge \frac{1}{xy} = \left(\frac{1}{y}\right)\left(\frac{1}{x}\right). \qquad \square$$

Because we are so used to working in *commutative* algebraic systems such as $(\mathbb{R}, \cdot)$, the "order inversion" in Part (ii) of Theorem 1.18 may come as a bit of a surprise. In a commutative monoid, it is irrelevant whether we write $(y^-) * (x^-)$ or $(x^-) * (y^-)$; however, the order inversion is quite essential in a noncommutative monoid.

1.20 DEFINITION

GROUP   A **group** is a monoid in which every element is invertible. A **commutative** group is a commutative monoid that is a group.

Commutative groups are often referred to as **abelian** groups in honor of the Norwegian mathematician Niels Henrik Abel (1802–1829), a leader in the development of several important branches of modern mathematics. The algebraic system $(\mathbb{R}, +)$ is a commutative, or abelian, group, and so is the algebraic system given by the table in Example 1.2 (Problem 25). The algebraic system given by the table in Example 1.8 is a *noncommutative* group (Problem 26). The algebraic system $(\mathbb{R}, \cdot)$ forms a commutative monoid but it is *not* a group because the element $0 \in \mathbb{R}$ has no multiplicative inverse.

If $(G, *)$ is a group and G is a finite set, then $(G, *)$ is called a **finite group**; otherwise, it is called an **infinite group**. For instance, $(\mathbb{R}, +)$ is an infinite group. If $(G, *)$ is a finite group, then $\#G$ is called the **order** of $(G, *)$. For instance, the group in Example 1.2 has order 4.

1.21 DEFINITION

CANCELLATION LAWS   Let $(S, *)$ be an algebraic system.

(i) $(S, *)$ is said to satisfy the **left-cancellation law** if, for all elements $a, x, y \in S$,

$$a * x = a * y \Rightarrow x = y.$$

(ii) $(S, *)$ is said to satisfy the **right-cancellation law** if, for all elements $a, x, y \in S$,

$$x * a = y * a \Rightarrow x = y.$$

Note that $(S, *)$ satisfies the left-cancellation law if and only if no element appears twice in any horizontal row of its operation table (Problem 29). Likewise, $(S, *)$ satisfies the right-cancellation law if and only if no element appears twice in any vertical column of its operation table. If an algebraic system $(S, *)$ satisfies both the left- and the right-cancellation laws, we simply say that it satisfies the **cancellation laws**.

1.22 THEOREM

> **CANCELLATION LAWS IN A GROUP**   A group satisfies the cancellation laws.

PROOF   Let $(S, *)$ be a group with neutral element e and suppose that $a, x, y \in S$ with

$$a * x = a * y.$$

Because a is invertible in S, a^- exists in S and we have

$$(a^-) * (a * x) = (a^-) * (a * y).$$

Therefore, since $*$ is associative,

$$[(a^-) * a] * x = [(a^-) * a] * y,$$

that is,

$$e * x = e * y,$$

so that

$$x = y.$$

This proves the left-cancellation law. The proof of the right-cancellation law is similar and is left as an exercise (Problem 31).   □

As a consequence of Theorem 1.22, no element appears more than once in any horizontal row or in any vertical column of the operation table for a group $(S, *)$. Moreover, each element appears *at least once* in each such row or column. Indeed, to say that the element $b \in S$ appears at least once in the row labeled by the element $a \in S$ is to say that the equation

$$a * x = b$$

can be solved for $x \in S$ (Problem 33), and this equation has the unique solution

$$x = (a^-) * b$$

(Problem 35). Likewise, to say that the element b appears at least once in the column labeled by the element a is to say that the equation

$$y * a = b$$

can be solved for $y \in S$ (Problem 33), and this equation has the unique solution

$$y = b * (a^{-})$$

(Problem 35). Therefore:

> In the operation table for a group, each element appears once and only once in each row and in each column.

PROBLEM SET 3.1

1. In the algebraic system $(S, *)$ of Example 1.8, find:
   (a) $a * b$ (b) $b * a$ (c) $a * d$ (d) $d * a$
   (e) $a * (b * c)$ (f) $(a * b) * c$ (g) $a * (c * b)$ (h) $(a * c) * b$

2. Let $S = \{a, b, c\}$, and define a binary operation $*$ on S by

$$x * y = \begin{cases} z \in S \text{ such that } x \neq z \text{ and } y \neq z & \text{if } x \neq y \\ x & \text{if } x = y \end{cases}$$

   for all $x, y \in S$.
   (a) Write out the operation table for $(S, *)$.
   (b) Is $(S, *)$ an associative algebraic system?
   (c) Is $(S, *)$ a commutative algebraic system?
   (d) Does $(S, *)$ have a neutral element? If so, what is it?

3. In each case, determine whether or not $*$, as given by the indicated formula for $x, y \in S$, is a binary operation on the set S. Give reasons for your answers.
   (a) $S =$ all the negative integers, $x * y = 4x - y^2$
   (b) $S =$ all positive integers, $x * y = 15x + y$
   (c) $S =$ all positive integers, $x * y = \sqrt{x^2 + y^2}$
   (d) $S =$ all real numbers, $x * y = (x + y)/(1 + x^2 + y^2)$
   (e) $S =$ all real numbers, $x * y = y^x$

4. If S is a finite set with $\#S = n$, show that there are precisely n^{n^2} different binary operations $*$ on S.

5. Define a binary operation $*$ on $\mathbb{R}$ by $x * y = x + y + xy$ for all $x, y \in \mathbb{R}$.
   (a) Prove that $(\mathbb{R}, *)$ is associative.
   (b) Prove that $(\mathbb{R}, *)$ is commutative.
   (c) Prove that $(\mathbb{R}, *)$ is a monoid.

6. Let $S = [0, 1]$ be the closed unit interval in $\mathbb{R}$, and define $x * y = (x + y + |x - y|)/2$ for $x, y \in S$.
   (a) Prove that $(S, *)$ is an algebraic system.
   (b) Prove that $(S, *)$ is associative.

(c) Prove that $(S, *)$ is commutative.

(d) Prove that $(S, *)$ is a monoid.

7. Let $(S, *)$ be an algebraic system with a neutral element e.

(a) If $x, y, z \in S$ and if at least one of the elements x, y, z is equal to e, show that $x * (y * z) = (x * y) * z$.

(b) If S is finite with $\#S = n$, use the result of Part (a) to show that the associativity of $(S, *)$ can be checked directly by verifying $(n - 1)^3$ equations.

8. Let $S = \{e, a\}$, and define $*: S \times S \to S$ by the following table:

$*$	e	a
e	e	a
a	a	e

(a) Prove that $(S, *)$ is a semigroup.

(b) Prove that $(S, *)$ is commutative.

(c) Prove that $(S, *)$ is a monoid.

(d) Prove that $(S, *)$ is a group.

9. Use the results in Problem 7 to verify the associativity of the algebraic system in Example 1.2 by checking the appropriate 27 equations.

10. Give an example of an algebraic system $(S, *)$ containing an element a such that $a * (a * a) \neq (a * a) * a$.

11. Let S be the set of all positive real numbers, and define $*: S \times S \to S$ by $x * y = x/y$ for all $x, y \in S$.

(a) Is $(S, *)$ an associative algebraic system?

(b) Is $(S, *)$ commutative?

(c) Does $(S, *)$ have a neutral element?

(d) Is $(S, *)$ a group?

12. Let $M = \{1, 2\}$ and let S be the set of all functions $f: M \to M$. Define $*: S \times S \to S$ by $f * g = f \circ g$ for all $f, g \in S$. Label the four functions in S as, say, $e, a, b,$ and c, where $e = \Delta_X$.

(a) Write the operation table for $(S, *)$.

(b) Is $(S, *)$ associative?

(c) Is $(S, *)$ commutative?

(d) Is $(S, *)$ a monoid?

(e) Is $(S, *)$ a group?

13. On page 151, we showed two ways in which the four elements $x, y, z,$ and w can be grouped under a binary composition $*$ without changing the order in which they occur. Find the remaining three ways in which these four elements can be grouped under the binary composition $*$ without changing this order.

14. Give an example of a semigroup that is not a monoid.

15. If $*$ is associative in Problem 13, prove that all five of the groupings produce the same result, namely, $x * y * z * w$.

16. In Problem 5:
(a) Find all of the invertible elements in the monoid $(\mathbb{R}, *)$.
(b) Is $(\mathbb{R}, *)$ a group? Why or why not?

17. Let S be a nonempty set, and define a binary operation $*$ on S by $x * y = x$ for all $x, y \in S$.
(a) Prove that $(S, *)$ is a semigroup.
(b) Is $(S, *)$ a monoid? Why or why not?
(c) Is $(S, *)$ a group? Why or why not?

18. In Problem 6:
(a) Find all of the invertible elements in the monoid $(S, *)$.
(b) Is $(S, *)$ a group? Why or why not?

19. Let S be a nonempty set and let a be a fixed element of S. Define a binary operation $*$ on S by $x * y = a$ for all $x, y \in S$.
(a) Prove that $(S, *)$ is a semigroup.
(b) Is $(S, *)$ a monoid? Why or why not?
(c) Is $(S, *)$ a group? Why or why not?

20. Let $\mathbb{N} = \{1, 2, 3, \ldots\}$ denote the set of all natural numbers (positive integers) and let S be the set of all mappings $f : \mathbb{N} \to \mathbb{N}$. Consider the monoid $(S, \circ)$, where $\circ$ denotes function composition. Let $s \in S$ be given by $s(n) = n + 1$ for all $n \in \mathbb{N}$.
(a) Show that s has a left-inverse in $(S, \circ)$.
(b) Show that s has no right-inverse in $(S, \circ)$.
(c) Show that s has infinitely many different left-inverses in $(S, \circ)$.

21. An element a in an algebraic system $(S, *)$ is called an **idempotent** if $a * a = a$.
(a) If $(S, *)$ is a monoid and if y is a left-inverse of x in S, show that $x * y$ is an idempotent in S.
(b) Show that the only idempotent element in a group is the neutral element.

22. Let i be the complex number with $i^2 = -1$. Show that $V = \{1, i, -1, -i\}$ forms a group $(V, \cdot)$ of order 4 under multiplication.

23. Complete the proof of Theorem 1.18 by showing that, if $z = (y^-) * (x^-)$, then $z * (x * y) = e$.

24. Let $(S, *)$ be a monoid and let G be the set of all invertible elements in $(S, *)$. If $\star : G \times G \to G$ is defined by $x \star y = x * y$ for all $x, y \in G$, show that $(G, \star)$ is a group. [We refer to $(G, \star)$ as the **group of invertible elements** of $(S, *)$.]

25. Show that the algebraic system in Example 1.2 is a group. [Hint: See Problem 9.]

26. It can be shown that the algebraic system in Example 1.8 is associative. Granting this, show that it is a noncommutative group.

27. Interpret Theorem 1.18 for the abelian group $(\mathbb{R}, +)$ of real numbers under the operation of addition.

28. Let $\mathcal{M}_2(\mathbb{R})$ denote the set of all two-by-two matrices with real number entries. Multiply two such matrices as follows:

$$\begin{bmatrix} a & b \\ c & d \end{bmatrix} \cdot \begin{bmatrix} x & y \\ z & w \end{bmatrix} = \begin{bmatrix} ax + bz & ay + bw \\ cx + dz & cy + dw \end{bmatrix}.$$

(a) Show that $(\mathcal{M}_2(\mathbb{R}), \cdot)$ is a monoid and identify its neutral element.

(b) Find necessary and sufficient conditions on the real numbers a, b, c, and d so that the matrix on the left in the display above is invertible.

29. Show that an algebraic system $(S, *)$ satisfies the left- (or the right-) cancellation law if and only if no element appears twice in any horizontal row (or any vertical column) of its operation table.

30. Show that the cancellation laws fail in the monoid $(\mathcal{M}_2(\mathbb{R}), \cdot)$ of Problem 28.

31. Complete the proof of Theorem 1.22 by showing that the right-cancellation law holds in a group.

32. (a) Does there exist a group of order 1? Explain.

(b) Write an operation table for a group $(G, *)$ of order 2, denoting the elements by $G = \{e, a\}$, where e is the neutral element. Explain why there is only one way to form such an operation table.

(c) Write an operation table for a group $(G, *)$ of order 3, denoting the elements by $G = \{e, a, b\}$, where e is the neutral element. Explain why there is only one way to form such an operation table.

33. Let $(S, *)$ be an algebraic system and let a, $b \in S$. Show that the equation $a * x = b$ (or $y * a = b$) can be solved for $x \in S$ (or $y \in S$) if and only if b appears at least once in the row (or column) labeled by a in the operation table for $(S, *)$.

34. Write two essentially different operation tables for a group $(G, *)$ of order 4, denoting the elements by $G = \{e, p, q, r\}$, where e is the neutral element.

35. Let $(S, *)$ be a group and let a, $b \in S$. Show that the equations $a * x = b$ and $y * a = b$ have unique solutions x, $y \in S$.

36. Suppose that $(S, *)$ is a semigroup and that there is an element $e \in S$ with the following two properties: (i) $e * x = x$ for every $x \in S$. (ii) For every $x \in S$, there exists $y \in S$ such that $y * x = e$. Prove that $(S, *)$ is a group with e as its neutral element.

37. Give an example of an infinite commutative monoid $(S, *)$ in which the cancellation laws hold but that is not a group.

38. Let $(S, *)$ be a semigroup and assume that, for every $a, b \in S$, the equations $a * x = b$ and $y * a = b$ can be solved for $x, y \in S$. Prove that $(S, *)$ is a group.

39. Let X be a nonempty set and let $\mathscr{P}(X)$ be the set of all subsets of X. Define a binary operation $*$ on $\mathscr{P}(X)$ by $M * N = (M \cup N) \setminus (M \cap N)$ for all $M, N \subseteq X$. Prove that $(\mathscr{P}(X), *)$ is a group.

40. Let $(S, *)$ be a commutative group and assume that $\#S$ is finite and odd. Prove that, for every element $a \in S$, there exists a unique element $b \in S$ such that $b * b = a$ (that is, every element has a unique "square root").

41. Let X be a nonempty set and let $\mathscr{B}(X)$ be the set of all bijections $\phi : X \to X$. Define a binary operation $*$ on $\mathscr{B}(X)$ by $\phi * \psi = \phi \circ \psi$ for all $\phi, \psi \in \mathscr{B}(X)$. Prove that $(\mathscr{B}(X), *)$ is a group.

42. Let A, B, C, D, E, and K be six fixed real numbers. Define a binary composition $*$ on the set $\mathbb{R}$ of all real numbers by $x * y = Ax^2 + Bxy + Cy^2 + Dx + Ey + K$ for all $x, y \in \mathbb{R}$. Find all possible values of A, B, C, D, E, and K for which the system $(\mathbb{R}, *)$ is a semigroup.

2

RINGS AND FIELDS

Until now, we have studied algebraic systems $(S, *)$ with a *single* binary operation $*$. *Rings* and *fields* are algebraic systems with *two* binary operations, called "addition" and "multiplication," governed by certain postulates suggested by the behavior of ordinary addition and multiplication in the real number system $\mathbb{R}$.

2.1 DEFINITION

RING A **ring** is an ordered triple $(R, +, \cdot)$ consisting of a nonempty set R and two binary operations $+$ and $\cdot$, called **addition** and **multiplication** on R, which satisfy the following conditions:

(i) $(R, +)$ is a commutative group.
(ii) $(R, \cdot)$ is a semigroup.
(iii) The **distributive laws** hold, so that, for all $x, y, z \in R$,

$$x \cdot (y + z) = (x \cdot y) + (x \cdot z) \quad \text{and} \quad (x + y) \cdot z = (x \cdot z) + (y \cdot z).$$

The first equation in Part (iii) of Definition 2.1 expresses the **left-distributive law** and the second equation expresses the **right-distributive law**. If $(R, +, \cdot)$ is a ring, then the neutral element of the group $(R, +)$ is written as 0 and called the **zero element** of the ring. If $x \in R$, then the inverse of x in the group $(R, +)$ is written as $-x$ rather than as x^- and it is called the **additive inverse** of x, or the **negative** of x in the ring. Of course, the real number system $(\mathbb{R}, +, \cdot)$ with the usual operations of addition and multiplication provides an example of a ring. However, it must not be supposed that the addition and multiplication operations in a general ring have anything to do with ordinary numerical addition and multiplication, except for the fact that they obey some of the same algebraic laws.

By abuse of language, we often speak of "the ring R" when what we really mean is the ring $(R, +, \cdot)$. Also, we frequently omit the symbol $\cdot$ for the ring multiplication and write simply xy rather than $x \cdot y$ for the "product" of the two elements x and y in R. We recall that, because of the associative laws, parentheses are not needed for iterated additions or multiplications, so that expressions such as

$$x + y + z, \qquad x + y + z + w, \qquad xyz, \qquad xyzw,$$

and so forth make perfectly good sense in a ring. Also, because $(R, +)$ is a *commutative*, or *abelian*, group, we are permitted to "shuffle" the terms in an iterated sum; for instance,

$$x + y + z + w = z + y + w + x.$$

However, in a general ring, there is no assumption that multiplication is commutative, so that the *order* of the factors in a product might be of significance.

Note that, in a ring, an expression such as

$$x + yz$$

could be ambiguous. Does it mean

$$x + (yz)$$

or does it mean

$$(x + y)z?$$

We resolve this ambiguity for a general ring, just as we do in ordinary algebra, by making the following notational agreement:

> *In writing expressions involving combinations of addition and multiplication in a ring, multiplication takes precedence over addition.*

In other words, you must carry out indicated multiplications *first*, before you carry out the additions. Thus, we understand that $x + yz$ means $x + (yz)$. With this notational agreement in mind, we can write the distributive laws as

$$x(y + z) = xy + xz \qquad \text{and} \qquad (x + y)z = xz + yz.$$

The following theorem is almost a straight transcription of Definition 2.1 (Problem 2).

2.2 THEOREM

> **POSTULATES FOR A RING**   Let R be a set containing a special object 0 and equipped with binary operations
>
> $$(x, y) \mapsto x + y \qquad \text{and} \qquad (x, y) \mapsto x \cdot y = xy.$$
>
> Then $(R, +, \cdot)$ is a ring if and only if the following conditions hold for all $x, y, z \in R$:
>
> (i) $x + (y + z) = (x + y) + z$ (ii) $x + y = y + x$
> (iii) $x + 0 = x$
> (iv) For every $x \in R$, there exists $-x \in R$ with $x + (-x) = 0$.
> (v) $x(yz) = (xy)z$
> (vi) $x(y + z) = xy + xz$ and $(x + y)z = xz + yz$

We call $(R, +, \cdot)$ a **finite ring** if $\#R$ is finite. For such a finite ring, the binary operations $+$ and $\cdot$ can be specified by operation tables. For instance, here are the operation tables for a ring called the **integers modulo 4** (see Problem 7) and denoted by $(\mathbb{Z}_4, +, \cdot)$, or simply by $\mathbb{Z}_4$, where $\mathbb{Z}_4 = \{0, 1, 2, 3\}$:

+	0	1	2	3
0	0	1	2	3
1	1	2	3	0
2	2	3	0	1
3	3	0	1	2

$\cdot$	0	1	2	3
0	0	0	0	0
1	0	1	2	3
2	0	2	0	2
3	0	3	2	1

Of course, 0, 1, 2, and 3 do not have their usual numerical meanings in $\mathbb{Z}_4$—they are just used as convenient symbols for the four elements of this ring! Nevertheless, portions of the two tables reproduce familiar numerical facts; for instance, in the multiplication table, $0 \cdot x = x \cdot 0 = 0$ holds for all elements $x \in \mathbb{Z}_4$. The following theorem shows that this is no accident:

2.3 THEOREM

> **MULTIPLICATION BY ZERO IN A RING**   If R is a ring and $x \in R$, then $0 \cdot x = x \cdot 0 = 0$.

PROOF

We prove that $0 \cdot x = 0$ and leave the analogous proof that $x \cdot 0 = 0$ as an exercise (Problem 6). By the right-distributive law, we have

$$(0 + 0) \cdot x = 0 \cdot x + 0 \cdot x.$$

Because 0 is the neutral element in $(R, +)$,

$$0 + 0 = 0,$$

and so we can rewrite the first equation as

$$0 \cdot x = 0 \cdot x + 0 \cdot x.$$

Again, because 0 is the additive neutral element,

$$0 \cdot x + 0 = 0 \cdot x.$$

Combining the last two equations, we have

$$0 \cdot x + 0 = 0 \cdot x + 0 \cdot x.$$

Because $(R, +)$ is a group, it satisfies the left-cancellation law (Theorem 1.22), and therefore, from the last equation, we infer that

$$0 = 0 \cdot x. \qquad \square$$

A number of additional properties of a ring R follow easily from the fact that $(R, +)$ is a commutative group with 0 as its neutral element. For instance, since the neutral element in a group is its own inverse, we have

$$-0 = 0.$$

Three more properties, all of which seem quite natural, appear in the following theorem:

2.4 THEOREM   **PROPERTIES OF NEGATION IN A RING**   Let R be a ring and let x, $y \in R$. Then:

   (i) $x + y = 0 \Rightarrow x = -y$
   (ii) $-(-x) = x$
   (iii) $-(x + y) = (-x) + (-y)$

PROOF   (i) If $x + y = 0$, then x is a left-inverse for y in $(R, +)$. Since $-y$ is a right-inverse for y in $(R, +)$, the fact that $x = -y$ follows from Theorem 1.14.
(ii) This follows directly from Part (i) of Theorem 1.18.
(iii) This follows from Part (ii) of Theorem 1.18 and the fact that $(R, +)$ is commutative. $\qquad \square$

The next theorem establishes some additional properties of negation in a ring, indicating how negation interacts with multiplication.

2.5 THEOREM   **PROPERTIES OF NEGATION AND MULTIPLICATION**   Let R be a ring and let x, $y \in R$. Then:

   (i) $(-x)y = -(xy)$
   (ii) $x(-y) = -(xy)$
   (iii) $(-x)(-y) = xy$

PROOF (i) By the right-distributive law,

$$(-x)y + xy = ((-x) + x)y.$$

Because $(-x) + x = 0$, we can rewrite the last equation as

$$(-x)y + xy = 0 \cdot y.$$

By Theorem 2.3, $0 \cdot y = 0$, and so we have

$$(-x)y + xy = 0.$$

Therefore, by Part (i) of Theorem 2.4,

$$(-x)y = -(xy).$$

(ii) The proof is analogous to the proof of Part (i) and is left as an exercise (Problem 11a).

(iii) Using Parts (i) and (ii) above, we have

$$(-x)(-y) = -(x(-y)) = -(-(xy)).$$

By Part (ii) of Theorem 2.4, we have

$$-(-(xy)) = xy,$$

and it follows that

$$(-x)(-y) = xy. \qquad \square$$

Note that Theorem 2.5 generalizes the well-known "rules of signs" in elementary algebra. However, here are some important words of warning: In a ring, the element $-x$ is often called "*minus x*" or "*the negative of x*," but *you must not make the mistake of supposing that $-x$ is necessarily negative!* In the first place, we are dealing with abstract rings here, and the idea of being "positive" or "negative" may not even make sense. For instance, in the ring $\mathbb{Z}_4$ discussed above, which elements are "positive" and which "negative"? (Answer: None—"positive" and "negative" make no sense in $\mathbb{Z}_4$.) In the second place, any element y in a ring R can be written in the form $y = -x$ for some $x \in R$. [Just take $x = -y$ and use Part (ii) of Theorem 2.4.] Thus, even in the ring $(\mathbb{R}, +, \cdot)$ of real numbers, where it makes sense to talk about *positive* and *negative* elements, $-x$ *need not be negative!*

2.6 DEFINITION

SUBTRACTION IN A RING   Let R be a ring and let $x, y \in R$. We define

$$x - y = x + (-y).$$

The element $x - y$ is read in English as "*x minus y*," and the binary operation $(x, y) \mapsto x - y$ is called **subtraction**. We leave the proof of the following lemma as an exercise (Problem 11b).

2.7 LEMMA | **TRANSPOSITION IN A RING**   Let R be a ring and let $x, y, z \in R$. Then

$$x + y = z \Leftrightarrow x = z - y.$$

Lemma 2.7 justifies the familiar procedure (called *transposition* in elementary algebra) of moving a term from one side of an equation to another at the expense of changing the algebraic sign of the term.

In a ring the operation of addition is, by definition, commutative; however, there is no such requirement for the operation of multiplication.

2.8 DEFINITION | **COMMUTATIVE RING**   If $(R, +, \cdot)$ is a ring and if $(R, \cdot)$ is a commutative semigroup, then we say that R is a **commutative** ring.

Both the ring $\mathbb{R}$ of real numbers and the ring $\mathbb{Z}_4$ of integers modulo 4 are commutative. An important example of a noncommutative ring is given in Problem 13.

The ring $\mathbb{R}$ of real numbers and the ring $\mathbb{Z}_4$ both contain an element 1 that is effective as a multiplicative neutral element. This observation leads us to the next definition.

2.9 DEFINITION | **RING WITH UNITY**   Let $(R, +, \cdot)$ be a ring. If $(R, \cdot)$ has a neutral element different from the additive neutral element 0, then this neutral element of $(R, \cdot)$ is called the **unity element** of the ring, and R is called a **ring with unity**.

In a ring R with unity, the unity element is ordinarily denoted by 1. Thus, 1 is the unique element in R such that $1 \neq 0$ and

$$1 \cdot x = x \cdot 1 = x$$

holds for all elements $x \in R$. If $(R, +, \cdot)$ is a ring with unity 1, then $(R, \cdot)$ is a monoid, and we can speak of the *invertible* elements in this monoid.

2.10 DEFINITION | **MULTIPLICATIVELY INVERTIBLE ELEMENTS IN A RING**   Let $(R, +, \cdot)$ be a ring with unity. If $x \in R$ and x is an invertible element of the monoid $(R, \cdot)$, then we call x a **multiplicatively invertible** element of the ring R and we write its inverse in $(R, \cdot)$ as x^{-1}.

In the ring $\mathbb{Z}_4$ of integers modulo 4, the multiplicatively invertible elements are 1 and 3 (Problem 7c). In the ring $\mathbb{R}$ of real numbers, every nonzero element is multiplicatively invertible. Thus, the real numbers form a *field* in the sense of the following definition:

2.11 DEFINITION

> **FIELD** A **field** is a commutative ring with unity in which every nonzero element is multiplicatively invertible.

The following theorem is almost a straight transcription of the definition of a field (Problem 21):

2.12 THEOREM

> **POSTULATES FOR A FIELD** Let F be a set containing two special objects 0 and 1 and equipped with two binary operations
>
> $$(x, y) \mapsto x + y \qquad \text{and} \qquad (x, y) \mapsto x \cdot y = xy.$$
>
> Then $(F, +, \cdot)$ is a field if and only if the following conditions hold for all $x, y, z \in F$:
>
> (i) Associative Laws
>
> $$x + (y + z) = (x + y) + z$$
>
> and
>
> $$x(yz) = (xy)z$$
>
> (ii) Commutative Laws
>
> $$x + y = y + x$$
>
> and
>
> $$xy = yx$$
>
> (iii) Neutral Elements
>
> $$x + 0 = x,$$
> $$x \cdot 1 = x,$$
>
> and
>
> $$0 \neq 1$$
>
> (iv) Inverses
>
> $$\forall x \in F, \qquad \exists (-x) \in F, \qquad x + (-x) = 0,$$
>
> and
>
> $$\forall x \in F, \qquad x \neq 0 \Rightarrow \exists x^{-1} \in F, \qquad x \cdot x^{-1} = 1$$
>
> (v) Distributive Law
>
> $$x(y + z) = xy + xz$$

Note that it is sufficient in Part (v) to write only the left-distributive law—the right-distributive law then follows from the commutativity of multiplication in Part (ii). Just as we did for rings, we often speak of "the field F" when what we really mean is the field $(F, +, \cdot)$. Below are some examples of fields:

1. $\mathbb{R}$, the field of *real numbers*.
2. $\mathbb{Q}$, the field of *rational numbers*, that is, all real numbers of the form m/n such that m and n are integers and $n \neq 0$.
3. $\mathbb{C}$, the field of *complex numbers*, that is, all numbers of the form $x + yi$, where $x, y \in \mathbb{R}$ and $i^2 = -1$.
4. $\mathbb{Z}_2$, the field of *integers modulo 2*, defined by $\mathbb{Z}_2 = \{0, 1\}$ with the operations given by the following tables:

+	0	1		$\cdot$	0	1
0	0	1		0	0	0
1	1	0		1	0	1

Note that 1–3 are *infinite* fields, whereas the field $\mathbb{Z}_2$ in 4 is a *finite* field. In general, a finite field is called a **Galois field** in honor of the French mathematician Evariste Galois (1811–1832), who used group theory to settle the classic problem of the solvability of equations by radicals.

2.13 THEOREM

MULTIPLICATIVELY INVERTIBLE ELEMENTS IN A FIELD
Let $(F, +, \cdot)$ be a field and let $x \in F$. Then x is multiplicatively invertible in F if and only if $x \neq 0$.

PROOF   By the definition of a field, if $x \neq 0$, then x is multiplicatively invertible in F. Conversely, if x is multiplicatively invertible in F, then x cannot be equal to 0. Indeed, if 0 were multiplicatively invertible in F, we would have

$$0 \cdot 0^{-1} = 1.$$

Moreover, by Theorem 2.3, we would also have

$$0 \cdot 0^{-1} = 0.$$

Consequently, we would have $1 = 0$, in contradiction with Part (iii) of Theorem 2.12.   □

The multiplicative inverse x^{-1} of a nonzero element x in a field is often called the **reciprocal** of x. The following corollary of Theorem 2.13 is obtained by applying the facts about invertible elements in a monoid (Section 3.1) to the monoid $(F, \cdot)$. We leave its routine verification as an exercise (Problem 22).

2.14 COROLLARY

> **PROPERTIES OF RECIPROCALS IN A FIELD** Let $(F, +, \cdot)$ be a field and let $x, y \in F$. Then:
>
> (i) $1^{-1} = 1$
> (ii) $x \neq 0 \Rightarrow x^{-1} \neq 0$ and $(x^{-1})^{-1} = x$
> (iii) $x \neq 0 \wedge y \neq 0 \Rightarrow xy \neq 0 \wedge (xy)^{-1} = x^{-1}y^{-1}$
> (iv) $xy = 1 \Rightarrow x, y \neq 0 \wedge y = x^{-1}$

As a consequence of Part (iii) of Corollary 2.14, we have

$$x \neq 0 \wedge y \neq 0 \Rightarrow xy \neq 0,$$

from which it follows that

$$xy = 0 \Rightarrow x = 0 \vee y = 0.$$

In words:

> *In a field, if the product of two elements is zero, then at least one of the elements must be zero.*

This is the justification for solving equations in elementary algebra by the method of factorization (Problem 23).

2.15 DEFINITION

> **DIVISION IN A FIELD** Let F be a field. If $x, y \in F$ and $y \neq 0$, we define $x/y \in F$ by
>
> $$x/y = xy^{-1}.$$

We read x/y in English as "x divided by y," or as "x over y," and we refer to the expression x/y as a **fraction** with **numerator** x and **denominator** y. Note that x/y is defined only if $y \neq 0$. Thus:

> *A fraction with a zero denominator is meaningless.*

Alternative notation for x/y is

$$x \div y \qquad \text{or} \qquad \frac{x}{y}.$$

As a particular case of Definition 2.15, we have, for $y \neq 0$,

$$\frac{1}{y} = 1 \cdot y^{-1} = y^{-1}.$$

Thus, the notation $1/y$ can be used rather than y^{-1} for the reciprocal of a nonzero element in a field.

The familiar process of "dividing both sides of an equation by a nonzero quantity" is justified by the following theorem, the proof of which we leave as an exercise (Problem 27).

2.16 THEOREM

> **DIVIDING AN EQUATION BY A NONZERO QUANTITY**   Let F be a field and suppose that $a, b \in F$ with $b \neq 0$. Then
>
> $$ab = c \Leftrightarrow a = \frac{c}{b}.$$

The next theorem justifies the rule that *the reciprocal of a fraction is obtained by inverting the fraction.*

2.17 THEOREM

> **THE RECIPROCAL OF A FRACTION**   Let F be a field and suppose that $a, b \in F$ with $a \neq 0$ and $b \neq 0$. Then $a/b \neq 0$ and
>
> $$\left(\frac{a}{b}\right)^{-1} = \frac{b}{a}.$$

PROOF   Let $x = a/b$ and let $y = b/a$. (Note that, by hypothesis, $b \neq 0$ and $a \neq 0$, so we can legitimately form these fractions.) By Definition 2.15, $x = ab^{-1}$ and $y = ba^{-1}$. Therefore:

$$xy = ab^{-1}ba^{-1} = aa^{-1}bb^{-1} = 1 \cdot 1 = 1.$$

By Part (iv) of Corollary 2.14, it follows that $a/b = x \neq 0$ and that

$$\frac{b}{a} = y = x^{-1} = \left(\frac{a}{b}\right)^{-1}. \qquad \square$$

The usual rule of arithmetic for multiplying two fractions is justified by the following theorem.

2.18 THEOREM

> **THE PRODUCT OF TWO FRACTIONS**   Let F be a field and suppose that $a, b, c, d \in F$ with $b, d \neq 0$. Then
>
> $$\left(\frac{a}{b}\right)\left(\frac{c}{d}\right) = \frac{ac}{bd}.$$

PROOF   By Definition 2.15, $a/b = ab^{-1}$ and $c/d = cd^{-1}$; hence,

$$\left(\frac{a}{b}\right)\left(\frac{c}{d}\right) = ab^{-1}cd^{-1} = acb^{-1}d^{-1}.$$

By Part (iii) of Corollary 2.14,

$$b^{-1}d^{-1} = (bd)^{-1},$$

and it follows that

$$\left(\frac{a}{b}\right)\left(\frac{c}{d}\right) = ac(bd)^{-1} = \frac{ac}{bd}$$

by Definition 2.15 again.   □

By a computation similar to the one in the proof of Theorem 2.18, you can prove the following theorem, which justifies the standard procedure of canceling a common factor in the numerator and denominator of a fraction (Problem 29).

2.19 THEOREM

> **CANCELLATION LAW FOR FRACTIONS**   Let F be a field and suppose that $a, b, c \in F$ with $b, c \neq 0$. Then
>
> $$\frac{ac}{bc} = \frac{a}{b}.$$

In a field F, the distributive law justifies the usual rule for adding fractions with a common denominator. Indeed, if $a, b, d \in F$ with $d \neq 0$, we have

$$\frac{a}{d} + \frac{b}{d} = ad^{-1} + bd^{-1} = (a + b)d^{-1} = \frac{a + b}{d}.$$

To add two fractions with different denominators, we use the cancellation law "in reverse" to obtain a common denominator and then add the fractions by the rule above.

2.20 THEOREM

> **THE SUM OF TWO FRACTIONS**   Let F be a field and suppose that $a, b, c, d \in F$ with $b, d \neq 0$. Then
>
> $$\frac{a}{b} + \frac{c}{d} = \frac{ad + bc}{bd}.$$

PROOF By Theorem 2.19 we can multiply the numerator and denominator of the first fraction by d, and we can multiply the numerator and denominator of the second fraction by b, to obtain

$$\frac{a}{b} + \frac{c}{d} = \frac{ad}{bd} + \frac{cb}{db} = \frac{ad}{bd} + \frac{bc}{bd} = \frac{ad + bc}{bd}. \qquad \square$$

We leave as an exercise the proof of the analogous rule for subtraction of fractions in a field (Problem 31).

The following theorem justifies the familiar rules for negation of a fraction.

2.21 THEOREM

> **NEGATION OF A FRACTION** Let F be a field and suppose that $a, b \in F$ with $b \neq 0$. Then
>
> $$-\left(\frac{a}{b}\right) = \frac{-a}{b} = \frac{a}{-b}.$$

PROOF By Part (i) of Theorem 2.5, we have

$$-\left(\frac{a}{b}\right) = -(ab^{-1}) = (-a)b^{-1} = \frac{-a}{b}.$$

If $-b$ were equal to 0, we would have $b = -(-b) = -0 = 0$, contradicting the hypothesis that $b \neq 0$. Therefore, $-b \neq 0$. Since $b \neq 0$ and $-b \neq 0$, it follows that $b(-b) \neq 0$. By Part (iii) of Theorem 2.5,

$$(-a)(-b) = ab;$$

hence,

$$\frac{(-a)(-b)}{b(-b)} = \frac{ab}{b(-b)} = \frac{ab}{(-b)b},$$

and, applying the cancellation law (Theorem 2.19) to the first and last fractions, we conclude that

$$\frac{-a}{b} = \frac{a}{-b}. \qquad \square$$

Integer exponents can be used in an arbitrary field F in the same way they are used in ordinary arithmetic. Thus, if n is a positive integer, and $a \in F$, we understand that

$$a^n = a \cdot a \cdot a \cdot \cdots \cdot a$$

where there are n factors on the right.

Positive integer exponents are characterized by the following two properties:

(i) $a^1 = a$

(ii) $a^{n+1} = a^n \cdot a$ for every positive integer n.

Properties (i) and (ii) are especially useful in proofs by mathematical induction of theorems involving exponents. If n and m are positive integers, it is easy to see that

$$a^{n+m} = a^n a^m$$

just by noting that there are $n + m$ factors on each side of the equation. (If desired, a more rigorous proof can be made by holding m constant and using mathematical induction on n.)

If $a \neq 0$, we follow the convention of ordinary arithmetic and define

$$a^0 = 1.$$

Note that 0^0 is undefined. Finally, if $a \neq 0$ and n is a positive integer, we define

$$a^{-n} = (a^{-1})^n$$

(noting that $-n$ is a negative integer). We leave it to you to check that all the usual rules for integer exponents hold in the field F (Problem 35).

PROBLEM SET 3.2

1. (a) If $R = \{0\}$, show that $(R, +, \cdot)$ is a ring if we define $0 + 0 = 0$ and $0 \cdot 0 = 0$.
   (b) If $(R, +, \cdot)$ is a ring such that 0 is a neutral element for the monoid $(R, \cdot)$, show that $R = \{0\}$.
   (c) In view of Parts (a) and (b), explain the requirement that $1 \neq 0$ in a ring with unity. (See Definition 2.9.)

2. Prove Theorem 2.2.

3. The usual rules for parity (that is, evenness or oddness) of integers can be written symbolically as

   $$\text{even} + \text{even} = \text{odd} + \text{odd} = \text{even},$$
   $$\text{even} + \text{odd} = \text{odd} + \text{even} = \text{odd},$$
   $$\text{even} \cdot \text{even} = \text{even} \cdot \text{odd} = \text{odd} \cdot \text{even} = \text{even},$$

   and

   $$\text{odd} \cdot \text{odd} = \text{odd}.$$

   Taking $R = \{\text{even}, \text{odd}\}$ and using the rules above for addition and multiplication, show that $(R, +, \cdot)$ is a commutative ring.

4. Let R be the set of all even integers (that is, all integer multiples of 2). If $+$ and $\cdot$ denote the ordinary addition and multiplication operations, show that $(R, +, \cdot)$ is a ring with no unity element.

5. Let $(R, +, \cdot)$ be a ring and let $a, b, c, d \in R$.
    (a) Prove that $(a + b)(c + d) = ac + ad + bc + bd$.
    (b) True or false: $(a + b)(a - b) = a \cdot a - b \cdot b$? Justify your answer.

6. Complete the proof of Theorem 2.3 by showing that $x \cdot 0 = 0$ in a ring.

7. Let n be a fixed integer greater than 1 and let $\mathbb{Z}_n = \{0, 1, 2, 3, \ldots, n - 1\}$. Define binary operations $+$ and $\cdot$ on $\mathbb{Z}_n$ as follows:

    $x + y =$ the remainder on division of the ordinary sum of x and y by n.

    $x \cdot y =$ the remainder on division of the ordinary product of x and y by n.

    It can be shown that $(\mathbb{Z}_n, +, \cdot)$ is a commutative ring with unity and that it is a field if and only if n is a prime number (that is, if ± 1 and $\pm n$ are the only integer divisors of n).
    (a) Show that the operation tables for $\mathbb{Z}_4$ and $\mathbb{Z}_2$ are, in fact, those given on pages 163 and 168.
    (b) Find the operation table for the field $\mathbb{Z}_3$.
    (c) Find all multiplicatively invertible elements in $\mathbb{Z}_4$.

8. Suppose that $(R, +, \cdot)$ is a system such that:
    (i) $(R, +)$ is a group with neutral element 0.
    (ii) $(R, \cdot)$ is a monoid with neutral element 1.
    (iii) The distributive laws hold.
    Prove that $(R, +, \cdot)$ is a ring with unity. [Hint: The only thing missing is the commutative law for addition. Prove that $x + y = y + x$ by beginning with $(1 + 1) \cdot (x + y)$ and expanding it in two ways using the distributive laws.]

9. Let $(G, +)$ be any commutative group with 0 as its neutral element. Define a binary operation $(x, y) \mapsto x \star y$ on G by $x \star y = 0$ for all $x, y \in G$. Is $(G, +, \star)$ a ring? Why or why not?

10. An element x in a ring R is said to be an **idempotent** if $x \cdot x = x$. (For instance, 0 and, if it exists, 1 are idempotents.) An **idempotent ring** is a ring in which *every* element is an idempotent.
    (a) Prove that, in an idempotent ring, the identity $x + x = 0$ holds, so that every element is its own negative. [Hint: Use the fact that $(x + x) \cdot (x + x) = x + x$.]
    (b) Prove that an idempotent ring is commutative. [Hint: Expand $(x + y) \cdot (x + y)$.]

11. (a) Prove Part (ii) of Theorem 2.5.
    (b) Prove Lemma 2.7.

12. Let X be a nonempty set and let $\mathcal{P}(X)$ denote the set of all subsets of X. For $M, N \in \mathcal{P}(X)$, define $M + N$, $M \cdot N \in \mathcal{P}(X)$ by $M + N = (M \cup N) \setminus (M \cap N)$ and $M \cdot N = M \cap N$. Show that $(\mathcal{P}(X), +, \cdot)$ is an idempotent ring (see Problem 10). Show that it is a ring with unity.

13. $\mathcal{M}_2(\mathbb{R})$ is defined to be the set of all two-by-two matrices with real number entries. Two such matrices are added by adding their corresponding entries:

$$\begin{bmatrix} a & b \\ c & d \end{bmatrix} + \begin{bmatrix} x & y \\ z & w \end{bmatrix} = \begin{bmatrix} a+x & b+y \\ c+z & d+w \end{bmatrix}.$$

They are multiplied by the row-by-column rule:

$$\begin{bmatrix} a & b \\ c & d \end{bmatrix} \cdot \begin{bmatrix} x & y \\ z & w \end{bmatrix} = \begin{bmatrix} ax+bz & ay+bw \\ cx+dz & cy+dw \end{bmatrix}.$$

(a) Show that $(\mathcal{M}_2(\mathbb{R}), +, \cdot)$ is a ring with unity.
(b) Show that this ring is *not* commutative.
Hint: Consider, for example, the products

$$\begin{bmatrix} 1 & 0 \\ 0 & 0 \end{bmatrix} \cdot \begin{bmatrix} 1 & 1 \\ 0 & 0 \end{bmatrix} \quad \text{and} \quad \begin{bmatrix} 1 & 1 \\ 0 & 0 \end{bmatrix} \cdot \begin{bmatrix} 1 & 0 \\ 0 & 0 \end{bmatrix}.$$

14. Let $(R, +, \cdot)$ be a commutative ring and let E denote the set of all idempotents in R (see Problem 10). For $a, b \in E$ define $a \oplus b = a + b - ab$ and $a \otimes b = a \cdot b$.
(a) If $a, b \in E$, show that $a \oplus b \in E$ and $a \otimes b \in E$.
(b) Show that $(E, \oplus, \otimes)$ is an idempotent ring.

15. Show that a matrix

$$\begin{bmatrix} a & b \\ c & d \end{bmatrix}$$

in the ring $\mathcal{M}_2(\mathbb{R})$ of Problem 13 is multiplicatively invertible if and only if its **determinant** $ad - bc$ is different from 0.

16. Let i denote the "imaginary" number with $i^2 = -1$ and let $\mathbb{C}$ be the **complex numbers**, $\mathbb{C} = \{x + yi \,|\, x, y \in \mathbb{R}\}$. Complex numbers are added and multiplied as follows:

$$(x + yi) + (u + vi) = (x + u) + (y + v)i,$$
$$(x + yi) \cdot (u + vi) = (xu - yv) + (xv + yu)i.$$

Show that $(\mathbb{C}, +, \cdot)$ is a field.

17. (a) Find two nonzero elements of $\mathcal{M}_2(\mathbb{R})$ whose product is the zero element of $\mathcal{M}_2(\mathbb{R})$. (See Problem 13.)
(b) An element x of a ring is said to be a **nilpotent of order 2** if $x \cdot x = 0$. Are there any nonzero nilpotents of order 2 in the ring $\mathcal{M}_2(\mathbb{R})$?

18. Show that two-by-two matrices having the form

$$\begin{bmatrix} \cos^2 \phi & \cos \phi \sin \phi \\ \cos \phi \sin \phi & \sin^2 \phi \end{bmatrix}$$

are idempotent elements in the ring $\mathcal{M}_2(\mathbb{R})$. (See Problems 10 and 13.)

19. A **polynomial function** $p:\mathbb{R} \to \mathbb{R}$ has the form

$$p(x) = a_n x^n + a_{n-1} x^{n-1} + \cdots + a_1 x + a_0,$$

where $x \in \mathbb{R}$ and the **coefficients** $a_n, a_{n-1}, \ldots, a_1, a_0$ are fixed real numbers. Let P denote the set of all polynomial functions $p:\mathbb{R} \to \mathbb{R}$.
(a) Show that P is a commutative ring with unity under the usual operations of addition and multiplication of functions as defined in calculus.
(b) What are the multiplicatively invertible elements in P?

20. (a) In a field, show that 0 and 1 are the only idempotents. (See Problem 10.)
(b) In a field, is it possible to have $1 = -1$? Explain.

21. Prove Theorem 2.12.

22. Prove Corollary 2.14.

23. Using suitable properties of the field $\mathbb{R}$, explain how to justify the usual technique in elementary algebra for solving equations by factoring.

24. (a) Write the addition and multiplication tables for the field $\mathbb{Z}_5$. (See Problem 7.)
(b) In the field $\mathbb{Z}_5$, show that $1^{-1} = 1$, $2^{-1} = 3$, $3^{-1} = 2$, and $4^{-1} = 4$.
(c) In the field $\mathbb{Z}_5$, show that $2/3 = 1/4$.

25. Let F be a field.
(a) Find all solutions in F of the equation $x^2 = 0$.
(b) Find all solutions in F of the equation $x = x^{-1}$ for $x \neq 0$.

26. In the field $\mathbb{R}$ of real numbers, the following condition holds:

$$x^2 + y^2 = 0 \Rightarrow x = y = 0.$$

Is this a special property of real numbers, or does it hold in any field F? Explain.

27. Prove Theorem 2.16.

28. Let $F = \{0, a, 1\}$ be a field with three elements.
(a) Show that $a \cdot a = 1$.
(b) Show that $1 + 1 = a$.
(c) Show that $1 + a = 0$.

29. Prove Theorem 2.19.

30. Let $\mathbb{C}$ be the field of complex numbers (Problem 16) and let $a \in \mathbb{C}$. Prove that the equation $z^2 = a$ has a solution $z \in \mathbb{C}$.

31. Let F be a field and suppose that $a, b, c, d \in F$ with $b, d \neq 0$. Prove that
$$\frac{a}{b} - \frac{c}{d} = \frac{ad - bc}{bd}.$$

32. (a) Show that the quadratic equation $x^2 + x + 1 = 0$ has no solution in the field $\mathbb{Z}_2$.
(b) Show that there exists a field $F = \{0, 1, a, b\}$ containing four distinct elements, in which 0 and 1 add and multiply as in $\mathbb{Z}_2$, and in which both of the elements a and b are solutions of $x^2 + x + 1 = 0$. Write the addition and multiplication tables for this field.

33. Prove the rule of **cross multiplication** in a field F: If $a, b, c, d \in F$ with $b, d \neq 0$, then
$$\frac{a}{b} = \frac{c}{d} \Rightarrow ad = bc.$$

34. Assume that F_1 and F_2 are two fields such that, as sets, $F_1 \subseteq F_2$, and such that the addition and multiplication operations in F_1 are just the restrictions to F_1 of the corresponding operations in F_2. Then F_1 is said to be a **subfield** of F_2, and F_2 is said to be an **extension field** of F_1.
(a) Show that the rational numbers $\mathbb{Q}$ form a subfield of $\mathbb{R}$.
(b) Show that the complex numbers $\mathbb{C}$ form an extension field of $\mathbb{R}$.

35. Let F be a field, let $a, b \in F$, and let m and n be integers. Prove each statement:
(a) $a^{n+m} = a^n a^m$ (b) $(a^n)^m = a^{nm}$ (c) $(ab)^n = a^n b^n$
(d) $\left(\dfrac{a}{b}\right)^n = \dfrac{a^n}{b^n}, \quad b \neq 0$ (e) $a^{n-m} = \dfrac{a^n}{a^m}, \quad a \neq 0$
[Hint: Assume to begin with that m and n are positive integers, then consider the case in which one or both are zero, and, finally, take care of the case in which one or both are negative.]

36. Let $\mathbb{R}^3$ denote the set of all vectors $\mathbf{v} = (x, y, z)$, $x, y, z \in \mathbb{R}$. Define the dot product $\mathbf{v} \cdot \mathbf{w}$ and the cross product $\mathbf{v} \times \mathbf{w}$ of vectors in $\mathbb{R}^3$ as usual. Let $\mathbb{H}$ be the Cartesian product $\mathbb{H} = \mathbb{R} \times \mathbb{R}^3$. Define addition and multiplication operations on $\mathbb{H}$ by
$$(a, \mathbf{v}) + (b, \mathbf{w}) = (a + b, \mathbf{v} + \mathbf{w})$$
and
$$(a, \mathbf{v})(b, \mathbf{w}) = (ab - \mathbf{v} \cdot \mathbf{w}, a\mathbf{w} + b\mathbf{v} + \mathbf{v} \times \mathbf{w}).$$
(a) Show that $(\mathbb{H}, +, \cdot)$ is a ring with unity.
(b) Show that every nonzero element in $\mathbb{H}$ is multiplicatively invertible.
[Hint: If $h = (a, \mathbf{v}) \neq 0$, so that $a \neq 0$ and $\mathbf{v} \neq \mathbf{0}$, then $h^{-1} = (ca, -c\mathbf{v})$, where $c = (a^2 + \mathbf{v} \cdot \mathbf{v})^{-1}$.]
(c) Show that $\mathbb{H}$ is not commutative and therefore is not a field.
[$\mathbb{H}$ is the ring of **real quaternions** discovered by the Irish mathematician W. R. Hamilton (1805–1865) in 1843.]

3 _____

ORDERED FIELDS AND THE REAL NUMBERS

By visualizing the real number field $\mathbb{R}$ in the usual way as points on a line, we see that there is a natural order relation $\leq$ on $\mathbb{R}$. Indeed, for $x, y \in \mathbb{R}$, $x < y$ means that the point corresponding to x lies to the left of the point corresponding to y on the line (Figure 3-1), and $x \leq y$ means that either $x < y$ or else $x = y$. The *positive* real numbers, whose corresponding points lie to the *right* of 0 on the line, are the numbers $x \in \mathbb{R}$ such that $0 < x$. Note that, if P denotes the set of all positive real numbers, then, for $x, y \in \mathbb{R}$,

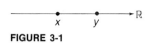

FIGURE 3-1

$$x < y \Leftrightarrow y - x \in P.$$

Thus, the order relation on $\mathbb{R}$ both determines and is determined by the set P of positive real numbers. These considerations suggest the following definition.

3.1 DEFINITION

> **ORDERED FIELD**   An **ordered field** is a field F together with a fixed subset $P \subseteq F$, called the set of **positive** elements of F, that satisfies the following conditions:
>
> (i) $0 \notin P$
> (ii) $x \in F, \quad x \neq 0, \quad x \notin P \Rightarrow -x \in P$
> (iii) $x, y \in P \Rightarrow x + y \in P$
> (iv) $x, y \in P \Rightarrow xy \in P$

Let F be an ordered field as in Definition 3.1. If $x \in P$, we say that x is **positive**. Condition (i) says that 0 is not positive. A nonzero element $x \in F$ is said to be **negative** if $-x$ is positive. Condition (ii) says that if a nonzero element is not positive, then it is negative. Conditions (iii) and (iv) require that the sum and the product of positive elements are positive; that is, the set P is **closed** under addition and multiplication.

3.2 THEOREM

> **LAW OF TRICHOTOMY**   Let F be an ordered field with P as its set of positive elements. Then, for $x \in F$, *exactly one* of the following conditions is true:
>
> (i) $x \in P$
> (ii) $x = 0$
> (iii) $-x \in P$

PROOF By Condition (ii) in Definition 3.1, at least one of the three statements (i), (ii), or (iii) must be true. We must show that *no two* of these three statements can be true simultaneously. Since $0 \notin P$, we cannot have both (i) and (ii). Since $-0 = 0$ and $0 \notin P$, we cannot have both (ii) and (iii). Finally, if (i) and (iii) were both true, then, since P is closed under addition, we would again have $0 = x + (-x) \in P$, contradicting Part (i) of Definition 3.1. □

The law of trichotomy can be paraphrased as follows:

> *In an ordered field, every element is positive, zero, or negative, and no element satisfies any two of these three conditions.*

3.3 THEOREM

SQUARES OF NONZERO ELEMENTS ARE POSITIVE Let F be an ordered field with P as its set of positive elements. Let $x \in F$ with $x \neq 0$. Then $x^2 \in P$.

PROOF Since $x \neq 0$, the law of trichotomy (Theorem 3.2) implies that either $x \in P$ or else $-x \in P$. If $x \in P$, then, since P is closed under multiplication, we have

$$x^2 = x \cdot x \in P.$$

Likewise, if $-x \in P$, then, using Part (iii) of Theorem 2.5, we have

$$x^2 = x \cdot x = (-x)(-x) \in P.$$

In any case, then, provided that $x \neq 0$, we have $x^2 \in P$. □

3.4 COROLLARY

POSITIVITY OF 1 In an ordered field F with P as its set of positive elements, it is always the case that

$$1 \in P.$$

PROOF $1 \neq 0$ and $1 = 1^2$. □

3.5 COROLLARY

POSITIVE PRODUCTS AND QUOTIENTS Let F be an ordered field and let $x, y \in F$ with $y \neq 0$. Then

$$xy \in P \Leftrightarrow \frac{x}{y} \in P.$$

PROOF   We prove the implication from left to right and leave the converse implication as an exercise (Problem 1). Suppose that $xy \in P$. By Theorem 3.3, $(1/y)^2 \in P$; hence, since the product of positive elements is positive,

$$\frac{x}{y} = xy\left(\frac{1}{y}\right)^2 \in P. \qquad \qquad \square$$

3.6 DEFINITION

> **THE RELATION < ON AN ORDERED FIELD**   Let F be an ordered field with P as its set of positive elements. We define the binary relation $<$ on F by
>
> $$x < y \Leftrightarrow y - x \in P$$
>
> for all $x, y \in F$.

If $x < y$ we say that x is **less than** y, or, equivalently, that y is **greater than** x. The statement $x < y$ is often written in the alternative form $y > x$. An expression of the form $x < y$, or $y > x$, is called a **strict inequality**. We define the **nonstrict inequality** $x \leq y$ by

$$x \leq y \Leftrightarrow x < y \lor x = y.$$

If $x \leq y$ we say that x is **less than or equal to** y, or, equivalently, that y is **greater than or equal to** x. The statement $x \leq y$ is often written in the alternative form $y \geq x$. Note that

$$x \leq x$$

holds for any element x in an ordered field F; that is, the *binary relation* $\leq$ *is reflexive on* F. (See Definition 2.9 in Chapter 2.)

We leave the proof of the following lemma as an exercise (Problem 3).

3.7 LEMMA

> Let F be an ordered field with P as its set of positive elements. Then, for $x, y \in F$:
>
> (i) $x \in P \Leftrightarrow 0 < x$
> (ii) $-x \in P \Leftrightarrow x < 0$
> (iii) $x < y \Rightarrow x \neq y$

Using Definition 3.6, we can restate the law of trichotomy (Theorem 3.2) in the following equivalent form (Problem 5).

3.8 THEOREM

> **LAW OF TRICHOTOMY (ALTERNATIVE VERSION)** If F is an ordered field, then, for $x, y \in F$, *exactly one* of the following is true:
>
> (i) $x < y$
> (ii) $x = y$
> (iii) $x > y$

The next theorem shows that both strict and nonstrict inequalities are transitive relations. (See Definition 2.9 in Chapter 2.)

3.9 THEOREM

> **TRANSITIVITY OF INEQUALITIES** Let F be an ordered field and let x, $y, z \in F$. Then:
>
> (i) $x < y \land y < z \Rightarrow x < z$
> (ii) $x \le y \land y \le z \Rightarrow x \le z$

PROOF We prove Part (i) and leave Part (ii) as an exercise (Problem 7). Thus, suppose that $x < y$ and $y < z$. Then, by Definition 3.6,

$$y - x \in P \qquad \text{and} \qquad z - y \in P.$$

Therefore, by Part (iii) of Definition 3.1,

$$z - x = (z - y) + (y - x) \in P;$$

hence, by Definition 3.6 again, $x < z$. □

We have noted that the relation $\le$ is reflexive on an ordered field F, and Part (ii) of Theorem 3.9 shows that it is transitive. As a consequence of the law of trichotomy, $\le$ is also *antisymmetric* (see Definition 2.11 in Chapter 2), and in fact, $(F, \le)$ *is a totally ordered set* (see Definition 2.24 in Chapter 2). We ask you to prove these facts in Problem 11.

In an ordered field F, we write the **compound inequality**

$$x < y < z$$

as an abbreviation for the statement that

$$x < y \land y < z.$$

It is considered to be good mathematical practice to arrange it so that all inequalities in a compound inequality "run in the same direction." Thus, writing an expression such as

$$x < y > w$$

to mean that $x < y$ and $w < y$ would be regarded as bad form. However, it is acceptable to abbreviate the statement that

$$x < y \wedge w < y$$

as

$$x, w < y.$$

Strict and nonstrict inequalities can be combined in a compound inequality, so that, for instance,

$$x < y \leq z$$

is understood to mean that

$$x < y \wedge y \leq z.$$

Note that, for such a "mixed" compound inequality, we have

$$x < y \wedge y \leq z \Rightarrow x < z$$

(Problem 13).

3.10 THEOREM

> **ADDITION OF INEQUALITIES** Let F be an ordered field and let $x, y, z, w \in F$. Then:
>
> (i) $x < y \Rightarrow x + z < y + z$
> (ii) $x < y$ and $z < w \Rightarrow x + z < y + w$

PROOF (i) Suppose that $x < y$. Then, by Definition 3.6, we have $y - x \in P$. But, then,

$$(y + z) - (x + z) = y - x \in P,$$

and, consequently,

$$x + z < y + z.$$

(ii) Suppose that $x < y$ and $z < w$. Then, by Part (i) above,

$$x + z < y + z$$

and, likewise,

$$z + y < w + y.$$

Therefore,

$$x + z < y + z = z + y < w + y,$$

and it follows from Part (i) of Theorem 3.9 (transitivity of the $<$ relation) that

$$x + z < w + y = y + w. \qquad \square$$

Suitably modified versions of Parts (i) and (ii) of Theorem 3.10 hold for nonstrict inequalities; for instance, we have

$$x \le y \Rightarrow x + z \le y + z$$

and

$$x < y \quad \text{and} \quad z \le w \Rightarrow x + z < y + w$$

(Problem 15a and b).

3.11 THEOREM

MULTIPLICATION OF INEQUALITIES   Let F be an ordered field and let $x, y, z, w \in F$. Then:

(i) $x < y$ and $0 < z \Rightarrow xz < yz$
(ii) $x < y$ and $z < 0 \Rightarrow yz < xz$
(iii) $0 < x < y$ and $0 < z < w \Rightarrow 0 < xz < yw$

PROOF

(i) Suppose that $x < y$ and $0 < z$. Then we have

$$y - x \in P \quad \text{and} \quad z \in P,$$

and it follows from Part (iv) of Definition 3.1 that

$$yz - xz = (y - x)z \in P;$$

hence, by Definition 3.6,

$$xz < yz.$$

(ii) Suppose that $x < y$ and $z < 0$. Then

$$y - x \in P \quad \text{and} \quad -z \in P,$$

so that

$$xz - yz = (y - x)(-z) \in P,$$

and, therefore,

$$yz < xz.$$

(iii) We leave the proof of Part (iii) as an exercise (Problem 9). $\square$

By Part (i) of Theorem 3.11, we are permitted to multiply both sides of an inequality by a *positive* element of F, but, if we multiply by a *negative* element of F, we must *reverse the inequality*. Suitably modified versions of Parts (i), (ii), and (iii) of the theorem hold for nonstrict inequalities; for instance, we have

$$x \leq y \quad \text{and} \quad 0 \leq z \Rightarrow xz \leq yz$$
$$x \leq y \quad \text{and} \quad z \leq 0 \Rightarrow yz \leq xz$$

and

$$0 \leq x \leq y \quad \text{and} \quad 0 \leq z \leq w \Rightarrow 0 \leq xz \leq yw$$

(Problem 19a, b, and c). Using the facts that $-1 < 0$ and $(-1)x = -x$, it is an easy matter to prove the following corollary (Problem 10).

3.12 COROLLARY

Let x and y be elements in an ordered field F. Then:

(i) $x < y \Leftrightarrow -y < -x$
(ii) $x \leq y \Leftrightarrow -y \leq -x$

3.13 THEOREM

SQUARING BOTH SIDES OF AN INEQUALITY   Let x and y be elements in an ordered field F and suppose that $0 < x, y$. Then

$$x < y \Leftrightarrow x^2 < y^2.$$

PROOF   If $0 < x < y$, then $x^2 = x \cdot x < y \cdot y = y^2$ follows from Part (iii) of Theorem 3.11. Conversely, suppose that $x^2 < y^2$, but that $x < y$ is false. Then, by the law of trichotomy, $y \leq x$. If $y = x$, then $y^2 = x^2$, contradicting $x^2 < y^2$. However, if $y < x$, then (by the same argument given above), $y^2 < x^2$, again contradicting $x^2 < y^2$. □

3.14 DEFINITION

SIGNUM FUNCTION   Let F be an ordered field. We define a function $\text{sgn}: F \to F$, called the **signum function** on F, by

$$\text{sgn}(x) = \begin{cases} 1, & \text{if } x > 0 \\ 0, & \text{if } x = 0 \\ -1, & \text{if } x < 0 \end{cases}.$$

Note that, by the law of trichotomy, sgn is a well-defined function on the ordered field F. The signum function has the property that

$$\text{sgn}(xy) = \text{sgn}(x) \cdot \text{sgn}(y)$$

for all $x, y \in F$ (Problem 21). Using the signum function, we can define the idea of absolute value as follows:

3.15 DEFINITION   **ABSOLUTE VALUE**   Let F be an ordered field. For $x \in F$, we define the **absolute value** of x, in symbols $|x|$, by

$$|x| = x \cdot \text{sgn}(x).$$

Note that, if $x \geq 0$, then $|x| = x$; however, if $x < 0$, then $|x| = -x$ (Problem 23b). In any case, then, $|x| \geq 0$.

3.16 LEMMA   **PRODUCTS AND ABSOLUTE VALUES**   If F is an ordered field and $x, y \in F$, then

$$|xy| = |x|\,|y|.$$

PROOF   Using the fact that

$$\text{sgn}(xy) = \text{sgn}(x) \cdot \text{sgn}(y)$$

and Definition 3.15, we have

$$|xy| = xy \cdot \text{sgn}(xy) = xy \cdot \text{sgn}(x) \cdot \text{sgn}(y) = x \cdot \text{sgn}(x) \cdot y \cdot \text{sgn}(y)$$
$$= |x|\,|y|. \qquad \square$$

Some of the properties of absolute value are most easily obtained by a *case analysis* as illustrated in the proof of the following lemma.

3.17 LEMMA   Let F be an ordered field. Then, for $x \in F$,

$$-|x| \leq x \leq |x|.$$

PROOF   We consider separately the cases in which $0 \leq x$ and in which $x < 0$. (By the law of trichotomy, these cases cover all possibilities.) If $0 \leq x$, then $x = |x|$ and

$$-|x| = -x \leq 0 \leq x = |x|.$$

If $x < 0$, then $-x = |x|$ and

$$-|x| = x < 0 \leq |x|.$$

Thus, in either case,

$$-|x| \leq x \leq |x|. \qquad \square$$

3.18 THEOREM

> **TRIANGLE INEQUALITY**   Let F be an ordered field. Then, for $x, y \in F$,
>
> $$|x + y| \leq |x| + |y|.$$

PROOF   By Lemma 3.17, we have

$$-|x| \leq x \leq |x| \qquad \text{and} \qquad -|y| \leq y \leq |y|. \qquad (1)$$

Adding these inequalities, we obtain

$$x + y \leq |x| + |y| \qquad (2)$$

and

$$-(|x| + |y|) \leq x + y. \qquad (3)$$

Multiplying the inequality in (3) by -1, and reversing it, we find that

$$-(x + y) \leq |x| + |y|. \qquad (4)$$

Now we consider first the case in which $0 \leq x + y$, and second the case in which $x + y < 0$. In the first case, we use (2) to conclude that

$$|x + y| = x + y \leq |x| + |y|. \qquad (5)$$

In the second case, we use (4) to obtain

$$|x + y| = -(x + y) \leq |x| + |y|. \qquad (6)$$

Thus, in either case,

$$|x + y| \leq |x| + |y|. \qquad (7)$$

$$\square$$

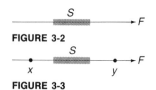

FIGURE 3-2

FIGURE 3-3

If we visualize an ordered field F as points on a line, then a subset S of F can be depicted by shading the portion of the line corresponding to the points in S (Figure 3-2).

Recall that an element $y \in F$ is called an *upper bound* for the set S if $s \leq y$ holds for all $s \in S$ (Definition 2.20 in Chapter 2). Similarly, an element $x \in F$ is called a *lower bound* for S if $x \leq s$ holds for all $s \in S$ (Figure 3-3).

Upper and lower bounds, when they exist, are not unique; in fact, any element of F that is larger than an upper bound for S is again an upper bound for S, and any element of F that is smaller than a lower bound for S is again a lower bound for S. When we say that a set *has a lower bound*, or that it *has an upper bound*, we do not mean to imply that this lower or upper bound necessarily belongs to the set. An upper bound a

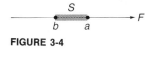

FIGURE 3-4

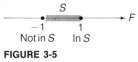

FIGURE 3-5

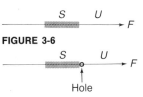

FIGURE 3-6

FIGURE 3-7

for S that actually *belongs* to S is called a **maximum, largest, greatest,** or **last element** of S, and a lower bound b for S that actually *belongs* to S is called a **minimum, smallest, least,** or **first element** of S (Figure 3-4).

A set $S \subseteq F$ may or may not have a greatest element, and it may or may not have a least element. For instance, the half-open interval

$$S = (-1, 1] = \{s \in F \,|\, -1 < s \leq 1\}$$

has 1 as a greatest element, but it has no least element. Indeed, -1 is the only possible candidate for the least element of S, but $-1 \notin S$ (Figure 3-5).

Now let S be any nonempty subset of F, suppose that S has an upper bound, and let U be the set of all upper bounds for S (Figure 3-6). Consider the question of whether or not U has a least element. Recall, from Section 2.2 of Chapter 2, that a least element of U, if it exists, is called a **least upper bound,** or a **supremum** of S. Suppose that, somehow, there is a "hole" in the field F, that all of the elements of S are to the left of the hole, and all of the elements of U are to the right of the hole (Figure 3-7). Then U would have no least element since the only possible least element of U would occupy the position where the hole is, and there is no element of F in this position.

One way to ensure that an ordered field F has no "holes" in it is to insist that every nonempty set S that has an upper bound has a least upper bound (that is, a supremum). This brings us to the following definition:

3.19 DEFINITION

> **COMPLETENESS**   An ordered field F is said to be **complete** if every nonempty subset S of F that has an upper bound has a least upper bound:
>
> $$a = \text{LUB}(S) = \sup(S).$$

Now we are ready to introduce the field $\mathbb{R}$ of real numbers on a more formal basis. We visualize $\mathbb{R}$ as a line with no "holes" in it, that is, as a complete ordered field. Therefore, we adopt the following axiom:

3.20 AXIOM

> **THE FIELD OF REAL NUMBERS** $\mathbb{R}$ is a complete ordered field.

Here we are postulating the existence of a complete ordered field, denoted by $\mathbb{R}$. We make no other assumptions about $\mathbb{R}$; indeed, all of the properties of the real number system can be derived deductively from the assumption that $\mathbb{R}$ forms a complete ordered field.

The idea of multiplying by -1 to reverse inequalities (Corollary 3.12) is exploited in the proof of the next lemma.

3.21 LEMMA

Let $S \subseteq \mathbb{R}$ and let

$$T = \{-s \mid s \in S\}.$$

Then:

(i) x is an upper bound for S if and only if $-x$ is a lower bound for T.

(ii) $a = \sup(S)$ if and only if $-a = \inf(T)$.

PROOF

(i) If x is an upper bound for S, then $s \leq x$ holds for all $s \in S$; hence, $-x \leq -s$ holds for all $s \in S$, that is, $-x$ is a lower bound for T. Conversely, if $-x$ is a lower bound for T, then $-x \leq -s$ holds for all $s \in S$; hence, $s \leq x$ holds for all $s \in S$, that is, x is an upper bound for S.

(ii) Suppose that $a = \sup(S)$ and let U be the set of all upper bounds of S. Then a is a lower bound for U and $a \in U$. Let

$$L = \{-x \mid x \in U\}.$$

By Part (i) above, L is the set of all lower bounds for T. Also, by Part (i), and the fact that a is a lower bound for U, we have $-a$ as an upper bound for L. Finally, since $a \in U$, we have $-a \in L$, and we conclude that $-a = \inf(T)$. A similar argument, which we leave as an exercise (Problem 37), shows that, conversely, if $-a = \inf(T)$, then $a = \sup(S)$. □

Lemma 3.21 is illustrated in Figure 3-8:

FIGURE 3-8

Pictorially, what is happening is just that *inequalities are reversed by a reflection through the origin.* We leave as an exercise (Problem 38) the proof of the following important corollary to Lemma 3.21:

3.22 COROLLARY

Every nonempty subset S of $\mathbb{R}$ that has a lower bound has a greatest lower bound:

$$b = \mathrm{GLB}(S) = \inf(S).$$

Although the proof of the next lemma is quite simple, you will find the lemma to be exceedingly useful when dealing with the supremum or infimum of a set $S \subseteq \mathbb{R}$.

3.23 LEMMA

Let $S \subseteq \mathbb{R}$ and let $p, q \in \mathbb{R}$. Then:

(i) If $a = \sup(S)$ and $p < a$, then there exists $s \in S$ with

$$p < s \le a.$$

(ii) If $b = \inf(S)$ and $b < q$, then there exists $s \in S$ with

$$b \le s < q.$$

PROOF

We prove Part (i) here and leave Part (ii) as an exercise (Problem 39). Because p is smaller than the least upper bound a of S, it follows that p cannot be an upper bound for S. Therefore, there must exist a number $s \in S$ such that

$$p < s$$

(for, otherwise, $s \le p$ would hold for all $s \in S$, and p would be an upper bound for S). Also, since a is an upper bound for S, $s \le a$, and we have

$$p < s \le a.$$ □

PROBLEM SET 3.3

1. Let F be an ordered field with P as its set of positive elements, and let $x, y \in F$ with $y \ne 0$. Complete the proof of Corollary 3.5 by showing that

$$\frac{x}{y} \in P \Rightarrow xy \in P.$$

2. Let F be an ordered field with P as its set of positive elements, and let K be a subfield of F. (See Problem 34 in Problem Set 3.2.) Show that K becomes an ordered field if we take its set of positive elements to be $P \cap K$.

3. Prove Lemma 3.7.

4. The rational numbers $\mathbb{Q}$ form a subfield of the real numbers $\mathbb{R}$. If $\mathbb{Q}$ is made into an ordered field as in Problem 2, show that the positive elements in $\mathbb{Q}$ are exactly those rational numbers that can be written in the form m/n, where

$$m, n \in \mathbb{N} = \{1, 2, 3, \ldots\}.$$

5. Prove Theorem 3.8.

6. Show that the field $\mathbb{C}$ of complex numbers cannot be made into an ordered field. [Hint: In $\mathbb{C}$, $i^2 = -1$.]

7. Prove Part (ii) of Theorem 3.9.

8. (a) Prove that, in an ordered field F, it is impossible to have $1 + 1 = 0$.
(b) Use the result in Part (a) to show that $\mathbb{Z}_2$ cannot be made into an ordered field.

9. Prove Part (iii) of Theorem 3.11.

10. Prove Corollary 3.12.

11. Let F be an ordered field.
(a) Prove that the relation $\leq$ on F is antisymmetric.
(b) Prove that F is totally ordered by the relation $\leq$.

12. True or false: In the ordered field $\mathbb{R}$ of real numbers

$$0 \leq x < y \Rightarrow \sqrt{x} < \sqrt{y}.$$

If true, prove it; if false, give a counterexample.

In Problems 13–34, suppose that F is an ordered field and that x, y, z, w ∈ F. Prove each statement.

13. $x < y \wedge y \leq z \Rightarrow x < z$ **14.** $0 < x \Rightarrow 0 < x^{-1}$

15. (a) $x \leq y \Rightarrow x + z \leq y + z$
(b) $x < y \wedge z \leq w \Rightarrow x + z < y + w$
(c) $x \leq y \wedge z \leq w \Rightarrow x + z \leq y + w$

16. $x < y \Rightarrow \exists a \in F, \quad x < a < y$ [Hint: Let $a = (x + y)/(1 + 1)$.]

17. (a) $x < y \Leftrightarrow x + z < y + z$ (b) $x \leq y \Leftrightarrow x + z \leq y + z$

18. (a) F contains infinitely many elements. [Hint: Use the result of Problem 16.]
(b) F cannot contain a largest element or a smallest element.

19. (a) $x \leq y \wedge 0 \leq z \Rightarrow xz \leq yz$
(b) $x \leq y \wedge z \leq 0 \Rightarrow yz \leq xz$
(c) $0 \leq x \leq y \wedge 0 \leq z \leq w \Rightarrow 0 \leq xz \leq yw$
(d) $0 < x < 1 \Rightarrow x < 1 < x^{-1}$
(e) $0 < x < 1 \Rightarrow x^2 < x$
(f) $1 < x \Rightarrow x < x^2$

20. $0 \leq x \leq y \Rightarrow (x - y)^2 \leq y^2 - x^2$ [Hint: Expand $(y^2 - x^2) - (x - y)^2$.]

21. $\operatorname{sgn}(xy) = \operatorname{sgn}(x) \cdot \operatorname{sgn}(y)$

22. $x = 0 \Leftrightarrow |x| = 0$

23. (a) $x \geq 0 \Rightarrow |x| = x$ (b) $x < 0 \Rightarrow |x| = -x$

24. $|x| \geq 0$ with equality if and only if $x = 0$

25. $|-x| = |x|$

26. (a) $|x| = x \Rightarrow x \geq 0$ (b) $|x| = -x \Rightarrow x \leq 0$

27. $y \neq 0 \Rightarrow \left| \dfrac{x}{y} \right| = \dfrac{|x|}{|y|}$

28. $y \neq 0 \Rightarrow \mathrm{sgn}(y) = \dfrac{y}{|y|}$

29. $|x - y| \leq |x| + |y|$

30. If $x \leq y$ and $-x \leq y$, then $|x| \leq y$.

31. $|x| < y \Leftrightarrow -y < x < y$

32. $||x| - |y|| \leq |x - y|$

33. $|x + y| = |x| + |y| \Leftrightarrow xy \geq 0$

34. $x, y \geq 0 \Rightarrow |x - y|^2 \leq |x^2 - y^2|$ [Hint: Use the result of Problem 20.]

35. (a) Prove that, in an ordered field F, the formula

$$\frac{x + y + |x - y|}{1 + 1}$$

gives the larger of the two elements x and y (or gives their common value if they are equal).

(b) Find a similar formula for the smaller of the two elements x and y (or their common value if they are equal).

36. Let $x, y \in \mathbb{R}$ with $x, y \geq 0$. Use the result of Problem 34 to prove that

$$|\sqrt{x} - \sqrt{y}| \leq \sqrt{|x - y|}.$$

37. (a) Complete the proof of Part (ii) of Lemma 3.21 by showing that, if $T = \{-s \mid s \in S\}$ and if $-a = \inf(T)$, then $a = \sup(S)$.

(b) Would the results of Lemma 3.21 hold in an arbitrary ordered field F? Explain.

38. Prove Corollary 3.22.

39. Prove Part (ii) of Lemma 3.23.

In Problems 40–48, assume that F is an ordered field.

40. The empty set $\varnothing$ is a subset of F.
(a) What is the set U of upper bounds of $\varnothing$?
(b) Does $\varnothing$ have a supremum in F? Explain.

41. A **ray** (opening to the left) in F is defined to be a subset R of F such that $R \neq \varnothing$, $R \neq F$, and, if $r, x \in F$ with $r \in R$ and $x \leq r$, then $x \in R$.
(a) Show that a ray in F must have an upper bound in F.
(b) Sketch a figure showing a ray.
(c) Suppose that $b \in F$ and define $R_b = \{x \in F \mid x \leq b\}$. Show that R_b is a ray. (A ray of the form R_b is called a **principal ray**.)
(d) Give an example of a ray in F that is not a principal ray.
(e) If $b, c \in F$, show that $b \leq c$ if and only if $R_b \subseteq R_c$.

42. Suppose that $S \subseteq T \subseteq F$, $a = \sup(S)$, $b = \inf(S)$, $c = \sup(T)$, and $d = \inf(T)$. Prove that $d \leq b \leq a \leq c$.

43. Let $\emptyset \neq S \subseteq F$, and define $R = \{r \in F \mid \exists s \in S, r \leq s\}$. Show that:
(a) R is a ray in F (see Problem 41).
(b) R and S have the same set of upper bounds in F.

44. A subset C of F is called a **Dedekind cut** in F if $C \neq \emptyset$, $C \neq F$, and C contains every lower bound in F of its own set of upper bounds in F.
(a) Show that every Dedekind cut in F is a ray in F (see Problem 41).
(b) Show that every principal ray in F is a Dedekind cut in F.
(c) Give an example of a ray in F that is not a Dedekind cut in F.
(d) If C is a Dedekind cut in F, show that $a \in F$ is the supremum of C if and only if every element $x \in F$ with $x < a$ is in C, but no element $y \in F$ with $a < y$ is in C.

45. Prove that F is complete if and only if every ray in F has a supremum (see Problems 41 and 43).

46. Prove that F is complete if and only if every Dedekind cut in F has a supremum (see Problem 44).

47. If F is complete and R is a ray in F, prove that either R is a principal ray or else there exists $b \in F$ such that $R = R_b \setminus \{b\}$.

48. Prove that F is complete if and only if every Dedekind cut in F is a principal ray in F (see Problems 41 and 44).

4

THE NATURAL NUMBERS AND MATHEMATICAL INDUCTION

In this section we are going to extract the system $\mathbb{N}$ of natural numbers from the complete ordered field $\mathbb{R}$ of real numbers. One of the consequences of this process is a justification of the *principle of mathematical induction*.

Intuitively, $\mathbb{N}$ consists of the numbers $1, 2, 3, \ldots$ with which we count finite nonempty sets. Thus, $\mathbb{N}$ should be a subset of $\mathbb{R}$ that contains 1 and has the additional property that

$$n \in \mathbb{N} \Rightarrow n + 1 \in \mathbb{N}.$$

Moreover, we should expect $\mathbb{N}$ to be the "smallest" subset of $\mathbb{R}$ that satisfies these conditions. In the next two definitions, we simply formalize these ideas.

4.1 DEFINITION

> **INDUCTIVE SET**   A subset I of $\mathbb{R}$ is said to be **inductive** if and only if it satisfies the following two conditions:
>
> (i) $1 \in I$
> (ii) For all $n \in \mathbb{R}$, $n \in I \Rightarrow n + 1 \in I$.

Note that there exists at least one inductive subset of $\mathbb{R}$; indeed, $\mathbb{R}$ is an inductive subset of itself. Of course, there are other inductive subsets of $\mathbb{R}$; for instance, the set P consisting of all positive real numbers is inductive (Problem 1).

4.2 DEFINITION

> **NATURAL NUMBERS**   A real number n is said to be a **natural number** if and only if n belongs to every inductive subset of $\mathbb{R}$. The set of all natural numbers is denoted by $\mathbb{N}$.

As our first result, we show that the natural numbers form the "smallest" inductive subset of $\mathbb{R}$.

4.3 LEMMA

> $\mathbb{N}$ is an inductive subset of $\mathbb{R}$ and, if I is an inductive subset of $\mathbb{R}$, then $\mathbb{N} \subseteq I$.

PROOF   By Definition 4.1, the real number 1 belongs to every inductive subset I of $\mathbb{R}$; hence, by Definition 4.2, we have $1 \in \mathbb{N}$. Now suppose that $n \in \mathbb{N}$. To finish the proof that $\mathbb{N}$ is inductive, we must show that $n + 1 \in \mathbb{N}$, that is, we must prove that $n + 1$ belongs to every inductive subset of $\mathbb{R}$. Thus, let I be an arbitrary inductive subset of $\mathbb{R}$. Because $n \in \mathbb{N}$, it follows that n belongs to every inductive subset of $\mathbb{R}$; hence, in particular, $n \in I$. Since I is inductive and $n \in I$, we conclude that $n + 1 \in I$. This shows that $n + 1$ belongs to every inductive subset of $\mathbb{R}$ and, hence, that $n + 1 \in \mathbb{N}$. Therefore, $\mathbb{N}$ is an inductive subset of $\mathbb{R}$.

To complete the proof, let I be an inductive subset of $\mathbb{R}$ and let $n \in \mathbb{N}$. Then n belongs to every inductive subset of $\mathbb{R}$, and, in particular, $n \in I$. Thus, $\mathbb{N} \subseteq I$.   □

Because $1 \in \mathbb{N}$ and $\mathbb{N}$ is an inductive subset of $\mathbb{R}$, it follows that $1 + 1 \in \mathbb{N}$. Naturally, we *define* $2 = 1 + 1$, so that $2 \in \mathbb{N}$. Again, since $\mathbb{N}$ is inductive and $2 \in \mathbb{N}$, it follows that $2 + 1 \in \mathbb{N}$. Naturally, we *define* $3 = 2 + 1$, so that $3 \in \mathbb{N}$. We can continue in this way as long as we please, *defining* $4 = 3 + 1, 5 = 4 + 1$, and so on, and showing that $4, 5, \ldots \in \mathbb{N}$. Note that, for all $n \in \mathbb{N}$, we have

$$n < n + 1.$$

(This follows on adding n to both sides of the inequality $0 < 1$, which is a consequence of Corollary 3.4.) Therefore,

$$1 < 2 < 3 < 4 < 5,$$

and so on.

Lemma 4.3 is used to prove the following theorem, which is the basis for proofs by mathematical induction.

4.4 THEOREM

> **PRINCIPLE OF MATHEMATICAL INDUCTION**   For each natural number n, let P_n be a proposition. Suppose that the following two conditions hold:
>
> (i) P_1 is true.
> (ii) For all $n \in \mathbb{N}$, if P_n is true, then P_{n+1} is true.
>
> Then P_n is true for all $n \in \mathbb{N}$.

PROOF   Let I be the subset of $\mathbb{N}$ consisting of all of the natural numbers n for which P_n is true. Conditions (i) and (ii) imply that I is an inductive subset of $\mathbb{R}$. By Lemma 4.3, it follows that $\mathbb{N} \subseteq I$, hence, that P_n is true for all $n \in \mathbb{N}$.   □

A proof that P_n is true for all $n \in \mathbb{N}$, obtained by using Theorem 4.4, is called a **proof by induction on** n. Such a proof involves two things: First, you must prove that P_1 is true. Second, you must prove that, for all $n \in \mathbb{N}$,

$$P_n \Rightarrow P_{n+1}.$$

Naturally, in proving this implication, you are entitled to *assume that P_n is true* for the purpose of showing that P_{n+1} follows from this assumption. In an inductive proof, the step in which you assume that P_n is true (for an arbitrary but fixed $n \in \mathbb{N}$) is called the **inductive hypothesis**. (The technique of making proofs by using the principle of mathematical induction is discussed in more detail and illustrated with further examples in Appendix I.)

4.5 Example   Prove that $n \geq 1$ holds for all $n \in \mathbb{N}$.

SOLUTION   For each $n \in \mathbb{N}$, let P_n be the proposition asserting that $n \geq 1$. We prove that P_n is true for all $n \in \mathbb{N}$ by using the principle of mathematical induction (Theorem 4.4). First, we must prove that P_1 is true. Second, we must prove that $P_n \Rightarrow P_{n+1}$. Because $1 \geq 1$, the proposition P_1 is true. To prove that $P_n \Rightarrow P_{n+1}$, assume that P_n is true for an arbitrary but fixed $n \in \mathbb{N}$; that is, assume that

$$n \geq 1. \tag{1}$$

(This is the induction hypothesis.) By adding 1 to both sides of the inequality in (1), we find that

$$n + 1 \geq 2. \tag{2}$$

By adding 1 to both sides of the inequality $1 > 0$, we also have

$$2 > 1. \tag{3}$$

By combining (2) and (3), we obtain

$$n + 1 \geq 1, \tag{4}$$

that is, P_{n+1} is true. This shows that $P_n \Rightarrow P_{n+1}$ and completes the proof. □

As the proof of the following lemma illustrates, it is not always necessary to invoke the induction hypothesis when you make a proof by mathematical induction. The proof is also a good review of some of the features of the implication connective.

4.6 LEMMA

> **SUBTRACTION OF 1 FROM A NATURAL NUMBER**   For every $n \in \mathbb{N}$,
>
> $n \neq 1 \Rightarrow n - 1 \in \mathbb{N}$.

PROOF   For each $n \in \mathbb{N}$, let P_n denote the proposition

$$n \neq 1 \Rightarrow n - 1 \in \mathbb{N}.$$

We prove, by the principle of mathematical induction, that P_n is true for all $n \in \mathbb{N}$. The proposition P_1 is true because its hypothesis, $1 \neq 1$, is false. We must prove that

$$P_n \Rightarrow P_{n+1}$$

holds for all $n \in \mathbb{N}$. Since an implication with a true conclusion is automatically true, it suffices to prove that P_{n+1} is true for all $n \in \mathbb{N}$. Thus, we must prove that

$$n + 1 \neq 1 \Rightarrow (n + 1) - 1 \in \mathbb{N}$$

holds for all $n \in \mathbb{N}$. In other words, we must prove that

$$n + 1 \neq 1 \Rightarrow n \in \mathbb{N}$$

holds for all $n \in \mathbb{N}$. But the last implication is true for all $n \in \mathbb{N}$ since its conclusion is true. □

If we visualize the real number system $\mathbb{R}$ as a number scale, then our intuition tells us that the natural numbers $1, 2, 3, \ldots$ are distributed along this number scale like "beads on a wire," starting at 1 and extending indefinitely to the right (Figure 3-9). The following theorem helps to confirm this.

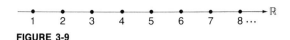

FIGURE 3-9

4.7 THEOREM

> **DISCRETENESS OF THE NATURAL NUMBERS**   If $n \in \mathbb{N}$, there exists no $k \in \mathbb{N}$ with
>
> $$n < k < n + 1.$$

PROOF   The proof is by induction on n. Let P_n be the proposition asserting that there is no natural number k strictly between n and $n + 1$. We begin by proving that P_1 is true. To this end, let

$$I = \{1\} \cup \{x \in \mathbb{R} \mid x \geq 2\}.$$

It is not difficult to check that I is an inductive subset of $\mathbb{R}$ (Problem 2); hence, by Lemma 4.3, $\mathbb{N} \subseteq I$. Thus, if $k \in \mathbb{N}$, then either $k = 1$ or else $k \geq 2$; hence, we cannot have $1 < k < 2$, and P_1 is proved.

To prove that $P_n \Rightarrow P_{n+1}$ holds for every $n \in \mathbb{N}$, we proceed by contradiction. Thus, suppose there exists $n \in \mathbb{N}$ such that P_n is true but P_{n+1} is false; that is, suppose there is *no* natural number strictly between n and $n + 1$, but there is a natural number $k \in \mathbb{N}$ that *is* strictly between $n + 1$ and $(n + 1) + 1$. Thus,

$$n + 1 < k < (n + 1) + 1. \tag{1}$$

By adding -1 to all members of the inequality in (1), we obtain

$$n < k - 1 < n + 1. \tag{2}$$

Now, k cannot be 1, for, if it were, then the first inequality in (2) would imply $n < 0$, contradicting Example 4.5. Since $k \in \mathbb{N}$ and $k \neq 1$, we can use Lemma 4.6 to conclude that

$$k - 1 \in \mathbb{N}. \tag{3}$$

By combining (2) and (3), we arrive at a contradiction to our supposition that there is no natural number strictly between n and $n + 1$, and our proof by contradiction is complete.   □

4.8 THEOREM

> **DIFFERENCE OF NATURAL NUMBERS**   If $m, n \in \mathbb{N}$ and $m > n$, then $m - n \in \mathbb{N}$.

PROOF Fix $m \in \mathbb{N}$. For each $n \in \mathbb{N}$, let P_n be the statement

$$m > n \Rightarrow m - n \in \mathbb{N}.$$

Then P_1 is the statement

$$m > 1 \Rightarrow m - 1 \in \mathbb{N},$$

which is true by Lemma 4.6.

Now we must prove that

$$P_n \Rightarrow P_{n+1}$$

holds for all $n \in \mathbb{N}$. We are going to prove this by contradiction, so suppose there is a $k \in \mathbb{N}$ for which $P_k \Rightarrow P_{k+1}$ is false. The only way that an implication can be false is for its hypothesis to be true and its conclusion false; hence, we have P_k true,

$$m > k \Rightarrow m - k \in \mathbb{N}, \tag{1}$$

and P_{k+1} false. To say that P_{k+1} is false is to say that its hypothesis is true,

$$m > k + 1, \tag{2}$$

and its conclusion is false,

$$m - (k + 1) \notin \mathbb{N}; \tag{3}$$

that is,

$$(m - k) - 1 \notin \mathbb{N}. \tag{4}$$

Now, $k + 1 > k$, and therefore, by (2), we have

$$m > k. \tag{5}$$

By (1), (5), and modus ponens,

$$m - k \in \mathbb{N}. \tag{6}$$

If $m - k \neq 1$, then, by (6) and Lemma 4.6, we would have $(m - k) - 1 \in \mathbb{N}$, contradicting (4). Consequently,

$$m - k = 1. \tag{7}$$

It follows that $m = k + 1$, contradicting (2).   □

After you become familiar with the technique of proving theorems "by induction," you can just keep in mind the proposition corresponding to each $n \in \mathbb{N}$ rather than denoting it explicitly by P_n. This is illustrated in the proof of the next theorem.

4.9 THEOREM
> **SUM OF NATURAL NUMBERS**   If $m, n \in \mathbb{N}$, then $m + n \in \mathbb{N}$.

PROOF   Fix $m \in \mathbb{N}$. We prove that $m + n \in \mathbb{N}$ by induction on n. For $n = 1$, we obtain the statement $m + 1 \in \mathbb{N}$, which is true because $m \in \mathbb{N}$ and $\mathbb{N}$ is an inductive set. Now we assume the inductive hypothesis

$m + n \in \mathbb{N}$

for some (arbitrary but fixed) $n \in \mathbb{N}$. Again, since $\mathbb{N}$ is inductive and $m + n \in \mathbb{N}$, we have $(m + n) + 1 \in \mathbb{N}$; hence,

$m + (n + 1) \in \mathbb{N}$.   □

We leave it as an exercise for you to prove the following theorem (Problem 11).

4.10 THEOREM
> **PRODUCT OF NATURAL NUMBERS**   If $m, n \in \mathbb{N}$, then $mn \in \mathbb{N}$.

Note carefully how the completeness of $\mathbb{R}$ is used in the proof of the following important theorem.

4.11 THEOREM
> **ARCHIMEDEAN PROPERTY OF $\mathbb{R}$**   Let $\varepsilon, a \in \mathbb{R}$ with $\varepsilon, a > 0$. Then there exists $n \in \mathbb{N}$ such that
>
> $n\varepsilon > a$.

PROOF   The proof is by contradiction. Thus, suppose that no such $n \in \mathbb{N}$ exists and let

$S = \{n\varepsilon \mid n \in \mathbb{N}\}$.

Then $\varepsilon = \varepsilon \cdot 1 \in S$, so $S \neq \varnothing$, and our supposition that $n\varepsilon > a$ is false for all $n \in \mathbb{N}$ means that a is an upper bound for S. Let $c = \sup(S)$. Using Part (i) of Lemma 3.23, with a replaced by c and p replaced by $c - \varepsilon/2$, we conclude that there exists $n \in \mathbb{N}$ such that

$$c - \frac{\varepsilon}{2} < n\varepsilon \leq c.$$

Consequently,

$$c < n\varepsilon + \frac{\varepsilon}{2} < n\varepsilon + \varepsilon = \varepsilon(n + 1).$$

But, since $n + 1 \in \mathbb{N}$, we have $\varepsilon(n + 1) \in S$; hence,

$\varepsilon(n + 1) \leq c = \sup(S)$.

Thus,

$$c < \varepsilon(n + 1) \le c,$$

which is a contradiction because we cannot have $c < c$.   □

In Theorem 4.11, it is most useful to think of ε as a "very small" positive real number, and to think of a as being "very large." The **Archimedean property** of the field $\mathbb{R}$ of real numbers, established by Theorem 4.11, can then be paraphrased as follows:

> *By adding a small positive real number ε to itself sufficiently many times, we can obtain an arbitrarily large positive real number.*

There are ordered fields that do not have the Archimedean property; however, *any subfield of a field with the Archimedean property inherits the Archimedean property* (Problem 18).

The following corollary of Theorem 4.11 shows that there are arbitrarily large natural numbers.

4.12 COROLLARY   If $x \in \mathbb{R}$, then there exists $n \in \mathbb{N}$ such that $x < n$.

PROOF   If $x \le 0$, just take $n = 1$. Thus, we can suppose that $x > 0$. In Theorem 4.11, take $\varepsilon = 1$ and $a = x$.   □

4.13 COROLLARY   Let $x \in \mathbb{R}$ with $1 \le x$. Then there exists a unique natural number n such that $n \le x < n + 1$.

PROOF   We prove the existence of n and leave the proof of its uniqueness as an exercise (Problem 17). Our proof is by contradiction, so suppose that

$$1 \le x, \tag{1}$$

but there does not exist $n \in \mathbb{N}$ such that $n \le x < n + 1$. In other words, for all $n \in \mathbb{N}$,

$$n \le x \Rightarrow n + 1 \le x. \tag{2}$$

By (1), (2), and the principle of mathematical induction, it follows that $n \le x$ holds for every natural number n, contradicting Corollary 4.12.

□

A totally ordered set $(X, \leq)$ is said to be **well-ordered** if every nonempty subset of X contains a smallest element. The next theorem tells us that the natural numbers, under the total order inherited from the real numbers, form a well-ordered set.

4.14 THEOREM

> **ℕ IS A WELL-ORDERED SET**   If M is a nonempty subset of $\mathbb{N}$, then M contains a smallest element. In fact, if $m = \inf(M)$, then m is the smallest element of M.

PROOF   By the result of Example 4.5, 1 is a lower bound for M; hence, by Corollary 3.22, M has an infimum. Let

$$m = \inf(M). \tag{1}$$

Because 1 is a lower bound for M, we have

$$1 \leq m. \tag{2}$$

At this point, we have no guarantee that m belongs to M, or even that it is a natural number. However, by Corollary 4.13, there exists $n \in \mathbb{N}$ such that

$$n \leq m < n + 1. \tag{3}$$

In particular, $m < n + 1$; hence, by using Part (ii) of Lemma 3.23 (with S replaced by M, b replaced by m, and q replaced by $n + 1$), we conclude that there exists $k \in M$ such that

$$m \leq k < n + 1. \tag{4}$$

Note that, by (3) and (4),

$$n \leq m \leq k. \tag{5}$$

As a consequence of (4) and (5), we have

$$n \leq k < n + 1. \tag{6}$$

Since $k \in M$ and $M \subseteq \mathbb{N}$, it follows that $k \in \mathbb{N}$; hence, by Theorem 4.7, we must have

$$n = k, \tag{7}$$

and so, by (5),

$$m = n = k \in M. \tag{8}$$

Since $m \in M$ and m is a lower bound for M, it follows that m is the smallest element in M. $\qquad\square$

Theorem 4.14 has the following very useful corollary:

4.15 COROLLARY

ALTERNATIVE FORM OF MATHEMATICAL INDUCTION For each natural number n, let P_n be a proposition. Suppose that the following two conditions hold:

(i) P_1 is true.
(ii) If m is a natural number, $1 < m$, and P_k is true for all natural numbers $k < m$, then P_m is true.

Then P_n is true for all $n \in \mathbb{N}$.

PROOF

The proof is by contradiction. Suppose that Parts (i) and (ii) hold, but that there exists $q \in \mathbb{N}$ such that P_q is false. Let

$$M = \{m \in \mathbb{N} \mid P_m \text{ is false}\}.$$

Then $q \in M$, so that M is a nonempty subset of $\mathbb{N}$. By Theorem 4.14, M contains a smallest element m. Thus, m is the smallest natural number for which P_m is false. Because P_1 is true, $m \neq 1$, and therefore $1 < m$. Furthermore, if k is a natural number and $k < m$, then P_k must be true (since m is the smallest natural number for which P_m is false). Therefore, by (ii), P_m is true, contradicting the fact that P_m is false.   □

Now we have nearly all of the elements of elementary arithmetic at our disposal in the system $\mathbb{N}$ of natural numbers. We can add and multiply natural numbers (Theorems 4.9 and 4.10), and we can subtract a smaller natural number from a larger (Theorem 4.8). We discuss the process of division (that is, the *division algorithm*) in Section 5.

PROBLEM SET 3.4

1. Prove that the set $P = \{x \in \mathbb{R} \mid x > 0\}$ is inductive.
2. Prove that the set $I = \{1\} \cup \{x \in \mathbb{R} \mid x \geq 2\}$ is inductive.
3. Prove that the set $J = \{x \in \mathbb{R} \mid x \geq 1\}$ is inductive.
4. If $\mathscr{I}$ denotes the set of all inductive subsets of $\mathbb{R}$, prove that $\mathbb{N} = \bigcap_{I \in \mathscr{I}} I$.
5. If $I \subseteq \mathbb{N}$ and I is an inductive set, prove that $I = \mathbb{N}$.
6. State and prove the converse of Lemma 4.6.
7. True or false: If I is an inductive set and $I \subseteq J \subseteq \mathbb{R}$, then J is an inductive set. If true, prove it; if false, give a counterexample.
8. Prove: If $n, m \in \mathbb{N}$, then $m > n$ if and only if there exists $k \in \mathbb{N}$ such that $m = n + k$.

9. Prove: If $n, m \in \mathbb{N}$, then $m > n \Rightarrow m \geq n + 1$. [Hint: Use Theorem 4.7.]

10. Prove the following generalized version of the principle of mathematical induction: *Let k be a fixed natural number. For each natural number $n \geq k$, let P_n be a proposition. Suppose P_k is true and suppose that, for every natural number $n \geq k$, $P_n \Rightarrow P_{n+1}$. Then P_n is true for every natural number $n \geq k$.*

11. Prove Theorem 4.10. [Hint: Fix m and prove that $mn \in \mathbb{N}$ by induction on n.]

12. Let F be a field (not necessarily ordered) and define a subset I of F to be **inductive** if $1 \in I$ and

$$x \in I \Rightarrow x + 1 \in I$$

for all $x \in F$. Define an analogue of $\mathbb{N}$ in F, call it $\mathbb{N}_F$, by the requirement that $x \in \mathbb{N}_F$ if and only if x belongs to every inductive subset of F. State and prove an analogue of Lemma 4.3 for $\mathbb{N}_F$.

13. Prove: Given any $\varepsilon \in \mathbb{R}$ with $\varepsilon > 0$, there exists $m \in \mathbb{N}$ such that, for all $n \in \mathbb{N}$, $n > m \Rightarrow 1/n < \varepsilon$.

14. As a consequence of Example 4.5, $0 \notin \mathbb{N}$. With the notation of Problem 12, show that there are fields F for which $0 \in \mathbb{N}_F$. [Hint: Try $F = \mathbb{Z}_2$.]

15. Prove: If $n, k \in \mathbb{N}$ and $n < k < n + 2$, then $k = n + 1$.

16. Prove that, if F is a subfield of $\mathbb{R}$, then $\mathbb{N} \subseteq F$. (See Problem 34 in Problem Set 3.2.)

17. Complete the proof of Corollary 4.13 by showing that, if $1 \leq x$, and if $n, m \in \mathbb{N}$ with $n \leq x < n + 1$ and $m \leq x < m + 1$, then $n = m$. [Hint: Begin by showing that $n < m + 1$ and $m < n + 1$, deduce that $n < m + 1 < n + 2$, and use the result of Problem 15.]

18. Prove that, if F is a subfield of $\mathbb{R}$, then F inherits the Archimedean property from $\mathbb{R}$. (See Problem 2 in Problem Set 3.3.)

19. A **Peano model**, named after the Italian mathematician Giuseppe Peano (1858–1932), is an ordered triple (N, a, s) consisting of a set N, a fixed element $a \in N$, and a mapping $s: N \to N$ satisfying the following three postulates:
(i) $s: N \to N$ is an injection.
(ii) $a \notin s(N)$.
(iii) If $a \in M \subseteq N$ and if $s(M) \subseteq M$, then $M = N$.
Define $s: \mathbb{N} \to \mathbb{N}$ by $s(n) = n + 1$ for all $n \in \mathbb{N}$ and show that $(\mathbb{N}, 1, s)$ is a Peano model.

20. Let P denote the set of all positive real numbers. With the usual ordering $\leq$ inherited from $\mathbb{R}$, is $(P, \leq)$ a well-ordered set? Explain.

21. If M is a nonempty subset of $\mathbb{N}$ and if M has an upper bound in $\mathbb{R}$, prove that M contains a largest element.

22. Suppose that $m, n \in \mathbb{N}$ with $n \leq m$. Let $K = \{k \in \mathbb{N} \mid kn \leq m\}$.
    (a) Show that K is a nonempty subset of $\mathbb{N}$.
    (b) Show that K has an upper bound in $\mathbb{R}$. [Hint: Use Theorem 4.11.]
    (c) Let q be the largest element of K (Problem 21). Show that
    $$qn \leq m < (q + 1)n.$$

23. A natural number n is said to be **even** if it can be written in the form $n = 2k$ for some natural number k. If $n \in \mathbb{N}$ is not even, it is said to be **odd**.
    (a) Show that, if n is even, then $n + 1$ is odd.
    (b) Show that, if n is odd, then $n + 1$ is even.
    (c) Show that every natural number is either even or odd.
    (d) Show that the product of two odd natural numbers is odd.
    (e) If n is a natural number and n^2 is even, show that n must be even.

24. Let n be a fixed natural number. Prove that every natural number m with $m < 2^n$ can be written in the form
    $$m = c_0 + 2c_1 + 2^2 c_2 + 2^3 c_3 + \cdots + 2^{n-1} c_{n-1}$$
    where $c_j \in \{0, 1\}$ for each $j = 0, 1, 2, \ldots, n - 1$.

25. Show that *there do not exist natural numbers n and m with $n^2 = 2m^2$* by carrying out the following: Suppose that such natural numbers exist.
    (a) Show then that there exists a smallest natural number n such that n^2 has the form $n^2 = 2m^2$ for some $m \in \mathbb{N}$.
    (b) Show that there exists $k \in \mathbb{N}$ such that $n = 2k$. [Hint: Use Part (e) of Problem 23.]
    (c) Show that $m^2 = 2k^2$, and conclude that there exists $r \in \mathbb{N}$ such that $m = 2r$.
    (d) Show that $k^2 = 2r^2$ and that $k < n$, thus contradicting Part (a).

5

THE INTEGERS AND THE RATIONAL NUMBERS

In the previous section, we showed that the system $\mathbb{N}$ of natural numbers is **closed** under addition and multiplication; that is, the sum and product of natural numbers are again natural numbers. However, $\mathbb{N}$ is not closed under subtraction because, if $m, n \in \mathbb{N}$ with $m \leq n$, then $m - n \notin \mathbb{N}$. If we enlarge $\mathbb{N}$ by appending 0 and the negative integers, we obtain the system $\mathbb{Z}$ of *integers* which is not only closed under subtraction but is closed under addition and multiplication as well.

5.1 DEFINITION | **INTEGERS** We define $\mathbb{Z} = \mathbb{N} \cup \{0\} \cup \{-n \mid n \in \mathbb{N}\}$. A real number is called an **integer** if it belongs to $\mathbb{Z}$.

Note that, if $k \in \mathbb{Z}$, then $k \in \mathbb{N}$ if and only if $k > 0$; in other words, $\mathbb{N}$ *is precisely the set of positive integers* (Problem 1). Also note that

$$k \in \mathbb{Z} \Leftrightarrow -k \in \mathbb{Z}$$

(Problem 3). We leave as an exercise the proof of the following lemma and its corollary (Problems 2 and 4).

5.2 LEMMA

> **CLOSURE OF $\mathbb{Z}$ UNDER SUMS, DIFFERENCES, AND PRODUCTS** Let m and n be integers. Then $m + n$, $m - n$, and mn are integers.

5.3 COROLLARY

> Under the operations of addition and multiplication inherited from $\mathbb{R}$, the system $(\mathbb{Z}, +, \cdot)$ forms a commutative ring with unity. Furthermore, if $x, y \in \mathbb{Z}$ with $xy = 0$, then $x = 0$ or $y = 0$, or both.

Like the natural numbers, the integers are distributed along the number line like "beads on a wire," but they extend indefinitely in both directions (Figure 3-10). In particular, we have the following analogue of Theorem 4.7:

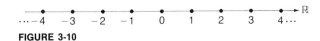

FIGURE 3-10

5.4 THEOREM

> **DISCRETENESS OF THE INTEGERS** If $n \in \mathbb{Z}$, there exists no $k \in \mathbb{Z}$ with $n < k < n + 1$.

PROOF   Suppose, on the contrary, that $n, k \in \mathbb{Z}$ with

$$n < k < n + 1.$$

By adding $1 - n$ to all members of this inequality, we obtain

$$n + (1 - n) < k + (1 - n) < n + 1 + (1 - n);$$

that is,

$$1 < k + 1 - n < 2.$$

Let $q = k + 1 - n$, so that

$$1 < q < 2.$$

By Lemma 5.2, $q \in \mathbb{Z}$ and, since $1 < q$, it follows that $q \in \mathbb{N}$. But then $1 < q < 2$ contradicts Theorem 4.7. ☐

We also have the following analogue for $\mathbb{Z}$ of Corollary 4.13:

5.5 THEOREM

> **GREATEST INTEGER LESS THAN OR EQUAL TO x** Let $x \in \mathbb{R}$. Then there exists a unique $n \in \mathbb{Z}$ such that
>
> $$n \leq x < n + 1.$$

PROOF By Corollary 4.12, there exists $q \in \mathbb{N}$ such that

$$1 - x < q.$$

Therefore,

$$1 < q + x,$$

and it follows from Corollary 4.13 that there exists $m \in \mathbb{N}$ such that

$$m \leq q + x < m + 1.$$

Let $n = m - q$ and add $-q$ to all members of the last inequality to obtain

$$n \leq x < n + 1.$$

By Lemma 5.2, $n \in \mathbb{Z}$. That n is uniquely determined by this condition is proved by arguing as in Problem 17 in Problem Set 3.4, and is left as an exercise (Problem 6). ☐

5.6 DEFINITION

> **GREATEST INTEGER FUNCTION**   Let $x \in \mathbb{R}$. Then the unique $n \in \mathbb{Z}$ such that
>
> $$n \leq x < n + 1$$
>
> is called the **greatest integer less than or equal to** x and is denoted by $[\![x]\!] = n$. The function from $\mathbb{R}$ into $\mathbb{Z}$ defined by $x \mapsto [\![x]\!]$ for all $x \in \mathbb{R}$ is called the **greatest integer function**.

Note that $[\![x]\!]$ is characterized by the two properties

$$[\![x]\!] \in \mathbb{Z}$$

and

$$[\![x]\!] \leq x < [\![x]\!] + 1.$$

In some computer languages, such as BASIC, the greatest integer function is written as INT, so that

$$\text{INT}(X) = [\![X]\!].$$

The effect of INT(X) is to round X *down* to the nearest integer. (In other computer languages, this is called the **floor** function.) A related function is the **closest integer function** CINT defined by

$$\text{CINT}(X) = \text{INT}(X + 0.5) = [\![X + 1/2]\!].$$

The effect of CINT(X) is to round off X to the nearest integer.

If n is a natural number, then the function

$$x \mapsto [\![x \cdot 10^n + 0.5]\!] \cdot 10^{-n}$$

has the effect of *rounding x off to n decimal places* (Problem 7). In BASIC, the corresponding formula

$$\text{CINT}(X*10\text{\textasciicircum}N)/10\text{\textasciicircum}N$$

is used to round X off to N decimal places.

Although $\mathbb{Z}$ is closed under addition, subtraction, and multiplication, *it is not closed under division.* For instance, $1, 2 \in \mathbb{Z}$, but $\frac{1}{2} \notin \mathbb{Z}$.

5.7 Example　Show that $\frac{1}{2} \notin \mathbb{Z}$.

SOLUTION　Since $0 < 1, 2$, we have

$$0 < \tfrac{1}{2}.$$

Also, since $0 < 1 < 2$, we have

$$\tfrac{1}{2} < 1.$$

Therefore,

$$0 < \tfrac{1}{2} < 1,$$

and it follows from Theorem 5.4 that $\frac{1}{2} \notin \mathbb{Z}$.　　　□

Let $m, n \in \mathbb{Z}$ with $n \neq 0$ and consider the fraction $m/n \in \mathbb{R}$. Although m/n need not be an integer, we can use long division to divide m by n and obtain an *integer quotient q* and a *remainder r* with $0 \leq r < |n|$:

$$\begin{array}{r} q \\ n\,\overline{\smash{\big)}\,m} \end{array} \quad \text{and a remainder } r.$$

That this is always possible, and that the quotient q and remainder r are uniquely determined by m and n, is the content of the following important theorem and corollary:

5.8 THEOREM

> **DIVISION ALGORITHM** Let $m, n \in \mathbb{Z}$ with $n > 0$. Then there exist unique integers q and r such that
>
> $$m = nq + r \quad \text{and} \quad 0 \le r < n.$$

PROOF We begin by proving the existence of q and r. Let $x = m/n$, let $q = [\![x]\!]$, and let $r = m - nq$. Then q and r are integers and

$$m = nq + r. \tag{1}$$

Also, since $q = [\![x]\!]$, we have

$$q \le x < q + 1. \tag{2}$$

Multiplying each member of (2) by the positive integer n, and using the fact that $x = m/n$, we find that

$$nq \le m < nq + n. \tag{3}$$

By adding $-nq$ to each member of (3) and using the fact that $r = m - nq$, we obtain

$$0 \le r < n. \tag{4}$$

To prove the uniqueness of q and r, suppose that we also have integers q' and r' with

$$m = nq' + r' \tag{5}$$

and

$$0 \le r' < n. \tag{6}$$

We must prove that $q' = q$ and that $r' = r$. From (5), we have

$$\frac{m}{n} = q' + \frac{r'}{n}, \tag{7}$$

and it follows from (6) that

$$0 \le \frac{r'}{n} < 1. \tag{8}$$

By adding q' to all members of the inequality in (8) and using (7), we obtain

$$q' \le \frac{m}{n} < q' + 1. \tag{9}$$

In view of (9) and Definition 5.6, we have

$$q' = \left[\!\!\left[\frac{m}{n}\right]\!\!\right] = q. \tag{10}$$

Finally, by combining (5) and (10), we find that

$$r' = m - nq' = m - nq = r. \tag{11}$$

$\square$

5.9 COROLLARY

> **GENERALIZED DIVISION ALGORITHM**   Let $m, n \in \mathbb{Z}$ with $n \neq 0$. Then there exist unique integers q and r such that
>
> $$m = nq + r \quad\text{and}\quad 0 \leq r < |n|.$$

PROOF   We prove the existence of q and r and leave the proof of their uniqueness as an exercise (Problem 8). By applying Theorem 5.8 to m and $|n|$, we find that there are unique integers k and r such that

$$m = |n|k + r \quad\text{and}\quad 0 \leq r < |n|.$$

Let $q = \operatorname{sgn}(n)k$ (see Definition 3.14). Then q is an integer and

$$m = |n|k + r = n \cdot \operatorname{sgn}(n)k + r = nq + r. \qquad\square$$

In Corollary 5.9, q is called the **integer quotient** and r is called the **remainder** on division of m by n, $n \neq 0$. Do not confuse the integer quotient q with the quotient m/n obtained by dividing m by n in the field $\mathbb{R}$. The relationship between these two types of quotients is expressed by the simple formula

$$\frac{m}{n} = q + \frac{r}{n}$$

obtained by dividing both sides of $m = nq + r$ by n. If $r = 0$, so that $m/n = q$, we say that the integer m is **exactly divisible** by the integer n, or, for short, that n **divides** m.

As we have seen, the system $\mathbb{Z}$ of integers is closed under addition, multiplication, negation, and subtraction, but it is not closed under division. Indeed, for $m, n \in \mathbb{Z}$ with $n \neq 0$, $m/n \in \mathbb{Z}$ if and only if n divides m. To obtain a number system that is closed under division (by nonzero numbers), we simply pass to the *rational numbers*.

5.10 DEFINITION

> **RATIONAL NUMBERS**   A real number x is called a **rational number** if it can be expressed in the form $x = m/n$, where $m, n \in \mathbb{Z}$ and $n \neq 0$. The set of all rational numbers is denoted by $\mathbb{Q}$.

Numbers of the form m/n with $m, n \in \mathbb{Z}$ and $n \neq 0$ are called *rational* because they are *ratios* of integers, not because they have anything to do with being "reasonable." The symbol $\mathbb{Q}$ is used for the set of all rational numbers to help remind us that these numbers are *quotients* (of integers). We leave it as an exercise for you to verify the following theorem (Problem 23).

5.11 THEOREM

> **$\mathbb{Q}$ IS AN ORDERED FIELD** $\mathbb{Q}$ is closed under addition, negation, subtraction, multiplication, and division by nonzero numbers. Furthermore, $(\mathbb{Q}, +, \cdot)$ is a field and, under the order relation $\leq$ inherited from $\mathbb{R}$, it is an ordered field.

Note that $\mathbb{Z} \subseteq \mathbb{Q}$ because every integer n can be expressed in the form $n = n/1$. Thus, we have

$$\mathbb{N} \subseteq \mathbb{Z} \subseteq \mathbb{Q} \subseteq \mathbb{R}.$$

5.12 DEFINITION

> **IRRATIONAL NUMBER** A real number that is not a rational number is called an **irrational number**.

Thus, $\mathbb{R} \setminus \mathbb{Q}$ is the set of all irrational numbers. There is no special symbol (apart from $\mathbb{R} \setminus \mathbb{Q}$) for the set of irrational numbers, and $\mathbb{R} \setminus \mathbb{Q}$ is not regarded as a "number system" because it is not closed under *any* of the arithmetic operations of addition, subtraction, multiplication, or division (Problem 21).

5.13 Example Show that $\sqrt{2}$ is an irrational number.[†]

SOLUTION Suppose, on the contrary, that $\sqrt{2}$ is rational. Then there exist integers m, n with $n \neq 0$ such that

$$\sqrt{2} = \frac{m}{n}. \tag{1}$$

By multiplying the numerator and denominator of the fraction m/n by -1 if necessary, we can assume without loss of generality that both m and n are natural numbers. Multiplying both sides (1) by n, we obtain

$$n\sqrt{2} = m. \tag{2}$$

[†] The symbol $\sqrt{2}$ denotes the unique positive real number whose square is 2. That such a number exists follows from Theorem 4.19 in Chapter 4.

After squaring both sides of (2), we find that

$$m^2 = 2n^2,$$ (3)

which contradicts Problem 25 in Problem Set 3.4.   □

Let $r \in \mathbb{Q}$ with $r \neq 0$. We say that r is expressed in **reduced form** if it is written as

$$r = \frac{m}{n}$$

where n and m have no common integer divisor, except for ± 1, and $n > 0$. It can be shown that every nonzero rational number can be written *uniquely* in reduced form (Problem 26).

As we have noted, the integers are spread out along the number line like "beads on a wire." How are the rational numbers distributed among the real numbers? Here the story is more complicated. To begin with, the rational numbers are **dense** in the real numbers in the sense that, *between any two real numbers, there is a rational number*. We devote the next theorem to a proof of this fact.

5.14 THEOREM

> **DENSITY OF Q IN R**   Let $a, b \in \mathbb{R}$ with $a < b$. Then there exists a rational number $r \in \mathbb{Q}$ such that $a < r < b$.

PROOF   Since $a < b$, it follows that $0 < 1/(b - a)$. By Corollary 4.12, there exists a natural number n such that

$$\frac{1}{b - a} < n,$$ (1)

that is,

$$0 < \frac{1}{n} < b - a.$$ (2)

As a consequence of (2), we have

$$a + \frac{1}{n} < b.$$ (3)

Now let

$$m = [\![na + 1]\!]$$ (4)

so that, by Definition 5.6, $m \in \mathbb{Z}$ and

$$m \leq na + 1 < m + 1.$$ (5)

The first inequality in (5) implies that

$$\frac{m}{n} \le a + \frac{1}{n}. \tag{6}$$

By combining (6) and (3), we obtain

$$\frac{m}{n} < b. \tag{7}$$

Also, from the second inequality in (5), $na < m$, that is,

$$a < \frac{m}{n}. \tag{8}$$

From (8) and (7),

$$a < \frac{m}{n} < b. \tag{9}$$

Let $r = m/n$. Since m and n are integers and $n \ne 0$, r is a rational number and (9) shows that $a < r < b$. $\square$

The fact that there is a rational number between any two real numbers might lead you to believe that "most" real numbers are rational and that the irrational numbers are comparatively rare. It turns out that this is not true—in fact, the situation is quite the contrary. As a first indication of this, we shall prove that *the irrational numbers are also dense in* $\mathbb{R}$.

5.15 THEOREM

> **DENSITY OF $\mathbb{R} \setminus \mathbb{Q}$ IN $\mathbb{R}$** Let $a, b \in \mathbb{R}$ with $a < b$. Then there exists an irrational number $x \in \mathbb{R} \setminus \mathbb{Q}$ such that $a < x < b$.

PROOF   Let u be any fixed positive irrational number, for instance, $u = \sqrt{2}$. By dividing both sides of the inequality $a < b$ by the positive number u, we have

$$\frac{a}{u} < \frac{b}{u}. \tag{1}$$

By Theorem 5.14, there exists a rational number $r \in \mathbb{Q}$ such that

$$\frac{a}{u} < r < \frac{b}{u}. \tag{2}$$

Without loss of generality, we can assume that

$$r \ne 0. \tag{3}$$

(If $r = 0$, just choose another rational number between 0 and b/u.) By multiplying all members of (2) by u, we obtain

$$a < ru < b. \tag{4}$$

Let $x = ru$. If x were rational, then $u = x/r$ would be rational by Theorem 5.2, contradicting our original choice of u as an irrational number. Hence, x is irrational and

$$a < x < b. \tag{5}$$

$\square$

PROBLEM SET 3.5

1. Prove that $\mathbb{N} = \{n \in \mathbb{Z} \mid n > 0\}$.

2. Prove Lemma 5.2.

3. Prove that, for $k \in \mathbb{R}$, $k \in \mathbb{Z} \Leftrightarrow -k \in \mathbb{Z}$.

4. Prove Corollary 5.3.

5. Suppose that $m, n \in \mathbb{Z}$ with $mn = 1$. Prove that either $m = n = 1$ or else $m = n = -1$.

6. Prove the uniqueness of n in Theorem 5.5.

7. Let $x, x^* \in \mathbb{R}$ and let $n \in \mathbb{N}$. We say that x^* approximates x **to n decimal places** if $|x - x^*| \le (10^{-n})/2$. Let $x^* = [\![x \cdot 10^n + 0.5]\!] \cdot 10^{-n}$. Show that x^* approximates x to n decimal places.

8. Prove the uniqueness of q and r in Corollary 5.9.

9. If $n, m \in \mathbb{Z}$, we write $n \mid m$ and say that n **divides** m if there exists $k \in \mathbb{Z}$ such that $kn = m$. (Note that this extends the definition of *divides* given in the text because there is no requirement here that $n \ne 0$.) For $n, m \in \mathbb{Z}$, prove the following:
   (a) $1 \mid m$ (b) $n \mid 0$ (c) $n \mid n$ (d) $-n \mid n$
   (e) $n \mid m \Leftrightarrow -n \mid m$ (f) $n \mid m \wedge m \mid n \Rightarrow n = m \vee n = -m$

10. If $n \in \mathbb{Z}$, define the subset $\mathbb{Z}n$ of $\mathbb{Z}$ by $\mathbb{Z}n = \{kn \mid k \in \mathbb{Z}\}$. In other words, $\mathbb{Z}n$ is the set of all *integer multiples of n*. For $n, m, k \in \mathbb{Z}$, prove the following:
    (a) $n \mid m \Leftrightarrow m \in \mathbb{Z}n$ (See Problem 9.)   (b) $n \mid m \Leftrightarrow \mathbb{Z}m \subseteq \mathbb{Z}n$
    (c) $m, k \in \mathbb{Z}n \Rightarrow m + k \in \mathbb{Z}n$ (d) $m \in \mathbb{Z}n \Rightarrow km \in \mathbb{Z}n$

11. Let $n, m, k \in \mathbb{Z}$. With the notation of Problem 9, prove the following:
    (a) $n \mid m \Rightarrow n \mid mk$ (b) $n \mid m \wedge n \mid k \Rightarrow n \mid (m + k)$
    (c) $n \mid m \wedge m \mid k \Rightarrow n \mid k$

12. Let $I \subseteq \mathbb{R}$, $I \ne \varnothing$, and suppose that I has the following two properties:
    (i) $m, n \in I \Rightarrow m + n \in I$.
    (ii) $k \in \mathbb{Z} \wedge m \in I \Rightarrow km \in I$.

Prove that there exists a unique $g \in \mathbb{Z}$ such that $g \geq 0$ and $I = \mathbb{Z}g$. (See Problem 10.) [Hint: First prove that $0 \in I$. Then take care of the case in which 0 is the only element in I. Finally, supposing that there is a nonzero element in I, prove that there is a positive element in I, and let g be the smallest positive element in I.]

13. Suppose that n is a fixed positive integer. Define the mapping $\rho : \mathbb{Z} \to \{0, 1, 2, 3, \ldots, n - 1\}$ by $\rho(m) = r$, where r is the remainder on division of m by n (Theorem 5.8). Let $h, k \in \mathbb{Z}$.
    (a) Prove that $\rho(h + k) = \rho(\rho(h) + \rho(k))$.
    (b) Prove that $\rho(hk) = \rho(\rho(h) \cdot \rho(k))$.

14. Let $m, n \in \mathbb{Z}$. With the notation of Problem 10, define $\mathbb{Z}m + \mathbb{Z}n = \{h + k \mid h \in \mathbb{Z}m, k \in \mathbb{Z}n\}$. Use the result of Problem 12 to prove that there exists a unique $g \in \mathbb{Z}$ such that $g \geq 0$ and $\mathbb{Z}m + \mathbb{Z}n = \mathbb{Z}g$. The integer g is called the **greatest common divisor** of m and n, in symbols, $g = \mathrm{GCD}(m, n)$.

15. Give an example of three nonzero integers $m, h,$ and k such that m divides hk, but m divides neither h nor k.

16. Let $m, n, k \in \mathbb{Z}$ and let $g = \mathrm{GCD}(m, n)$ as in Problem 14. Prove the following:
    (a) $g \mid m$ and $g \mid n$
    (b) There exist $a, b \in \mathbb{Z}$ such that $g = am + bn$.
    (c) $k \mid m \wedge k \mid n \Rightarrow k \mid g$

17. Let $m, n \in \mathbb{Z}$. If $m, n \neq 0$, we define the **least common multiple** of m and n, in symbols $\mathrm{LCM}(m, n)$, to be the smallest natural number that is an integer multiple of both m and n. If either $m = 0$ or $n = 0$, we define $\mathrm{LCM}(m, n) = 0$. Let $c = \mathrm{LCM}(m, n)$ and suppose that k is an integer multiple of both m and n. Prove that $c \mid k$ (see Problem 9). [Hint: By the division algorithm (Theorem 5.8) there are integers q and r with $k = cq + r$ and $0 \leq r < c$. Prove that $m \mid r$ and $n \mid r$.]

18. If $m, n \in \mathbb{Z}$, we say that m and n are **relatively prime** if $\mathrm{GCD}(m, n) = 1$. (See Problem 14.) If $m, n \neq 0$, prove that m and n are relatively prime if and only if there are integers a, b such that $am + bn = 1$. [Hint: Use Problem 16b.]

19. If $m, n \in \mathbb{Z}$ and $n \mid m$, we say that n is a **proper divisor** of m if $n \neq \pm 1$ and $n \neq \pm m$ (see Problem 9). A natural number p is called a **prime** if $p \neq 1$ and p has no proper divisors. If n is a natural number and $n > 1$, prove that n can be factored as a product of finitely many primes. [Hint: Assume that there is a natural number $n > 1$ that cannot be so factored and consider the smallest such number.]

20. Prove **Euclid's Lemma:** If $m, n, k \in \mathbb{Z}$, m and n are relatively prime, and $m \mid nk$, then $m \mid k$. (See Problems 9 and 18.)

21. Show that the set $\mathbb{R} \setminus \mathbb{Q}$ of irrational numbers is not closed under addition, subtraction, multiplication, or division. [Hint: If $u \in \mathbb{R}$ is irrational, so are $-u$ and $1/u$.]

22. Let $m, n \in \mathbb{Z}$ with $m, n \neq 0$ and let $g = \text{GCD}(m, n)$ (see Problem 14). Prove that the integers m/g and n/g are relatively prime (see Problem 18).

23. Prove Theorem 5.11.

24. Prove the **Fundamental Theorem of Arithmetic**: If n is an integer greater than 1, then n can be factored as a product of finitely many distinct primes raised to positive integer powers, and, apart from the order in which these factors are written, this factorization is unique.

25. Show that the field $\mathbb{Q}$ of rational numbers has no proper subfields, that is, show that, if $F \subseteq \mathbb{Q}$, $1 \in F$, and F is closed under addition, subtraction, multiplication, and division by nonzero elements, then $F = \mathbb{Q}$.

26. Prove that every nonzero rational number can be written uniquely in reduced form. [Hint: Use Problems 22 and 20.]

27. Let x be a positive real number with a **terminating** decimal expansion, that is, for some $n \in \mathbb{N}$, all of the digits after the nth decimal place are zeros. Prove that x is a rational number.

28 A **proper fraction** is a rational number of the form m/n where $0 \leq m < n$. Show that every nonnegative rational number r can be written uniquely in the form $r = k + (m/n)$ where k is a nonnegative integer and either $m = 0$ or m/n is a proper fraction in reduced form.

29. A rational approximation x^* to a positive real number x can be obtained by dropping all digits after the nth digit in the decimal expansion of x. (This procedure is called **chopping**.) If x^* is obtained by chopping x in this way, show that $x^* = [\![10^n x]\!] \cdot 10^{-n}$.

30. Let x be a positive real number with a **repeating** decimal expansion, that is, for some $n \in \mathbb{N}$, all of the digits from the nth decimal place and beyond repeat in blocks of k digits. Let $y = 10^{n-1}x - [\![10^{n-1}x]\!]$. Show that $0 \leq y < 1$ and that, in the decimal expansion of y, the digits to the right of the decimal point repeat in blocks of k digits.

31. Let x be a positive real number and let $x^* = [\![10^n x]\!] \cdot 10^{-n}$ as in Problem 29. Prove that $0 \leq x - x^* < 10^{-n}$.

32. Suppose that $0 \leq y < 1$ and that, in the decimal expansion of y, the digits to the right of the decimal point repeat in blocks of k digits. Let $m = [\![10^k y]\!]$ and let $n = 10^k - 1$. Show that $y = m/n$.

33. Let $0 < x < 1$. Show that the nth digit in the decimal expansion of x is given by $[\![10^n x]\!] - 10[\![10^{n-1}x]\!]$.

34. By combining the results of Problems 30 and 32, show that a positive real number x with a repeating decimal expansion is a rational number.

35. Let $x = 3.142\overline{446}$. (The overbar indicates that the block of digits 446 repeats forever.) Express x as a rational number in reduced form. [Hint: Let $y = 1000x$, $z = y - [\![y]\!]$, and $n = 1000z - z$. Show that n is a positive integer.]

36. Show that the decimal expansion of a positive rational number x either terminates as in Problem 27 or repeats as in Problem 30. [Hint: Let $x = m/n$ in reduced form, and consider the successive remainders produced by long division of m by n. If one of these remainders is zero, the decimal expansion of x terminates. Otherwise each of the remainders is a positive integer less than n; hence, after at most n steps, one of these remainders must be repeated.]

HISTORICAL NOTES

The modern trend in mathematics to define and use abstract structures probably can be credited to the work of Évariste Galois (1811–1832), who used groups and fields to solve one of the great unsolved problems of the day: to show that the fifth-degree polynomial equation, unlike those of lower degree, cannot in general be solved in terms of sums, differences, products, quotients, and roots of the coefficients. Galois's solution involved associating a group with each equation, so that solvability "by radicals" would be equivalent to the group's having certain properties. This problem had been engaging the efforts of mathematicians since the mid-seventeenth century, when methods for solving "by radicals" the fourth-degree polynomial equation had been discovered. Earlier attempts at a solution by Paulo Ruffini (1765–1822) and by Joseph Louis Lagrange (1736–1813), though ingenious, had failed.

In fact, the idea of a group had been around earlier, although an abstract formulation had not been developed. Augustin Louis Cauchy (1789–1857), as early as 1815, had written about groups of permutations. Groups in number theory had certainly appeared in the work of Euler and Gauss, although they had not specifically referred to group structure. Later in the nineteenth century, Felix Klein (1849–1925) used groups to classify geometries in his celebrated *Erlanger Programm*, and Marius Sophus Lie (1842–1899) developed a whole new area of mathematics concerned with groups of transformations. Klein and Jules Henri Poincaré (1854–1912) also contributed in this area. The axiomatic treatment was first formulated in the mid-nineteenth century by Arthur Cayley (1821–1895), by Walther Franz von Dyck (1856–1934), and finally, in modern form, in 1893, by Heinrich Weber (1842–1913) in his text, *Lehrbuch der Algebra*.

Rings and fields were studied later, in the 1920s and 1930s, primarily in Germany. The great algebraist, Emmy Noether (1882–1935), did pioneering work in these areas prior to moving to the United States where she taught at Bryn Mawr. Integral domains and fields had been known, at least informally, since antiquity, but their formal axiomatization was a relatively late affair. It was found too that there are fields, such as finite fields, which had not occurred to earlier generations but which satisfied all the standard properties expected of the fields of rational or real numbers.

Records exist that indicate that the real numbers were used by the Babylonians, but a formal study of real numbers did not occur until the late nineteenth century. The integers were used, of course, early on, and the rational numbers posed no problems. Unfortunately, the irrational numbers *did* pose a problem for the Greeks, who objected to "incommensurable" numbers. They turned their attention to geometry in order to avoid such numbers, thereby setting a style and standard that unfortunately prevailed in Western culture for many centuries. The advent of Arabic algebra into Europe around 1200 began the flowering of European mathematics. Decimals were introduced into Europe by Simon Stevin (1548–1620), allowing the use of irrational numbers.

In the nineteenth century, with the need to clarify and provide a solid foundation for calculus, and hence for more advanced analysis, it was seen that it would be necessary to characterize properly the real numbers. Two important conditions in this quest were the least upper bound principle and the Cauchy completeness criterion. In the context of an Archimedean ordered field, each of these concepts implies the other. This led Karl Weierstrass (1815–1897), Julius Wilhelm Richard Dedekind (1831–1916), Georg Cantor, and Leopold Kronecker (1823–1891) to try to construct the real numbers from the rational numbers by some systematic process. Dedekind chose the least upper bound principle to be the defining property of the real numbers, and he defined a real number to be a cut (now called a Dedekind cut, or *Schnitt*). Cantor's approach used the Cauchy criterion.

Kronecker viewed the whole problem differently. He rejected the real number system as too imprecise, and when, in 1882, Ferdinand Lindemann (1852–1939) showed that π is transcendental (that is, not the root of any polynomial equation with rational coefficients), Kronecker remarked: "Very interesting. Then π does not exist." A more famous remark of Kronecker's was that "God made the positive integers, all else is the work of man." Kronecker would allow humans to use only algebraic processes in the manufacture of new numbers. Thus, π, e, and the great majority of other real numbers, which can be defined only by means of limiting processes, lay beyond the pale for Kronecker. He claimed that it is nonsense to talk about numbers no one could compute. Of course, he had a point. The computation of π continues to occupy computers and their programmers, with millions of decimal places now calculated. But it gets one nowhere nearer π in a sense. Providing only better and better rational approximations is little more than a curiosity (though with some spin-off uses in testing computers).

In the early twentieth century, Kronecker's ideas were taken up again, this time by the so-called intuitionists led by Luitzen E. J. Brouwer (1881–1966). The intuitionists insisted that a mathematical object exists only if it can be constructed in a finite number of steps. This limits mathematics to the kind of things that can be done on a digital computer. So the dissidents of the early twentieth century have in some sense been vindicated by subsequent events.

ANNOTATED BIBLIOGRAPHY

Birkhoff, Garrett, and T. C. Bartee. *Modern Applied Algebra* (New York: McGraw-Hill), 1970.

> *One of the earliest and best surveys of applications of algebra to modern problems of engineering and technology.*

Birkhoff, Garrett, and Saunders MacLane. *A Survey of Modern Algebra*, 3rd ed. (New York: Macmillan), 1965.

> *This is the book that introduced a whole generation of mathematicians to abstract algebra. First published in 1941, it had enormous impact in bringing to the United States ideas then prevalent in Europe and in introducing these ideas at the undergraduate level. Further, the text is the result of a collaboration of two U.S. mathematicians of international reputation.*

Fraleigh, John B. *A First Course in Abstract Algebra*, 3rd ed. (Reading, Mass.: Addison-Wesley), 1982.

> *A popular text in algebra that has many nice problems and takes up topics of some difficulty, albeit often in the direction of applications in algebraic topology.*

Gardiner, A. *Infinite Processes: Background to Analysis* (New York: Springer-Verlag), 1982.

> *An interesting book with a variety of pre-analysis topics, and an exceptionally good treatment of number systems.*

Hardy, G. H., and E. M. Wright, *An Introduction to the Theory of Numbers*, 5th ed. (Oxford: Clarendon Press), 1979.

> *The classic book on the theory of numbers, concerned primarily with the integers, but with some sections on irrational numbers as well.*

Herstein, I. N. *Abstract Algebra* (New York: Macmillan), 1986.

> *An elementary text in algebra by a first-class algebraist.*

Hillman, A. P., and G. L. Alexanderson. *A First Undergraduate Course in Abstract Algebra*, 4th ed. (Belmont, Ca.: Wadsworth), 1988.

> *A widely used text in abstract algebra treating groups, rings, and fields in some depth.*

Kleiner, Israel. "The Evolution of Group Theory. A Brief Survey," *Mathematics Magazine* 59 (1986), 195–215.

A fresh look at the history of group theory. The article won the Allendoerfer Prize for expository writing in 1987.

McCoy, Neal H. *Rings and Ideals* (Carus Monograph No. 8) (Washington, D.C.: The Mathematical Association of America), 1948.

Now somewhat old, this nevertheless remains an interesting and valuable treatment of the subject by a well-known ring theorist.

Niven, Ivan. *Irrational Numbers* (Carus Monograph No. 11) (Washington, D.C.: The Mathematical Association of America), 1956.

A beautiful treatment of some of the classic problems of deciding whether certain numbers are irrational and, if so, whether they are algebraic or transcendental. Somewhat mathematically demanding, but rewarding nonetheless.

4

TOPOLOGY OF THE REAL NUMBER SYSTEM

As we mentioned in Section 3 of Chapter 3, all of the properties of the real number system $\mathbb{R}$ follow from the assumption that it forms a complete ordered field. We also observed that $\mathbb{R}$ can be thought of as a number line; that is, we can visualize a one-to-one correspondence between $\mathbb{R}$ and the points on an infinite straight line. Intuitively, we picture a line as a uniform whole, containing no vacant spaces or gaps. In English, such an entity is called a *continuum*. Thus, to emphasize those features of the real numbers that result from viewing them as points on a line, $\mathbb{R}$ is often referred to as a **linear continuum**.

In this chapter, we study certain properties of $\mathbb{R}$, called *topological* properties, which are closely related to its structure as a linear continuum. A **topological** property of $\mathbb{R}$ is, by definition, a property that can be formulated purely in terms of set theory and the notion of an open set.

1

OPEN AND CLOSED SETS

If $a, b \in \mathbb{R}$ and $a < b$, we use the usual notation[†] (a, b) for the set of all real numbers strictly between a and b:

$$(a, b) = \{x \in \mathbb{R} \mid a < x < b\}.$$

Note that

$$a = \inf((a, b)) \qquad \text{and} \qquad b = \sup((a, b))$$

[†] Unfortunately, this is the same notation used for ordered pairs—however, it is always possible to tell from the context what is intended.

(Problem 1). In particular, (a, b) has both a lower bound and an upper bound. We refer to a subset of $\mathbb{R}$ having the form (a, b) as a **bounded open interval**. Similarly, the **bounded closed interval** $[a, b]$ is defined for $a < b$ by

$$[a, b] = \{x \in \mathbb{R} \mid a \leq x \leq b\},$$

so that

$$[a, b] = (a, b) \cup \{a, b\}.$$

Again we have

$$a = \inf([a, b]) \qquad \text{and} \qquad b = \sup([a, b])$$

(Problem 2).

1.1 DEFINITION

> **NEIGHBORHOOD**   Let $c \in \mathbb{R}$. By a **neighborhood** of c, we mean a bounded open interval of the form
>
> $$(c - \varepsilon, c + \varepsilon)$$
>
> with $\varepsilon \in \mathbb{R}$, $\varepsilon > 0$. We say that c is the **center** of the neighborhood $(c - \varepsilon, c + \varepsilon)$, and we refer to ε as its **radius**.

The idea of a neighborhood allows us to formulate the notion of the *boundary* of a set. Intuitively, the *boundary* of a set lies between the set and its complement. The following definition makes this intuitive notion more precise.

1.2 DEFINITION

> **BOUNDARY POINT**   Let $M \subseteq \mathbb{R}$ and let $b \in \mathbb{R}$. We say that b is a **boundary point** of M if every neighborhood of b has a nonempty intersection with both M and its complement $\mathbb{R} \setminus M$. The set of all boundary points of M is called the **boundary** of M and is denoted by $bd(M)$.

For instance, if $a, b \in \mathbb{R}$ with $a < b$, then the boundary of the open interval (a, b) is the set $\{a, b\}$ (Problem 4a). Likewise, the boundary of the closed interval $[a, b]$ is $\{a, b\}$ (Problem 4b).

The next lemma, which further confirms our intuitive notion of a boundary, is an immediate consequence of the fact that the complement of the complement of a set is the set itself. In Problem 3, we ask you to supply the simple proof.

1.3 LEMMA	**BOUNDARIES AND COMPLEMENTS** If $M \subseteq \mathbb{R}$, then $$bd(M) = bd(\mathbb{R} \setminus M).$$

Note that an open interval (a, b) contains neither of its boundary points, whereas a closed interval $[a, b]$ contains both of its boundary points. This observation suggests the following definition.

1.4 DEFINITION	**OPEN AND CLOSED SETS** Let $U, F \subseteq \mathbb{R}$. (i) If $U \cap bd(U) = \varnothing$, we say that U is an **open** set. (ii) If $bd(F) \subseteq F$, we say that F is a **closed** set.

Observe that, in mathematics, the word *closed* is used in two different senses: the *algebraic sense* as in the statement that the rational numbers are closed under addition, and the *topological sense* as in Part (ii) of Definition 1.4. However, it is usually easy to tell from the context which sense is intended.

The next theorem establishes a fundamental connection between open and closed sets.

1.5 THEOREM	**OPEN SETS, CLOSED SETS, AND THEIR COMPLEMENTS** A subset of $\mathbb{R}$ is open if and only if its complement is closed.

PROOF   Let $U \subseteq \mathbb{R}$, let $F = \mathbb{R} \setminus U$, and let $B = bd(U)$. By Lemma 1.3, $B = bd(F)$. By elementary set algebra,

$$U \cap B = \varnothing \Leftrightarrow B \subseteq F,$$

that is,

$$U \text{ is open} \Leftrightarrow F \text{ is closed.} \qquad \square$$

Some authors use the characterization of open sets given in the following theorem as their definition of an open set.

1.6 THEOREM	**CHARACTERIZATION OF OPEN SETS** A set $U \subseteq \mathbb{R}$ is open if and only if, for every point $u \in U$, there exists a neighborhood N of u such that $N \subseteq U$.

PROOF   Suppose that U is open and that $u \in U$. Then $u \notin bd(U)$, so there exists a neighborhood N of u such that

$$N \cap U \neq \varnothing \wedge N \cap (\mathbb{R} \setminus U) \neq \varnothing \quad \text{is false.}$$

Therefore, by De Morgan's law,

$$N \cap U = \varnothing \vee N \cap (\mathbb{R} \setminus U) = \varnothing.$$

Since $u \in N \cap U$, we cannot have $N \cap U = \varnothing$, and it follows that

$$N \cap (\mathbb{R} \setminus U) = \varnothing.$$

But, by elementary set algebra, the last condition implies that

$$N \subseteq U.$$

Conversely, assume that, for every point $u \in U$, there exists a neighborhood N of u such that $N \subseteq U$. Let $b \in bd(U)$. We have to show that $b \notin U$. Suppose, on the contrary, that $b \in U$. Then, by our hypothesis, there exists a neighborhood N of b such that $N \subseteq U$. However, since $b \in bd(U)$, we must have

$$N \cap (\mathbb{R} \setminus U) \neq \varnothing,$$

which (again by elementary set algebra) contradicts $N \subseteq U$. □

Our definition of an open set was originally suggested by the idea of an open interval, so it should come as no surprise that an open interval is an open set. Nevertheless, it is necessary to prove that this is so, and we do this in the next lemma. In the proof, we use the notation

$$\min(p, q)$$

for the *minimum* of two real numbers p, q; that is, $\min(p, q)$ denotes the smaller of the two numbers p, q (or their common value if they are equal). Note that

$$\min(p, q) = \text{GLB}(\{p, q\}) = \inf(\{p, q\})$$

(Problem 5a).

1.7 LEMMA | If $a, b \in \mathbb{R}$ with $a < b$, then (a, b) is an open set.

PROOF   Let $c \in (a, b)$, so that

$$a < c < b.$$

By Theorem 1.6, it is sufficient to prove that there exists a neighborhood $N = (c - \varepsilon, c + \varepsilon)$ of c such that $N \subseteq (a, b)$. Let

$$\varepsilon = \min(b - c, c - a),$$

noting that, since $b - c$ and $c - a$ are positive, so is ε. Since $\varepsilon \leq c - a$, it follows that

$$a \leq c - \varepsilon$$

and, since $\varepsilon \leq b - c$, we also have

$$c + \varepsilon \leq b.$$

To prove that $(c - \varepsilon, c + \varepsilon) \subseteq (a, b)$, suppose that

$$x \in (c - \varepsilon, c + \varepsilon).$$

Then

$$a \leq c - \varepsilon < x < c + \varepsilon \leq b,$$

and it follows that $x \in (a, b)$. $\qquad \square$

1.8 THEOREM

> **THE UNION OF OPEN SETS**   The union of a family of open subsets of $\mathbb{R}$ is an open set.

PROOF   Let $(V_j)_{j \in J}$ be a family of open sets and let

$$U = \bigcup_{j \in J} V_j.$$

Suppose that $u \in U$. By Theorem 1.6, it will be sufficient to prove that there exists a neighborhood N of u such that $N \subseteq U$. Because $u \in U$, there exists $i \in J$ such that

$$u \in V_i.$$

Since V_i is open and $u \in V_i$, there exists (by Theorem 1.6) a neighborhood N of u such that

$$N \subseteq V_i.$$

But

$$V_i \subseteq \bigcup_{j \in J} V_j = U,$$

and it follows that $N \subseteq U$. $\qquad \square$

1.9 LEMMA

> Let $c \in \mathbb{R}$ and let J and K be neighborhoods of c. Then $J \cap K$ is a neighborhood of c.

PROOF   We outline the proof and leave the details as an exercise (Problem 7). Let α be the radius of J and let β be the radius of K. Then

$$J = (c - \alpha, c + \alpha) \qquad \text{and} \qquad K = (c - \beta, c + \beta).$$

Let $\varepsilon = \min(\alpha, \beta)$. Then

$$J \cap K = (c - \varepsilon, c + \varepsilon).$$

□

1.10 COROLLARY

Let U and V be open sets. Then $U \cap V$ is an open set.

PROOF   Let $c \in U \cap V$. By Theorem 1.6, it will be sufficient to prove that there exists a neighborhood N of c such that $N \subseteq U \cap V$. Because $c \in U$ and U is open, there exists a neighborhood J of c such that $J \subseteq U$. Likewise, because $c \in V$ and V is open, there exists a neighborhood K of c such that $K \subseteq V$. Let $N = J \cap K$. By Lemma 1.9, N is a neighborhood of c, and we have

$$N = J \cap K \subseteq U \cap V.$$

□

1.11 THEOREM

THE INTERSECTION OF OPEN SETS   The intersection of a finite family of open subsets of $\mathbb{R}$ is an open set.

PROOF   Let $V_1, V_2, V_3, \ldots, V_n$ be open sets. We use induction on n to prove that

$$V_1 \cap V_2 \cap V_3 \cap \cdots \cap V_n$$

is an open set. This is clear for $n = 1$, so we assume as our induction hypothesis that it is true for an arbitrary but fixed value of n. Now let

$$V_1, V_2, V_3, \ldots, V_n, V_{n+1}$$

be open sets and let

$$U = V_1 \cap V_2 \cap V_3 \cap \cdots \cap V_n.$$

By the induction hypothesis, U is an open set. Therefore, by Lemma 1.10, $U \cap V_{n+1}$ is an open set, and it follows that

$$V_1 \cap V_2 \cap V_3 \cap \cdots \cap V_n \cap V_{n+1} = U \cap V_{n+1}$$

is an open set.

□

Theorems 1.8 and 1.11 are often paraphrased as follows:

> *The collection of all open subsets of $\mathbb{R}$ is closed under arbitrary unions and finite intersections.*

As a direct consequence of Definition 1.2, the empty set has an empty boundary,

$$bd(\varnothing) = \varnothing.$$

Therefore, by Definition 1.4, $\varnothing$ is both open and closed. Because the complement of $\varnothing$ is $\mathbb{R}$, it follows from Theorem 1.5 that $\mathbb{R}$ is also both open and closed. To summarize:

> *The empty set and the set $\mathbb{R}$ of all real numbers are sets that are both open and closed.*

By using Theorem 1.5 and applying the generalized De Morgan laws to the results of Theorems 1.8 and 1.11, we obtain the following theorem.

1.12 THEOREM

THE UNION AND INTERSECTION OF CLOSED SETS

(i) The intersection of a family of closed subsets of $\mathbb{R}$ is a closed set.
(ii) The union of a finite family of closed subsets of $\mathbb{R}$ is a closed set.

PROOF   We prove Part (i) here and leave the analogous proof of Part (ii) as an exercise (Problem 11). For simplicity, we denote the complement of a subset M of $\mathbb{R}$ by $M' = \mathbb{R} \setminus M$. Let $(F_j)_{j \in J}$ be a family of closed subsets of $\mathbb{R}$ and let

$$F = \bigcap_{j \in J} F_j.$$

To prove that F is closed, it is sufficient to prove that F' is open (Theorem 1.5). But, by the generalized De Morgan law [Part (ii) of Theorem 6.13 in Chapter 1],

$$F' = \left(\bigcap_{j \in J} F_j \right)' = \bigcup_{j \in J} (F_j)'.$$

However, by Theorem 1.5, the set $(F_j)'$ is open for every $j \in J$, and it follows from Theorem 1.8 that F' is open.   □

If $a \in \mathbb{R}$, we use the usual notation[†] (a, ∞) and $(-\infty, a)$ for the intervals

$$(a, \infty) = \{x \in \mathbb{R} \,|\, a < x\} \qquad \text{and} \qquad (-\infty, a) = \{x \in \mathbb{R} \,|\, x < a\}.$$

Both (a, ∞) and $(-\infty, a)$ can be written as unions of bounded open intervals,

$$(a, \infty) = \bigcup_{x \in I}(a, x), \qquad I = (a, \infty),$$

and

$$(-\infty, a) = \bigcup_{x \in J}(x, a), \qquad J = (-\infty, a),$$

(Problem 13), and it follows from Lemma 1.7 and Theorem 1.8 that (a, ∞) and $(-\infty, a)$ are open sets. We also use the standard notation $[a, \infty)$ and $(-\infty, a]$ for the intervals

$$[a, \infty) = \{x \in \mathbb{R} \,|\, a \le x\} \qquad \text{and} \qquad (-\infty, a] = \{x \in \mathbb{R} \,|\, x \le a\}.$$

We note that

$$[a, \infty) = \mathbb{R} \setminus (-\infty, a) \qquad \text{and} \qquad (-\infty, a] = \mathbb{R} \setminus (a, \infty),$$

so, in view of the fact that $(-\infty, a)$ and (a, ∞) are open sets, it follows from Theorem 1.5 that $[a, \infty)$ and $(-\infty, a]$ are closed sets.

Note that

$$\{a\} = (-\infty, a] \cap [a, \infty)$$

and, since the intersection of finitely many closed sets is a closed set [Part (ii) of Theorem 1.12], it follows that $\{a\}$ is a closed set. In other words:

> *Singleton subsets of $\mathbb{R}$ are closed sets.*

Therefore, since any finite set M is the union of finitely many singleton sets (one for each point in M), we can conclude that:

> *Finite subsets of $\mathbb{R}$ are closed sets.*

Although a singleton set $\{a\} \subseteq \mathbb{R}$ is closed, it is certainly not open. Indeed, if $\{a\}$ were open, there would be a positive number ε such that

$$(a - \varepsilon, a + \varepsilon) \subseteq \{a\}$$

which is impossible. $\left[\text{For instance,} \, a + \dfrac{\varepsilon}{2} \in (a - \varepsilon, a + \varepsilon), \text{ but } a + \dfrac{\varepsilon}{2} \notin \{a\}. \right]$

[†] The symbol ∞ does not represent a real number. It is used only for convenience in writing unbounded intervals.

However, note that

$$\{a\} = \bigcap_{\varepsilon > 0}(a - \varepsilon, a + \varepsilon)$$

(Problem 15), which shows that the intersection of infinitely many open sets need not be open. Similarly, the union of infinitely many closed sets need not be closed (Problem 16).

1.13 DEFINITION

> **ACCUMULATION POINT**   Let $M \subseteq \mathbb{R}$, and let $a \in \mathbb{R}$. We say that a is an **accumulation point** of the set M if, for every neighborhood N of a,
>
> $(N \setminus \{a\}) \cap M \neq \emptyset.$

Definition 1.13 can be paraphrased as follows: a is an accumulation point of M provided that every neighborhood of a contains a point of M *other than a itself*. Intuitively, the elements of a set M are "piling up" near such an accumulation point. (Some authors refer to an accumulation point of a set as a *limit point* of the set.) Note that an accumulation point of a set may or may not belong to the set. For instance, the endpoints a and b of a half-open interval,

$$[a, b) = \{x \in \mathbb{R} \mid a \leq x < b\},$$

are accumulation points of the interval, $a \in [a, b)$, but $b \notin [a, b)$.

1.14 LEMMA

> If $V, M \subseteq \mathbb{R}$, $V \cap M = \emptyset$, and V is open, then no accumulation point of M belongs to V.

PROOF   Suppose, on the contrary, that a is an accumulation point of M and $a \in V$. By Theorem 1.6, there exists a neighborhood N of a such that

$$N \subseteq V.$$

By Definition 1.13,

$$(N \setminus \{a\}) \cap M \neq \emptyset,$$

from which it follows that

$$N \cap M \neq \emptyset.$$

But, since $N \subseteq V$ and $V \cap M = \emptyset$, we also have

$$N \cap M = \emptyset,$$

contradicting $N \cap M \neq \emptyset$.   □

1.15 THEOREM | **ACCUMULATION POINTS AND CLOSED SETS**   Let $M \subseteq \mathbb{R}$ and let A be the set of all accumulation points of M. Then M is closed if and only if $A \subseteq M$.

PROOF   Let

$$V = \mathbb{R} \setminus M, \tag{1}$$

noting that, by Theorem 1.5,

$$M \text{ is closed} \Leftrightarrow V \text{ is open}. \tag{2}$$

As a consequence of (1), we have

$$V \cap M = \varnothing. \tag{3}$$

Also, by elementary set algebra,

$$A \subseteq M \Leftrightarrow V \cap A = \varnothing. \tag{4}$$

If M is closed, then V is open by (2), so by (3) and Lemma 1.14, $V \cap A = \varnothing$; hence, by (4), $A \subseteq M$.

Conversely, suppose that

$$A \subseteq M. \tag{5}$$

On the basis of (5), we propose to prove that V is open. Thus, let

$$v \in V, \tag{6}$$

and note that, by Theorem 1.6, it is sufficient to show that there exists a neighborhood N of v such that $N \subseteq V$. From (3) and (6), we have

$$v \notin M. \tag{7}$$

By combining (5) and (7), we find that

$$v \notin A. \tag{8}$$

Therefore, by Definition 1.13, there exists a neighborhood N of v such that

$$(N \setminus \{v\}) \cap M = \varnothing. \tag{9}$$

Because of (7), condition (9) implies that

$$N \cap M = \varnothing. \tag{10}$$

By combining (1) and (10), we have

$$N \subseteq V, \tag{11}$$

which shows that V is an open set. Therefore, by (2), M is a closed set.   □

Theorem 1.15 suggests the following definition:

1.16 DEFINITION

CLOSURE OF A SET   Let $M \subseteq \mathbb{R}$ and let A be the set of all accumulation points of M. We define the **closure** of M, in symbols $\bar{M}$, by

$$\bar{M} = M \cup A.$$

In other words, the closure of a set is obtained by appending to the set all of its own accumulation points.

1.17 THEOREM

PROPERTIES OF CLOSURE   Let $M, F, K \subseteq \mathbb{R}$. Then:

(i) $M \subseteq \bar{M}$ and $\bar{M}$ is a closed set.
(ii) M is closed if and only if $M = \bar{M}$.
(iii) $M \subseteq K \Rightarrow \bar{M} \subseteq \bar{K}$.
(iv) If F is closed and $M \subseteq F$, then $\bar{M} \subseteq F$.
(v) $\overline{M \cup K} = \bar{M} \cup \bar{K}$.

PROOF   We leave the proofs of Parts (i) and (ii) as an exercise (Problem 22). Part (iii) follows from the obvious fact that, if $M \subseteq K$, then every accumulation point of M is an accumulation point of K.

To prove Part (iv), assume that F is closed and that $M \subseteq F$. Then, by Part (ii), $F = \bar{F}$; hence, by Part (iii),

$$\bar{M} \subseteq \bar{F} = F.$$

To prove Part (v), note that $M, K \subseteq M \cup K$; hence, by Part (iii),

$$\bar{M}, \bar{K} \subseteq \overline{M \cup K},$$

from which it follows that

$$\bar{M} \cup \bar{K} \subseteq \overline{M \cup K}.$$

To prove the opposite inclusion, let

$$F = \bar{M} \cup \bar{K}.$$

By Part (i), both $\bar{M}$ and $\bar{K}$ are closed sets; hence, by Part (ii) of Theorem 1.12, F is a closed set. Because

$$M \subseteq \bar{M} \quad \text{and} \quad K \subseteq \bar{K},$$

we have

$$M \cup K \subseteq F.$$

Therefore, by Part (iv),

$$\overline{M \cup K} \subseteq F = \bar{M} \cup \bar{K},$$

and the proof is complete.   □

1.18 DEFINITION

> **ISOLATED POINT**   Let $M \subseteq \mathbb{R}$. A point $p \in M$ is called an **isolated point** of M if it is not an accumulation point of M.

A subset D of $\mathbb{R}$ such that every point of D is an isolated point is said to be **discrete**. (This accounts for the terminology in Theorems 4.7 and 5.4 of Chapter 3—see Problem 27.) We leave the proof of the following lemma as an exercise (Problem 25).

1.19 LEMMA

> If $M \subseteq \mathbb{R}$ and $p \in M$, then p is an isolated point of M if and only if there is a neighborhood N of p such that
>
> $$N \cap M = \{p\}.$$

1.20 DEFINITION

> **INTERIOR POINT AND INTERIOR OF A SET**   Let $M \subseteq \mathbb{R}$. A point $p \in M$ is called an **interior point** of M if there is a neighborhood N of p such that
>
> $$N \subseteq M.$$
>
> The set of all interior points of M is called the **interior** of M and is denoted by M^o.

The following lemma provides a connection between interiors and closures. Again, we leave the proof as an exercise (Problem 29).

1.21 LEMMA

> If $M \subseteq \mathbb{R}$, then
>
> $$M^o = \mathbb{R} \setminus (\overline{\mathbb{R} \setminus M}).$$

PROBLEM SET 4.1

1. If $a, b \in \mathbb{R}$ with $a < b$, prove that $a = \inf((a, b))$ and $b = \sup((a, b))$.
2. If $a, b \in \mathbb{R}$ with $a < b$, prove that $a = \inf([a, b])$ and $b = \sup([a, b])$.
3. Prove Lemma 1.3.

4. Let $a, b \in \mathbb{R}$ with $a < b$.
   (a) Prove that $bd((a, b)) = \{a, b\}$.
   (b) Prove that $bd([a, b]) = \{a, b\}$.

5. Let $p, q \in \mathbb{R}$.
   (a) Prove that $\min(p, q) = \inf(\{p, q\})$.
   (b) If $p, q > 0$, prove that $\inf(\{p, q\}) > 0$.

6. Let M be a finite nonempty subset of $\mathbb{R}$.
   (a) Prove that M has a supremum and that $\sup(M) \in M$.
   (b) Prove that M has an infimum and that $\inf(M) \in M$. [Hint: Use mathematical induction on $\#M$.]

7. Fill in the details in the proof of Lemma 1.9.

8. Let $M \subseteq \mathbb{R}$, $b \in \mathbb{R}$. Prove that b is a boundary point of M if and only if, for every open set U, $b \in U$ implies that $U \cap M \neq \emptyset$ and that $U \cap (\mathbb{R} \setminus M) \neq \emptyset$.

9. Prove that a subset of $\mathbb{R}$ is open if and only if it is a union of bounded open intervals.

10. True or false: If $P \subseteq Q \subseteq \mathbb{R}$, then $bd(P) \subseteq bd(Q)$. If true, prove it; if false, give a counterexample.

11. Prove Part (ii) of Theorem 1.12.

12. If $a \in \mathbb{R}$, prove that

$$\{a\} = \bigcap_{n \in \mathbb{N}} \left(a - \frac{1}{n}, a + \frac{1}{n} \right).$$

[Hint: Use the Archimedean property of $\mathbb{R}$.]

13. Let $a \in \mathbb{R}$, $I = (a, \infty)$, and $J = (-\infty, a)$.
    (a) Prove that $I = \bigcup_{x \in I}(a, x)$.
    (b) Prove that $J = \bigcup_{x \in J}(x, a)$.

14. If $a \in \mathbb{R}$, prove that $\{a\}$ is closed by making a direct proof (using Theorem 1.6) that $\mathbb{R} \setminus \{a\}$ is an open set.

15. If $a \in \mathbb{R}$, prove that $\{a\} = \bigcap_{\varepsilon > 0}(a - \varepsilon, a + \varepsilon)$.

16. Give an example to show that the union of infinitely many closed sets need not be closed.

17. If $a, b \in \mathbb{R}$ with $a < b$, prove that the closed bounded interval $[a, b]$ is a closed set. [Hint: $\mathbb{R} \setminus [a, b] = (-\infty, a) \cup (b, \infty)$.]

18. Let $M \subseteq \mathbb{R}$ and let $a \in \mathbb{R}$. Prove that a is an accumulation point of M if and only if, for every open set U, $a \in U$ implies that $(U \setminus \{a\}) \cap M \neq \emptyset$.

19. (a) If $M \subseteq \mathbb{R}$, $b = \sup(M)$, and $b \notin M$, prove that b is an accumulation point of M.
    (b) State and prove a similar result for the infimum.

20. If $M \subseteq \mathbb{R}$ and A is the set of all accumulation points of M, prove that A is a closed set.

21. Let $M \subseteq \mathbb{R}$ be a nonempty closed set and suppose that M has both an upper bound and a lower bound, so that $a = \inf(M)$ and $b = \sup(M)$ exist by the completeness of $\mathbb{R}$.
   (a) Prove that $a, b \in M$. [Hint: Use the results of Problem 19 and the fact that M is closed.]
   (b) Prove that a is the smallest element in M and that b is the largest element in M.

22. (a) Prove Part (i) of Theorem 1.17.
   (b) Prove Part (ii) of Theorem 1.17. [Hint: Use Definition 1.16 and Theorem 1.15.]

23. (a) If $M \subseteq \mathbb{R}$, prove that $bd(M) = \bar{M} \cap (\overline{\mathbb{R} \setminus M})$.
   (b) If $M \subseteq \mathbb{R}$, prove that $bd(M)$ is a closed set.

24. If $M \subseteq \mathbb{R}$, prove that $bd(M) = \bar{M} \setminus M^o$.

25. Prove Lemma 1.19.

26. Let $M \subseteq \mathbb{R}$, let A be the set of all accumulation points of M, and let B be the set of all accumulation points of A. Prove that $B \subseteq A$.

27. Prove that $\mathbb{N}$ and $\mathbb{Z}$ are discrete subsets of $\mathbb{R}$.

28. Prove that $\{1/n \,|\, n \in \mathbb{N}\}$ is a discrete subset of $\mathbb{R}$.

29. Prove Lemma 1.21.

30. True or false: If P and Q are discrete subsets of $\mathbb{R}$, then $P \cup Q$ is a discrete subset of $\mathbb{R}$. If true, prove it; if false, give a counterexample.

31. Let $M \subseteq \mathbb{R}$.
   (a) Prove that $\bar{M}$ is the intersection of all closed subsets of $\mathbb{R}$ that contain M.
   (b) Prove that M^o is the union of all open subsets of $\mathbb{R}$ that are contained in M.

32. (a) If $P, Q \subseteq \mathbb{R}$, show that $\overline{P \cap Q} \subseteq \bar{P} \cap \bar{Q}$.
   (b) Give a counterexample to show that the inclusion in Part (a) cannot be strengthened to an equality.

33. If $M \subseteq \mathbb{R}$, prove that $bd(bd(M)) \subseteq bd(M)$.

34. Let $F \subseteq \mathbb{R}$ be closed, and let $U \subseteq \mathbb{R}$ be open. Prove:
   (a) $\overline{F^o} \subseteq F$ (b) $U \subseteq (\bar{U})^o$
   (c) If $U = F^o$, then $U = (\bar{U})^o$.   (d) If $F = \bar{U}$, then $\overline{F^o} = F$.

35. Let $P, Q \subseteq \mathbb{R}$. Prove:
   (a) $bd(P \cup Q) \subseteq bd(P) \cup bd(Q)$
   (b) $bd(P \cup Q) \cap P \subseteq bd(P)$

36. A set $U \subseteq \mathbb{R}$ is said to be a **regular open** set if $U = (\bar{U})^o$.
   (a) Show that every open interval of the form (a, b), (a, ∞), or $(-\infty, a)$ is a regular open set.
   (b) Give an example of a set $U \subseteq \mathbb{R}$ that is open but not regular open.

(c) If U is an open set, prove that U is a regular open set if and only if $bd(U) \subseteq bd(\bar{U})$.

37. True or false: If $P, Q \subseteq \mathbb{R}$ and a is an accumulation point of $P \cup Q$, then a is an accumulation point of at least one of the sets P, Q. If true, prove it; if false, give a counterexample.

38. Let $M \subseteq \mathbb{R}$, and let $U = (\bar{M})^\circ$.
(a) Prove that U is a regular open set (see Problem 36).
(b) If V is a regular open set and $M \subseteq V$, then $U \subseteq V$; that is, with respect to set inclusion, U is the "smallest" regular open set containing M.

39. An idea or property is said to be **topological** if it can be expressed, with the aid of set theory, purely in terms of open sets. Which of the following are topological, and why?
(a) The idea of a neighborhood of a point
(b) The idea of a boundary point of a set
(c) The property of a set being closed
(d) The idea of an accumulation point
(e) The idea of the closure of a set
(f) The idea of the interior of a set
(g) The idea of an isolated point of a set
(h) The property of a set being discrete

40. (a) Prove that the intersection of two regular open sets is a regular open set. (See Problems 36 and 38.)
(b) Give an example to show that the union of regular open sets need not be regular open.
(c) If U and V are regular open sets and if $U \cap V \neq \varnothing$, prove that $U \cup V$ is a regular open set.

2

LIMITS

By a **real-valued function of a real variable** we mean a function whose domain and range are nonempty subsets of $\mathbb{R}$. Calculus may be defined as the study of such functions with the aid of the notion of *limit*. In elementary calculus, however, much of this study is conducted on an intuitive, nonrigorous basis. In the branch of mathematics known as **mathematical analysis**, the ideas of elementary calculus are reexamined with the care and precision that they deserve—in short, mathematical analysis begins with *careful calculus*. We launch our study of mathematical analysis by carefully defining the concept of a limit of a real-valued function of a real variable. To avoid repetition in what follows, we refer to a real-valued function of a real variable simply as a **function**.

2.1 DEFINITION

> **LIMIT OF A FUNCTION** Let f be a function with domain D, and let $a, L \in \mathbb{R}$. We say that L is a **limit of** f **at** a if the following two conditions hold:
>
> (i) a is an accumulation point of D.
> (ii) For every $\varepsilon \in \mathbb{R}$ with $\varepsilon > 0$, there exists $\delta \in \mathbb{R}$ with $\delta > 0$ such that, for every $x \in D$,
> $$0 < |x - a| < \delta \Rightarrow |f(x) - L| < \varepsilon.$$

Definition 2.1 is essentially the same as the definition given in most elementary calculus textbooks, the only real difference being the requirement that a be an accumulation point of the domain of f. We begin our study of limits by recasting this definition in the language of neighborhoods.

2.2 LEMMA

> Let $a, b, x, y, L \in \mathbb{R}$, and let $\varepsilon, \delta \in \mathbb{R}$ with $\varepsilon, \delta > 0$. Then:
>
> (i) $|b| < \varepsilon \Leftrightarrow -\varepsilon < b < \varepsilon$
> (ii) $|y - L| < \varepsilon \Leftrightarrow y \in (L - \varepsilon, L + \varepsilon)$
> (iii) $0 < |x - a| < \delta \Leftrightarrow x \in (a - \delta, a + \delta) \setminus \{a\}$

PROOF

(i) The condition $|b| < \varepsilon$ is equivalent to the requirement that both $b < \varepsilon$ and $-b < \varepsilon$ hold; that is, $b < \varepsilon$ and $b > -\varepsilon$ (see Problem 4). Therefore, $|b| < \varepsilon$ is equivalent to $-\varepsilon < b < \varepsilon$.

(ii) By replacing b in Part (i) by $y - L$, we find that
$$|y - L| < \varepsilon \Leftrightarrow -\varepsilon < y - L < \varepsilon.$$

Evidently,
$$-\varepsilon < y - L < \varepsilon \Leftrightarrow L - \varepsilon < y < L + \varepsilon.$$

(To go from left to right, add L to each member; to go from right to left, subtract L from each member.) Statement (ii) now follows from the transitivity of logical equivalence.

(iii) By replacing y by x, L by a, and ε by δ in (ii), we find that
$$|x - a| < \delta \Leftrightarrow x \in (a - \delta, a + \delta).$$

Statement (iii) now follows from the observation that
$$0 < |x - a| \Leftrightarrow x \neq a. \qquad \square$$

By using Parts (ii) and (iii) of Lemma 2.2 to rewrite the conclusion and hypothesis of the implication in Definition 2.1, denoting the neighborhood

$(L - \varepsilon, L + \varepsilon)$ of L by N, and denoting the neighborhood $(a - \delta, a + \delta)$ of a by J, we obtain the following corollary:

2.3 COROLLARY

Let f be a function with domain D, and let $a, L \in \mathbb{R}$. Then L is a limit of f at a if and only if the following two conditions hold:

(i) a is an accumulation point of D.
(ii) For every neighborhood N of L there exists a neighborhood J of a such that, for every $x \in \mathbb{R}$,

$$x \in (J \setminus \{a\}) \cap D \Rightarrow f(x) \in N.$$

Corollary 2.3 makes it plain why it is desirable to require that a be an accumulation point of D. Indeed, without this requirement, $(J \setminus \{a\}) \cap D$ might be empty, in which case the hypothesis of the implication in Part (ii) would be false, and the implication would be true automatically.

The next lemma will be used to help prove that the number L in Definition 2.1 and Corollary 2.3 is uniquely determined.

2.4 LEMMA

Let $p, q \in \mathbb{R}$ with $p \neq q$. Then there exist neighborhoods N and K of p and q, respectively, such that

$$N \cap K = \varnothing.$$

PROOF   Since $p \neq q$, we have $|p - q| > 0$. Let

$$\varepsilon = \tfrac{1}{2}|p - q|,$$

noting that $\varepsilon > 0$ and $2\varepsilon = |p - q|$. Let

$$N = (p - \varepsilon, p + \varepsilon),$$

and let

$$K = (q - \varepsilon, q + \varepsilon).$$

We claim that $N \cap K = \varnothing$. Suppose not. Then there exists a real number x with

$$x \in N \cap K.$$

By Part (ii) of Lemma 2.2, we have

$$|x - p| < \varepsilon \qquad \text{and} \qquad |x - q| < \varepsilon.$$

Since $p - x = -(x - p)$, it follows that $|p - x| = |x - p|$, and so we have

$$|p - x| < \varepsilon \qquad \text{and} \qquad |x - q| < \varepsilon.$$

By adding these inequalities, we find that

$$|p - x| + |x - q| < \varepsilon + \varepsilon = 2\varepsilon = |p - q|.$$

Therefore, by the triangle inequality (Theorem 3.18 in Chapter 3),

$$|p - q| = |(p - x) + (x - q)| \le |p - x| + |x - q| < |p - q|,$$

which is a contradiction since $|p - q|$ cannot be strictly less than itself.
$\square$

It is customary to say that the neighborhoods N of p and K of q **separate** the points p and q if, as in Lemma 2.4, $N \cap K = \varnothing$. That distinct points can be separated by neighborhoods, or more generally by open sets, is known as the **Hausdorff property** [in honor of the German mathematician, Felix Hausdorff (1868–1942)]. It is this property that guarantees the uniqueness of limits.

2.5 THEOREM

> **UNIQUENESS OF LIMITS**   Let f be a function, let $a \in \mathbb{R}$, and suppose that L and L' are limits of f at a. Then $L = L'$.

PROOF   Suppose $L \ne L'$. By Lemma 2.4, there exist neighborhoods N and K of L and L', respectively, such that

$$N \cap K = \varnothing. \tag{1}$$

Let $D = \text{domain}(f)$; by Corollary 2.3, there exists a neighborhood J_1 of a such that, for all $x \in \mathbb{R}$,

$$x \in (J_1 \setminus \{a\}) \cap D \Rightarrow f(x) \in N. \tag{2}$$

Similarly, by Corollary 2.3, with L replaced by L', there exists a neighborhood J_2 of a such that, for all $x \in \mathbb{R}$,

$$x \in (J_2 \setminus \{a\}) \cap D \Rightarrow f(x) \in K. \tag{3}$$

Now let

$$J = J_1 \cap J_2, \tag{4}$$

noting that, by Lemma 1.9, J is a neighborhood of a. Since a is an accumulation point of D, there exists $x \in \mathbb{R}$ such that

$$x \in (J \setminus \{a\}) \cap D. \tag{5}$$

As a consequence of (4) and (5), we have

$$x \in (J_1 \setminus \{a\}) \cap D \qquad \text{and} \qquad x \in (J_2 \setminus \{a\}) \cap D; \qquad (6)$$

hence, in view of (2) and (3),

$$f(x) \in N \qquad \text{and} \qquad f(x) \in K, \qquad (7)$$

contradicting (1). $\qquad\qquad\qquad\qquad\qquad\qquad\qquad\qquad\qquad\qquad\qquad\quad$ □

2.6 DEFINITION

> **NOTATION FOR LIMITS** If f is a function, $a \in \mathbb{R}$, and f has a limit L at a, then this unique limit is denoted by
>
> $$\lim_{x \to a} f(x) = L \qquad \text{or} \qquad \lim_{x \to a} f(x) = L$$

Many elementary calculus textbooks (and even some textbooks on mathematical analysis) use the notation of Definition 2.6 before proving the uniqueness of limits. This is a logical error because the uniqueness of the limit is already implicit in the notation. (After all, if

$$\lim_{x \to a} f(x) = L \qquad \text{and} \qquad \lim_{x \to a} f(x) = L',$$

then $L = L'$ simply by substitution of the first equation into the second.) From now on, if we write

$$\lim_{x \to a} f(x) = L,$$

we mean that the limit of f at a exists and that its value is L.

Note that the symbol x in the notation

$$\lim_{x \to a} f(x) = L$$

is a dummy variable. Indeed, if f has the limit L at a, it is a matter of indifference whether we write

$$\lim_{x \to a} f(x) = L$$

or, for instance,

$$\lim_{t \to a} f(t) = L.$$

As we mentioned earlier, an idea or a property is said to be *topological* if it can be expressed, with the aid of set theory, purely in terms of open sets. The following theorem shows that the idea of limit is a topological idea. The proof, which we leave as an exercise (Problem 7), is a consequence of Corollary 2.3 and Theorem 1.6.

2.7 THEOREM

> **LIMITS ARE TOPOLOGICAL**   Let f be a function with domain D, and let $a \in \mathbb{R}$. Then $\lim_{x \to a} f(x)$ exists and equals L if and only if the following two conditions hold:
>
> (i) For every open set V with $a \in V$, $(V \setminus \{a\}) \cap D \neq \varnothing$.
> (ii) For every open set U with $L \in U$, there exists an open set V with $a \in V$ such that, for all $x \in \mathbb{R}$,
>
> $$x \in (V \setminus \{a\}) \cap D \Rightarrow f(x) \in U.$$

The requirement that $\lim_{x \to a} f(x) = L$ is a **local condition** in that it depends only on the behavior of the function f in a neighborhood of the point a (or, more generally, in an open set containing a). Furthermore, this condition is independent of the value of the function at a, or even whether a belongs to the domain of the function. The next theorem, which makes this more precise, depends on the idea of *restricting* a function to a subset of its domain.

If f is a function and $\varnothing \neq M \subseteq \text{dom}(f)$, then we define the **restriction** of f to M, in symbols $f|_M$, to be the function with domain M defined by

$$(f|_M)(x) = f(x)$$

for all $x \in M$.

2.8 THEOREM

> **LOCAL NATURE OF LIMITS**   Let f be a function, let $a \in \mathbb{R}$, suppose W is an open subset of $\mathbb{R}$ with $a \in W$, let $M = (W \setminus \{a\}) \cap \text{dom}(f)$, assume that $M \neq \varnothing$, and let
>
> $$g = f|_M.$$
>
> Then $\lim_{x \to a} f(x)$ exists if and only if $\lim_{x \to a} g(x)$ exists, and the two limits, if they exist, are equal.

PROOF   Assume the hypotheses of the theorem and suppose

$$L = \lim_{x \to a} f(x).$$

To prove that L is the limit of g at a, we propose to use Theorem 2.7. Thus, let V be an open set with $a \in V$. Then $a \in V \cap W$ and, by Theorem

1.11, $V \cap W$ is an open set. Therefore, since L is the limit of f at a, Part (i) of Theorem 2.7 implies that

$$((V \cap W)\setminus\{a\}) \cap \mathrm{dom}(f) \neq \varnothing.$$

Consequently,

$$(V\setminus\{a\}) \cap (W\setminus\{a\}) \cap \mathrm{dom}(f) \neq \varnothing,$$

so, since $(W\setminus\{a\}) \cap \mathrm{dom}(f) = M = \mathrm{dom}(g)$, we have

$$(V\setminus\{a\}) \cap \mathrm{dom}(g) \neq \varnothing.$$

Hence, the condition in Part (i) of Theorem 2.7 is satisfied.

Now let U be an open set with $L \in U$. By Part (ii) of Theorem 2.7, there exists an open set V with $a \in V$ such that, for all $x \in \mathbb{R}$,

$$x \in (V\setminus\{a\}) \cap \mathrm{dom}(f) \Rightarrow f(x) \in U.$$

Suppose that

$$x \in (V\setminus\{a\}) \cap \mathrm{dom}(g).$$

Then, since $\mathrm{dom}(g) \subseteq \mathrm{dom}(f)$, we have

$$x \in (V\setminus\{a\}) \cap \mathrm{dom}(f),$$

and therefore,

$$f(x) \in U.$$

Also, since $x \in \mathrm{dom}(g)$, we have $g(x) = f(x)$, and it follows that

$$g(x) \in U.$$

This shows that, for all $x \in \mathbb{R}$,

$$x \in (V\setminus\{a\}) \cap \mathrm{dom}(g) \Rightarrow g(x) \in U.$$

Hence, g also satisfies the condition in Part (ii) of Theorem 2.7, and we can conclude that

$$L = \lim_{x \to a} g(x).$$

The proof of the converse is quite similar and is left as an exercise (Problem 9). □

As a consequence of Theorem 2.8, if two functions agree everywhere, except possibly at one point, and if one of the functions has a limit at that point, then so does the other, and the two limits are the same. More precisely, we have the following corollary of Theorem 2.8:

2.9 COROLLARY

Let f and g be functions, and let $a \in \mathbb{R}$. Suppose that

$$\text{dom}(f) \setminus \{a\} = \text{dom}(g) \setminus \{a\} \neq \varnothing$$

and that

$$f(x) = g(x)$$

holds for all $x \in \text{dom}(f) \setminus \{a\}$. Then, if $\lim_{x \to a} f(x)$ exists, so does $\lim_{x \to a} g(x)$, and

$$\lim_{x \to a} f(x) = \lim_{x \to a} g(x).$$

PROOF   Let $W = \mathbb{R}$ in Theorem 2.8, so that $M = \text{dom}(f) \setminus \{a\}$, and $f|_M = g|_M$. Thus, if f has the limit L at a, so does $g|_M$; hence, g also has the limit L at a.   □

We leave the proof of the next lemma as an exercise (Problem 11).

2.10 LEMMA

Let $D \subseteq \mathbb{R}$, let $c \in \mathbb{R}$, and let a be an accumulation point of D.

(i) If $f : D \to \mathbb{R}$ is defined by $f(x) = c$ for all $x \in D$, then $\lim_{x \to a} f(x) = c$.

(ii) If $g : D \to \mathbb{R}$ is defined by $g(x) = x$ for all $x \in D$, then $\lim_{x \to a} g(x) = a$.

The results of Lemma 2.10 are often written in the abbreviated forms

$$\lim_{x \to a} c = c \qquad \text{and} \qquad \lim_{x \to a} x = a.$$

The following lemma can be interpreted as stating that the binary operation $+ : \mathbb{R} \times \mathbb{R} \to \mathbb{R}$ is *continuous* (Problem 19). (We study the idea of continuity in the next section.)

2.11 LEMMA

CONTINUITY OF ADDITION   Let $c, d \in \mathbb{R}$ and suppose that N is a neighborhood of $c + d$. Then there exist neighborhoods J and K of c and d, respectively, such that

$$u \in J, \, v \in K \Rightarrow u + v \in N.$$

PROOF Denote by ε the radius of N and let $\delta = \varepsilon/2$. Let J and K be neighborhoods of c and d, respectively, both with radius δ. Let $u \in J, v \in K$. Then, by Part (ii) of Lemma 2.2,

$$|u - c| < \delta \qquad \text{and} \qquad |v - d| < \delta.$$

Thus, by the triangle inequality,

$$|(u + v) - (c + d)| = |(u - c) + (v - d)| \leq |u - c| + |v - d| < \delta + \delta$$
$$= 2\delta = \varepsilon,$$

and it follows from Part (ii) of Lemma 2.2 that

$$u + v \in N. \qquad \qquad \square$$

2.12 THEOREM

> **LIMIT OF A SUM** Let f and g be functions, suppose that $D \subseteq \text{dom}(f) \cap \text{dom}(g)$, and let a be an accumulation point of D. Let
>
> $$\lim_{x \to a} f(x) = L \qquad \text{and} \qquad \lim_{x \to a} g(x) = L'.$$
>
> Define $h : D \to \mathbb{R}$ by
>
> $$h(x) = f(x) + g(x)$$
>
> for all $x \in D$. Then
>
> $$\lim_{x \to a} h(x) = L + L'.$$

PROOF Let N be a neighborhood of $L + L'$. By Lemma 2.11, there exist neighborhoods J and K of L and L', respectively, such that

$$u \in J, \quad v \in K \Rightarrow u + v \in N.$$

Since $\lim_{x \to a} f(x) = L$ and J is a neighborhood of L, there exists a neighborhood I_1 of a such that, for all $x \in \mathbb{R}$,

$$x \in (I_1 \setminus \{a\}) \cap \text{dom}(f) \Rightarrow f(x) \in J.$$

Likewise, since $\lim_{x \to a} g(x) = L'$ and K is a neighborhood of L', there exists a neighborhood I_2 of a such that, for all $x \in \mathbb{R}$,

$$x \in (I_2 \setminus \{a\}) \cap \text{dom}(g) \Rightarrow g(x) \in K.$$

Let

$$I = I_1 \cap I_2,$$

noting that I is a neighborhood of a by Lemma 1.9. Since

$$\text{dom}(h) = D \subseteq \text{dom}(f) \cap \text{dom}(g),$$

we have, for all $x \in \mathbb{R}$,

$$x \in (I \setminus \{a\}) \cap \operatorname{dom}(h) \Rightarrow f(x) \in J, \quad g(x) \in K.$$

However,

$$f(x) \in J, \quad g(x) \in K \Rightarrow h(x) = f(x) + g(x) \in N,$$

and it follows that, for all $x \in \mathbb{R}$,

$$x \in (I \setminus \{a\}) \cap \operatorname{dom}(h) \Rightarrow h(x) \in N.$$

Therefore,

$$\lim_{x \to a} h(x) = L + L'. \qquad \square$$

The result of Theorem 2.12 is often written in the abbreviated form

$$\lim_{x \to a} (f(x) + g(x)) = \lim_{x \to a} f(x) + \lim_{x \to a} g(x)$$

and paraphrased by the statement that *the limit of a sum is the sum of the limits*.

The following lemma can be interpreted as stating that the binary composition $\cdot : \mathbb{R} \times \mathbb{R} \to \mathbb{R}$ is continuous.

2.13 LEMMA

CONTINUITY OF MULTIPLICATION   Let $c, d \in \mathbb{R}$ and suppose that N is a neighborhood of cd. Then there exist neighborhoods J and K of c and d, respectively, such that

$$u \in J, \quad v \in K \Rightarrow uv \in N.$$

PROOF   Denote by ε the radius of N. Our problem is to find positive numbers δ and η (the radii of J and K, respectively) so that, if $|u - c| < \delta$ and $|v - d| < \eta$, then

$$|uv - cd| < \varepsilon. \qquad (*)$$

Note that

$$\begin{aligned}
uv - cd &= uv + (-ud + ud) - cd \\
&= (uv - ud) + (ud - cd) \\
&= u(v - d) + (u - c)d.
\end{aligned}$$

Therefore, by the triangle inequality,

$$|uv - cd| \le |u(v - d)| + |(u - c)d| = |u| |v - d| + |u - c| |d|.$$

Hence, in order to ensure that $(*)$ holds, it will be sufficient to find δ and η so that, if $|u - c| < \delta$ and $|v - d| < \eta$, then

$$|u| |v - d| < \frac{\varepsilon}{2} \qquad (**)$$

and

$$|u - c||d| < \frac{\varepsilon}{2} \qquad\qquad (***)$$

both hold.

If $d \neq 0$, let

$$\delta = \frac{\varepsilon}{2|d|},$$

and, if $d = 0$, let δ be any fixed positive number (say, $\delta = 1$). Let

$$\eta = \frac{\varepsilon}{2(\delta + |c|)}.$$

Now suppose that

$$|u - c| < \delta \qquad \text{and} \qquad |v - d| < \eta.$$

If $d = 0$, then (***) is automatic, otherwise we have

$$|u - c||d| < \delta|d| = \frac{\varepsilon}{2},$$

and, in either case, (***) holds. By the triangle inequality,

$$|u| = |(u - c) + c| \leq |u - c| + |c| < \delta + |c|.$$

Therefore,

$$|u||v - d| < (\delta + |c|)\eta = \frac{\varepsilon}{2},$$

and (**) also holds. $\qquad\qquad\qquad\qquad\qquad\qquad\qquad\qquad \square$

The proof of the next theorem follows from Lemma 2.13 in much the same way as the proof of Theorem 2.12 followed from Lemma 2.11, and we leave it as an exercise (Problem 13).

2.14 THEOREM **LIMIT OF A PRODUCT** Let f and g be functions, suppose that $D \subseteq \text{dom}(f) \cap \text{dom}(g)$, and let a be an accumulation point of D. Let

$$\lim_{x \to a} f(x) = L \qquad \text{and} \qquad \lim_{x \to a} g(x) = L'.$$

Define $h : D \to \mathbb{R}$ by

$$h(x) = f(x) \cdot g(x)$$

for all $x \in D$. Then

$$\lim_{x \to a} h(x) = L \cdot L'.$$

The result of Theorem 2.14 is often written in the abbreviated form

$$\lim_{x \to a} (f(x) \cdot g(x)) = \left(\lim_{x \to a} f(x)\right) \cdot \left(\lim_{x \to a} g(x)\right)$$

and paraphrased by the statement that *the limit of a product is the product of the limits.*

The following lemma can be used to prove the theorem usually paraphrased by the statement that *the limit of a reciprocal is the reciprocal of the limit, provided that the limit is nonzero* (Problem 23).

2.15 LEMMA

> **CONTINUITY OF RECIPROCATION**   Let $c \in \mathbb{R}$ with $c \neq 0$, and let N be a neighborhood of $1/c$. Then there exists a neighborhood J of c such that $0 \notin J$ and, for every $x \in \mathbb{R}$,
>
> $$x \in J \Rightarrow \frac{1}{x} \in N.$$

PROOF   Denote by ε the radius of N. Our problem is to find a positive number δ (the radius of J) such that, if $|x - c| < \delta$, then

$$x \neq 0 \qquad \text{and} \qquad \left|\frac{1}{x} - \frac{1}{c}\right| < \varepsilon. \qquad (*)$$

Now

$$\left|\frac{1}{x} - \frac{1}{c}\right| = \left|\frac{c - x}{cx}\right| = \frac{|c - x|}{|c||x|},$$

so (*) is equivalent to

$$x \neq 0 \qquad \text{and} \qquad |x - c| < \varepsilon|c||x|. \qquad (**)$$

Let

$$\delta = \min(\tfrac{1}{2}|c|, \tfrac{1}{2}\varepsilon|c|^2)$$

and suppose that $|x - c| < \delta$. Then, by the triangle inequality,

$$|c| = |x + (c - x)| \leq |x| + |c - x|$$
$$= |x| + |x - c| < |x| + \delta \leq |x| + \tfrac{1}{2}|c|,$$

from which it follows that

$$0 < \tfrac{1}{2}|c| < |x|.$$

Therefore, $x \neq 0$ and

$$|x - c| < \delta \leq \tfrac{1}{2}\varepsilon|c|^2 = \varepsilon|c| \cdot \tfrac{1}{2}|c| < \varepsilon|c||x|,$$

that is, (**) holds.   □

PROBLEM SET 4.2

1. Take $\mathbb{R}$ as the domain of discourse and use the convention that ε and δ represent positive real numbers to write the condition in Part (ii) of Definition 2.1 as a quantified statement in predicate calculus.

2. Given that a is an accumulation point of the domain D of a function f, show that a real number b is *not* a limit of f at a if and only if there exists a positive real number ε such that, for every positive real number δ, there is a number $x \in D$ satisfying the conditions $0 < |x - a| < \delta$ and $|f(x) - b| \geq \varepsilon$.

3. Let $a \in \mathbb{R}$, suppose that J is a neighborhood of a, and let δ be the radius of J.
   (a) Prove that $x \in J \Rightarrow |a| - \delta < |x| < |a| + \delta$. [Hint: Use the triangle inequality.]
   (b) Let N be the neighborhood of $|a|$ with radius δ. Prove that $x \in J \Rightarrow |x| \in N$.

4. In the proof of Part (i) of Lemma 2.2, it is asserted that the condition $|b| < \varepsilon$ is equivalent to the requirement that both $b < \varepsilon$ and $-b < \varepsilon$ hold. Make a rigorous proof of this fact. [Hint: Use Lemma 3.17 in Chapter 3.]

5. Suppose that $\lim_{x \to a} f(x) = L$ and that $B \in \mathbb{R}$ with $L > B$. Prove that there exists a neighborhood J of a such that, for all $x \in \mathbb{R}$, $x \in J \cap \mathrm{dom}(f) \Rightarrow f(x) > B$.

6. Let f be a function, and let a be an accumulation point of the domain of f. Prove that a real number L is the limit of f at a if and only if, for every open set U with $L \in U$, there exists an open set V with $a \in V$ such that $f(V \setminus \{a\}) \subseteq U$. [See Part (i) of Definition 4.10 in Chapter 2.]

7. Prove Theorem 2.7.

8. In a calculus textbook, we find the calculation
   $$\lim_{x \to 7} \frac{x^2 - 49}{x - 7} = \lim_{x \to 7} (x + 7).$$
   How is this calculation justified? [Hint: See Corollary 2.9.]

9. Finish the proof of Theorem 2.8 by showing that, under the hypotheses of the theorem, if $L = \lim_{x \to a} g(x)$, then $L = \lim_{x \to a} f(x)$.

10. Prove the **"squeezing property"** of limits: Let g, f, and h be functions such that $\mathrm{dom}(g) \setminus \{a\} = \mathrm{dom}(f) \setminus \{a\} = \mathrm{dom}(h) \setminus \{a\}$ and suppose that $g(x) \leq f(x) \leq h(x)$ holds for all $x \in \mathrm{dom}(g) \setminus \{a\}$. Then, if $\lim_{x \to a} g(x) = L = \lim_{x \to a} h(x)$, it follows that $\lim_{x \to a} f(x) = L$.

11. Prove Lemma 2.10.

12. Let f be a function, suppose that $a \in \mathbb{R}$, and let $D = \{h \,|\, a + h \in \mathrm{dom}(f)\}$. Define $g : D \to \mathbb{R}$ by $g(h) = f(a + h)$ for all $h \in D$. Prove that $\lim_{x \to a} f(x) = L$ if and only if $\lim_{h \to 0} g(h) = L$. [This result is often written in the abbreviated form $\lim_{x \to a} f(x) = \lim_{h \to 0} f(a + h)$.]

13. Prove Theorem 2.14.

14. In elementary calculus, the derivative $f'(a)$ is written either as

$$\lim_{x \to a} \frac{f(x) - f(a)}{x - a} \quad \text{or as} \quad \lim_{h \to 0} \frac{f(a + h) - f(a)}{h}.$$

Use the result of Problem 12 to show that, under appropriate conditions, these two limits are, in fact, the same.

15. If $\lim_{x \to a} f(x) = L$ and if $g:\text{dom}(f) \to \mathbb{R}$ is defined by $g(x) = |f(x)|$ for all $x \in \text{dom}(f)$, prove that $\lim_{x \to a} g(x) = |L|$. [Hint: See Problem 3.]

16. Let f be a function with domain D and suppose that $a, L \in \mathbb{R}$. Define the function $g:D \to \mathbb{R}$ by $g(x) = |f(x) - L|$. Prove that $\lim_{x \to a} f(x) = L$ if and only if $\lim_{x \to a} g(x) = 0$. [This result is often written in the abbreviated form $\lim_{x \to a} f(x) = L$ if and only if $\lim_{x \to a} |f(x) - L| = 0$.]

17. If $\lim_{x \to a} f(x) = L$, n is a positive integer, and $g:\text{dom}(f) \to \mathbb{R}$ is defined by $g(x) = (f(x))^n$ for every $x \in \text{dom}(f)$, prove that $\lim_{x \to a} g(x) = L^n$.

18. Define $f:\mathbb{R} \to \mathbb{R}$ by

$$f(x) = \begin{cases} 1, & \text{if } x \in \mathbb{Q} \\ 0, & \text{if } x \in \mathbb{R} \setminus \mathbb{Q} \end{cases}.$$

If $a \in \mathbb{R}$, prove that f has no limit at a.

19. By definition, a binary operation $*:\mathbb{R} \times \mathbb{R} \to \mathbb{R}$ is said to be **continuous** if and only if the following condition holds: If $c, d \in \mathbb{R}$ and N is a neighborhood of $c * d$, then there exist neighborhoods J and K of c and d, respectively, such that $u \in J, v \in K \Rightarrow u * v \in N$ (for instance, see Lemmas 2.11 and 2.13). Prove that subtraction is a continuous binary operation on $\mathbb{R}$.

20. Prove the following **substitution property of limits**: Let f and g be functions and let a and c be real numbers such that $\lim_{y \to c} f(y)$ exists and $\lim_{x \to a} g(x) = c$. Suppose there is an open set W such that $a \in W, W \setminus \{a\} \subseteq \text{dom}(f \circ g)$, and $c \notin g(W \setminus \{a\})$. Then $\lim_{x \to a}(f \circ g)(x) = \lim_{y \to c} f(y)$.

21. Make a careful statement of the result usually paraphrased as *the limit of a difference is the difference of the limits*, and prove the resulting theorem. [Hint: Use the result in Problem 19.]

22. In a calculus textbook, we find the following argument: To find $\lim_{x \to 0} [(\sin 5x)/x]$, we let $y = 5x$ and note that $y \to 0$ when $x \to 0$; hence,

$$\lim_{x \to 0} \frac{\sin 5x}{x} = \lim_{y \to 0} \frac{\sin y}{y/5} = 5 \lim_{y \to 0} \frac{\sin y}{y} = 5.$$

Assume that $\lim_{y \to 0} [(\sin y)/y] = 1$, and provide a rigorous argument to justify this calculation. [Hint: Among other things, you will find the result of Problem 20 useful.]

23. Make a careful statement of the theorem usually paraphrased as *the limit of a reciprocal is the reciprocal of the limit*, and prove the resulting theorem using Lemma 2.15.

24. Suppose that f, g, and h are functions with the same domain D and that $f(x) \leq g(x) \leq h(x)$ holds for all $x \in D$.
(a) If f, g, and h have limits at a, prove that

$$\lim_{x \to a} f(x) \leq \lim_{x \to a} g(x) \leq \lim_{x \to a} h(x).$$

(b) Does the result in Part (a) still hold if all inequalities are replaced by strict inequalities?

25. Make a careful statement of the theorem usually paraphrased as *the limit of a quotient is the quotient of the limits*, and prove the resulting theorem.

26. Let f and g be functions and let $a \in \mathbb{R}$. Suppose that W is an open set, $W \setminus \{a\} \subseteq \text{dom}(f) \cap \text{dom}(g)$, $\lim_{x \to a} f(x) = 0$, and that there is a positive real number B such that $|g(x)| \leq B$ holds for all $x \in W \setminus \{a\}$. Let $h : W \setminus \{a\} \to \mathbb{R}$ be defined by $h(x) = f(x)g(x)$ for all $x \in W \setminus \{a\}$. Prove that $\lim_{x \to a} h(x) = 0$.

27. Let f be a function with domain D, let $a \in \mathbb{R}$, let $A = D \cap (a, \infty)$, and let $B = D \cap (-\infty, a)$. If $A \neq \emptyset$, $f|_A$ denotes the restriction of f to A as in Problem 9, and $\lim_{x \to a}(f|_A)(x) = L^+$ exists, then we refer to L^+ as the **limit on the right of f at a** and write $\lim_{x \to a^+} f(x) = L^+$. Similarly, if $B \neq \emptyset$ and $\lim_{x \to a}(f|_B)(x) = L^-$ exists, then we refer to L^- as the **limit on the left of f at a** and write $\lim_{x \to a^-} f(x) = L^-$. If both of these **one-sided limits** exist and have the same value L, prove that $\lim_{x \to a} f(x) = L$.

28. Use the result of Problem 26 to prove that

$$\lim_{x \to 0} \left(x \sin \frac{1}{x} \right) = 0.$$

29. Let f be a function, let W be an open set such that $W \setminus \{a\} \subseteq \text{dom}(f)$, and suppose that $\lim_{x \to a} f(x)$ exists and equals L. Prove that both of the one-sided limits $\lim_{x \to a^+} f(x)$ and $\lim_{x \to a^-} f(x)$ exist and equal L. (See Problem 27.)

30. If f is the function of Problem 18 and if $g : \mathbb{R} \to \mathbb{R}$ is defined by $g(x) = x \cdot f(x)$ for all $x \in \mathbb{R}$, investigate the question of which numbers $a \in \mathbb{R}$ satisfy the condition that g has a limit at a.

31. Let f be a function and let $L \in \mathbb{R}$. We say that **L is a limit of f at $+\infty$** if (i) dom(f) has no upper bound, and (ii) for every neighborhood N of L, there exists $b \in \mathbb{R}$ such that, for all $x \in \mathbb{R}$, $x \in \text{dom}(f) \cap (b, \infty) \Rightarrow f(x) \in N$. If f has a limit L at $+\infty$, prove that this limit is unique, thus justifying the usual notation $\lim_{x \to +\infty} f(x) = L$.

32. Let f be a function and let $a \in \mathbb{R}$. We say that **f has a limit of $+\infty$ at a** and write $\lim_{x \to a} f(x) = +\infty$ if (i) a is an accumulation point of dom(f), and (ii) for every positive number B, there exists a neighborhood J of a such that, for all $x \in \mathbb{R}$, $x \in (J \setminus \{a\}) \cap \operatorname{dom}(f) \Rightarrow f(x) > B$. Define $D = \{x \in \mathbb{R} \mid f(x) > 0\}$, suppose $D \neq \varnothing$, and define $g : D \to \mathbb{R}$ by $g(x) = 1/f(x)$ for all $x \in D$. Prove that $\lim_{x \to a} f(x) = +\infty$ if and only if $\lim_{x \to a} g(x) = 0$.

33. Let f be a function and let $L \in \mathbb{R}$. Let $D = \{x \in \mathbb{R} \mid 1/x \in \operatorname{dom}(f)\}$, and suppose that $D \neq \varnothing$. Define $g : D \to \mathbb{R}$ by $g(x) = f(1/x)$ for all $x \in D$. With the notation of Problems 29 and 31, prove that $\lim_{x \to +\infty} f(x) = L$ if and only if $\lim_{x \to 0^+} g(x) = L$.

34. Formulate a definition for $\lim_{x \to a} f(x) = -\infty$ patterned after the definition in Problem 32 for $\lim_{x \to a} f(x) = +\infty$, and prove that $\lim_{x \to a} f(x) = -\infty$ holds if and only if $\lim_{x \to a} (-f(x)) = +\infty$.

35. Formulate a definition for $\lim_{x \to -\infty} f(x) = L$ patterned after the definition in Problem 31 for $\lim_{x \to +\infty} f(x) = L$, and prove that $\lim_{x \to -\infty} f(x) = L$ if and only if $\lim_{x \to +\infty} f(-x) = L$.

36. Formulate suitable definitions for:
(a) $\lim_{x \to +\infty} f(x) = +\infty$ (b) $\lim_{x \to +\infty} f(x) = -\infty$
(c) $\lim_{x \to -\infty} f(x) = +\infty$ (d) $\lim_{x \to -\infty} f(x) = -\infty$

3
_____

CONTINUITY

The intuitive idea of *continuity*, familiar from elementary calculus, is roughly that a function f is continuous at a point a if $f(x)$ is close to $f(a)$ when x is close to a. More accurately:

3.1 DEFINITION

CONTINUITY OF A FUNCTION AT A POINT   Let f be a function with domain D. We say that f is **continuous at the point** $a \in \mathbb{R}$ if the following two conditions hold:

(i) $a \in D$.
(ii) For every $\varepsilon \in \mathbb{R}$ with $\varepsilon > 0$, there exists $\delta \in \mathbb{R}$ with $\delta > 0$ such that, for every $x \in D$,

$$|x - a| < \delta \Rightarrow |f(x) - f(a)| < \varepsilon.$$

To prove that f is continuous at $a \in \operatorname{dom}(f)$ by a direct use of Definition 3.1, we must show that, for an *arbitrary* positive ε, there exists a positive δ (usually depending on ε) such that the condition in Part (ii) holds. Doing this by actually determining a δ that works is called *finding the δ*, and proofs in which we "find the δ" are called ε–δ **proofs**.

3.2 Example   Let $f(x) = x^2$ for all $x \in \mathbb{R}$. If $a \in \mathbb{R}$, make an ε–δ proof that f is continuous at a.

SOLUTION   Since $\text{dom}(f) = \mathbb{R}$, the condition $a \in \text{dom}(f)$ is automatic. Thus, let ε be an arbitrary positive number and let

$$\delta = \min\left(1, \frac{\varepsilon}{2|a| + 1}\right).$$

Suppose that

$$|x - a| < \delta.$$

Then we have

$$|x - a| < 1 \qquad \text{and} \qquad \delta(2|a| + 1) \le \varepsilon.$$

Therefore, using the triangle inequality (twice), we have

$$
\begin{aligned}
|x^2 - a^2| &= |(x - a)(x + a)| \\
&= |x - a|\,|x + a| \\
&< \delta|x + a| \\
&\le \delta(|x| + |a|) \\
&= \delta(|a + (x - a)| + |a|) \\
&\le \delta(|a| + |x - a| + |a|) \\
&= \delta(2|a| + |x - a|) \\
&< \delta(2|a| + 1) \\
&\le \varepsilon.
\end{aligned}
$$

$\square$

Although the proof in Example 3.2 is *logically* flawless, you may be left with an uneasy feeling even after checking all of the algebra. It is *psychologically* difficult to accept the validity of the proof in the face of the nagging question: Where did the δ come from, anyway? A mathematician would say that the proof is "not motivated." Although the lack of motivation is probably on the part of the reader of the proof, rather than the proof itself, this somewhat ungrammatical parlance is accepted in mathematical circles.

When mathematicians ask for a proof to be *motivated*, they are asking for some insight into how the prover conceived of the proof in the first place. This insight is often provided by a so-called *heuristic* argument— that is, an argument involving suggestive or persuasive reasoning that is not necessarily logically impeccable. For instance, heuristic arguments often commit the logical transgression of proceeding "backward" from conclusion to hypotheses.

3.3 Example   Motivate the argument in Example 3.2.

SOLUTION   Given $\varepsilon > 0$, our problem is to find $\delta > 0$ such that, if

$$|x - a| < \delta,$$

then

$$|x^2 - a^2| < \varepsilon.$$

Since

$$|x^2 - a^2| = |(x - a)(x + a)| = |x - a|\,|x + a|,$$

what we want is

$$|x - a|\,|x + a| < \varepsilon.$$

If $|x - a| < \delta$, then the last condition will certainly hold provided that

$$\delta|x + a| < \varepsilon.$$

Now the critical question is: If $|x - a| < \delta$, then how large can $|x + a|$ be? Questions like this can often be settled by using the triangle inequality.
  By the triangle inequality, we have

$$|x + a| \le |x| + |a|,$$

so the question of how large $|x + a|$ can be hinges on the question of how large $|x|$ can be. A second use of the triangle inequality shows that

$$|x| = |a + (x - a)| \le |a| + |x - a|,$$

and, therefore, if $|x - a| < \delta$, it follows that

$$|x| < |a| + \delta.$$

Consequently, if $|x - a| < \delta$, then we have

$$|x + a| \le |x| + |a| < (|a| + \delta) + |a| = 2|a| + \delta.$$

For instance, if we require that $\delta \le 1$, then the condition that $|x - a| < \delta$ will ensure that

$$|x + a| < 2|a| + 1$$

and, therefore, that

$$\delta|x - a| < \delta(2|a| + 1).$$

Hence, if we make sure that $\delta \le 1$ and *also* that $\delta \le \varepsilon/(2|a| + 1)$, we will have

$$\delta|x - a| < \varepsilon.$$

This suggests that we let

$$\delta = \min\left(1, \frac{\varepsilon}{2|a| + 1}\right). \qquad \square$$

In mathematical writing, and especially in ε–δ proofs, the heuristics are often suppressed, usually in the interest of brevity, but (one suspects) sometimes for the effect when a δ that works is brought forth as if by magic. Whenever you feel unsatisfied with a proof, in spite of being convinced that it is logically correct, you should try to pinpoint the source of your discomfort. If your uneasiness is caused by a failure to understand the proof writer's motivation, you might try to supply the missing heuristics yourself.

The following lemma is obtained by translating Definition 3.1 into the language of neighborhoods (Problem 7).

3.4 LEMMA

> Let f be a function with domain D. Then f is continuous at a point $a \in \mathbb{R}$ if and only if the following two conditions hold:
>
> (i) $a \in D$.
> (ii) For every neighborhood N of $f(a)$, there exists a neighborhood J of a such that, for every $x \in \mathbb{R}$,
>
> $$x \in J \cap D \Rightarrow f(x) \in N.$$

A comparison of Corollary 2.3 with Lemma 3.4 reveals a close connection between limits and continuity. Indeed, we have the following theorem:

3.5 THEOREM

> **LIMITS AND CONTINUITY**   Let f be a function with domain D, and let $a \in D$. Then:
>
> (i) If a is an isolated point of D, then f is continuous at a.
> (ii) If a is an accumulation point of D, then f is continuous at a if and only if
>
> $$\lim_{x \to a} f(x) = f(a).$$

PROOF   Note that Part (i) of Lemma 3.4 holds by assumption.

(i) Suppose that a is an isolated point of D, and let N be a neighborhood of $f(a)$. By Lemma 1.19, there exists a neighborhood J of a such that $J \cap D = \{a\}$. Therefore, for every $x \in \mathbb{R}$,

$$x \in J \cap D \Rightarrow x = a \Rightarrow f(x) = f(a) \in N,$$

so Part (ii) of Lemma 3.4 holds.

(ii) Suppose that a is an accumulation point of D, so that Part (i) of Corollary 2.3 holds. Let $L = f(a)$ in Corollary 2.3. Then Part (ii) of

Lemma 3.4 implies Part (ii) of Corollary 2.3 because, if $x \in (J \setminus \{a\}) \cap D$, then $x \in J \cap D$. Conversely, suppose that Part (ii) of Corollary 2.3 holds, and let N be a neighborhood of $f(a)$. Then there exists a neighborhood J of a such that, for all $x \in \mathbb{R}$,

$$x \in (J \setminus \{a\}) \cap D \Rightarrow f(x) \in N.$$

Therefore, for all $x \in \mathbb{R}$ with $x \neq a$, we have

$$x \in J \cap D \Rightarrow f(x) \in N.$$

However, if $x = a$, then $f(x) = f(a) \in N$ is automatic, and it follows that Part (ii) of Lemma 3.4 holds.   □

In elementary calculus, it is unusual to define continuity in such a way that a function is automatically continuous at an isolated point of its domain—indeed, the question of continuity at such a point is rarely considered. In mathematical analysis, however, Definition 3.1 is almost universally accepted, and, consequently, functions are regarded as being continuous at isolated points of their domains.

Another discrepancy in terminology between elementary calculus and mathematical analysis concerns the word *discontinuous*. In elementary calculus, this word is used as a synonym for *not continuous*; for instance, it is usual to say that the function $f(x) = 1/x$ is "discontinuous at 0." This practice is avoided in mathematical analysis simply because 0 does not belong to the domain of $f(x) = 1/x$. Thus, in this textbook, we speak about a discontinuity only for points *in the domain* of a function. More precisely:

3.6 DEFINITION   **POINT OF DISCONTINUITY**   Let f be a function. A point $a \in \mathbb{R}$ is called a **point of discontinuity** of f, and f is said to be **discontinuous** at a, if $a \in \text{dom}(f)$ and f is not continuous at a.

3.7 Example   Show that the function sgn is discontinuous at 0.

SOLUTION   The signum function, abbreviated sgn, was introduced in Definition 3.14 of Chapter 3. To show that sgn is discontinuous at 0, we must show that $0 \in \text{dom}(\text{sgn})$, but that sgn is not continuous at 0. By the definition of sgn, $0 \in \text{dom}(\text{sgn})$, and, in fact, $\text{sgn}(0) = 0$. Thus, to show that sgn is discontinuous at 0, we need only to prove that the condition in Part (ii) of Definition 3.1 fails.

As a quantified statement in predicate calculus, Part (ii) of Definition 3.1 reads as follows:

$$(\forall \varepsilon)(\exists \delta)(\forall x)(|x - 0| < \delta \Rightarrow |\text{sgn}(x) - \text{sgn}(0)| < \varepsilon),$$

that is,

$$(\forall\varepsilon)(\exists\delta)(\forall x)(|x| < \delta \Rightarrow |\mathrm{sgn}(x)| < \varepsilon),$$

with the understanding that $\mathbb{R}$ is the domain of discourse and that ε and δ represent positive real numbers. The denial of this statement is

$$\sim(\forall\varepsilon)(\exists\delta)(\forall x)(|x| < \delta \Rightarrow |\mathrm{sgn}(x)| < \varepsilon),$$

that is,

$$(\exists\varepsilon)(\forall\delta)(\exists x)(|x| < \delta \wedge |\mathrm{sgn}(x)| \geq \varepsilon). \tag{*}$$

To prove (*), we must find a positive number ε such that, for every positive number δ, there is a number $x \in \mathbb{R}$ such that

$$|x| < \delta \qquad \text{and} \qquad |\mathrm{sgn}(x)| \geq \varepsilon. \tag{**}$$

Note that

$$x \neq 0 \Rightarrow |\mathrm{sgn}(x)| = 1,$$

so, if we take $\varepsilon = 1$, then

$$x \neq 0 \Rightarrow |\mathrm{sgn}(x)| \geq \varepsilon.$$

Thus, let

$$\varepsilon = 1,$$

and suppose δ is an arbitrary positive number. To prove that there is a number $x \in \mathbb{R}$ such that (**) holds, it will suffice to prove that there is a number $x \in \mathbb{R}$ such that

$$|x| < \delta \qquad \text{and} \qquad x \neq 0.$$

But this is easy—for instance, just take $x = \delta/2$. $\qquad\square$

The following theorem is useful for showing that various combinations (sums, differences, products, and so forth) of continuous functions are continuous. (For the definition of a *continuous binary operation*, see Problem 19 in Problem Set 4.2.)

3.8 THEOREM

CONTINUITY OF COMBINATIONS OF FUNCTIONS Let f and g be functions and let $D = \mathrm{dom}(f) \cap \mathrm{dom}(g)$. Suppose that $a \in D$ and that both f and g are continuous at a. Let $*:\mathbb{R} \times \mathbb{R} \to \mathbb{R}$ be a continuous binary operation, and define $h:D \to \mathbb{R}$ by

$$h(x) = f(x) * g(x) \qquad \text{for all } x \in D.$$

Then h is continuous at a.

PROOF   We verify Parts (i) and (ii) of Lemma 3.4. Part (i) holds by hypothesis, so we focus on Part (ii). Thus, let N be a neighborhood of $h(a) = f(a) * g(a)$. Then, by the definition in Problem 19 of Problem Set 4.2, there exist neighborhoods J and K of $f(a)$ and $g(a)$, respectively, such that

$$u \in J, \quad v \in K \Rightarrow u * v \in N. \tag{1}$$

Since J is a neighborhood of $f(a)$ and f is continuous at a, there is a neighborhood J_1 of a such that, for every $x \in \mathbb{R}$,

$$x \in J_1 \cap \mathrm{dom}(f) \Rightarrow f(x) \in J. \tag{2}$$

Since K is a neighborhood of $g(a)$ and g is continuous at a, there is a neighborhood K_1 of a such that, for every $x \in \mathbb{R}$,

$$x \in K_1 \cap \mathrm{dom}(g) \Rightarrow g(x) \in J. \tag{3}$$

Let $I = J_1 \cap K_1$, noting that I is a neighborhood of a by Lemma 1.9. Furthermore, if $x \in I \cap D$, then the hypotheses of the implications in (2) and (3) both hold, and it follows that $f(x)$, $g(x) \in J$. But, if $f(x)$, $g(x) \in J$, then $h(x) = f(x) * g(x) \in N$ by (1). Therefore,

$$x \in I \cap D \Rightarrow h(x) \in N. \tag{4}$$

□

By combining Theorem 3.8 with Lemmas 2.11 and 2.13 and Problem 19 in Problem Set 4.2, we obtain the following corollary (Problem 11):

3.9 COROLLARY

> **CONTINUITY OF SUMS, DIFFERENCES, AND PRODUCTS**   Let f and g be functions and let $D = \mathrm{dom}(f) \cap \mathrm{dom}(g)$. Suppose that $a \in D$ and that both f and g are continuous at a. Define $s:D \to \mathbb{R}, d:D \to \mathbb{R}$, and $p:D \to \mathbb{R}$ by
>
> $$s(x) = f(x) + g(x),$$
> $$d(x) = f(x) - g(x),$$
>
> and
>
> $$p(x) = f(x)g(x)$$
>
> for all $x \in D$. Then s, d, and p are continuous at a.

The functions s, d, and p in Corollary 3.9 are called the **sum**, the **difference**, and the **product** of the functions f and g. Thus, the corollary can be paraphrased as follows:

> *The sum, difference, and product of two functions are continuous at each point of continuity of the functions.*

Because many of the functions that arise in practice are built up from simpler functions by composition, the next theorem provides an important method for establishing continuity.

3.10 THEOREM

CONTINUITY OF COMPOSITIONS OF FUNCTIONS Let f and g be functions and suppose that g is continuous at a and that f is continuous at $g(a)$. Then the composite function $f \circ g$ is continuous at a.

PROOF   Since g is continuous at a, we have $a \in \operatorname{dom}(g)$. Likewise, since f is continuous at $g(a)$, we have $g(a) \in \operatorname{dom}(f)$, and it follows that $a \in \operatorname{dom}(f \circ g)$. Let N be a neighborhood of $(f \circ g)(a) = f(g(a))$. Since f is continuous at $g(a)$, there exists a neighborhood K of $g(a)$ such that, for every $y \in \mathbb{R}$,

$$y \in K \cap \operatorname{dom}(f) \Rightarrow f(y) \in N. \tag{1}$$

Since K is a neighborhood of $g(a)$ and g is continuous at a, there exists a neighborhood J of a such that, for every $x \in \mathbb{R}$,

$$x \in J \cap \operatorname{dom}(g) \Rightarrow g(x) \in K. \tag{2}$$

We propose to finish the proof by showing that

$$x \in J \cap \operatorname{dom}(f \circ g) \Rightarrow (f \circ g)(x) \in N. \tag{*}$$

Thus, assume the hypothesis of (*), so that

$$x \in J, \quad x \in \operatorname{dom}(g), \quad \text{and} \quad g(x) \in \operatorname{dom}(f). \tag{3}$$

Then, by (2), we have

$$g(x) \in K. \tag{4}$$

From (4) and (3), we have

$$g(x) \in K \cap \operatorname{dom}(f). \tag{5}$$

Therefore, by (1) with y replaced by $g(x)$, we have

$$(f \circ g)(x) = f(g(x)) \in N, \tag{6}$$

which is the conclusion of (*).   □

If g is a function and

$$D = \{x \in \operatorname{dom}(g) \mid g(x) \neq 0\},$$

then, provided that $D \neq \emptyset$, the function $r:D \to \mathbb{R}$, defined by

$$r(x) = \frac{1}{g(x)},$$

for each $x \in D$ is called the **reciprocal** of g.

3.11 COROLLARY

> **CONTINUITY OF THE RECIPROCAL**   Let g be a function, suppose $a \in \text{dom}(g)$, $g(a) \neq 0$, and g is continuous at a. Let r be the reciprocal of g. Then r is continuous at a.

PROOF   Define $f:\mathbb{R}\setminus\{0\} \to \mathbb{R}$ by $f(x) = 1/x$ for all $x \neq 0$. By Lemma 2.15, f is continuous at each point $c \in \mathbb{R}\setminus\{0\}$. Evidently, $r = f \circ g$, and, by Theorem 3.10, it follows that r is continuous at a.   □

If f and g are functions and

$$D = \{x \in \text{dom}(f) \cap \text{dom}(g) \,|\, g(x) \neq 0\},$$

then, provided that $D \neq \emptyset$, the function $q:D \to \mathbb{R}$ defined by

$$q(x) = \frac{f(x)}{g(x)}$$

for each $x \in D$ is called the **quotient** of f and g. The next lemma establishes the result usually paraphrased by the following statement:

> The quotient of two functions is continuous at each point for which the numerator is continuous and the denominator is nonzero and continuous.

3.12 LEMMA

> **CONTINUITY OF QUOTIENTS**   Let f and g be functions and let
>
> $$D = \{x \in \text{dom}(f) \cap \text{dom}(g) \,|\, g(x) \neq 0\}.$$
>
> Then, if $a \in D$, f and g are continuous at a, and q is the quotient of f and g, it follows that q is continuous at a.

PROOF   We sketch the proof and leave the details as an exercise (Problem 19). Let r be the reciprocal of g. Note that q is the product of f and r. Therefore, q is continuous by Corollaries 3.9 and 3.11.   □

The next theorem is an analogue for continuity of Theorem 2.7 for limits. Its proof, which follows easily from Theorem 1.6, is left as an exercise (Problem 21).

3.13 THEOREM

CONTINUITY AT A POINT IS TOPOLOGICAL Let f be a function with domain D. Then f is continuous at a point $a \in \mathbb{R}$ if and only if the following two conditions hold:

(i) $a \in D$.
(ii) For every open set U with $f(a) \in U$, there exists an open set V with $a \in V$ such that, for all $x \in \mathbb{R}$,

$$x \in V \cap D \Rightarrow f(x) \in U.$$

Just as in elementary calculus, we define a function to be *continuous* if it is continuous at every point of its domain.

3.14 DEFINITION

CONTINUOUS FUNCTION A function f is continuous if it is continuous at each point $a \in \text{dom}(f)$.

3.15 Example Prove that the **absolute-value function** $f: \mathbb{R} \to \mathbb{R}$, defined by $f(x) = |x|$ for every $x \in \mathbb{R}$, is continuous.

SOLUTION Let $a \in \text{dom}(f) = \mathbb{R}$, let N be a neighborhood of $f(a) = |a|$, and let ε be the radius of N. Let J be the neighborhood of a of the same radius, ε. By Problem 3 in Problem Set 4.2, we have

$$x \in J \Rightarrow f(x) = |x| \in N,$$

and, therefore, f is continuous at a. □

We leave the proofs of the following lemma and its corollary as exercises (Problems 23 and 24).

3.16 LEMMA

(i) Let $c \in \mathbb{R}$. Then the constant function f defined by $f(x) = c$ for all $x \in \mathbb{R}$ is continuous.
(ii) The identity function j defined on $\mathbb{R}$ by $j(x) = x$ for all $x \in \mathbb{R}$ is continuous.
(iii) If n is a positive integer, then the power function p defined on $\mathbb{R}$ by $p(x) = x^n$ for all $x \in \mathbb{R}$ is continuous.

3.17 COROLLARY

> If $f : \mathbb{R} \to \mathbb{R}$ is a polynomial function of degree n, that is, if n is a positive integer and there are constants $c_n, c_{n-1}, \ldots, c_1, c_0$ in $\mathbb{R}$ such that
>
> $$f(x) = c_n x^n + c_{n-1} x^{n-1} + \cdots + c_1 x + c_0$$
>
> for all $x \in \mathbb{R}$, then f is continuous.

A very useful topological characterization of continuity can be made on the basis of the next definition.

3.18 DEFINITION

> **RELATIVELY OPEN AND CLOSED SETS**   Let $M \subseteq \mathbb{R}$.
>
> (i) A subset H of M is said to be a **relatively open subset of** M, or simply **open in** M, if there is an open subset V of $\mathbb{R}$ such that
>
> $$H = V \cap M.$$
>
> (ii) A subset K of M is said to be a **relatively closed subset of** M, or simply **closed in** M, if there is a closed subset F of $\mathbb{R}$ such that
>
> $$K = F \cap M.$$

The topological characterization of continuity depends on the notion of relatively open sets and the idea of the inverse image of a set under a function [Part (ii) of Definition 4.10 in Chapter 2].

3.19 THEOREM

> **TOPOLOGICAL CHARACTERIZATION OF CONTINUITY**   A function f is continuous if and only if, for every open set $U \subseteq \mathbb{R}$, $f^{-1}(U)$ is relatively open in $\mathrm{dom}(f)$.

PROOF   Suppose that $f^{-1}(U)$ is relatively open in $\mathrm{dom}(f)$ for every open set $U \subseteq \mathbb{R}$. Let $a \in \mathrm{dom}(f)$. To prove that f is continuous at a, we use Theorem 3.13. Thus, suppose that $U \subseteq \mathbb{R}$ is open and that $f(a) \in U$. Then $a \in f^{-1}(U)$. Since $f^{-1}(U)$ is relatively open in $\mathrm{dom}(f)$, there exists an open set $V \subseteq \mathbb{R}$ such that

$$f^{-1}(U) = V \cap \mathrm{dom}(f).$$

Because $a \in f^{-1}(U)$, it follows that $a \in V$. Furthermore,

$$x \in V \cap \mathrm{dom}(f) \Rightarrow x \in f^{-1}(U) \Rightarrow f(x) \in U,$$

and so f is continuous at a by Theorem 3.13.

Conversely, suppose that f is continuous and let U be an open subset of $\mathbb{R}$. We must prove that $f^{-1}(U)$ is relatively open in dom(f). Note that $f^{-1}(U) \subseteq \text{dom}(f)$. If $a \in f^{-1}(U)$, then $a \in \text{dom}(f)$ and $f(a) \in U$; hence, by Theorem 3.13, there exists an open set $V_a \subseteq \mathbb{R}$ (possibly depending on a) such that $a \in V_a$ and, for all $x \in \mathbb{R}$,

$$x \in V_a \cap \text{dom}(f) \Rightarrow f(x) \in U \Rightarrow x \in f^{-1}(U).$$

Consequently,

$$a \in f^{-1}(U) \Rightarrow a \in V_a \cap \text{dom}(f) \subseteq f^{-1}(U).$$

Let

$$V = \bigcup\nolimits_{a \in f^{-1}(U)} V_a,$$

noting that V is open by Theorem 1.8. By the generalized distributive law (Theorem 6.6 in Chapter 1),

$$V \cap \text{dom}(f) = \bigcup\nolimits_{a \in f^{-1}(U)}(V_a \cap \text{dom}(f)).$$

Because each point $a \in f^{-1}(U)$ belongs to $V_a \cap \text{dom}(f)$, it follows that

$$f^{-1}(U) \subseteq V \cap \text{dom}(f).$$

Because, for each point $a \in f^{-1}(U)$, $V_a \cap \text{dom}(f) \subseteq f^{-1}(U)$, it also follows that

$$V \cap \text{dom}(f) \subseteq f^{-1}(U).$$

Therefore,

$$f^{-1}(U) = V \cap \text{dom}(f),$$

so $f^{-1}(U)$ is relatively open in dom(f).   □

Theorem 3.19 is fundamental for the topological study of the notion of continuity. (In fact, it provides a basis for the general definition of continuity in the theory of topological spaces.) In particular, the theorem will have a critical role to play in our work in the next two sections of this textbook.

PROBLEM SET 4.3

1. Find $\delta > 0$ such that, if $|x - 2| < \delta$, then $|x^2 - 4| < 0.007$. [Hint: See the solution of Example 3.2.]

2. Let $f(x) = x^3$ for all $x \in \mathbb{R}$. If $a \in \mathbb{R}$, make an ε–δ proof that f is continuous at a. [Hint: Use the fact that $x^3 - a^3 = (x - a)(x^2 + ax + a^2)$.]

3. If $f(x) = 2x - 3$ for all $x \in \mathbb{R}$, make an ε–δ proof that f is continuous at each point $a \in \mathbb{R}$.

4. If $f(x) = x|x|$ for all $x \in \mathbb{R}$, make an ε–δ proof that f is continuous at each point $a \in \mathbb{R}$.

5. Lemma 2.15 actually shows that the function g defined for $x \in \mathbb{R} \setminus \{0\}$ by $g(x) = 1/x$ is continuous. Motivate the given proof; that is, give an explanation of how one might be led to select the particular value of δ used in the proof.

6. Motivate the selection of δ and η in the proof of Lemma 2.13.

7. Prove Lemma 3.4.

8. Show that the function f in Problem 18 of Problem Set 4.2 is discontinuous at every point $a \in \mathbb{R}$.

9. Let $f:\mathbb{R} \to \mathbb{R}$ be defined by $f(x) = \text{sgn}|x|$ for every $x \in \mathbb{R}$.
(a) Sketch the graph of f.
(b) Show that f is discontinuous at 0.
(c) Show that f is continuous at every real number $a \neq 0$.

10. The function $\text{sinr}:\mathbb{R} \to \mathbb{R}$ defined for $x \in \mathbb{R}$ by

$$\text{sinr}(x) = \begin{cases} \sin(1/x), & \text{if } x \neq 0 \\ 0, & \text{if } x = 0 \end{cases}$$

plays an important role in the construction of counterexamples in mathematical analysis. (We have labeled this function *sinr* because, for $x \neq 0$, it is the *sine* of the *reciprocal* of x.) The graph of sinr is shown in Figure 4-1. Assuming that the sine function has the standard properties developed in elementary calculus, prove that:
(a) sinr is continuous at every point $a \neq 0$.
(b) sinr is discontinuous at 0.

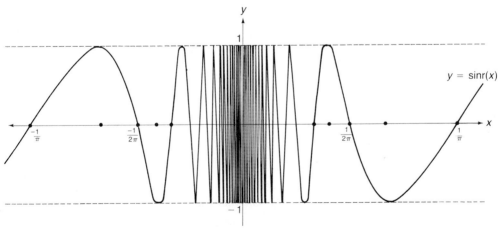

FIGURE 4-1

11. Prove Corollary 3.9.

12. Let $f : \mathbb{R} \to \mathbb{R}$ be defined by $f(x) = x \, \text{sinr}(x)$ for all $x \in \mathbb{R}$ (see Problem 10). The graph of f is shown in Figure 4-2. Prove that f is continuous.

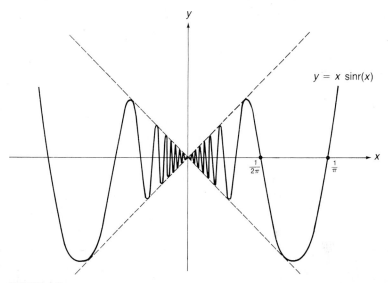

$$y = x \, \text{sinr}(x)$$

FIGURE 4-2

13. Use Theorems 2.12 and 3.5 to give an alternative proof of the part of Corollary 3.9 corresponding to the statement that *the sum of continuous functions is continuous*. (This is the proof usually given in elementary calculus.)

14. If f and g are functions, $D = \text{dom}(f) \cap \text{dom}(g)$, $a \in D$, f is continuous at a, and g is discontinuous at a, prove that the function $s : D \to \mathbb{R}$ defined by $s(x) = f(x) + g(x)$ for $x \in D$ is discontinuous at a. [Hint: Make an indirect proof.]

15. If $* : \mathbb{R} \times \mathbb{R} \to \mathbb{R}$ is a continuous binary operation (Problem 19 in Problem Set 4.2) and $g : \mathbb{R} \to \mathbb{R}$ is a continuous function, prove that the binary operation $\star : \mathbb{R} \times \mathbb{R} \to \mathbb{R}$ defined by $x \star y = g(x * y)$ for all $x, y \in \mathbb{R}$ is a continuous binary operation.

16. Let $\alpha : \mathbb{R} \times \mathbb{R} \to \mathbb{R}$, $\beta : \mathbb{R} \times \mathbb{R} \to \mathbb{R}$, and $\gamma : \mathbb{R} \times \mathbb{R} \to \mathbb{R}$ be continuous binary operations, and define a binary operation $* : \mathbb{R} \times \mathbb{R} \to \mathbb{R}$ by $x * y = (x \, \alpha \, y) \, \gamma \, (x \, \beta \, y)$. Prove that $*$ is a continuous binary operation.

17. Define the binary operation $\rho : \mathbb{R} \times \mathbb{R} \to \mathbb{R}$ by $x \, \rho \, y = |x - y|$ for all $x, y \in \mathbb{R}$. Prove that ρ is a continuous binary operation. [Hint: Use the result of Problem 15.]

18. Define the binary operation $\min: \mathbb{R} \times \mathbb{R} \to \mathbb{R}$ as usual by $\min(x, y) = \inf(\{x, y\})$ for $x, y \in \mathbb{R}$. Prove that min is a continuous binary operation on $\mathbb{R}$. [Hint: Use the fact that $\min(x, y) = \frac{1}{2}(x + y + |x - y|)$.]

19. Supply the details of the proof of Lemma 3.12.

20. Let f and g be functions, $D = \text{dom}(f) \cap \text{dom}(g)$, and suppose that both f and g are continuous at $a \in D$. Define $h: D \to \mathbb{R}$ by $h(x) = \min(f(x), g(x))$ for all $x \in D$. Prove that h is continuous at a. [Hint: See Problem 18.]

21. Prove Theorem 3.13.

22. A function $f: \mathbb{R} \to \mathbb{R}$ is said to be **additive** if $f(x + y) = f(x) + f(y)$ holds for all $x, y \in \mathbb{R}$.
  (a) Prove that an additive function is continuous if it is continuous at 0.
  (b) Prove that an additive function is continuous if it is continuous at any one point in $\mathbb{R}$.

23. Prove Lemma 3.16. [Hint: Use induction on n to prove Part (iii).]

24. Prove Corollary 3.17. [Hint: Use induction on n.]

25. If n is a positive integer, prove that $f: \mathbb{R} \setminus \{0\} \to \mathbb{R}$ defined by $f(x) = x^{-n}$ is continuous. [Hint: $f = p \circ r$ where $r(x) = 1/x$ for $x \neq 0$ and $p(x) = x^n$ for all x.]

26. A **rational function** is a function R for which there are polynomial functions P and Q, with Q not identically zero, such that $R(x) = P(x)/Q(x)$ for all $x \in \mathbb{R}$ with $Q(x) \neq 0$. Prove that every rational function is continuous.

27. A function f is said to be **bounded** on a set M if $f(M)$, the image of M under f, is a bounded set [that is, $f(M)$ has both a lower bound and an upper bound]. If f is continuous at a point $a \in \mathbb{R}$, prove that there exists a neighborhood J of a such that f is bounded on J.

28. If f is a continuous function and M is a bounded set, is it true that $f(M)$ is a bounded set? If true, prove it; if false, give a counterexample.

29. A function f is said to be **bounded away from zero** on a set M if there is a neighborhood N of 0 such that $f(M) \cap N = \emptyset$. If $a \in \text{dom}(f)$, f is continuous at a, and $f(a) \neq 0$, prove that there is a neighborhood J of a such that f is bounded away from zero on J.

30. If $M \subseteq \mathbb{R}$, prove that the union of a set of relatively open subsets of M is again a relatively open subset of M.

31. If $M \subseteq \mathbb{R}$, prove that a subset H of M is relatively open in M if and only if its relative complement $M \setminus H$ is relatively closed in M.

32. If $M \subseteq \mathbb{R}$, prove that the intersection of a finite set of relatively open subsets of M is again a relatively open subset of M.

33. Prove that a function f is continuous if and only if, for every closed set $F \subseteq \mathbb{R}$, $f^{-1}(F)$ is relatively closed in $\text{dom}(f)$.

34. If f is a function, prove that f is continuous if and only if $f(\bar{M}) \subseteq \overline{f(M)}$ holds for every $M \subseteq \mathbb{R}$.

35. If f is a function, prove that f is continuous if and only if the condition $\overline{f^{-1}(M)} \cap \text{dom}(f) \subseteq f^{-1}(\bar{M})$ holds for every $M \subseteq \mathbb{R}$.

36. If $M \subseteq \mathbb{R}$, prove that M is discrete if and only if every subset of M is relatively open in M.

37. If $f : \mathbb{N} \to \mathbb{R}$, prove that f is continuous.

38. Let f be a function, and suppose that Y is a set containing the range of f. Prove that f is continuous if and only if, for every set A that is relatively open (or, relatively closed) in Y, $f^{-1}(A)$ is relatively open (or, relatively closed) in the domain of f.

39. Let f be a continuous function and suppose that M is a nonempty subset of $\text{dom}(f)$. Prove that $f|_M$, the restriction of f to M, is a continuous function.

4
_____

CONNECTED SETS AND THE INTERMEDIATE VALUE THEOREM

An open or closed interval, (a, b) or $[a, b]$, is a *connected* set in the intuitive sense that it contains no vacant spaces or gaps. In this section, we are going to show that the idea of connectedness can be made quite precise, that it is a topological idea, and that it is an idea of some importance in mathematical analysis. However, we begin by studying the opposite idea—the notion of being *disconnected* or *separated*.

4.1 DEFINITION

> **SEPARATED SETS** Two subsets A and B of $\mathbb{R}$ are said to be **separated** if they are disjoint and no point of either set is an accumulation point of the other set.

For instance, the two open intervals $(1, 2)$ and $(2, 3)$ are separated. (Note that 2 is an accumulation point of both sets, but it belongs to neither set.)

4.2 LEMMA

> Let $A, B \subseteq \mathbb{R}$. Then A and B are separated if and only if
> $$\bar{A} \cap B = \varnothing \qquad \text{and} \qquad A \cap \bar{B} = \varnothing.$$

PROOF We sketch the proof and leave the details as an exercise (Problem 1). Let A^* be the set of all accumulation points of A, and let B^* be the set of all accumulation points of B. Then

$$\bar{A} = A \cup A^* \qquad \text{and} \qquad \bar{B} = B \cup B^*.$$

The condition that no point of A is an accumulation point of B is equivalent to $A \cap B^* = \varnothing$; the condition that no point of B is an accumulation point of A is equivalent to $A^* \cap B = \varnothing$; and the condition that A and B are disjoint is equivalent to $A \cap B = \varnothing$. Thus, the lemma follows from the observations that

$$\bar{A} \cap B = (A \cap B) \cup (A^* \cap B) \quad \text{and} \quad A \cap \bar{B} = (A \cap B) \cup (A \cap B^*). \quad \square$$

4.3 DEFINITION

> **SEPARATION OF A SET** A **separation** of a set $D \subseteq \mathbb{R}$ is a pair of sets $A, B \subseteq \mathbb{R}$ such that $A \neq \varnothing$, $B \neq \varnothing$, A and B are separated, and
>
> $$D = A \cup B.$$

In words, a separation of a set is a decomposition of the set into two nonempty separated pieces.

4.4 LEMMA

> Suppose that the pair of sets A, B forms a separation of the set D. Then both A and B are relatively open in D and relatively closed in D.

PROOF   By hypothesis, we have $A \neq \varnothing$, $B \neq \varnothing$,

$$D = A \cup B, \tag{1}$$

$$A \cap B = \varnothing, \tag{2}$$

$$\bar{A} \cap B = \varnothing, \tag{3}$$

and

$$A \cap \bar{B} = \varnothing. \tag{4}$$

From (1) and (4), it follows that

$$A \subseteq D \cap (\mathbb{R} \setminus \bar{B}). \tag{5}$$

Since $B \subseteq \bar{B}$, we have $\mathbb{R} \setminus \bar{B} \subseteq \mathbb{R} \setminus B$, and so, by (1) and (2), we have

$$D \cap (\mathbb{R} \setminus \bar{B}) \subseteq D \cap (\mathbb{R} \setminus B) = D \setminus B = A. \tag{6}$$

Combining (5) and (6), we conclude that

$$A = D \cap (\mathbb{R} \setminus \bar{B}). \tag{7}$$

By Part (i) of Theorem 1.17, $\bar{B}$ is a closed set; hence, by Theorem 1.5, $\mathbb{R} \setminus \bar{B}$ is an open set, and it follows from (7) and Part (i) of Definition 3.18 that A is a relatively open subset of D.

By (1) and the distributive law, we have

$$D \cap \bar{A} = (A \cup B) \cap \bar{A} = (A \cap \bar{A}) \cup (B \cap \bar{A}). \tag{8}$$

Since $A \subseteq \bar{A}$, it follows that

$$A \cap \bar{A} = A. \tag{9}$$

Combining (9), (3), and (8), we obtain

$$A = D \cap \bar{A}. \tag{10}$$

Since $\bar{A}$ is a closed set, it follows from Part (ii) of Definition 3.18 that A is a relatively closed subset of D. Thus, A is both relatively open and relatively closed in D. A symmetric argument shows that B is both relatively open and relatively closed in D. □

4.5 COROLLARY

> Suppose that the pair of sets A, B forms a separation of an open set U. Then both A and B are open sets.

PROOF   The proof follows from Lemma 4.4 and the observation that a relatively open subset of an open set is the intersection of two open sets and therefore is an open set by Theorem 1.11. □

4.6 DEFINITION

> **DISCONNECTED AND CONNECTED SETS**
>
> (i) A set D is **disconnected** if it admits a separation.
> (ii) A set C is **connected** if it is not disconnected.

If our definition of a connected set is reasonable, we ought to be able to prove that a bounded open interval is a connected set. However, to make the proof, we are going to need the next lemma.

4.7 LEMMA

> If $U \subseteq \mathbb{R}$ is a nonempty open set and $k = \sup(U)$, then
>
> $$k \in \bar{U} \setminus U.$$

PROOF   It is sufficient to prove that $k \notin U$ and that k is an accumulation point of U. If $k \in U$, then, by Theorem 1.6, there exists a positive number δ such that $(k - \delta, k + \delta) \subseteq U$. But then

$$k < k + \frac{\delta}{2} \in U,$$

contradicting the fact that k is an upper bound for U. Therefore, $k \notin U$.

To prove that k is an accumulation point of U, let N be a neighborhood of k. If ε is the radius of N, we have $k - \varepsilon < k$, and, therefore, by Part (i) of Lemma 3.23 in Chapter 3, there exists $u \in U$ such that $k - \varepsilon < u \leq k$. Since $k \notin U$, we cannot have $u = k$, and so

$$k - \varepsilon < u < k < k + \varepsilon.$$

Therefore,

$$u \in (k - \varepsilon, k + \varepsilon) = N \qquad \text{and} \qquad u \neq k.$$

It follows that

$$u \in (N \setminus \{k\}) \cap U,$$

and, hence, that

$$(N \setminus \{k\}) \cap U \neq \varnothing.$$

Therefore, k is an accumulation point of U. □

4.8 THEOREM	**BOUNDED OPEN INTERVALS ARE CONNECTED** Let $c, d \in \mathbb{R}$ with $c < d$. Then the bounded open interval (c, d) is a connected set.

PROOF   Suppose, on the contrary, that (c, d) is disconnected, so there exists a pair of sets A, B forming a separation of (c, d). Since the sets A and B are non-empty, we can choose points $a \in A$ and $b \in B$. The sets A and B are disjoint, so $a \neq b$, and, therefore, either $a < b$ or $b < a$. We suppose that $a < b$. (A symmetric argument takes care of the case in which $b < a$.) Thus,

$$c < a < b < d.$$

By Corollary 4.5, A is an open set. Let

$$U = (c, b) \cap A,$$

noting that

$$a \in U$$

and that U, being the intersection of two open sets, is an open set. Also, b is an upper bound for U, so U is a nonempty set with an upper bound. By the completeness of $\mathbb{R}$, the set U has a supremum. Let

$$k = \sup(U).$$

Since $a \in U$, b is an upper bound for U and k is the least upper bound for U, we have

$$a \leq k \leq b.$$

Therefore,

$$c < a \leq k \leq b < d,$$

and it follows that

$$k \in (c, d) = A \cup B.$$

By Lemma 4.7,

$$k \notin U = (c, b) \cap A.$$

Now we propose to show that $k \in B$. If $k = b$, this is clear. Thus, suppose $k \neq b$. Then $c < k < b$, so $k \in (c, b)$. But, if $k \in (c, b)$, the fact that $k \notin U = (c, b) \cap A$ implies that $k \notin A$, which (in view of the fact that $k \in A \cup B$)

implies that $k \in B$. In any case then,

$k \in B$.

By Lemma 4.7 again,

$k \in \bar{U}$.

However, $U \subseteq A$, and it follows that $\bar{U} \subseteq \bar{A}$, and, consequently,

$k \in \bar{A}$.

Therefore,

$k \in \bar{A} \cap B$,

contradicting the fact that

$\bar{A} \cap B = \varnothing$. □

4.9 LEMMA

Let $C \subseteq \mathbb{R}$ be a connected set, suppose that A and B are separated subsets of $\mathbb{R}$, and let

$C \subseteq A \cup B$.

Then

$C \subseteq A$ or $C \subseteq B$.

PROOF Since $C \subseteq A \cup B$ and $A \cap B = \varnothing$, it will be sufficient to prove that at least one of the two sets $C \cap A$ or $C \cap B$ is empty. (For instance, if $C \cap A = \varnothing$, it will follow that $C \subseteq B$.) Suppose, on the contrary, that both $C \cap A$ and $C \cap B$ are nonempty. Since $C \subseteq A \cup B$, we have

$C = (C \cap A) \cup (C \cap B)$.

Because $C \cap A \subseteq A$, it follows that

$\overline{C \cap A} \subseteq \bar{A}$,

and so

$\overline{C \cap A} \cap (C \cap B) \subseteq \bar{A} \cap B = \varnothing$.

Therefore,

$\overline{C \cap A} \cap (C \cap B) = \varnothing$.

A symmetric argument shows that

$\overline{C \cap B} \cap (C \cap A) = \varnothing$;

hence, by Lemma 4.2, the pair $C \cap A$, $C \cap B$ forms a disconnection of C, contradicting the supposition that C is connected. □

4.10 COROLLARY

> Let $C \subseteq \mathbb{R}$ be a connected set and suppose that
>
> $$C \subseteq M \subseteq \bar{C}.$$
>
> Then M is a connected set.

PROOF   Suppose the contrary, and let the pair A, B of sets form a separation of M. Then

$$C \subseteq M = A \cup B,$$

and it follows from Lemma 4.9 that $C \subseteq A$ or $C \subseteq B$. Suppose that $C \subseteq A$. (A symmetric argument takes care of the case in which $C \subseteq B$.) Then

$$M \subseteq \bar{C} \subseteq \bar{A},$$

and it follows from the fact that $\bar{A} \cap B = \varnothing$ that

$$B = M \cap B = \varnothing,$$

contradicting the supposition that the pair A, B forms a separation of M.
□

As an immediate consequence of Corollary 4.10:

> *The closure of a connected set is connected.*

Another consequence of Corollary 4.10 is that *bounded half-open intervals and bounded closed intervals are connected.* Indeed, if $a, b \in \mathbb{R}$, then

$$(a, b) \subseteq (a, b], [a, b) \subseteq [a, b] = \overline{(a, b)}$$

(Problem 7), and (a, b) is connected by Theorem 4.8.

In general, the union of connected sets need not be connected. For instance, the open intervals $(1, 2)$ and $(2, 3)$ are connected, but their union, $(1, 2) \cup (2, 3)$, is not connected. However, as the next theorem shows, under suitable conditions, the union of connected sets will be connected.

4.11 THEOREM

> **THE UNION OF CONNECTED SETS**   Let $\mathscr{C}$ be a set of connected sets such that
>
> $$H, K \in \mathscr{C} \Rightarrow H \cap K \neq \varnothing.$$
>
> Let
>
> $$C = \bigcup \mathscr{C}.$$
>
> Then C is a connected set.

PROOF   Suppose, on the contrary, that C is disconnected, and let the pair A, B of sets form a separation of C. Choose and fix any $K \in \mathscr{C}$. Then $K \subseteq C = A \cup B$, and it follows from Lemma 4.9 that $K \subseteq A$ or $K \subseteq B$. Assume that $K \subseteq A$. (The case in which $K \subseteq B$ is handled by a similar argument.) Now, for any $H \in \mathscr{C}$, $H \subseteq C = A \cup B$, and again we have $H \subseteq A$ or $H \subseteq B$. But, since $H \cap K \neq \varnothing$ and $A \cap B = \varnothing$, we cannot have $K \subseteq A$ and $H \subseteq B$. Consequently, $H \subseteq A$ for all $H \in \mathscr{C}$, and it follows that

$$B \subseteq A \cup B = C = \bigcup \mathscr{C} \subseteq A,$$

and, therefore, since $A \cap B = \varnothing$, that $B = \varnothing$, contradicting the hypothesis that A, B provides a separation of C. $\square$

A set $C \subseteq \mathbb{R}$ is said to be *convex* if, whenever $a, b \in C$, every point c between a and b belongs to C. More precisely:

4.12 DEFINITION

CONVEX SUBSET OF $\mathbb{R}$ A set $C \subseteq \mathbb{R}$ is said to be **convex** if, for every $a, b \in C$ with $a < b$, we have $[a, b] \subseteq C$.

Evidently, the empty set, all singleton sets, all bounded intervals (open, closed, or half-open), all unbounded intervals (open or closed), and $\mathbb{R}$ itself are convex sets (Problem 11). We leave the proof of the next lemma as an exercise (Problem 13).

4.13 LEMMA

If $C \subseteq \mathbb{R}$ is a convex set, C contains at least two points, and $c \in C$, then C is the union of the set of all closed bounded intervals $[a, b]$ such that $c \in [a, b] \subseteq C$.

Theorem 4.11 and Lemma 4.13 yield the following corollary.

4.14 COROLLARY

If $C \subseteq \mathbb{R}$ is convex, then C is connected.

PROOF   If $C = \varnothing$ or C is a singleton, then C is connected (Problem 5). Therefore, we can assume that C contains at least two points. Choose and fix a point $c \in C$. By Lemma 4.13, C is a union of a set of closed bounded intervals, each of which contains the point c; hence, by Theorem 4.11, C is connected. $\square$

4.15 THEOREM

CONVEX SETS AND CONNECTED SETS   A subset of $\mathbb{R}$ is convex if and only if it is connected.

PROOF

Let $C \subseteq \mathbb{R}$. By Corollary 4.14, if C is convex, then C is connected. To finish the proof, we show that if C is not convex, then C is disconnected. Suppose C is not convex. Then there exist points $a, b \in C$ with $a < b$ such that $[a, b] \not\subseteq C$. Hence, there exists $d \in [a, b]$ with $d \notin C$. Since $a, b \in C$, it follows that $a < d < b$. Let

$$A = (-\infty, d) \cap C$$

and let

$$B = (d, \infty) \cap C,$$

noting that $a \in A$ and $b \in B$, so $A \neq \varnothing$ and $B \neq \varnothing$. Also, since $d \notin C$, we have

$$C = A \cup B.$$

Now $A \subseteq (-\infty, d)$, so

$$\bar{A} \subseteq \overline{(-\infty, d)} = (-\infty, d],$$

and it follows that

$$\bar{A} \cap B = \varnothing.$$

A similar argument shows that

$$A \cap \bar{B} = \varnothing,$$

and it follows that the pair A, B provides a separation of the set C. Therefore, C is disconnected.   □

If $C \subseteq \mathbb{R}$ and f is a continuous function with $C \subseteq \text{dom}(f)$, we refer to the set $f(C)$ as a **continuous image** of C.

4.16 THEOREM

A CONTINUOUS IMAGE OF A CONNECTED SET IS CONNECTED   If f is a continuous function, $C \subseteq \text{dom}(f)$, and C is a connected set, then $f(C)$ is a connected set.

PROOF

We prove that if $f(C)$ is disconnected, then C is disconnected. Suppose the pair of sets A, B forms a separation of $f(C)$. Let

$$H = C \cap f^{-1}(A) \quad \text{and} \quad K = C \cap f^{-1}(B).$$

We are going to show that the pair H, K forms a separation of C. Since $\varnothing \neq A$, $B \subseteq f(C)$, it follows that H, $K \neq \varnothing$ (Problem 23). By using the distributive law and Problem 33b in Problem Set 2.4, we obtain

$$H \cup K = C \cap (f^{-1}(A) \cup f^{-1}(B)) = C \cap f^{-1}(A \cup B)$$
$$= C \cap f^{-1}(f(C)).$$

But, because $C \subseteq \mathrm{dom}(f)$, Problem 31e in Problem Set 2.4 shows that $C \subseteq f^{-1}(f(C))$, and, therefore,

$$C \cap f^{-1}(f(C)) = C.$$

Consequently,

$$H \cup K = C.$$

Since $H \subseteq f^{-1}(A)$, we have

$$\bar{H} \subseteq \overline{f^{-1}(A)}.$$

Therefore,

$$\bar{H} \cap K \subseteq \overline{f^{-1}(A)} \cap K = \overline{f^{-1}(A)} \cap C \cap f^{-1}(B)$$
$$\subseteq \overline{f^{-1}(A)} \cap \mathrm{dom}(f) \cap f^{-1}(B).$$

But, by Problem 35 in Problem Set 4.3,

$$\overline{f^{-1}(A)} \cap \mathrm{dom}(f) \subseteq f^{-1}(\bar{A}),$$

and it follows that

$$\bar{H} \cap K \subseteq f^{-1}(\bar{A}) \cap f^{-1}(B).$$

By applying Problem 33c in Problem Set 2.4 to the right side of the last inclusion, we obtain

$$f^{-1}(\bar{A}) \cap f^{-1}(B) = f^{-1}(\bar{A} \cap B) = f^{-1}(\varnothing) = \varnothing,$$

from which we can conclude that

$$\bar{H} \cap K = \varnothing.$$

A similar argument shows that

$$H \cap \bar{K} = \varnothing,$$

and completes the proof that the pair H, K provides a separation of C.

$\square$

The following theorem, which is a consequence of Theorem 4.16, is sometimes known as **Bolzano's theorem** because it was first proved analytically in a pamphlet published in 1817 by Bernhard Bolzano (1781–1848). Bolzano, who lived in Prague all his life, was a scholar, philosopher, theologian, and mathematician who made a number of important contributions to mathematical analysis. Unfortunately, his ideas were so far ahead of their time that his contemporaries did not always comprehend their full significance.

4.17 THEOREM

> **THE INTERMEDIATE VALUE THEOREM**   Let f be a continuous function, let $a, b \in \mathbb{R}$, and suppose that $[a, b] \subseteq \mathrm{dom}(f)$. Then, if y is any real number between $f(a)$ and $f(b)$, there exists $x \in [a, b]$ such that $y = f(x)$.

PROOF   Because $[a, b]$ is connected, it follows from Theorem 4.16 that $f([a, b])$ is connected; hence, by Theorem 4.15, that $f([a, b])$ is convex. Therefore, since $f(a), f(b) \in f([a, b])$ and y is between $f(a)$ and $f(b)$, it follows that $y \in f([a, b])$; hence, there exists $x \in [a, b]$ such that $y = f(x)$.   ☐

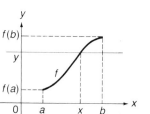

FIGURE 4-3

As Figure 4-3 illustrates, Bolzano's intermediate value theorem has a very simple geometric interpretation. The figure shows the graph of a continuous function on the interval $[a, b]$ and the theorem says that the graph of f must intersect every horizontal line between the levels of $f(a)$ and $f(b)$. Intuitively, it reflects the fact that the graph of a continuous function contains no jumps or gaps.

Bolzano's intermediate value theorem is often paraphrased as follows:

> *A continuous function defined on an interval takes on all values between any two of its values on the interval.*

If f is a function and $z \in \mathrm{dom}(f)$, we call z a **zero** of f if

$$f(z) = 0.$$

The following corollary of the intermediate value theorem, which is called the **change-of-sign property** of continuous functions, is particularly useful for finding such zeros. We leave the simple proof as an exercise (Problem 27).

4.18 COROLLARY

> **CHANGE-OF-SIGN PROPERTY OF CONTINUOUS FUNCTIONS**   Suppose that f is a continuous function, that $a, b \in \mathbb{R}$ with $a \neq b$, and that the closed interval with endpoints a and b is contained in $\mathrm{dom}(f)$. Then, if $f(a)$ and $f(b)$ have opposite algebraic signs, it follows that there is a zero z of f strictly between a and b.

We submit the next theorem as a further indication of the utility of the intermediate value theorem.

4.19 THEOREM

> **EXISTENCE OF *n*th ROOTS** Let n be a positive integer and let y be a real number with $y \geq 0$. Then there exists a unique real number x with $x \geq 0$ such that $y = x^n$.

PROOF   Define $f : \mathbb{R} \to \mathbb{R}$ by $f(x) = x^n$ for every $x \in \mathbb{R}$. Let $y \geq 0$. Choose a real number $b \geq 0$ such that $b^n \geq y$. (For instance, if $y \leq 1$, choose $b = 1$, and if $y > 1$, choose $b = y$.) Because $f(0) = 0$, we have

$$f(0) \leq y \leq f(b).$$

By Part (iii) of Lemma 3.16, f is continuous; hence, by Theorem 4.17, there exists $x \in [0, b]$ such that

$$y = f(x) = x^n.$$

This proves the existence of $x \in \mathbb{R}$ with $x \geq 0$ such that $y = x^n$. To prove the uniqueness, suppose that $z \in \mathbb{R}$ with $z \geq 0$ and $y = z^n$. We must prove that $x = z$. Suppose not. Then either $x < z$ or $z < x$. Suppose $x < z$. (The case in which $z < x$ is handled similarly.) Then

$$x^n < z^n,$$

(Problem 28), contradicting $x^n = y = z^n$.   □

In Theorem 4.19, the unique real number x is called the **principal *n*th root** of y and is written as

$$x = y^{1/n}.$$

If n is *odd*, this notation is extended to negative real numbers by defining, for $y > 0$,

$$(-y)^{1/n} = -(y^{1/n}).$$

Note that (when working within the real number system), *n*th *roots of negative numbers are undefined for values of n that are even*. Of course, the usual square root notation is used for $y^{1/2}$, so that, if $y \geq 0$, then, by definition

$$\sqrt{y} = y^{1/2}.$$

PROBLEM SET 4.4

1. Supply the details of the proof of Lemma 4.2.

2. If A and B are separated sets, $A_1 \subseteq A$, and $B_1 \subseteq B$, prove that A_1 and B_1 are separated sets.

3. Prove that two disjoint sets are separated if and only if no point of either set is a boundary point of the other set.

4. Let $D = A \cup B$ and suppose that $A \neq \varnothing$, $B \neq \varnothing$, and $A \cap B = \varnothing$.
    (a) Prove that the pair A, B forms a separation of D if and only if both A and B are relatively open in D.
    (b) Prove that the pair A, B forms a separation of D if and only if both A and B are relatively closed in D.

5. (a) Show that the empty set is connected.
    (b) Show that every singleton set is connected.

6. Prove that a set is connected if and only if it contains no proper subset that is both relatively open and relatively closed.

7. If $a, b \in \mathbb{R}$ with $a < b$, prove that $\overline{(a, b)} = [a, b]$ and thus conclude, using Corollary 4.10, that $(a, b]$, $[a, b)$, and $[a, b]$ are connected sets.

8. Prove that an open set is connected if and only if it cannot be written as the union of two disjoint, nonempty, open subsets.

9. If $D \subseteq \mathbb{R}$, D contains at least two points, and D contains an isolated point, prove that D is disconnected.

10. (a) Prove that the empty set and $\mathbb{R}$ itself are the only subsets of $\mathbb{R}$ that are both open and closed.
    (b) Prove that every nontrivial subset of $\mathbb{R}$ has a nonempty boundary.

11. Prove that the empty set, all singleton sets, all bounded intervals, all unbounded intervals, and $\mathbb{R}$ itself are convex sets.

12. If D is a finite subset of $\mathbb{R}$ and $\#D > 1$, prove that D is disconnected.

13. Prove Lemma 4.13.

14. Let f be continuous, $C \subseteq \mathrm{dom}(f)$, and $A \subseteq f(C)$.
    (a) If A is relatively open in $f(C)$, prove that $C \cap f^{-1}(A)$ is relatively open in C.
    (b) Prove that $C \setminus (C \cap f^{-1}(A)) = C \cap f^{-1}(f(C) \setminus A)$.
    (c) Using Parts (a) and (b), prove that, if A is relatively closed in $f(C)$, then $C \cap f^{-1}(A)$ is relatively closed in C.
    (d) If A is a proper subset of $f(C)$, prove that $C \cap f^{-1}(A)$ is a proper subset of C.

15. If M is a set and $x \in M$, prove that the union C of all the connected subsets of M that contain the point x is a connected subset of M. The set C is called the **connected component** of x in M. A subset of M is called a

connected component of M if it is the connected component of a point $x \in M$.

16.   Combine the results of Problems 6 and 14 to give an alternative proof of Theorem 4.16.

17.   If C is a connected subset of M, prove that C is a connected component of M if and only if C is a maximal connected subset of M in the sense that if $C \subseteq K \subseteq M$ and K is connected, then $C = K$. (See Problem 15.)

18.   Prove that a set $D \subseteq \mathbb{R}$ is disconnected if and only if there exists $c \in \mathbb{R} \setminus D$ such that $(-\infty, c) \cap D \neq \varnothing$ and $(c, \infty) \cap D \neq \varnothing$.

19.   Let $M \subseteq \mathbb{R}$.
       (a) If C_1 and C_2 are connected components of M and $C_1 \neq C_2$, prove that $C_1 \cap C_2 = \varnothing$.
       (b) Prove that M is the union of all its own connected components. (See Problem 15.)

20.   Let C and K be connected subsets of $\mathbb{R}$ with $C \subseteq K$. Suppose that $K \setminus C = A \cup B$, where A and B are separated sets. Prove that both $C \cup A$ and $C \cup B$ are connected sets.

21.   (a) Prove that every connected component of a nonempty open subset of $\mathbb{R}$ is an open interval. (See Problem 15.)
       (b) Prove that every open subset of $\mathbb{R}$ is a union of disjoint open intervals. [Hint: Use Part (a) and Problem 19.]

22.   In Problem 20, suppose that $C = \{x\}$. Prove that $\bar{A} = A \cup \{x\}$ and $\bar{B} = B \cup \{x\}$.

23.   In the proof of Theorem 4.16, prove that the sets H and K are nonempty.

24.   Let C be a connected set with $C = A \cup B$, and assume that both A and B are relatively closed in C. If $A \cap B$ is connected, prove that both A and B are connected.

25.   If $D \subseteq \mathbb{R}$, prove that D is disconnected if and only if there exists a continuous function $f : D \to \mathbb{R}$ such that $f(D)$ consists of exactly two points. [Hint: For the "if" part of the proof, use Theorem 4.16. For the "only if" part, suppose that A, B provides a separation of D and define $f(x) = 0$ for $x \in A$ and $f(x) = 1$ for $x \in B$.]

26.   Let D be a discrete subset of $\mathbb{R}$.
       (a) If K is a connected subset of D, prove that $K = \{k\}$ for some $k \in D$.
       (b) If C is a connected subset of $\mathbb{R}$ and $f : C \to D$ is continuous, prove that f is a constant function.

27.   Prove Corollary 4.18.

28.   In the proof of Theorem 4.19, it is asserted that, if n is a positive integer and $0 \leq x < z$, then $x^n < z^n$. Use mathematical induction on n to prove this.

29. Show that the function f defined for all $x \in \mathbb{R}$ by $f(x) = x^3 + x - 1$ has a zero between 0 and 1.

30. If C is a nonempty connected subset of $\mathbb{R}$ and $f: C \to \mathbb{Q}$ is a continuous function, prove that f is a constant function. [Hint: Use Theorem 5.15 in Chapter 3.]

31. Show that there is a real number x with $1 < x < 2$ such that $x^5 - 2x^3 = 9$.

32. Give a direct proof of Corollary 4.18 as follows: Suppose for definiteness that $a < b$, so that $[a, b] \subseteq \text{dom}(f)$. Assume that $f(a) < 0 < f(b)$. [The case in which $f(b) < 0 < f(a)$ is handled by a similar argument.] Let $A = \{x \in [a, b] \mid f(x) \leq 0\}$. Show that A is a nonempty set and that A has an upper bound. Conclude that A has a supremum. If $c = \sup(A)$, prove that $f(c) = 0$.

33. If f is a function and $f(c) = c$, then c is called a **fixed point** of f. A subset C of $\mathbb{R}$ is said to have the **fixed point property** if every continuous function $f: C \to C$ has a fixed point. Prove that the closed unit interval $[0, 1]$ has the fixed point property. [Hint: If $f: [0, 1] \to [0, 1]$ is continuous, consider the function $g: [0, 1] \to \mathbb{R}$ defined by $g(x) = f(x) - x$ for all $x \in [0, 1]$.]

34. Give an alternative proof of Theorem 4.17 based on the result of Problem 32. [Hint: Suppose that f is continuous and that $[a, b] \subseteq \text{dom}(f)$. In the case in which $f(a) < k < f(b)$, apply the result of Problem 32 to the function $g: \text{dom}(f) \to \mathbb{R}$ given by $g(x) = f(x) - k$ for all $x \in \text{dom}(f)$.]

35. (a) Prove that every closed bounded interval $[a, b]$ has the fixed point property. (See Problem 33.)

(b) Prove that the open unit interval $(0, 1)$ does not have the fixed point property.

36. Make an ε–δ proof to show that the function $f: [0, \infty) \to \mathbb{R}$ defined by $f(x) = \sqrt{x}$ for all $x \in [0, \infty)$ is continuous at every point $a \in [0, \infty)$. [Hint: Use Problem 36 in Problem Set 3.3.]

37. (a) Assume that the exponential function $x \mapsto e^x$ has the usual properties developed in elementary calculus. Prove that the equation $e^x = 3x$ has a solution $x \in [0, 1]$.

(b) Show that $e^x = 3x$ has a second solution $x \in [1, 2]$.

(c) Does this equation have any more solutions? Why or why not?

(d) Use a calculator and Newton's method from elementary calculus to find all solutions of $e^x = 3x$ (correct to four decimal places).

38. Prove that every polynomial function of odd degree has at least one zero in $\mathbb{R}$.

39. Let f be a function and let C be a connected set with $\text{dom}(f) \subseteq C$. Let F be the set of points in C where f is undefined, discontinuous, or takes the value zero, and suppose that F is finite.

(a) Prove that $C \setminus F$ is a union of finitely many open intervals.

(b) Prove that f maintains a constant algebraic sign on each of these open intervals. (This is the basis for the so-called **cut-point method** for determining the intervals on which a function is either positive or negative.)

40. Use the cut-point method (Problem 39) to determine the intervals on which the function

$$f(x) = \frac{x^2 + x - 6}{x^2 - 1}$$

is positive or negative.

5

COMPACT SETS AND THE EXTREME VALUE THEOREM

In our study of the system $\mathbb{R}$ of real numbers, we have seen that several important ideas and properties can be expressed topologically, that is, in terms of open sets. Our purpose in this section is to show that, in $\mathbb{R}$, the concept of *boundedness* can also be captured topologically. This is done by introducing and developing the idea of *compactness*.

The notion of compactness, in various guises, has a long and involved history. One of the seeds of the idea is found in the same 1817 pamphlet in which Bolzano proved the intermediate value theorem. Bolzano's work was extended and incorporated into lectures given in Berlin in the 1860s by the German mathematician, Karl Weierstrass (1815–1897), who is widely regarded as the "father of modern mathematical analysis." In the course of these lectures, Weierstrass proved a fundamental result, now known as the *Bolzano–Weierstrass theorem* (see Theorem 5.12), which helped to lead later investigators (Heine, Borel, Lebesgue, and others) to the modern concept of compactness.

The notion of compactness is formulated in terms of the idea of an *open cover*, defined as follows:

5.1 DEFINITION

> **COVER AND OPEN COVER**   If M is a set and $\mathscr{C}$ is a set of sets, we say that $\mathscr{C}$ **covers** M if $M \subseteq \bigcup \mathscr{C}$. If $\mathscr{C}$ covers M and every set $V \in \mathscr{C}$ is open, we say that $\mathscr{C}$ is an **open cover** of M.

Often, in the interest of euphony, we refer to a set of sets as a *collection* of sets. Note that a collection of sets $\mathscr{C}$ covers a set M if and only if every point $x \in M$ belongs to at least one set $V \in \mathscr{C}$. When we say that $\mathscr{C}$ is an open cover of M, we do not mean to imply that $\mathscr{C}$ itself is an open set— this would not even make sense because $\mathscr{C}$ is not a subset of $\mathbb{R}$, it is a collection of subsets of $\mathbb{R}$. The adjective *open* in the phrase *open cover* refers not to the collection of sets but rather to the sets in the collection. Perhaps it would be clearer to speak of a *cover by open sets*, but the terminology *open cover* is well-established mathematical parlance.

5.2 Example Let $f: D \to \mathbb{R}$ be a continuous function and let $\varepsilon > 0$. Then, for each $a \in D$, there is a $\delta_a > 0$ such that, for all $x \in D$,

$$|a - x| < \delta_a \Rightarrow |f(a) - f(x)| < \varepsilon.$$

For each $a \in D$, let N_a denote the neighborhood with center a and radius δ_a. If

$$\mathscr{C} = \{N_a | a \in D\},$$

show that $\mathscr{C}$ is an open cover of D.

SOLUTION If $a \in D$, then $a \in N_a \in \mathscr{C}$, so $\mathscr{C}$ covers D. By Lemma 1.7, every $N_a \in \mathscr{C}$ is an open set; hence, $\mathscr{C}$ is an open cover of D. ☐

5.3 DEFINITION **SUBCOVER AND FINITE SUBCOVER** Let $\mathscr{C}$ be a cover of the set M. A subset $\mathscr{F}$ of $\mathscr{C}$ is called a **subcover** of M if it too is a cover of M. If $\mathscr{F}$ is such a subcover of M, and if there are only a finite number of sets in $\mathscr{F}$, then we refer to $\mathscr{F}$ as a **finite subcover** of M.

Thus, if $\mathscr{C}$ is a cover of M, then to say that $\mathscr{F}$ is a finite subcover of M means that all three of the following conditions hold:

(i) $\#\mathscr{F}$ is finite.
(ii) $\mathscr{F} \subseteq \mathscr{C}$.
(iii) $\mathscr{F}$ covers M.

Note that the adjective *finite* in the phrase *finite subcover* refers to the number of sets in the collection $\mathscr{F}$ itself and not to the individual sets in the collection. Even though there are only a finite number of sets in $\mathscr{F}$, each set $V \in \mathscr{F}$ may well be infinite.

5.4 DEFINITION **COMPACT SET** A set $M \subseteq \mathbb{R}$ is said to be **compact** if every open cover $\mathscr{C}$ of M contains a finite subcover $\mathscr{F}$ of M.

The definition of compactness is often paraphrased as follows:

A set is compact if every open cover of the set can be reduced to a finite subcover.

For instance, every finite set M is compact (Problem 2).

5.5 LEMMA | In $\mathbb{R}$, every closed bounded interval is compact.

PROOF Let $a, b \in \mathbb{R}$ with $a < b$ and suppose that $\mathscr{C}$ is an open cover of $[a, b]$. Consider the closed subintervals $[a, k]$ of $[a, b]$, for $a \leq k \leq b$, where, for purposes of this proof, we understand that $[a, a]$ denotes the singleton $\{a\}$. We define K to be the set of all $k \in [a, b]$ such that there exists a finite subset $\mathscr{F}$ of $\mathscr{C}$ that covers the subinterval

$$[a, k]$$

of $[a, b]$. To prove the theorem, it is sufficient to show that $b \in K$.

Since $a \in [a, b]$ and $\mathscr{C}$ covers $[a, b]$, there exists $U \in \mathscr{C}$ such that $a \in U$. Now $\{U\} \subseteq \mathscr{C}$, $\{U\}$ is finite, and $\{U\}$ covers the set

$$\{a\} = [a, a],$$

and so, by the definition of K, we have $a \in K$. In particular, $K \neq \varnothing$. Because $K \subseteq [a, b]$, it follows that b is an upper bound for K. Therefore, by the completeness of $\mathbb{R}$, K has a supremum. Let

$$d = \sup(K).$$

Next, we are going to prove that $d \in K$. Since $a \in K$, we have $a \leq d$, and, since b is an upper bound for k, we also have $d \leq b$. Therefore

$$d \in [a, b],$$

so there exists an open set $V \in \mathscr{C}$ such that $d \in V$. By Theorem 1.6, there exists $\varepsilon > 0$ such that

$$(d - \varepsilon, d + \varepsilon) \subseteq V.$$

Because

$$d - \varepsilon < d = \sup(K),$$

Part (i) of Lemma 3.23 in Chapter 3 implies the existence of $k \in K$ such that

$$d - \varepsilon < k \leq d.$$

If $d = k$, then $d \in K$, as we wanted to show. Thus, suppose that $k < d$. Then

$$(k, d] \subseteq (d - \varepsilon, d + \varepsilon) \subseteq V.$$

By the definition of K, there exists a finite subset $\mathscr{F}$ of $\mathscr{C}$ that covers the subinterval $[a, k]$ of $[a, b]$. Since $(k, d] \subseteq V$, it follows that $\mathscr{F} \cup \{V\}$ is a finite subset of $\mathscr{C}$ that covers

$$[a, k] \cup (k, d] = [a, d].$$

Therefore, by the definition of K, we have $d \in K$. In any case, then, we have

$$d \in K.$$

If $d = b$, we have $b \in K$, and our proof is finished. Suppose, on the contrary, that $d < b$. By the definition of K and the fact that $d \in K$, there exists a finite subset $\mathscr{F}$ of $\mathscr{C}$ that covers $[a, d]$. In particular, there exists an open set $W \in \mathscr{F}$ such that $d \in W$. By Theorem 1.6, there exists $\delta > 0$ such that

$$(d - \delta, d + \delta) \subseteq W.$$

Let

$$\eta = \min\left(\frac{\delta}{2}, b - d\right),$$

and let

$$c = d + \eta.$$

Now

$$a \leq d < c = d + \eta \leq d + (b - d) = b,$$

so $c \in [a, b]$. Also,

$$d < c = d + \eta \leq d + \frac{\delta}{2} < d + \delta,$$

so

$$[d, c] \subseteq (d - \delta, d + \delta) \subseteq W \in \mathscr{F}.$$

This shows that $\mathscr{F}$ covers not only $[a, d]$ but also that it actually covers the larger interval

$$[a, d] \cup [d, c] = [a, c].$$

Consequently, $c \in K$. But $d < c$, contradicting the fact that d is the supremum of K.   □

5.6 LEMMA | If $F \subseteq \mathbb{R}$ is closed, $K \subseteq \mathbb{R}$ is compact, and $F \subseteq K$, then F is compact.

PROOF   Let $\mathscr{C}$ be an open cover of F. We must show that there is a finite subset $\mathscr{F}$ of $\mathscr{C}$ that still covers F. Let

$$U = \mathbb{R} \setminus F,$$

noting that U is open and that

$K \setminus F \subseteq U$.

Let

$\mathscr{E} = \mathscr{C} \cup \{U\}$.

Then, since $\mathscr{C}$ covers F and $\{U\}$ covers $K \setminus F$, it follows that $\mathscr{E}$ covers K. Because K is compact, there exists a finite subset $\mathscr{G}$ of $\mathscr{E}$ that covers K.
Let

$\mathscr{F} = \mathscr{G} \setminus \{U\}$,

noting that $\mathscr{F}$ is a finite subset of $\mathscr{C}$. Let $x \in F$. Then $x \in K$, and it follows that there exists $V \in \mathscr{G}$ such that $x \in V$. Since $x \in F$, we cannot have $x \in U = \mathbb{R} \setminus F$, and so $V \neq U$. Therefore, $V \in \mathscr{F}$. This shows that for each point $x \in F$ there exists $V \in \mathscr{F}$ such that $x \in V$; that is, $\mathscr{F}$ covers F. $\quad\square$

A set $M \subseteq \mathbb{R}$ is said to be **bounded** if it is bounded both above and below; that is, if it has both an upper bound b and a lower bound a. In other words, M is bounded if and only if it is contained in a closed bounded interval.

By combining Lemmas 5.5 and 5.6, we obtain the following theorem of classical analysis, named in honor of the German mathematician Heinrich Eduard Heine (1821–1881) and the French mathematician Félix Édouard Émile Borel (1871–1956), both of whom were pioneers in the development of modern mathematical analysis.

5.7 THEOREM

> **HEINE–BOREL THEOREM** Every closed and bounded subset of $\mathbb{R}$ is compact.

PROOF Let F be a closed and bounded subset of $\mathbb{R}$. Since F is bounded, there exist $a, b \in \mathbb{R}$ such that $F \subseteq [a, b]$. By Lemma 5.5, $[a, b]$ is compact, hence, by Lemma 5.6, F is compact. $\quad\square$

It turns out that the converse of the Heine–Borel theorem is also true, so that the compact subsets of $\mathbb{R}$ are precisely the sets that are closed and bounded. To prove this, we begin with a lemma and a corollary to the lemma.

5.8 LEMMA

> If M is a compact subset of $\mathbb{R}$ and $a \in \mathbb{R} \setminus M$, then there exist open subsets U and V of $\mathbb{R}$ such that $M \subseteq U$, $a \in V$, and $U \cap V = \varnothing$.

PROOF   For each $x \in M$, we have $x \neq a$, and it follows from the Hausdorff property of $\mathbb{R}$ (Lemma 2.4) that there exist neighborhoods N_x of x and K_x of a such that

$$N_x \cap K_x = \varnothing.$$

If $x \in M$, then N_x and K_x are open sets by Lemma 1.7; hence,

$$\{N_x | x \in M\}$$

is an open cover of the compact set M. Consequently, there exists a finite subset F of M such that

$$\{N_x | x \in F\}$$

covers F. Therefore, if we let

$$U = \bigcup_{x \in F} N_x,$$

then U is an open set by Theorem 1.8 and we have

$$M \subseteq U.$$

Also, if we let

$$V = \bigcap_{x \in F} K_x,$$

then V is an open set by Theorem 1.11 and we have

$$x \in V.$$

To prove that $U \cap V = \varnothing$, suppose $q \in U \cap V$. Since $q \in U$, there exists $x \in F$ such that

$$q \in N_x.$$

But since $q \in V$, we also have

$$q \in K_x,$$

contradicting the fact that $N_x \cap K_x = \varnothing$.   □

5.9 COROLLARY | If M is a compact subset of $\mathbb{R}$, then M is closed.

PROOF   It is sufficient to prove that $\mathbb{R} \setminus M$ is open. If $a \in \mathbb{R} \setminus M$, then, by Lemma 5.8, there exist open sets $U, V \subseteq \mathbb{R}$ such that $M \subseteq U$, $a \in V$, and $U \cap V = \varnothing$. In particular, then, $V \cap M = \varnothing$, so

$$a \in V \subseteq \mathbb{R} \setminus M.$$

By Theorem 1.6, there is a neighborhood N of a with $N \subseteq V$. Thus, for every point $a \in \mathbb{R} \setminus M$, there is a neighborhood N of a such that $N \subseteq \mathbb{R} \setminus M$. Consequently, $\mathbb{R} \setminus M$ is open by Theorem 1.6.   □

5.10 THEOREM

> **CHARACTERIZATION OF COMPACT SUBSETS OF $\mathbb{R}$**   A subset M of $\mathbb{R}$ is compact if and only if it is closed and bounded.

PROOF   By the Heine–Borel theorem (Theorem 5.7), M is compact if it is closed and bounded. Conversely, suppose that M is compact. By Corollary 5.9, M is closed, and we have only to show that it is bounded. Let

$$\mathscr{C} = \{(-r, r) | r \in \mathbb{R}, r > 0\}.$$

Since $\mathscr{C}$ covers $\mathbb{R}$, it certainly covers M. Therefore, $\mathscr{C}$ is an open cover of M, and it follows from the compactness of M that there is a finite set F of positive real numbers such that

$$M \subseteq \{(-r, r) | r \in F\}.$$

Let b denote the largest of the positive numbers in the finite set F. Then

$$M \subseteq (-b, b) \subseteq [-b, b],$$

so M is bounded. $\square$

5.11 COROLLARY

> **CHARACTERIZATION OF BOUNDED SUBSETS OF $\mathbb{R}$**   A subset M of $\mathbb{R}$ is bounded if and only if its closure is compact.

PROOF   If M is bounded, then there exists a closed bounded interval $[a, b]$ such that

$$M \subseteq [a, b].$$

Therefore,

$$\bar{M} \subseteq \overline{[a, b]} = [a, b].$$

By Lemma 5.5, $[a, b]$ is compact, and, since $\bar{M}$ is closed, it follows from Lemma 5.6 that $\bar{M}$ is compact. Conversely, suppose that $\bar{M}$ is compact. Then, by Theorem 5.10, $\bar{M}$ is bounded. Because $M \subseteq \bar{M}$, it follows that M is bounded. $\square$

5.12 THEOREM

> **BOLZANO–WEIERSTRASS THEOREM**   If M is a bounded infinite subset of $\mathbb{R}$, then M has at least one accumulation point in $\mathbb{R}$.

PROOF   It is sufficient to prove that, if M is a bounded subset of $\mathbb{R}$ and M has no accumulation points in $\mathbb{R}$, then M is finite. Thus, assume that M is bounded and has no accumulation points. Since M has no accumulation points, it follows that $M = \bar{M}$. Thus, by Corollary 5.11, M is compact. If $x \in M$, then, since x is not an accumulation point of M, it is an isolated point of M, and it follows from Lemma 1.19 that there exists a neighborhood N_x of x such that

$$N_x \cap M = \{x\}.$$

Since

$$\{N_x | x \in M\}$$

is an open cover of M and M is compact, there is a finite subset F of M such that

$$M \subseteq \bigcup_{x \in F} N_x.$$

Therefore,

$$M = \left(\bigcup_{x \in F} N_x \right) \cap M = \bigcup_{x \in F} (N_x \cap M) = \bigcup_{x \in F} \{x\} = F,$$

so M is a finite set.   □

5.13 THEOREM

> **A CONTINUOUS IMAGE OF A COMPACT SET IS COMPACT**   If f is a continuous function, $C \subseteq \text{dom}(f)$, and C is a compact set, then $f(C)$ is a compact set.

PROOF   Let $\mathscr{C}$ be an open cover of $f(C)$. For the proof, it will be convenient to index the open sets in $\mathscr{C}$ so that, say,

$$\mathscr{C} = \{U_\alpha | \alpha \in A\},$$

where A is the indexing set. Since f is continuous, $f^{-1}(U_\alpha)$ is relatively open in $\text{dom}(f)$ for every $\alpha \in A$ (Theorem 3.19). Therefore, for each $\alpha \in A$, there exists an open set $V_\alpha \subseteq \mathbb{R}$ such that

$$f^{-1}(U_\alpha) = V_\alpha \cap \text{dom}(f).$$

We claim that

$$\{V_\alpha | \alpha \in A\}$$

is an open cover of C. Indeed, if $x \in C$, then $f(x) \in f(C)$, so there exists $\alpha \in A$ such that $f(x) \in U_\alpha$; therefore, $x \in f^{-1}(U_\alpha) \subseteq V_\alpha$. Consequently, since C is compact, there is a finite subset F of A such that

$$\{V_\alpha | \alpha \in F\}$$

covers C.

We finish the proof by showing that

$$\{U_\alpha \,|\, \alpha \in F\}$$

covers $f(C)$. Indeed, suppose $x \in f(C)$. Then there exists $c \in C$ such that $x = f(c)$, and there exists $\alpha \in F$ such that $c \in V_\alpha$. Thus, $c \in C \subseteq \text{dom}(f)$, so

$$c \in V_\alpha \cap \text{dom}(f) = f^{-1}(U_\alpha);$$

hence,

$$x = f(c) \in U_\alpha,$$

and it follows that $\{U_\alpha \,|\, \alpha \in F\}$ covers $f(C)$. $\square$

5.14 THEOREM

EXTREME VALUE THEOREM   Let f be a continuous function, and let C be a nonempty compact set contained in $\text{dom}(f)$. Then there exist $p, q \in C$ such that, for all $x \in C$,

$$f(p) \le f(x) \le f(q).$$

PROOF   By Theorem 5.13, $f(C)$ is a compact set. Hence, by Theorem 5.10, $f(C)$ is a closed and bounded set. Since $\varnothing \ne C \subseteq \text{dom}(f)$, it follows that $f(C)$ is nonempty. By the completeness of $\mathbb{R}$, there exist $a, b \in \mathbb{R}$ with

$$a = \inf(f(C)) \qquad \text{and} \qquad b = \sup(f(C)).$$

Because $f(C)$ is closed, $a, b \in f(C)$ (see Problem 21a in Problem Set 4.1). Therefore, there exist $p, q \in C$ such that

$$f(p) = a \qquad \text{and} \qquad f(q) = b.$$

Because a is a lower bound for $f(C)$ and b is an upper bound for $f(C)$, we have, for all $x \in C$,

$$f(p) = a \le f(x) \le b = f(q).$$ $\square$

In the proof of Theorem 5.14, $a = \inf(f(C))$ and $b = \sup(f(C))$ are, respectively, the *minimum* and the *maximum* values $f(x)$ of the function f for $x \in C$. Therefore, the extreme value theorem is often paraphrased as:

On a compact set, a continuous function attains its minimum and maximum values.

Of course, a closed bounded interval $[a, b]$ is a compact set, and so, in particular, on a closed bounded interval, a continuous function attains its minimum and maximum values.

Compactness is often used to help prove existence theorems. The prototype of such theorems is the following:

5.15 THEOREM

> **NESTED INTERVAL THEOREM**   Let $\mathscr{B}$ be a nonempty collection of closed and bounded intervals with the property that, for $I, J \in \mathscr{B}$, either $I \subseteq J$ or $J \subseteq I$. Then there exists $p \in \mathbb{R}$ such that
>
> $p \in I$ holds for every $I \in \mathscr{B}$.

PROOF   Choose and fix an interval $K \in \mathscr{B}$. By Theorem 5.10, K is compact. Let

$$\mathscr{C} = \{\mathbb{R} \setminus J \mid J \in \mathscr{B}\}.$$

Note that each set $\mathbb{R} \setminus J$ in $\mathscr{C}$ is open.

Suppose $\mathscr{C}$ covers K. Then, by the compactness of K, there exists a finite subset

$$\{J_1, J_2, J_3, \ldots, J_n\} \subseteq \mathscr{B}$$

such that

$$\{\mathbb{R} \setminus J_1, \mathbb{R} \setminus J_2, \mathbb{R} \setminus J_3, \ldots, \mathbb{R} \setminus J_n\}$$

covers K; that is,

$$K \subseteq \mathbb{R} \setminus J_1 \cup \mathbb{R} \setminus J_2 \cup \mathbb{R} \setminus J_3 \cup \cdots \cup \mathbb{R} \setminus J_n.$$

By hypothesis, for $1 \leq i, j \leq n$, we have

$$J_i \subseteq J_j \qquad \text{or} \qquad J_j \subseteq J_i,$$

and therefore one of the sets

$$J_1, J_2, J_3, \ldots, J_n$$

must be contained in all of the others (Problem 4). By relabeling, if necessary, we can suppose that

$$J_1 \subseteq J_1, J_2, J_3, \ldots, J_n.$$

Therefore,

$$\mathbb{R} \setminus J_1, \mathbb{R} \setminus J_2, \mathbb{R} \setminus J_3, \ldots, \mathbb{R} \setminus J_n \subseteq \mathbb{R} \setminus J_1.$$

Consequently,

$$K \subseteq \mathbb{R} \setminus J_1 \cup \mathbb{R} \setminus J_2 \cup \mathbb{R} \setminus J_3 \cup \cdots \cup \mathbb{R} \setminus J_n \subseteq \mathbb{R} \setminus J_1,$$

and it follows that

$$K \cap J_1 = \varnothing.$$

But, by hypothesis, $K \subseteq J_1$ or $J_1 \subseteq K$, and therefore $K = \emptyset$ or $J_1 = \emptyset$, contradicting the fact that K and J_1 are closed bounded intervals.

From the argument above, we conclude that $\mathscr{C}$ does not cover K. Therefore, there exists a point $p \in K$ such that, for all $J \in \mathscr{B}$, $p \notin \mathbb{R} \setminus J$; that is, $p \in J$. $\qquad\qquad\square$

Another important application of the idea of compactness appears in the following lemma, which bears the name of the French mathematician Henri Léon Lebesgue (1875–1941).

5.16 LEMMA

> **LEBESGUE COVERING LEMMA**   If $D \subseteq \mathbb{R}$ is compact and $\mathscr{C}$ is an open cover of D, then there exists $\delta > 0$ such that, for every $a, b \in D$ with $|a - b| < \delta$, there exists $U \in \mathscr{C}$ with $a, b \in U$.

PROOF

For each $d \in D$, there is an open set $U_d \in \mathscr{C}$ with $d \in U_d$. Since U_d is open, there is a neighborhood K_d of d such that $K_d \subseteq U_d$. For each $d \in D$, denote by ε_d the radius of K_d and let N_d be the neighborhood of d whose radius is $\varepsilon_d/2$. Then, for each $d \in D$,

$$d \in N_d \subseteq K_d \subseteq U_d \in \mathscr{C}. \tag{1}$$

Since $\{N_d \mid d \in D\}$ is an open cover of D, and D is compact, there is a finite set $F \subseteq D$ such that

$$\{N_p \mid p \in F\} \text{ covers } D. \tag{2}$$

Let ε be the smallest number in the finite set $\{\varepsilon_p \mid p \in F\}$, so that

$$0 < \varepsilon \leq \varepsilon_p \qquad \text{for all } p \in F, \tag{3}$$

and let

$$\delta = \frac{\varepsilon}{2}. \tag{4}$$

Then, by (3) and (4),

$$0 < \delta \leq \frac{\varepsilon_p}{2} \qquad \text{for all } p \in F. \tag{5}$$

Now suppose that $a, b \in D$ and

$$|a - b| < \delta. \tag{6}$$

By (2), there exists $c \in F$ such that

$$a \in N_c. \tag{7}$$

Since $a \in N_c$ and the radius of N_c is $\varepsilon_c/2$, we have

$$\left|c - a\right| < \frac{\varepsilon_c}{2}. \tag{8}$$

By the triangle inequality, (8), and (5), we have

$$\left|c - b\right| \leq \left|c - a\right| + \left|a - b\right| < \frac{\varepsilon_c}{2} + \delta$$

$$\leq \frac{\varepsilon_c}{2} + \frac{\varepsilon_c}{2} = \varepsilon_c, \tag{9}$$

and it follows that

$$b \in K_c. \tag{10}$$

Finally, from (1), (7), and (10), we conclude that

$$a, b \in U_c \in \mathscr{C}. \tag{11}$$

$\square$

If $f : D \to \mathbb{R}$ is a continuous function and $\varepsilon > 0$, then, for each point $d \in D$, there is a $\delta > 0$ (possibly depending on d) such that, for every $x \in D$,

$$\left|x - d\right| < \delta \Rightarrow \left|f(x) - f(d)\right| < \varepsilon. \tag{*}$$

In general, however, it may be necessary to use *different values of δ for different points $d \in D$.* If, given an $\varepsilon > 0$, a single $\delta > 0$ can be found so that (*) holds for all d and all x in D, then we say that f is *uniformly continuous.* More precisely:

5.17 DEFINITION   **UNIFORM CONTINUITY**   Let f be a function and let $D \subseteq \mathrm{dom}(f)$. We say that f is **uniformly continuous on D** if, for every $\varepsilon > 0$, there exists $\delta > 0$ such that, for all $a, b \in D$,

$$\left|a - b\right| < \delta \Rightarrow \left|f(a) - f(b)\right| < \varepsilon.$$

The following theorem is fundamental in modern mathematical analysis.

5.18 THEOREM   **UNIFORM CONTINUITY ON COMPACT SETS**   Let f be a function, let $D \subseteq \mathrm{dom}(f)$, and suppose that f is continuous at each point in D. Then, if D is compact, it follows that f is uniformly continuous on D.

PROOF   Let $\varepsilon > 0$. For each point $d \in D$, there is a neighborhood N_d of d such that, for all $x \in D$,

$$x \in N_d \Rightarrow |f(x) - f(d)| < \frac{\varepsilon}{2}.$$

The collection $\mathscr{C} = \{N_d | d \in D\}$ is an open cover of the compact set D; hence, by the Lebesgue covering lemma (Lemma 5.16) there exists $\delta > 0$ such that, if $a, b \in D$ with $|a - b| < \delta$, then there exists $d \in D$ such that $a, b \in N_d$. However, if $a, b \in N_d$, we have

$$|f(a) - f(d)| < \frac{\varepsilon}{2} \quad \text{and} \quad |f(b) - f(d)| < \frac{\varepsilon}{2},$$

and, therefore, by the triangle inequality,

$$|f(a) - f(b)| < \frac{\varepsilon}{2} + \frac{\varepsilon}{2} = \varepsilon. \qquad \square$$

PROBLEM SET 4.5

1. Make a direct proof that the open unit interval $(0, 1)$ is not compact by showing that $\mathscr{C} = \{(0, 1 - \varepsilon) | 0 < \varepsilon < 1\}$ is an open cover of $(0, 1)$ that has no finite subcover.

2. If F is a finite subset of $\mathbb{R}$ and $\mathscr{C}$ is an open cover of F, make a direct argument (not using the results in this section) to show that $\mathscr{C}$ can be reduced to a finite subcover of F.

3. Make a direct proof that $\mathbb{N} = \{1, 2, 3, \ldots\}$ is not a compact subset of $\mathbb{R}$ by exhibiting an open cover that has no finite subcover.

4. In the proof of Theorem 5.15, it is asserted that if n is a positive integer and $J_1, J_2, J_3, \ldots, J_n \subseteq \mathbb{R}$ are such that, for $1 \le i, j \le n$, $J_i \subseteq J_j$ or $J_j \subseteq J_i$, then one of the sets J_k must be contained in all of the others. Prove this by mathematical induction on n.

5. Prove that the union of a finite collection of compact subsets of $\mathbb{R}$ is a compact set.

6. Prove that a subset of $\mathbb{R}$ that is both discrete and compact must be finite.

7. Let $M \subseteq \mathbb{R}$. Prove that M is compact if and only if every cover of M by sets that are relatively open in M can be reduced to a finite subcover.

8. Prove the following extension of Lemma 5.8: If A and B are compact subsets of $\mathbb{R}$ and $A \cap B = \varnothing$, then there exist open subsets U and V of $\mathbb{R}$ such that $A \subseteq U$, $B \subseteq V$, and $U \cap V = \varnothing$.

9. Prove that the boundary of a compact set is a compact set.

10. Let $\mathcal{B}$ be a nonempty collection of sets. We say that $\mathcal{B}$ is a **nested** collection of sets, or that $\mathcal{B}$ forms a **chain** if, for all $A, B \in \mathcal{B}$, $A \subseteq B$ or $B \subseteq A$. We say that $\mathcal{B}$ has the **finite intersection property** if, for every finite nonempty subcollection $\mathcal{F} \subseteq \mathcal{B}$, $\bigcap \mathcal{F} \neq \varnothing$. If $\mathcal{B}$ is a nested collection of nonempty sets, prove that $\mathcal{B}$ has the finite intersection property.

11. Show that Theorem 5.15 remains true if the word *intervals* is replaced by the words *nonempty sets*.

12. Prove that a set $M \subseteq \mathbb{R}$ is compact if and only if every collection $\mathcal{B}$ of relatively closed subsets of M that has the finite intersection property (see Problem 10) has a nonempty intersection. [Hint: Look at the complements of the sets in $\mathcal{B}$, use the De Morgan laws, and use the result of Problem 7.]

13. If f is a continuous function and I is a closed bounded interval contained in $\text{dom}(f)$, prove that $f(I)$ is a singleton set or a closed bounded interval.

14. If $\varnothing \neq M \subseteq \mathbb{R}$ and the set $D = \{|a - b|\,|\,a, b \in M\}$ has an upper bound, then $\sup(D)$ is called the **diameter** of M. Show that M is bounded if and only if it has a finite diameter.

15. Let f be a continuous function and suppose that $\text{dom}(f)$ is a compact set. If $M \subseteq \mathbb{R}$ is closed, show that $f(M)$ is closed.

16. With the terminology of Problem 14, show that, if M is bounded, then the diameter of M is $\sup(M) - \inf(M)$.

17. Let $X, Y \subseteq \mathbb{R}$, suppose that X is compact and nonempty, and let $f : X \to Y$ be a continuous bijection. Prove that $f^{-1} : Y \to X$ is continuous. [Hint: Use the result in Problem 15.]

18. If $\mathcal{C}$ is an open cover of the set D, then a positive number δ is called a **Lebesgue number** for $\mathcal{C}$ if every subset A of D such that the diameter of A is less than δ is contained in at least one set $U \in \mathcal{C}$ (see Problems 14 and 16). Prove that every open cover $\mathcal{C}$ of a compact set $D \subseteq \mathbb{R}$ has a Lebesgue number. [Hint: Argue as in the proof of Lemma 5.16.]

19. If, in Problem 17, X is not compact, then $f^{-1} : Y \to X$ is not necessarily continuous. For instance, let $X = [0, 1] \cup [2, 3)$, $Y = [0, 2]$, and define $f : X \to Y$ by $f(x) = x$ for $x \in [0, 1]$ and $f(x) = 4 - x$ for $x \in [2, 3)$.
(a) Show that $f : X \to Y$ is a bijection.
(b) Sketch graphs of both f and f^{-1}.
(c) Prove that f is continuous.
(d) Prove that f^{-1} is discontinuous at $1 \in Y$.

20. If D is a bounded (but not necessarily closed) set and $f : D \to \mathbb{R}$ is uniformly continuous on D, prove that $f(D)$ is a bounded set. [Hint: Find $\delta > 0$ such that, for $a, b \in D$, $|a - b| < \delta \Rightarrow |f(a) - f(b)| < 1$ and use the fact that D is bounded to cover it with a finite number of open intervals each of radius δ.]

21. Let $f:D \to \mathbb{R}$. In Parts (a) and (b) write a quantified expression in predicate calculus equivalent to the condition that:
    (a) f is continuous on D.
    (b) f is uniformly continuous on D.
    (c) Explain why the uniform continuity of f on D implies its continuity on D.

22. Use the result of Problem 36 in Problem Set 3.3 to prove that the function $f(x) = \sqrt{x}$ is uniformly continuous on $[0, \infty)$.

23. Use the results of Problem 21 to write a quantified expression in predicate calculus equivalent to the condition that f is not uniformly continuous on D.

24. Making free use of suitable results of elementary calculus, give a direct ε–δ proof that the sine function is uniformly continuous on $\mathbb{R}$.

25. Let $f:[0, \infty) \to \mathbb{R}$ be defined by $f(x) = x^2$ for $x \geq 0$. Prove that f is not uniformly continuous on $[0, \infty)$. [Hint: Show that if δ is any positive real number, there exist $x, y \geq 0$ such that $|x - y| < \delta$ but $|x^2 - y^2| \geq 1$. Try letting $y = x + (\delta/2)$, so that $|x - y| < \delta$ is automatic, and then find a suitable value of x so that $|x^2 - y^2| \geq 1$.]

26. Let sinr$:\mathbb{R} \to \mathbb{R}$ be the function defined in Problem 10 in Problem Set 4.3. Show that sinr is not uniformly continuous on the interval $(0, 1)$.

27. Make a direct ε–δ argument to show that $f(x) = x^2$ is uniformly continuous on $[0, 1]$.

28. With the notation of Problem 26, define $f:(0, 1) \to \mathbb{R}$ by $f(x) = x$ sinr(x). Show that f is uniformly continuous on $(0, 1)$.

29. If a function f is uniformly continuous on a set D, show that it is uniformly continuous on every subset of D.

30. True or false: If two functions are uniformly continuous on a set D, then their sum is uniformly continuous on D. If true, prove it; if false, give a counterexample.

31. True or false: If two functions are uniformly continuous on a set D, then their product is uniformly continuous on D. If true, prove it; if false, give a counterexample.

32. If $f:A \to B$ is uniformly continuous on A and $g:B \to C$ is uniformly continuous on B, prove that $g \circ f:A \to C$ is uniformly continuous on A.

33. Let D be a subset of the domain of the function f, let a be an accumulation point of D, and suppose that κ is a positive real number such that $f((a - \kappa, a + \kappa) \cap D)$ is a bounded set.
    (a) If $0 < \eta \leq \kappa$, prove that $f((a - \eta, a + \eta) \cap D)$ is a nonempty bounded set.
    (b) If $0 < \eta \leq \kappa$, define
    $$s_\eta = \sup(f((a - \eta, a + \eta) \cap D)).$$

Explain why this supremum exists in $\mathbb{R}$ and prove that, if $0 < \lambda \leq \eta \leq \kappa$, then $s_\lambda \leq s_\eta$.

(c) Prove that $\{s_\eta | 0 < \eta \leq \kappa\}$ has a lower bound and conclude that the infimum

$$L = \inf(\{s_\eta | 0 < \eta \leq \kappa\})$$

exists in $\mathbb{R}$.

(d) Prove that, if ε is a positive real number, there exists $\eta \in \mathbb{R}$ with $0 < \eta \leq \kappa$ and there exists $d \in D$ with $|d - a| < \eta$ such that $|f(d) - L| < \varepsilon$. [Hint: Apply Part (ii) of Lemma 3.23 in Chapter 3 with the set $S = \{s_\eta | 0 < \eta \leq \kappa\}$, $b = L$, and $q = L + \varepsilon$ to produce $\eta \in \mathbb{R}$ with $0 < \eta \leq \kappa$ and $L \leq s_\eta < L + \varepsilon$. Then note that $L - \varepsilon < s_\eta$, and apply Part (i) of the same lemma.]

34. Suppose that the function f is uniformly continuous on a set $D \subseteq \text{dom}(f)$ and let a be an accumulation point of D.

(a) If κ is positive, prove that $f((a - \kappa, a + \kappa) \cap D))$ is a nonempty bounded set.

(b) With the notation of Problem 33, prove that $L = \lim_{x \to a} f(x)$.

35. If f is a function, $D \subseteq \text{dom}(f)$, C is a compact set with $D \subseteq C$, and there exists a continuous function $g : C \to \mathbb{R}$ such that $f(d) = g(d)$ for all $d \in D$, prove that f is uniformly continuous on D.

36. Suppose that the function f is uniformly continuous on the nonempty set $D \subseteq \text{dom}(f)$. Define $g : \bar{D} \to \mathbb{R}$ for all $a \in \bar{D}$ by

$$g(a) = \begin{cases} f(a), & \text{if } a \in D \\ \lim_{x \to a} f(x), & \text{if } a \in \bar{D} \setminus D \end{cases},$$

using the result of Problem 34 to guarantee the existence of the limit. Prove that g is uniformly continuous on $\bar{D}$.

37. Let C be a compact set and let $f : C \to \mathbb{R}$ be a continuous function. If M is a closed subset of $\mathbb{R}$, prove that $f^{-1}(M)$ is a compact set.

38. Let $f : D \to \mathbb{R}$ be a function and suppose that D is a bounded set. Show that f is uniformly continuous on D if and only if f can be extended to a continuous function $g : \bar{D} \to \mathbb{R}$.

6

COUNTABLE AND UNCOUNTABLE SETS

How do we actually count a finite set M? We point to the elements of M, one at a time, in some definite order. As we point to each element, we name a natural number, starting from 1 and proceeding with the successive natural numbers until we have exhausted the set M. At the end of this counting process, a certain natural number n is assigned to the last element of M, and we conclude that $\#M = n$. The effect of the counting process is to set up a bijection

$$f:\{1, 2, 3, \ldots, n\} \to M$$

from an **initial segment** $\{1, 2, 3, \ldots, n\}$ of the set $\mathbb{N}$ of natural numbers onto M. If $n \in \mathbb{N}$, we denote the initial segment $\{1, 2, 3, \ldots, n\}$ by I_n, so that

$$I_n = \{m \in \mathbb{N} \mid m \leq n\}.$$

Using this notation, we can give a formal definition of the counting process.

6.1 DEFINITION

COUNTING A FINITE SET　A set M is said to be **finite** if it is empty or if there exists $n \in \mathbb{N}$ and a bijection $f:I_n \to M$. To **count** the set M means to find such a bijection.

As the German mathematician and physicist Hermann Weyl (1885–1955) remarked in his influential book, *Theory of Groups and Quantum Mechanics* (New York: Dover, 1950, p. 131), the fact that when a finite set M is counted in two different ways one always obtains the same number is "perhaps the most fundamental theorem of mathematics." Thus, we have the following theorem:

6.2 THEOREM

FUNDAMENTAL THEOREM OF COUNTING　Let M be a set, and let $m, n \in \mathbb{N}$. If there exist bijections

$$f:I_n \to M \qquad \text{and} \qquad g:I_m \to M,$$

then $n = m$.

Although this theorem is intuitively obvious, its proof is not as simple as one might think. [A proof can be found in D. J. Foulis and M. A. Munem, *After Calculus: Algebra* (San Francisco: Dellen/Macmillan, 1988), p. 313.] Theorem 6.2 provides a foundation for the following basic definition.

6.3 DEFINITION

THE CARDINAL NUMBER OF A FINITE SET　Let M be a finite, non-empty set. The unique natural number n such that there is a bijection $f:I_n \to M$ is called the **cardinal number** of M, written

$$n = \#M.$$

Using Definition 6.3, and defining $\#\varnothing = 0$, it is relatively easy to provide rigorous proofs for the properties of finite sets given in Section 5 of Chapter 1, pages 63–65. (Again, see *After Calculus: Algebra*, pp. 313–317, for the details.) Similarly, rigorous proofs can be given for the following two theorems, both of which are intuitively obvious (Problems 1 and 2).

6.4 THEOREM

> **EQUALITY OF CARDINAL NUMBERS**   If M and N are finite sets, then $\#M = \#N$ if and only if there exists a bijection $f:M \to N$.

6.5 THEOREM

> **ORDERING OF CARDINAL NUMBERS**   If M and N are finite sets, then $\#M \le \#N$ if and only if there exists an injection $f:M \to N$.

A set that is not finite is said to be **infinite**. For instance, the set $\mathbb{N}$ of all natural numbers is an infinite set (Problem 3a). A set that contains an infinite subset must itself be infinite (Problem 4b), and therefore the sets $\mathbb{Z}$, $\mathbb{Q}$, and $\mathbb{R}$ are also infinite. Georg Cantor (1845–1918), the founder of set theory, realized that the ideas implicit in Theorems 6.4 and 6.5 can be extended to infinite sets and that the result is a useful and logically self-consistent theory of *transfinite numbers*. Our purpose in this section is to develop a small fragment of this theory.

6.6 DEFINITION

> **EQUINUMEROUS SETS**   If M and N are sets, we say that M and N are **equinumerous**, or that they have the same **cardinality**, and write
>
> $$\#M = \#N$$
>
> if there exists a bijection $f:M \to N$.

The relation of being equinumerous is reflexive, symmetric, and transitive (Problem 5). Note that, if M and N are infinite sets, Definition 6.6 does not specify what $\#M$ and $\#N$ *are*, it merely establishes the notation $\#M = \#N$ as shorthand for the statement that there is a bijection $f:M \to N$. In more advanced textbooks, a suitable meaning is assigned to the symbol $\#M$ for an infinite set M, and $\#M$ is referred to as a **transfinite cardinal number**. For instance, the symbol $\aleph_0$, called **aleph-subscript-zero**, is used to denote the cardinal number of the set $\mathbb{N}$ of all natural numbers, so that

$$\#\mathbb{N} = \aleph_0.$$

(Aleph, written $\aleph$, is the first letter of the Hebrew alphabet.) This leads to our next definition.

6.7 DEFINITION

> **COUNTABLE AND UNCOUNTABLE SETS**
>
> A set M is said to be **countably infinite** if there is a bijection
>
> $$f : \mathbb{N} \to M.$$
>
> We say that M is **countable** if it is either finite or countably infinite. If M is not countable, it is said to be **uncountable**.

The notation $\# M = \aleph_0$ may be used to express the condition that the set M is countably infinite. Some authors prefer to use the word **denumerable** rather than *countable* in Definition 6.7. A set that is not denumerable is called **nondenumerable**.

6.8 Example   If E is the set of all even positive integers, show that $\# E = \aleph_0$.

SOLUTION   The mapping $f : \mathbb{N} \to E$ given by $f(n) = 2n$ for all $n \in \mathbb{N}$ is a bijection. $\qquad\square$

6.9 Example   Show that the set $\mathbb{Z}$ of all integers is countably infinite.

SOLUTION   Define $f : \mathbb{N} \to \mathbb{Z}$ by

$$f(n) = \begin{cases} n/2, & \text{if } n \text{ is even} \\ (1 - n)/2, & \text{if } n \text{ is odd} \end{cases}$$

for all $n \in \mathbb{N}$. In Problem 7, we ask you to show that $f : \mathbb{N} \to \mathbb{Z}$ is a bijection. $\qquad\square$

6.10 Example   Show that the set $\mathscr{P}(\mathbb{N})$ of all subsets of $\mathbb{N}$ is uncountable.

SOLUTION   Clearly, $\mathscr{P}(\mathbb{N})$ is not a finite set (Problem 9). We must show that $\mathscr{P}(\mathbb{N})$ is not countably infinite. Suppose, on the contrary, that there exists a bijection

$$f : \mathbb{N} \to \mathscr{P}(\mathbb{N}). \tag{1}$$

Then, for each $k \in \mathbb{N}$, $f(k)$ is a subset of $\mathbb{N}$. Let

$$K = \{ k \in \mathbb{N} \mid k \notin f(k) \}. \tag{2}$$

Because $K \in \mathscr{P}(\mathbb{N})$ and $f : \mathbb{N} \to \mathscr{P}(\mathbb{N})$ is surjective, there exists $n \in \mathbb{N}$ such that

$$f(n) = K. \tag{3}$$

If $n \in K$, then, by (2), $n \notin f(n) = K$, a contradiction. Therefore, $n \notin K$. But then $n \notin f(n)$, and it follows from (2) that $n \in K$, contradicting $n \notin K$.   □

To show that a set X is countably infinite, it is sufficient to produce a bijective mapping from $\mathbb{N}$ onto X. A mapping

$$s : \mathbb{N} \to X$$

(bijective or not) is called a **sequence** in X and, for $n \in \mathbb{N}$, $s(n)$ is called the **nth term** of the sequence. Often the nth term $s(n)$ is written in the alternative form

$$s_n = s(n),$$

and it is imagined that these terms are arranged in an endless progression

$$s_1, s_2, s_3, \ldots.$$

Although this helps us to think about sequences, it is essential to keep in mind that a sequence is really a mapping $s : \mathbb{N} \to X$.

If a sequence $s : \mathbb{N} \to X$ is defined by a specific formula, we say that we have a description of the sequence in **closed form**. For instance, the formula

$$s_n = \sqrt{n}, \qquad \forall n \in \mathbb{N},$$

provides a closed-form description of a sequence $s : \mathbb{N} \to \mathbb{R}$.

Another important technique for constructing a sequence $s_1, s_2, s_3, \ldots$ is by **recursion**, that is, by specifying the first k terms $s_1, s_2, s_3, \ldots, s_k$ for some $k \in \mathbb{N}$ and giving a rule whereby, for each $n \in \mathbb{N}$ with $n \geq k$, s_{n+1} is determined by the preceding terms $s_1, s_2, s_3, \ldots, s_k, \ldots, s_n$. The case in which $k = 1$ and a rule is given that determines s_{n+1} from s_n for each $n \geq 1$ is called **simple recursion**. Thus, for simple recursion, the first term s_1 is given and the rule for determining s_{n+1} from s_n takes the form

$$s_{n+1} = g(s_n),$$

where g is a specified function. That sequences can be defined by simple recursion is the content of the next theorem.

6.11 THEOREM

> **PRINCIPLE OF SIMPLE RECURSION**   Let X be a set, let $a \in X$, and let $g : X \to X$. Then there exists a unique sequence $s : \mathbb{N} \to X$ such that $s(1) = a$ and $s(n + 1) = g(s(n))$ for all $n \in \mathbb{N}$.

Although the principle of simple recursion is intuitively clear, its rigorous proof is somewhat involved and is omitted here. (For a proof, see D. J. Foulis and M. A. Munem, *After Calculus: Algebra*, pp. 326–329.)

6.12 LEMMA

> Let $X \subseteq \mathbb{N}$, and suppose that X is an infinite set. Then $\#X = \aleph_0$.

PROOF   Our proof makes repeated use of the fact that $\mathbb{N}$ is well-ordered (Theorem 4.14 in Chapter 3). Thus, if M is a nonempty subset of $\mathbb{N}$, then $\inf(M)$ exists, belongs to M, and is the smallest element of M.

For each natural number $x \in X$, define

$$M_x = \{y \in X \mid x < y\}.$$

If $x \in X$, then $M_x \neq \varnothing$, for otherwise we would have $X \subseteq I_x$, contradicting the hypothesis that X is an infinite set. (Note that the initial segment I_x of $\mathbb{N}$ is a finite set with $\#I_x = x$.) Define $g : X \to X$ by

$$g(x) = \inf(M_x)$$

for all $x \in X$. Thus, for every $x \in X$, $g(x)$ is the smallest element of X that is larger than x. In particular,

$$x < g(x)$$

and there exists no element $y \in X$ with $x < y < g(x)$. [Intuitively, if $x \in X$, then $g(x)$ is the "next" natural number in X.]

Now we define a sequence $s : \mathbb{N} \to X$ by simple recursion (Theorem 6.11), so that

$$s(1) = \inf(X)$$

and, for every $n \in \mathbb{N}$,

$$s(n + 1) = g(s(n)).$$

Thus, $s(1)$ is the smallest element of X and, for every $n \in \mathbb{N}$,

$$s(n) < s(n + 1).$$

Also, there exists no element $y \in X$ with $s(n) < y < s(n + 1)$. [Intuitively, the natural numbers $s(1), s(2), s(3), \ldots$ are the successive elements of X in increasing order.]

The fact that $s(n) < s(n + 1)$ for all $n \in \mathbb{N}$ implies that $s : \mathbb{N} \to X$ is an injection and that

$$n \leq s(n)$$

holds for all $n \in \mathbb{N}$ (Problem 13). If $s : \mathbb{N} \to X$ is also a surjection, then it is a bijection, and our proof is complete. Thus, suppose $a \in X$. We must show that there exists $n \in \mathbb{N}$ such that $a = s(n)$. Since $s(1) = \inf(X)$ and $a \in X$, we have

$$s(1) \leq a.$$

If $a = s(1)$, our proof is finished, so we can suppose that

$$s(1) < a.$$

Because $a \le s(a)$, the set $\{m \in \mathbb{N} \,|\, a \le s(m)\}$ is nonempty. Let

$$n = \inf(\{m \in \mathbb{N} \,|\, a \le s(m)\}).$$

Thus, n is the smallest natural number such that

$$a \le s(n).$$

Since $s(1) < a$, it follows that $n \ne 1$; so that $n - 1 \in \mathbb{N}$. Because $n - 1$ is smaller than n, we cannot have $a \le s(n - 1)$, and therefore we must have

$$s(n - 1) < a \le s(n).$$

But, there is no element in X strictly between $s(n - 1)$ and $s(n)$, and it follows that $a = s(n)$. □

6.13 THEOREM

> **INJECTIONS INTO A COUNTABLE SET** Let M and N be sets, let $h: M \to N$ be an injection, and suppose that N is countable. Then M is countable.

PROOF   If M is finite, then it is countable and we are done. Thus, we can suppose that M is infinite. Let

$$J = h(M) \tag{1}$$

and note that

$$h: M \to J \tag{2}$$

is a bijection, so that

$$\# M = \# J. \tag{3}$$

Therefore, J is infinite, and, since $J \subseteq N$, it follows that N is infinite. By the hypothesis that N is countable, we conclude that N is countably infinite and therefore there exists a bijection $f: \mathbb{N} \to N$. Let $g = f^{-1}$, so that

$$g: N \to \mathbb{N} \tag{4}$$

is a bijection, and let

$$X = g(J). \tag{5}$$

Note that $X \subseteq \mathbb{N}$. If $g|_J$ denotes the restriction of g to J, then

$$(g|_J): J \to X \tag{6}$$

is a bijection (Problem 15), and it follows that

$$\#J = \#X. \tag{7}$$

Because J is infinite, X is an infinite subset of $\mathbb{N}$, so, by Lemma 6.12,

$$\#X = \aleph_0. \tag{8}$$

By using Relations (3), (7), and (8) and the fact that being equinumerous is a transitive relation, we conclude that

$$\#M = \aleph_0. \tag{9}$$

$\square$

We leave as an exercise the proof of the following corollary to Theorem 6.13 (Problem 17a).

6.14 COROLLARY

A set M is countable if and only if there exists an injection $h : M \to \mathbb{N}$.

Our next theorem is an analogue for surjections of Theorem 6.13 for injections.

6.15 THEOREM

SURJECTIONS FROM A COUNTABLE SET Let N and M be sets, let $f : N \to M$ be a surjection, and suppose that N is countable. Then M is countable.

PROOF By Corollary 6.14, there exists an injection

$$h : N \to \mathbb{N}.$$

Since $f : N \to M$ is a surjection, $f^{-1}(y)$ is a nonempty subset of N for each $y \in M$. Therefore, for each $y \in M$, $h(f^{-1}(y))$ is a nonempty subset of $\mathbb{N}$. Define

$$g : M \to \mathbb{N}$$

by

$$g(y) = \inf(h(f^{-1}(y)))$$

for each $y \in M$. Since $\mathbb{N}$ is well-ordered, $g(y)$ exists and is the smallest element in $h(f^{-1}(y))$.

We are going to prove that $g : M \to \mathbb{N}$ is an injection. Thus, suppose that $y, z \in M$ and that

$$g(y) = g(z).$$

Since $g(y) \in h(f^{-1}(y))$, there exists $u \in f^{-1}(y)$ such that $g(y) = h(u)$. Likewise, there exists $v \in f^{-1}(z)$ such that $g(z) = h(v)$. Consequently,

$$h(u) = g(y) = g(z) = h(v),$$

and the injectivity of h implies that

$$u = v.$$

But, since $u \in f^{-1}(y)$ and $v \in f^{-1}(z)$, we have

$$y = f(u) = f(v) = z,$$

and we conclude that $g : M \to \mathbb{N}$ is an injection. Thus, M is countable by Corollary 6.14.   □

We leave as an exercise the proof of the following corollary to Theorem 6.15 (Problem 17b).

6.16 COROLLARY

> A nonempty set M is countable if and only if there exists a surjection $f : \mathbb{N} \to M$.

The scheme for setting up a bijection $f : \mathbb{N} \times \mathbb{N} \to \mathbb{N}$ indicated in the informal proof of the next theorem is called the **Cantor diagonal process** in honor of its discoverer, Georg Cantor. With a bit of work, a formal proof of the theorem can be made, based on the idea of the Cantor diagonal process (Problem 16).

6.17 THEOREM

> **COUNTABILITY OF $\mathbb{N} \times \mathbb{N}$**
>
> $\#(\mathbb{N} \times \mathbb{N}) = \aleph_0$.

INFORMAL PROOF   Arrange the elements of $\mathbb{N} \times \mathbb{N}$ in an array as follows:

$$(1,1) \quad (1,2) \quad (1,3) \quad (1,4) \quad (1,5) \quad \cdots$$
$$(2,1) \quad (2,2) \quad (2,3) \quad (2,4) \quad (2,5) \quad \cdots$$
$$(3,1) \quad (3,2) \quad (3,3) \quad (3,4) \quad (3,5) \quad \cdots$$
$$(4,1) \quad (4,2) \quad (4,3) \quad (4,4) \quad (4,5) \quad \cdots$$
$$\vdots \qquad \vdots \qquad \vdots \qquad \vdots \qquad \vdots$$

We "count" the elements in this array by following the successive diagonals as indicated, so that the ordered pairs in $\mathbb{N} \times \mathbb{N}$ are rearranged in a sequence: $(1, 1), (2, 1), (1, 2), (3, 1), (2, 2), (1, 3), (4, 1), (3, 2), (2, 3), (1, 4)$, and so on. In other words, we have a bijection

$$f : \mathbb{N} \to \mathbb{N} \times \mathbb{N}$$

such that

$$f(1) = (1, 1), \quad f(2) = (2, 1), \quad f(3) = (1, 2), \quad f(4) = (3, 1),$$

and so on.   □

6.18 COROLLARY

> If M and N are countable sets, then $M \times N$ is a countable set.

PROOF   By Corollary 6.14, there exist injections

$$p : M \to \mathbb{N} \quad \text{and} \quad q : N \to \mathbb{N}.$$

Define

$$h : M \times N \to \mathbb{N} \times \mathbb{N}$$

by

$$h((m, n)) = (p(m), q(n))$$

for all $(m, n) \in M \times N$. Then $h : M \times N \to \mathbb{N} \times \mathbb{N}$ is an injection (Problem 19) and, by Theorem 6.17, $\mathbb{N} \times \mathbb{N}$ is a countable set, so it follows from Theorem 6.13 that $M \times N$ is a countable set.   □

6.19 THEOREM

> **COUNTABILITY OF $\mathbb{Q}$**   The set $\mathbb{Q}$ of all rational numbers is countably infinite; that is,
>
> $$\# \mathbb{Q} = \aleph_0.$$

PROOF   By Example 6.9, $\mathbb{Z}$ is a countable set; hence, by Corollary 6.18, $\mathbb{Z} \times \mathbb{N}$ is also a countable set. Define

$$f : \mathbb{Z} \times \mathbb{N} \to \mathbb{Q}$$

by

$$f((n, m)) = \frac{n}{m}$$

for each $(n, m) \in \mathbb{Z} \times \mathbb{N}$. Then $f : \mathbb{Z} \times \mathbb{N} \to \mathbb{Q}$ is a surjection from the countable set $\mathbb{Z} \times \mathbb{N}$ onto $\mathbb{Q}$, and it follows from Theorem 6.15 that $\mathbb{Q}$ is countable. Since $\mathbb{Q}$ is infinite, it is countably infinite.   □

6.20 LEMMA Let $(D_j)_{j \in J}$ be a pairwise disjoint family of sets indexed by the nonempty set J. Suppose that J is countable and that D_j is countable for every $j \in J$. Then $\bigcup_{j \in J} D_j$ is a countable set.

PROOF   We outline the proof and leave the details as an exercise (Problem 21). Using Corollary 6.14, we choose,[†] for each $j \in J$, an injection

$$h_j : D_j \to \mathbb{N}.$$

Define

$$\alpha : \bigcup_{j \in J} D_j \to J$$

for each $x \in \bigcup_{j \in J} D_j$ by

$$\alpha(x) = j,$$

where j is the unique element of J for which $x \in D_j$. [Here we use the hypothesis that the family $(D_j)_{j \in J}$ is pairwise disjoint.] Define

$$f : \bigcup_{j \in J} D_j \to J \times \mathbb{N}$$

for each $x \in J$ by

$$f(x) = (\alpha(x), h_{\alpha(x)}(x)).$$

Then $f : \bigcup_{j \in J} D_j \to J \times \mathbb{N}$ is an injection. Since both J and $\mathbb{N}$ are countable, it follows from Corollary 6.18 that $J \times \mathbb{N}$ is countable; hence, by Theorem 6.13, $\bigcup_{j \in J} D_j$ is countable.   □

6.21 THEOREM **COUNTABLE UNIONS OF COUNTABLE SETS ARE COUNTABLE**   Let $(C_j)_{j \in J}$ be a family of sets indexed by the nonempty set J. Suppose that J is countable and that C_j is countable for each $j \in J$. Then $\bigcup_{j \in J} C_j$ is a countable set.

PROOF   For each $j \in J$, let

$$D_j = \{j\} \times C_j.$$

Since both $\{j\}$ and C_j are countable sets, D_j is a countable set by Corollary 6.18. Furthermore, if $j, k \in J$ with $j \neq k$, then $D_j \cap D_k = \emptyset$ (Problem 23), and it follows from Lemma 6.20 that $\bigcup_{j \in J} D_j$ is a countable set. An element of $\bigcup_{j \in J} D_j$ belongs to $\{j\} \times C_j$ for some $j \in J$, and therefore has the

[†] Here we use the so-called axiom of choice. [See D. J. Foulis and M. A. Munem, *After Calculus: Algebra* (San Francisco: Dellen/Macmillan, 1988), p. 330.]

form (j, x) where $x \in C_j$. The mapping

$$f: \bigcup_{j \in J} D_j \to \bigcup_{j \in J} C_j$$

defined by

$$f((j, x)) = x$$

for each element $(j, x) \in \bigcup_{j \in J} D_j$ is surjective (Problem 25), and it follows from Theorem 6.15 that $\bigcup_{j \in J} C_j$ is countable. □

6.22 LEMMA

The unit interval $[0, 1] \subseteq \mathbb{R}$ is an uncountable set.

PROOF Suppose, on the contrary, that $[0, 1]$ is a countable set. Because $[0, 1]$ is an infinite set (Problem 3b), it is then countably infinite, so there exists a bijection

$$s: \mathbb{N} \to [0, 1].$$

We use the notation

$$s_n = s(n)$$

for each $n \in \mathbb{N}$. Then, for $n, m \in \mathbb{N}$,

$$n \neq m \Rightarrow s_n \neq s_m$$

and, if $x \in [0, 1]$, there exists $n \in \mathbb{N}$ such that

$$x = s_n.$$

If $[a, b]$ is a closed and bounded interval in $\mathbb{R}$ and $x \in \mathbb{R}$, we construct a closed and bounded interval I in $\mathbb{R}$ such that

$$I \subseteq [a, b] \quad \text{and} \quad x \notin I$$

as follows: On the one hand, if $x \notin [a, b]$, we just let $I = [a, b]$. On the other hand, if $x \in [a, b]$, we let

$$I = \begin{cases} \left[a, \dfrac{a + b}{2} \right], & \text{if } x = b \\ \left[\dfrac{x + b}{2}, b \right], & \text{if } x < b \end{cases}.$$

(See Problem 27.)

By applying the construction above with $x = s_1$ and $[a, b] = [0, 1]$, we obtain a closed and bounded interval I_1 with

$$I_1 \subseteq [0, 1] \quad \text{and} \quad s_1 \notin I_1.$$

By applying the construction again, this time with $x = s_2$ and $[a, b] = I_1$, we obtain a closed and bounded interval I_2 with

$$I_2 \subseteq I_1 \quad \text{and} \quad s_2 \notin I_2.$$

Continuing in this way (Problem 36), we recursively construct a sequence of closed and bounded intervals $I_1, I_2, I_3, \ldots$ with

$$[0, 1] \supseteq I_1 \supseteq I_2 \supseteq I_3 \supseteq \cdots$$

such that, for all $n \in \mathbb{N}$,

$$s_n \notin I_n.$$

By the nested interval theorem (Theorem 5.15), there exists $x \in \mathbb{R}$ such that

$$x \in I_n$$

holds for every $n \in \mathbb{N}$. In particular, $x \in I_1 \subseteq [0, 1]$, so there exists $n \in \mathbb{N}$ with $x = s_n$. But then $s_n \in I_n$, contradicting the fact that $s_n \notin I_n$. $\square$

6.23 THEOREM

> **$\mathbb{R}$ IS UNCOUNTABLE**   The set $\mathbb{R}$ of all real numbers is an uncountable set.

PROOF   Let $h:[0, 1] \to \mathbb{R}$ be defined by $h(x) = x$ for all $x \in [0, 1]$. Then h is an injection. If $\mathbb{R}$ were countable, then $[0, 1]$ would be countable by Theorem 6.13. However, this would contradict Lemma 6.22. $\square$

PROBLEM SET 4.6

1. Prove Theorem 6.4.

2. Prove Theorem 6.5.

3. (a) Prove that the set $\mathbb{N}$ of all natural numbers is an infinite set.
(b) Prove that the unit interval $[0, 1] \subseteq \mathbb{R}$ is an infinite set.

4. (a) Prove that every subset of a finite set is a finite set.
(b) Prove that, if a set contains an infinite subset, it must be an infinite set.

5. If L, M, and N are sets, prove:
(a) $\#L = \#L$.
(b) $\#L = \#M \Rightarrow \#M = \#L$.
(c) $\#L = \#M$ and $\#M = \#N \Rightarrow \#L = \#N$.
[Caution: The symbolism $\#M = \#N$ is an abbreviation for the statement that there exists a bijection $f:M \to N$. In spite of its appearance, it does not assert that anything is *equal* to anything else.]

6. Prove the **Dirichlet pigeonhole principle**: If M and N are finite sets with $\#M > \#N$, then there exists no injection $f:M \to N$. (Think of the elements of M as pigeons, the elements of N as pigeonholes, and $f:M \to N$ as a placement of pigeons in holes. If there are more pigeons than pigeonholes, at least two pigeons must be in the same hole.)

7. In Example 6.9, prove that the function $f:\mathbb{N} \to \mathbb{Z}$ is a bijection.

8. If N is a finite set, M is a subset of N, and $\#M = \#N$, prove that $M = N$. [Hint: Use the result of Problem 6.]

9. If X is a set, prove that X is finite if and only if $\mathscr{P}(X)$ is finite.

10. If M and N are sets, M is finite, and $f:M \to N$ is a surjection, prove that N is finite and that $\#M \geq \#N$.

11. A sequence $s:\mathbb{N} \to \mathbb{N}$ is defined by simple recursion as follows: $s_1 = 1$ and for all $n \in \mathbb{N}$, $s_{n+1} = 2s_n$. Find a closed-form description of this sequence.

12. If M and N are finite sets with $\#M = \#N$ and $f:M \to N$, prove that f is an injection if and only if it is a surjection.

13. Let $s:\mathbb{N} \to \mathbb{N}$ be a sequence such that $s(n) < s(n+1)$ holds for all $n \in \mathbb{N}$.
    (a) If $n, m \in \mathbb{N}$ with $n < m$, prove that $s(n) < s(m)$. [Hint: Fix n and use mathematical induction on m.]
    (b) Use the result of Part (a) to prove that $s:\mathbb{N} \to \mathbb{N}$ is an injection.
    (c) Prove that $n < s(n)$ holds for all $n \in \mathbb{N}$. [Hint: Use mathematical induction and Theorem 4.7 in Chapter 3.]

14. If X is a set, prove that there exists no bijection $f:X \to \mathscr{P}(X)$. [Hint: Argue as in Example 6.10.]

15. Let $g:N \to Y$ be a bijection, let $J \subseteq N$, and let $X = g(J)$. Denote the restriction of g to J by $g|_J$, and prove that $(g|_J):J \to X$ is a bijection.

16. Let the mapping $g:\mathbb{N} \times \mathbb{N} \to \mathbb{N}$ be defined for $(x, y) \in \mathbb{N} \times \mathbb{N}$ by $g(x, y) = y + (x + y - 1)(x + y - 2)/2$. Prove that g is a bijection and that $g^{-1}:\mathbb{N} \to \mathbb{N} \times \mathbb{N}$ is the bijection $f:\mathbb{N} \to \mathbb{N} \times \mathbb{N}$ in the informal proof of Theorem 6.17.

17. (a) Prove Corollary 6.14.
    (b) Prove Corollary 6.16.

18. If M and N are sets, define $\#M \leq \#N$ to mean that there exists an injection $f:M \to N$ (see Theorem 6.5). If L, M, N, and P are sets, prove:
    (a) $\#L \leq \#L$.   (b) $\#L = \#M \Rightarrow \#L \leq \#M$.
    (c) If $\#L \leq \#M$ and $\#M \leq \#N$, then $\#L \leq \#N$.
    (d) If $\#L \leq \#M$, $\#L = \#P$, and $\#M = \#N$, then $\#P \leq \#N$.

19. Prove that the mapping $h:M \times N \to \mathbb{N} \times \mathbb{N}$ defined in the proof of Corollary 6.18 is an injection.

20. Suppose that M, N, A, B, C, and D are sets with $M = A \cup B$, $N = C \cup D$, $A \cap B = \varnothing$, $C \cap D = \varnothing$, $\#A = \#C$, and $\#B = \#D$. Prove that $\#M = \#N$.

21. Supply the details for the proof of Lemma 6.20.

22. Let M be a set and let $\phi:\mathscr{P}(M) \to \mathscr{P}(M)$ be a mapping such that, for $S, T \in \mathscr{P}(M), S \subseteq T \Rightarrow \phi(S) \subseteq \phi(T)$. Prove that there exists $A \in \mathscr{P}(M)$ such that $\phi(A) = A$. [Hint: Let $\mathscr{F} = \{S \in \mathscr{P}(M)|\phi(S) \subseteq S\}$ and let $A = \bigcap \mathscr{F}$.]

23. In the proof of Theorem 6.21, show that, if $i, k \in J$ with $i \neq k$, then $D_i \cap D_k = \varnothing$.

24. Suppose that M and N are sets and that $f:M \to N$ and $g:N \to M$ are injections. Define $\phi:\mathscr{P}(M) \to \mathscr{P}(M)$ by $\phi(S) = M \setminus g(N \setminus f(S))$ for all $S \in \mathscr{P}(M)$. Prove that there exists $A \in \mathscr{P}(M)$ such that $\phi(A) = A$. [Hint: Use the result of Problem 22.]

25. Show that the mapping $f:\bigcup_{j \in J} D_j \to \bigcup_{j \in J} C_j$ defined in the proof of Theorem 6.21 is a surjection.

26. With the notation and hypotheses of Problem 24, let $C = f(A), D = N \setminus C$, and $B = M \setminus A$. Prove that $(f|_A):A \to C$ and $(g|_D):D \to B$ are bijections.

27. With the notation used in the proof of Lemma 6.22, prove that:
    (a) I is a closed and bounded interval.
    (b) $I \subseteq [a, b]$.   (c) $x \notin I$.

28. By combining the results of Problems 24 and 20 and using the notation established in Problem 18, prove the **Cantor–Schroeder–Bernstein theorem**:   If M and N are sets such that $\#M \leq \#N$ and $\#N \leq \#M$, then $\#M = \#N$.

29. Use the argument in the proof of Lemma 6.22 to show that, if C is a connected subset of $\mathbb{R}$ and C contains at least two distinct points, then C is an uncountable set.

30. Let X be a set, suppose that $a \in X$, and let $g:\mathbb{N} \times X \to X$. Define $G:\mathbb{N} \times X \to \mathbb{N} \times X$ by $G((n, x)) = (n + 1, g((n, x)))$ for all $(n, x) \in \mathbb{N} \times X$. Define $S:\mathbb{N} \to \mathbb{N} \times X$ by simple recursion in such a way that $S(1) = (1, a)$ and, for all $n \in \mathbb{N}$, $S(n + 1) = G(S(n))$. Define $\pi:\mathbb{N} \times X \to X$ by $\pi((n, x)) = x$ for all $(n, x) \in \mathbb{N} \times X$, and define $s:\mathbb{N} \to X$ by $s(n) = \pi(S(n))$ for all $n \in \mathbb{N}$. Use mathematical induction to prove that $S(n + 1) = (n + 1, g((n, s(n))))$ holds for all $n \in \mathbb{N}$.

31. If M is an uncountable set, $M = A \cup B$, and A is a countable set, prove that B is an uncountable set. [Hint: Make an indirect argument using Theorem 6.21.]

32. Use the result of Problem 30 to prove the following **extended principle of simple recursion**:   Let X be a set, let $a \in X$, and let $g:\mathbb{N} \times X \to X$. Then there exists a unique sequence $s:\mathbb{N} \to X$ such that $s(1) = a$ and $s(n + 1) = g((n, s(n)))$ for all $n \in \mathbb{N}$.

33. Prove that the set $\mathbb{R} \setminus \mathbb{Q}$ of all irrational numbers is uncountable. [Hint: Use Theorem 6.23 and the result of Problem 31.]

34. Let $s: \mathbb{N} \to \mathbb{N}$ be the sequence obtained by using the extended principle of simple recursion (see Problem 32) as follows: $s(1) = 1$ and $s(n + 1) = (n + 1) \cdot s(n)$ for all $n \in \mathbb{N}$. Find a closed-form description for this sequence.

35. If $n \in \mathbb{N}$, let P_n denote the set of all polynomials of degree n with integer coefficients. A real number is said to be **algebraic** if it is a root of a polynomial in P_n for some $n \in \mathbb{N}$, otherwise it is said to be **transcendental**.
(a) Prove that $\# P_n = \aleph_0$ for each $n \in \mathbb{N}$. [Hint: Use Theorem 6.21.]
(b) Prove that the set of all algebraic numbers is countably infinite.
(c) Prove that the set of all transcendental numbers is uncountable. [Hint: Use Part (b), Theorem 6.23, and the result of Problem 31.]

36. Use the extended principle of simple recursion (see Problem 32) to make a rigorous proof of the existence of the sequence $I_1, I_2, I_3, \ldots$ of closed and bounded intervals used in the proof of Lemma 6.22.

37. Show that $\#(0, 1) = \#(0, \infty)$. [Hint: Use the mapping $f: (0, 1) \to (0, \infty)$ defined by $f(x) = x/(1 - x)$ for $x \in (0, 1)$.]

38. Prove that $\#(0, 1) = \# \mathbb{R}$. [Hint: See Problem 37.]

39. Prove that every nonempty open subset U of $\mathbb{R}$ is the union of a countable set of pairwise disjoint open intervals. [Hint: By Problem 21b in Problem Set 4.4, U is a union of a collection $\mathscr{C}$ of pairwise disjoint open intervals. Let $J = \mathbb{Q} \cap U$. Since $\mathbb{Q}$ is countable (Theorem 6.19), so is J. Define $f: J \to \mathscr{C}$ by $f(q) = (a, b)$, where (a, b) is the unique open set in $\mathscr{C}$ such that $q \in (a, b)$. Prove that f is a well-defined surjection, and use Theorem 6.15 to conclude that $\mathscr{C}$ is countable.]

7

CLASSIFICATION OF FUNCTIONS

Mathematical analysis is often referred to as the **theory of functions**—functions are, quite literally, the name of the game. Our work in this and the following chapters is restricted to functions $f: D \to \mathbb{R}$ with $D \subseteq \mathbb{R}$; that is, we are studying **real analysis**. Another branch of mathematical analysis, called **complex analysis**, focuses on functions $f: D \to \mathbb{C}$ with $D \subseteq \mathbb{C}$; yet another branch, called **functional analysis**, studies functions $f: V \to F$ where V is a vector space and F is its field of scalars; and so on.

Powerful techniques of classification and construction that enable mathematicians to cope with an immense variety of functions have been developed in all branches of mathematical analysis. Functions are often classified in terms of important properties that some functions have and others lack, such as continuity, monotonicity, differentiability, or integrability. We have studied the property of continuity in the earlier sections of the present chapter, and, in the next chapter, we turn our attention to differentiability and integrability. Later in this section, we have a brief look at the property of monotonicity.

After a particular class of functions has been singled out and studied, new functions may be constructed from the functions in this class, and thus new classes of functions may emerge. Some of the more important techniques used in real analysis for constructing new functions from old ones may be categorized as follows:

1. *Set-theoretic.* Among the set-theoretic techniques are *composition, inversion, restriction, extension,* and *amalgamation.* Thus, two functions f and g may be composed to obtain a new function $f \circ g$; an injective function f may be inverted to obtain a new function f^{-1}; a new function $f|_M$ may be obtained from f by restricting its domain to the set M; and, vice versa, a new function h (called an extension of f) may be constructed so that $h|_{\mathrm{dom}(f)} = f$. Also, if f and g agree on $\mathrm{dom}(f) \cap \mathrm{dom}(g)$, a new function h (called an amalgamation of f and g) may be constructed with $\mathrm{dom}(h) = \mathrm{dom}(f) \cup \mathrm{dom}(g)$ in such a way that h is an extension of both f and g.

2. *Algebraic.* Among the algebraic techniques are *formation of algebraic combinations* and *solution of equations.* Thus, if f and g are functions, we can form the sum $f + g$, the product $f \cdot g$, the negative $-f$, the reciprocal $1/f$, the difference $f - g$, the quotient f/g, and linear combinations $af + bg$ (where a, b are constants). Furthermore, if n is a positive integer, we can form the (principal) nth root $(f)^{1/n}$. Also, new functions may be constructed from old ones by solving algebraic equations or systems of equations.

3. *Analytic.* Among the analytic techniques are *taking limits, differentiating,* and *forming (indefinite) integrals.* Thus, we may form a new function by taking a suitable limit of a sequence $(f_n)_{n \in \mathbb{N}}$ of functions. (There are a number of different ways in which this can be done.) Also, if f is a differentiable function, the derivative f' is a new function, whereas if f is a continuous function, the indefinite integral provides a new function $F(x) = \int_a^x f(t)\, dt$.

We studied composition and inversion of functions in Chapter 2, and we introduced the idea of function restriction in Section 2 of the present chapter (page 238). By Theorem 3.10, the composition of continuous functions is again continuous, and the inverse of a continuous injection with a compact domain is continuous (Problem 17 in Problem Set 4.5). In general, however, discontinuous functions can arise from the inversion of continuous injections (Problem 19 in Problem Set 4.5). It is easily seen that the restriction of a continuous function remains continuous (Problem 39 in Problem Set 4.3); however, extensions of continuous functions need not be continuous. Some of the key ideas in mathematical analysis involve the question of whether a function admits an extension that has a particular desirable property, for instance, continuity or differentiability.

If f and g are functions that agree on the intersection of their domains, so that

$$x \in \text{dom}(f) \cap \text{dom}(g) \Rightarrow f(x) = g(x),$$

then f and g can be combined to form a new function defined by

$$h(x) = \begin{cases} f(x), & \text{if } x \in \text{dom}(f) \\ g(x), & \text{if } x \in \text{dom}(g) \end{cases}$$

for all $x \in \text{dom}(f) \cup \text{dom}(g)$. If f, g, and h are regarded as sets of ordered pairs, then

$$h = f \cup g$$

(Problem 1). We say that h is obtained from f and g by **amalgamation**. In general, even if f and g are continuous, h need not be continuous. (However, see Problem 2.)

Let $\mathscr{C}$ be a class of functions, for instance, the class of all continuous functions. A function obtained by amalgamating finitely many functions, each of which is a restriction of a function in the class $\mathscr{C}$ to an interval of real numbers, is said to be **piecewise defined** from the functions in $\mathscr{C}$. For instance, a function is **piecewise continuous** if its domain is a union of finitely many intervals and the restriction of the function to each of these intervals is continuous.

A **constant function** f satisfies the condition that there is a fixed real number c such that

$$f(x) = c$$

for all $x \in \text{dom}(f)$. For the case in which $\text{dom}(f) = \mathbb{R}$, the same symbol c is customarily used to denote both the fixed real number and the constant function f that it determines. For instance, in elementary calculus, when we write

$$\frac{d}{dx}(x^2 + 1) = \frac{d}{dx}x^2 + \frac{d}{dx}1 = 2x + 0 = 2x,$$

we understand that 1 denotes the constant function

$$x \mapsto 1, \quad \forall x \in \mathbb{R},$$

and 0 denotes the constant function

$$x \mapsto 0, \quad \forall x \in \mathbb{R}.$$

However, since this practice can lead to confusion, it is often avoided in more formal mathematical work.

A piecewise constant function whose domain is an interval is called a **step function**. For instance, the function

$$f(x) = \begin{cases} 0, & \text{if } -1 \le x < 0 \\ 2, & \text{if } 0 \le x \le 1 \\ -3, & \text{if } 1 < x < 4 \end{cases}$$

is a step function defined on the interval $[-1, 4)$. The definite (Riemann) integral of elementary calculus is based on the idea of approximating a function by a step function.

A **linear function** f, defined on $\mathbb{R}$, satisfies the condition that there are fixed real numbers a and b such that

$$f(x) = ax + b$$

for all $x \in \mathbb{R}$. A function whose domain is an interval is said to be **piecewise linear** if it is obtained from two or more linear functions by restricting their domains to intervals and amalgamating the resulting functions. For instance, the function defined for all $x \in \mathbb{R}$ by

$$f(x) = \begin{cases} 2x, & \text{if } x < 1 \\ 3x - 1, & \text{if } x \ge 1 \end{cases}$$

is piecewise linear. Note that the **absolute-value** function defined by

$$|x| = \begin{cases} -x, & \text{if } x < 0 \\ x, & \text{if } x \ge 0 \end{cases},$$

for all $x \in \mathbb{R}$, is piecewise linear. The trapezoidal rule of elementary calculus is based on the idea of approximating a function by a piecewise linear function.

Algebraic combinations of functions are understood to be formed *pointwise*. For instance, if f and g are functions, then the function $f + g$ is defined only if $\text{dom}(f) \cap \text{dom}(g) \ne \varnothing$, in which case, $\text{dom}(f + g) = \text{dom}(f) \cap \text{dom}(g)$ and

$$(f + g)(x) = f(x) + g(x)$$

for all $x \in \text{dom}(f) \cap \text{dom}(g)$. Likewise, the function $1/f$ is defined only if $\{x \in \mathbb{R} \,|\, f(x) \ne 0\} \ne \varnothing$, in which case, $\text{dom}(1/f) = \{x \in \mathbb{R} \,|\, f(x) \ne 0\}$ and

$$\frac{1}{f}(x) = \frac{1}{f(x)}$$

for all $x \in \text{dom}(1/f)$. (These are the standard definitions given in elementary calculus textbooks.)

A nonempty class of functions is called a **vector space of functions**, or simply a **function space**, if it is closed under the formation of linear combinations. For instance, the class of all linear functions forms a function space (Problem 9). If I is an interval, then the class of all step functions with domain I is also a function space (Problem 10). So is the class of all piecewise linear functions with domain I (Problem 12).

7.1 DEFINITION

> **C(X)** If $\varnothing \neq X \subseteq \mathbb{R}$, we denote by $C(X)$ the set of all continuous functions $f: X \to \mathbb{R}$.

We leave it to you to prove that $C(X)$ forms a vector space of functions (Problem 11).

A function space that is closed under multiplication is called an **algebra of functions**. For instance, if $\varnothing \neq X \subseteq \mathbb{R}$, then $C(X)$ is an algebra of functions (Problem 13). Note that an algebra of functions forms a commutative ring in the algebraic sense (Section 2 of Chapter 3) and that $C(X)$ is, in fact, a commutative ring with unity.

Because the product of linear functions is not necessarily linear, the class of all linear functions does not form an algebra. However, the class of all **polynomial functions**, that is, functions defined on $\mathbb{R}$ and having the form

$$f(x) = c_n x^n + c_{n-1} x^{n-1} + c_{n-2} x^{n-2} + \cdots + c_2 x^2 + c_1 x + c_0,$$

where the **coefficients** $c_n, c_{n-1}, c_{n-2}, \ldots, c_2, c_1,$ and c_0 are fixed real numbers, does form an algebra of functions (Problem 15). Indeed, the class of polynomial functions is the "smallest" algebra of functions that contains all of the linear functions. The zero element in this algebra is the constant function 0, and the unity element is the constant function 1.

Piecewise polynomial functions are often used to approximate more complicated functions; for instance, Simpson's (parabolic) rule of elementary calculus is based on the idea of approximating a function by a piecewise quadratic function. A differentiable piecewise polynomial function is called a **spline**, and the problem of approximating functions by splines arises in many branches of applied mathematics.

Although the algebra of polynomial functions is closed under composition, addition, subtraction, and multiplication, it is not closed under division. However, the class of all **rational functions**, that is, functions of the form

$$f(x) = \frac{p(x)}{q(x)},$$

where $p(x), q(x)$ are polynomials and $q(x)$ is not the constant zero function, is an algebra of functions that is closed under division (by functions other than the constant zero function) as well as under function composition (Problem 16).

Although the rational functions form a commutative ring with unity, they do not, as such, form a field (see Definition 2.11 in Chapter 3). Here is the difficulty: Consider a rational function f defined by

$$f(x) = \frac{p(x)}{q(x)},$$

where $q(x)$ is not the constant zero polynomial. The domain of f is understood to be the set of all real numbers x such that $q(x) \neq 0$. Suppose that f is not the constant zero function so that $p(x)$ is not the constant zero polynomial. Then the only possible candidate for the multiplicative inverse of f in the ring of all rational functions is the rational function g defined by

$$g(x) = \frac{q(x)}{p(x)}.$$

However, because the domain of 1 is $\mathbb{R}$ and the domain of fg consists of all real numbers x for which $p(x) \cdot q(x) \neq 0$, the product function fg is not necessarily equal to the constant function 1. (If two functions are equal, they must have the same domain!)

The difficulty mentioned above can be circumvented by introducing the following idea: Two rational functions r and s are said to be **equivalent** if $r(x) = s(x)$ holds for all but possibly finitely many values of $x \in \mathbb{R}$. Then, by dealing with *equivalence classes* of rational functions (see Section 5 of Chapter 2), rather than with the functions themselves, a bona fide field is obtained. In the interest of simplicity, we sometimes write

$$r = s$$

when what we really mean is that the functions r and s are equivalent. For instance, in elementary calculus, when we write

$$\frac{x^2 - 1}{x^2 - 3x + 2} = \frac{x + 1}{x - 2},$$

we do not really mean that the rational function on the left (which is not defined for $x = 1$) is equal to the rational function on the right (which is defined for $x = 1$); we mean that they are equivalent functions. Needless to say, this abuse of the equality symbol should be avoided in more formal mathematical situations.

A **monomial function of two variables** is a function

$$M: \mathbb{R} \times \mathbb{R} \to \mathbb{R}$$

of the form

$$M(x, y) = Cx^m y^n$$

for all $(x, y) \in \mathbb{R} \times \mathbb{R}$, where m and n are nonnegative integers and the **coefficient** C is a fixed real number. A **polynomial function of two variables** is a sum of finitely many such monomial functions. A **polynomial equation in two variables** is an equation of the form

$$P(x, y) = 0,$$

where P is a polynomial function of two variables. By an **algebraic function**, we mean—roughly speaking[†]—a function f obtained by solving such a polynomial equation for y in terms of x, so that

$$P(x, f(x)) = 0$$

holds for all $x \in \mathrm{dom}(f)$. For instance, a rational function

$$f(x) = \frac{p(x)}{q(x)}$$

can be obtained by solving the polynomial equation

$$q(x)y - p(x) = 0$$

for y in terms of x; hence, every rational function is an algebraic function. More generally, any function built by forming algebraic combinations of finitely many polynomial functions $p(x)$, $q(x)$, $r(x)$, ... with the aid of addition, subtraction, multiplication, division, and the extraction of nth roots is an algebraic function. For instance, the function

$$f(x) = \frac{\sqrt{x^2 - 3x + 2}}{4x - 1}$$

provides a solution $y = f(x)$ of the polynomial equation

$$(4x - 1)^2 y^2 - (x^2 - 3x + 2) = 0$$

and is therefore an algebraic function.

A function that is not algebraic is said to be **transcendental** because it transcends (that is, goes beyond) purely algebraic methods. The logarithmic, exponential, trigonometric, and inverse trigonometric functions of elementary calculus are examples of transcendental functions.

Many well-behaved classes of functions—for instance, linear functions, polynomial functions, rational functions, and algebraic functions—are closed under differentiation. Thus, the derivative of a polynomial is again a polynomial, the derivative of a rational function is again a rational function, and so on. Integration is quite another matter. Although an

---

[†] The formal definition of an algebraic function depends on concepts developed in complex analysis.

(indefinite) integral of a polynomial is again a polynomial, indefinite integrals of rational functions are often transcendental. For instance,

$$\ln x = \int_1^x \frac{1}{t}\, dt.$$

In elementary calculus, the theory of partial fractions gives a complete account of the types of functions that can arise from integrating rational functions (Problem 18). The fact that

$$\mathrm{Sin}^{-1}(x) = \int_0^x \frac{1}{\sqrt{1 - t^2}}\, dt$$

indicates how familiar transcendental functions can arise from integration of algebraic functions. Note, however, that transcendental functions not encountered in elementary calculus, such as

$$f(x) = \int_0^x \sqrt{1 + t^4}\, dt,$$

can also arise as integrals of algebraic functions.

Monotone functions, and linear combinations of such functions, play a crucial role in real analysis. We devote the remainder of this section to a brief study of such functions.

7.2 DEFINITION

> **INCREASING, DECREASING, AND MONOTONE FUNCTIONS**   Let f be a function. We say that f is **increasing** if, for all $x, y \in \mathrm{dom}(f)$,
>
> $$x \le y \Rightarrow f(x) \le f(y).$$
>
> We say that f is **decreasing** if, for all $x, y \in \mathrm{dom}(f)$,
>
> $$x \le y \Rightarrow f(x) \ge f(y).$$
>
> If f is either increasing or decreasing, it is said to be **monotone**.

If, for all $x, y \in \mathrm{dom}(f)$,

$$x < y \Rightarrow f(x) < f(y),$$

then we say that f is **strictly increasing**. Similarly, if, for all $x, y \in \mathrm{dom}(f)$,

$$x < y \Rightarrow f(x) > f(y),$$

then we say that f is **strictly decreasing**. Since multiplication by -1 reverses inequalities, it is clear that f is increasing (respectively, strictly increasing) if and only if $-f$ is decreasing (respectively, strictly decreasing). A function that is either strictly increasing or strictly decreasing is said to be **strictly monotone**.

7.3 THEOREM

> **CONTINUOUS INJECTIONS ON CONNECTED SETS**   Let $f : C \to \mathbb{R}$ be a continuous injection and suppose that C is a connected subset of $\mathbb{R}$. Then f is strictly monotone.

PROOF   Since f is injective, we have

$$a < b \Rightarrow f(a) \neq f(b)$$

for all $a, b \in C$. Since C is connected, Theorem 4.15 implies that

$$a, b \in C \quad \text{with } a < b \Rightarrow [a, b] \subseteq C = \text{dom}(f).$$

Our proof is by contradiction. Suppose f is not strictly monotone. Then either there exist $a, b, c \in C$ such that

$$a < b < c \quad \text{and} \quad f(a), f(c) < f(b)$$

or else there exist $a, b, c \in C$ such that

$$a < b < c \quad \text{and} \quad f(a), f(c) > f(b)$$

(Problem 19). By replacing f by $-f$ if necessary, we can suppose without loss of generality that the last condition is the one that holds. Now either $f(a) < f(c)$ or else $f(c) < f(a)$. We suppose that $f(a) < f(c)$, leaving the opposite case as an exercise (Problem 20). Thus, we have

$$a < b < c \quad \text{and} \quad f(b) < f(a) < f(c).$$

Choose and fix a k such that

$$f(b) < k < f(a).$$

Since $f(a) < f(c)$, it follows that

$$f(b) < k < f(c).$$

By applying the intermediate value theorem (Theorem 4.17) twice, we conclude that there is a number p with

$$a \leq p \leq b \quad \text{and} \quad f(p) = k$$

and there is a number q with

$$b \leq q \leq c \quad \text{and} \quad f(q) = k.$$

Since f is injective, it follows that

$$p = q$$

and therefore, since $p \leq b \leq q$, we must have

$$p = b.$$

Therefore,

$$k = f(p) = f(b),$$

contradicting the fact that $f(b) < k$.   □

7.4 LEMMA

> Let f be an increasing function, let $a \in \mathrm{dom}(f)$, define $A = (-\infty, a) \cap \mathrm{dom}(f)$, and suppose that a is an accumulation point of A. Then $\lim_{x \to a}(f|_A)$ exists, $\sup(f(A))$ exists, and
>
> $$\lim_{x \to a} (f|_A) = \sup(f(A)) \le f(a).$$

PROOF   If $x \in A$, then $x < a$; hence, since f is increasing, $f(x) \le f(a)$. Consequently, $f(a)$ is an upper bound for $f(A)$. Because a is an accumulation point of A, it follows that $A \ne \varnothing$; hence, since $A \subseteq \mathrm{dom}(f)$, $f(A) \ne \varnothing$. Therefore, by the completeness of $\mathbb{R}$, $f(A)$ has a supremum. Let

$$L = \sup(f(A)). \tag{1}$$

Since $f(a)$ is an upper bound for $f(A)$, it follows that

$$L \le f(a). \tag{2}$$

To finish the proof, we must prove that $\lim_{x \to a}(f|_A) = L$. Thus, let ε be a positive real number. By Part (i) of Lemma 3.23 in Chapter 3, there exists $b \in A$ such that

$$L - \varepsilon < f(b) \le L. \tag{3}$$

Since $b \in A = (-\infty, a) \cap \mathrm{dom}(f)$, we have $b \in \mathrm{dom}(f)$ and

$$b < a. \tag{4}$$

Let

$$\delta = a - b, \tag{5}$$

noting that $\delta > 0$, and suppose that $x \in \mathrm{dom}(f|_A) = A$ with

$$0 < |x - a| < \delta. \tag{6}$$

We have to prove that $|(f|_A)(x) - L| < \varepsilon$. Since $x \in A$, we have $(f|_A)(x) = f(x)$, so it will be sufficient to prove that $|f(x) - L| < \varepsilon$. Now, $x \in A = (-\infty, a) \cap \mathrm{dom}(f)$, and therefore,

$$x < a. \tag{7}$$

By combining (5), (6), and (7), we find that

$$b < x. \tag{8}$$

Since f is increasing, (8) implies that

$$f(b) \le f(x). \tag{9}$$

By combining (3) and (9), we have

$$L - \varepsilon < f(x). \tag{10}$$

Also, since $x \in A$, we have $f(x) \in f(A)$, and it follows from the fact that $L = \sup(f(A))$ that

$$f(x) \le L < L + \varepsilon. \tag{11}$$

Finally, as a consequence of (10) and (11), we have

$$|f(x) - L| < \varepsilon. \tag{12}$$

$\square$

The limit as $x \to a$ of the restricted function $f|_A$ in Lemma 7.4 is called the *limit on the left of f at a* and is written as

$$\lim_{x \to a^-} f(x) = \lim_{x \to a}(f|_A)(x)$$

(see Problem 27 in Problem Set 4.2). Thus, provided that $a \in \mathrm{dom}(f)$ and that a is an accumulation point of $(-\infty, a) \cap \mathrm{dom}(f)$, the conclusion of Lemma 7.4 may be written as

$$\lim_{x \to a^-} f(x) = \sup_{x < a} f(x) \le f(a)$$

where $\sup_{x<a} f(x)$ is an abbreviation for $\sup(\{x \in \mathrm{dom}(f) \,|\, x < a\})$. We use similar notation in the statement of the following theorem.

7.5 THEOREM **INCREASING FUNCTIONS** Let f be an increasing function, let $a \in \mathrm{dom}(f)$, and suppose that a is an accumulation point of both of the sets $(-\infty, a) \cap \mathrm{dom}(f)$ and $(a, \infty) \cap \mathrm{dom}(f)$. Then

$$\lim_{x \to a^-} f(x) = \sup_{x < a} f(x) \le f(a) \le \inf_{a < x} f(x) = \lim_{x \to a^+} f(x).$$

PROOF That

$$\lim_{x \to a^-} f(x) = \sup_{x < a} f(x) \le f(a)$$

is the content of Lemma 7.4. That

$$f(a) \le \inf_{a < x} f(x) = \lim_{x \to a^+} f(x)$$

follows from an argument similar to the proof of Lemma 7.4 and is left as an exercise (Problem 23). $\square$

7.6 LEMMA

> Let f be an increasing function, let $a, b \in \mathrm{dom}(f)$, suppose that a is an accumulation point of $(a, \infty) \cap \mathrm{dom}(f)$, and suppose that b is an accumulation point of $(-\infty, b) \cap \mathrm{dom}(f)$. Then
>
> $$a < b \Rightarrow \lim_{x \to a^+} f(x) \le \lim_{x \to b^-} f(x).$$

PROOF   By Lemma 7.4, $\lim_{x \to b^-} f(x)$ exists and equals $\sup_{x < b} f(x)$. By Problem 23, $\lim_{x \to a^+} f(x)$ exists and equals $\inf_{a < x} f(x)$. This will be sufficient to prove that

$$a < b \Rightarrow \inf_{a < x} f(x) \le \sup_{x < b} f(x). \tag{*}$$

Suppose $a < b$ and let

$$\delta = \frac{b - a}{2}.$$

Since a is an accumulation point of $(a, \infty) \cap \mathrm{dom}(f)$, there exists $p \in \mathrm{dom}(f)$ such that

$$a < p < a + \delta.$$

Since b is an accumulation point of $(-\infty, b) \cap \mathrm{dom}(f)$, there exists $q \in \mathrm{dom}(f)$ such that

$$b - \delta < q < b.$$

By our choice of δ, we have

$$a + \delta = b - \delta,$$

and it follows that

$$a < p < q < b.$$

Therefore,

$$\inf_{a < x} f(x) \le f(p) \le f(q) \le \sup_{x < b} f(x),$$

from which (*) follows.   □

7.7 THEOREM

> **DISCONTINUITIES OF A MONOTONE FUNCTION**   Let C be an open, connected set of real numbers, let $f : C \to \mathbb{R}$ be a monotone function, and let D be the set of points in C at which f is discontinuous. Then D is a countable set.

PROOF   We sketch the proof and leave the details as an exercise (Problem 25). We can assume without loss of generality that f is increasing. (If f is decreasing, just replace f by $-f$.) By Theorem 4.15, the hypotheses of Theorem 7.5 hold for every $a \in C = \text{dom}(f)$. By Theorem 7.5 and Problem 27 in Problem Set 4.2,

$$a \in D \Leftrightarrow a \in C \quad \text{and} \quad \lim_{x \to a^-} f(x) < \lim_{x \to a^+} f(x).$$

For each $a \in D$, let I_a be the open interval

$$I_a = \left(\lim_{x \to a^-} f(x), \lim_{x \to a^+} f(x) \right).$$

As a consequence of Lemma 7.6,

$$a, b \in D \quad \text{with } a \neq b \Rightarrow I_a \cap I_b = \varnothing.$$

Define

$$J = \mathbb{Q} \cap \bigcup_{a \in D} I_a$$

and let $q \in J$. Then there is a unique $a \in D$ such that $q \in I_a$. If we denote this unique a by $a = \phi(q)$, then

$$\phi : J \to D$$

is a well-defined surjection. Since $J \subseteq \mathbb{Q}$, it is countable, and it follows from Theorem 6.15 that D is countable.   □

PROBLEM SET 4.7

1. Let f and g be functions such that $f(x) = g(x)$ for all $x \in \text{dom}(f) \cap \text{dom}(g)$. Define $h: \text{dom}(f) \cup \text{dom}(g) \to \mathbb{R}$ by

$$h(x) = \begin{cases} f(x), & \text{if } x \in \text{dom}(f) \\ g(x), & \text{if } x \in \text{dom}(g) \end{cases}.$$

Prove that, as sets of ordered pairs, $h = f \cup g$.

2. With the notation of Problem 1, assume that f and g are continuous.
    (a) If both $\text{dom}(f)$ and $\text{dom}(g)$ are closed, prove that h is continuous.
    (b) If both $\text{dom}(f)$ and $\text{dom}(g)$ are open, prove that h is continuous.
    (c) Give an example to show that, in general, h need not be continuous.

3. Prove that the class of all linear functions on $\mathbb{R}$ is closed under function composition.

4. Let $f : [a, b] \to \mathbb{R}$ and let $a < c < b$. Show that there exist constants A, B, and C such that the polynomial function $g(x) = Ax^2 + Bx + C$ satisfies $f(a) = g(a)$, $f(b) = g(b)$, and $f(c) = g(c)$.

5. If f and g are functions such that $D = \mathrm{dom}(f) \cap \mathrm{dom}(g) \neq \varnothing$, define functions $f \vee g$ and $f \wedge g$, both with domain D, by $(f \vee g)(x) = \max(f(x), g(x))$ and $(f \wedge g)(x) = \min(f(x), g(x))$. Prove that $(f \vee g) = -((-f) \wedge (-g))$.

6. Let $f : [a, b] \to \mathbb{R}$ and let n be an even positive integer. Let $\Delta x = (b - a)/n$ and define $x_j = a + j \cdot \Delta x$ for all integers j with $0 \leq j \leq n$. Use the result of Problem 4 to prove that there exists a continuous, piecewise quadratic function g such that $f(x_j) = g(x_j)$ for $j = 0, 1, 2, \ldots, n$.

7. If f is a function with domain D, define $|f|$ to be the function with domain D such that $|f|(x) = |f(x)|$ for all $x \in D$.
   (a) Prove that $|f| = (f \cdot f)^{1/2}$.
   (b) With the notation of Problem 5, prove that

$$f \vee g = \frac{f + g + |f - g|}{2}.$$

8. With the notation of Problems 5 and 7, prove that an algebra of functions is closed under the formation of absolute values if and only if it is closed under $\vee$ and $\wedge$.

9. Prove that the class of all linear functions forms a vector space of functions.

10. If I is an interval of real numbers, prove that the class of all step functions $f : I \to \mathbb{R}$ forms a vector space of functions.

11. If $\varnothing \neq X \subseteq \mathbb{R}$, prove that $C(X)$ forms a vector space of functions.

12. If I is an interval of real numbers, prove that the class of all piecewise linear functions $f : I \to \mathbb{R}$ forms a vector space of functions.

13. With the notation of Problem 9, prove that $C(X)$ forms an algebra of functions and that, as a ring, it is a commutative ring with unity.

14. If I is an interval of real numbers, prove that the ring $C(I)$ is not a field.

15. Prove that the class of all polynomial functions forms an algebra of functions and that it is closed under function composition.

16. Prove that the class of all rational functions forms an algebra of functions and that it is closed under function composition.

17. If f is a strictly increasing function, show that f is injective and that f^{-1} is a strictly increasing function.

18. Give a complete account of the various types of functions that can arise from indefinite integration of rational functions.

19. Let $f : C \to \mathbb{R}$ be an injection and suppose that f is not monotone. Prove that either there exist $a, b, c \in C$ such that $a < b < c$ and $f(a)$, $f(c) < f(b)$, or else there exist $a, b, c \in C$ such that $a < b < c$ and $f(a)$, $f(c) > f(b)$.

20. Complete the proof of Theorem 7.3 by considering the case in which $f(c) < f(a)$.

21. If C is a connected set, then a function $f:C \to \mathbb{R}$ is said to have the **intermediate value property** if, whenever $a, b \in C$ with $a < b$ and k is a number between $f(a)$ and $f(b)$, then there exists $p \in [a, b]$ such that $f(p) = k$. Show that, if C is connected and $f:C \to \mathbb{R}$ is an injection with the intermediate value property, then f is strictly monotone. [Hint: Examine carefully the proof of Theorem 7.3.]

22. Let f be the function obtained by restricting the function sinr of Problem 10 in Problem Set 4.3 to the interval $[-1, 1]$. Show that, although f is not continuous, it does have the intermediate value property (see Problem 21 above).

23. Let f be an increasing function, let $a \in \mathrm{dom}(f)$, define the set $A = (a, \infty) \cap \mathrm{dom}(f)$, and suppose that a is an accumulation point of A. Prove that $\lim_{x \to a^+} f(x)$ exists, $\inf_{a < x} f(x)$ exists, and that

$$f(a) \le \inf_{a < x} f(x) = \lim_{x \to a^+} f(x).$$

24. True or false: If $f:A \to B$ and $g:B \to C$ are monotone functions, then $(g \circ f):A \to C$ is a monotone function. If true, prove it; if false, give a counterexample.

25. Supply the details for the proof of Theorem 7.7.

26. Prove that the conclusion of Theorem 7.7 holds even if the connected set C is not open.

27. Let C be an interval of real numbers. A function $f:C \to \mathbb{R}$ is said to be of **bounded variation** on C if there are increasing functions $g, h:C \to \mathbb{R}$ such that $f = g - h$. [A different, but equivalent, definition of bounded variation is usually given—see H. L. Royden, *Real Analysis*, 3rd ed. (New York: Macmillan, 1988), pp. 102–103.] Show that the class of all functions of bounded variation on C forms a vector space of functions. [Hint: Note that the constant zero function on C is increasing.]

28. If $f:[a, b] \to \mathbb{R}$ is injective and has the intermediate value property (see Problem 21), prove that both f and f^{-1} are continuous.

29. Suppose that C is an interval of real numbers and that $f:C \to \mathbb{R}$ is of bounded variation (see Problem 27). Prove that f can have at most countably many discontinuities on C.

30. Let C be an open, connected subset of $\mathbb{R}$ and suppose that $f:C \to \mathbb{R}$ is continuous and has the property that, for every open set $U \subseteq C$, $f(U)$ is open. Prove that f is monotone.

31. Let C be an open, connected set and let $f:C \to \mathbb{R}$ be increasing. Define $g:C \to \mathbb{R}$ by $g(x) = \lim_{t \to x^+} f(t)$ for each $x \in C$. Show that g is increasing

and that, for every $a \in C$,

$$g(a) = \inf_{a < x} g(x) = \lim_{x \to a^+} g(x).$$

32. Let I and J be intervals and suppose that $f : I \to J$ is strictly increasing and surjective. Prove that f is continuous and that $f^{-1} : J \to I$ is strictly increasing and surjective.

HISTORICAL NOTES

In the eighteenth century, Leonhard Euler (1707–1783) wrote two books on calculus that for many years were the models for texts: the *Institutiones Calculi Differentialis* (1787) and the *Institutionum Calculi Integralis* (1768–1770). But books on calculus and analysis today all reflect the treatment given the subject in a series of three books between 1821 and 1829 by Augustin Louis Cauchy (1789–1857), primarily in his famous text of 1821, *Cours d'Analyse de l'École Polytechnique*. Cauchy used the limit approach pioneered by Jean Le Rond d'Alembert (1717–1783), which gave it an arithmetic character, in contrast to the earlier dependence on geometry. Cauchy's definition reads: "When the successive values attributed to a variable approach indefinitely a fixed value so as to end by differing from it by as little as one wishes, this last is called the limit of all the others."

Cauchy's version of continuity, using this limit definition, comes close to the version we use today. Treatises on calculus in the eighteenth century had treated integration as the inverse of differentiation, which it is, of course, but Cauchy treated integration as the limit of sums. In this he was taking a more modern approach; this is the approach used in modern texts. Functions fail to have derivatives at points of discontinuity, but discontinuous curves may define areas, and hence, it makes sense to talk about integrals of such functions. Cauchy recognized this and used the mean value theorem to establish the fundamental theorem.

Cauchy's concern for precision in the definitions of calculus possibly is a result of his being a pioneer in the development of the theory of functions of a complex variable. Euler and d'Alembert had used complex variables in problems arising in physics. And Gauss, along with Jean Robert Argand (1768–1822) and Caspar Wessel (1745–1818), had placed the complex numbers in the plane and developed the first fundamental properties of complex functions. But it was Cauchy who wrote extensively on the subject and developed the powerful theorems in this area that bear his name. Since the derivative of a complex function no longer has the intuitive interpretation as a slope of a tangent line to a curve, Cauchy was forced to think of these ideas in more general terms.

Cauchy, who was extremely conservative in matters of religion and politics, went into exile with King Charles X rather than stay in France during the republican years. He accepted a position in Turin, but also spent some time in Prague, where Bernhard Bolzano (1781–1848), a Czechoslovakian priest, was writing his ideas on the definitions of limit, derivative, continuity, and convergence, which were quite similar to Cauchy's (though there is no reason to believe that they ever met). Bolzano's important book on the subject, *Rein analytischer Beweis*, was published in 1817. Bolzano early on discovered that there could be a one-to-one correspondence between the elements of an infinite set and a proper subset of that set, thus anticipating by many years Cantor's work on countable and uncountable sets and mappings between sets.

While Cauchy's definition of limit went a long way in the direction of eliminating geometrical intuition, there was still some reliance on geometrical ideas. It was only later, in 1872, that Eduard Heine (1821–1881), a student of Karl Weierstrass, gave the following definition: "If, given any ε, there is an η_0 such that for $0 < \eta < \eta_0$ the difference $f(x_0 \pm \eta) - L$ is less in absolute value than ε, then L is the limit of $f(x)$ for $x = x_0$." Finally we have arrived at a definition that fully uses what is now called the ε–δ approach to calculus, the bane of beginning calculus students but now a standard approach to proofs in analysis.

No account of the "arithmetization of analysis" could exclude Weierstrass, one of the giant figures in this whole area. Oddly enough, he started out as a secondary school teacher, and it was only some years into his career, at the age of 40 and after he had published a number of startling and important papers, that he received the call to become a professor at Berlin, perhaps at that time the best department of mathematics in Europe (and hence in the world). Weierstrass is, therefore, a counterexample to that oft-stated metatheorem in mathematics that only the young produce good mathematics!

Heine had proved what is now called the Heine–Borel theorem in 1872, but it had been ignored by the mathematical community. It was not until 1895 that it was proved again by Félix Edouard Émile Borel (1871–1956), who had worked extensively in the application of set theory to the theory of functions.

ANNOTATED BIBLIOGRAPHY

Calinger, Ronald, ed. *Classics of Mathematics* (Oak Park, Ill.: Moore), 1982.
  A source book with commentaries. Includes texts not only by Cauchy, Weierstrass, Dedekind, and Cantor on the arithmetization of analysis, but also excerpts from modern foundational works by Russell, Brouwer, and Gödel.

Fauvel, John, and Jeremy Gray, ed. *The History of Mathematics/A Reader* (London: Macmillan), 1987.

   An up-to-date look at a number of the original sources in the history of mathematics, in particular the nineteenth century attempts to make calculus rigorous. It reflects the thinking of recent scholars.

Grabiner, Judith V. *The Origins of Cauchy's Rigorous Calculus* (Cambridge, Mass.: MIT Press), 1981.

   A scholarly treatment of the contributions of Cauchy by one of the most respected historians of mathematics in our time.

MacLane, Saunders. *Mathematics: Form and Function* (New York: Springer-Verlag), 1986.

   A wide-ranging book that covers many areas of mathematics, but also frequently touches on the development of topological notions motivated by the need for more care in calculus.

Rudin, Walter. *Principles of Mathematical Analysis*, 3rd ed. (New York: McGraw-Hill), 1976.

   A classic text on post-calculus analysis by a widely respected mathematician.

5

DIFFERENTIATION
AND INTEGRATION

The idea of approximation is a theme that unifies all of mathematical analysis: We seek the true value τ of a quantity and use the symbolism

$$\tau \approx \alpha$$

to express the idea that α is an *approximation* to τ. The *error* $\mathscr{E}$ involved in the approximation is defined to be the difference between the true value τ and the approximate value α:

$$\mathscr{E} = \tau - \alpha.$$

By using the concept of error, we can rewrite the relation $\tau \approx \alpha$ in the form of the equation

$$\tau = \alpha + \mathscr{E}.$$

The smaller the error $\mathscr{E}$, the more accurate the approximation.

In this chapter, we use the idea of approximation to approach both the derivative and the integral. The derivative arises from the problem of approximating a function by a linear function, and the integral arises from the problem of approximating the area under the graph of a function.

1

LINEAR APPROXIMATION AND THE DERIVATIVE

Let f be a function, let $\mathscr{C}$ be a class of functions, and consider the problem of approximating f by a function α in the class $\mathscr{C}$, so that

$$f = \alpha + \mathscr{E},$$

where the error function

$$\mathscr{E} = f - \alpha$$

is small in some appropriate sense. We distinguish two special cases of this general problem:

1. *Interval approximation.* An interval $I \subseteq \mathrm{dom}(f)$ is specified and it is required that $I \subseteq \mathrm{dom}(\alpha)$ and that, in some sense, $\mathscr{E}$ is small over the entire interval I. For instance, if ε is a given positive real number, it may be required that

$$\sup_{x \in I} |\mathscr{E}(x)| < \varepsilon,$$

or, if $I = [a, b]$, the stipulation might be that

$$\int_a^b |\mathscr{E}(x)|\, dx < \varepsilon.$$

2. *Local approximation near a point.* A point $a \in \mathrm{dom}(f)$ is specified and it is required that $a \in \mathrm{dom}(\alpha)$ and that, in some sense, $\mathscr{E}(x)$ is small when $x \in \mathrm{dom}(f) \cap \mathrm{dom}(\alpha)$ and $|x - a|$ is small.

In what follows, we concentrate on the idea of local approximation and attempt to fix the meaning of the word *small*. A minimal requirement would seem to be that

$$\lim_{x \to a} \mathscr{E}(x) = 0. \qquad\qquad (*)$$

Note that $(*)$ alone places no condition on the value of $\mathscr{E}(a)$; it simply requires that the approximation $f(x) \approx \alpha(x)$ gets "better and better" as x approaches a. Thus, it seems reasonable to augment requirement $(*)$ by stipulating that

$$\mathscr{E}(a) = 0, \qquad\qquad (**)$$

so there is no error at all when $x = a$.

Although it may seem that $(*)$ and $(**)$ capture the idea that f is approximated by α near a, it turns out that these conditions by themselves are too weak. To see what the difficulty is, suppose that f and α are continuous at a. If we grant Condition $(**)$, which merely requires that $f(a) = \alpha(a)$, then (by the continuity of f and α at a) we have

$$\lim_{x \to a} \mathscr{E}(x) = \lim_{x \to a} [f(x) - \alpha(x)] = \lim_{x \to a} f(x) - \lim_{x \to a} \alpha(x)$$
$$= f(a) - \alpha(a) = 0,$$

and Condition $(*)$ is automatically satisfied. What is needed is a stronger criterion for the smallness of the error. The appropriate idea is furnished by the following definition.

1.1 DEFINITION

RELATIVE RATES OF APPROACH TO ZERO Let $\mathscr{E}$ and w be functions and suppose that a is an accumulation point of $\operatorname{dom}(\mathscr{E}/w)$. We say that $\mathscr{E}$ **approaches zero at** a **more rapidly than** w does if

$$\lim_{x \to a} \frac{\mathscr{E}(x)}{w(x)} = 0.$$

Suppose that $\mathscr{E}$ approaches zero at a more rapidly than w, so that

$$\lim_{x \to a} \frac{\mathscr{E}(x)}{w(x)} = 0.$$

Suppose also that

$$\lim_{x \to a} w(x) = 0.$$

The latter condition would normally tend to make the fraction $\mathscr{E}(x)/w(x)$ large in absolute value when x is close to a. The former condition obliges $\mathscr{E}(x)$ to approach zero so rapidly that this tendency is wiped out!

1.2 LEMMA

Let $\mathscr{E}$ and w be functions and suppose that a is an accumulation point of $\operatorname{dom}(\mathscr{E}/w)$. Then $\mathscr{E}$ approaches zero at a more rapidly than w does if and only if there exists a function ε with the following four properties:

(i) $\operatorname{dom}(\varepsilon) = \operatorname{dom}(\mathscr{E}/w) \cup \{a\}$.
(ii) $\mathscr{E}(x) = \varepsilon(x)w(x)$ for all $x \in \operatorname{dom}(\mathscr{E}/w) \setminus \{a\}$.
(iii) $\varepsilon(a) = 0$.
(iv) $\lim_{x \to a} \varepsilon(x) = 0$.

PROOF   Suppose that such a function ε exists. Then, by Conditions (i) and (ii), we have

$$\frac{\mathscr{E}(x)}{w(x)} = \varepsilon(x) \quad \text{for all } x \in \operatorname{dom}\left(\frac{\mathscr{E}}{w}\right) \setminus \{a\};$$

hence, by Condition (iii) and Corollary 2.9 in Chapter 4,

$$\lim_{x \to a} \frac{\mathscr{E}(x)}{w(x)} = 0.$$

Conversely, suppose that $\mathscr{E}$ approaches zero at a more rapidly than w does, and define $\varepsilon:\mathrm{dom}(\mathscr{E}/w) \cup \{a\} \to \mathbb{R}$ by

$$\varepsilon(x) = \begin{cases} \dfrac{\mathscr{E}(x)}{w(x)}, & \text{if } x \neq a \\ 0, & \text{if } x = a \end{cases}$$

for all $x \in \mathrm{dom}(\mathscr{E}/w) \cup \{a\}$. Then Conditions (i), (ii), and (iii) are obviously satisfied, and Condition (iv) follows from Corollary 2.9 in Chapter 4. $\square$

The idea of a derivative arises from the following **problem of local linear approximation**: We are given a function f and an accumulation point a of $\mathrm{dom}(f)$ such that $a \in \mathrm{dom}(f)$. We seek a *linear* function α that approximates f near a in such a way that

$$f(a) = \alpha(a)$$

and the error

$$\mathscr{E}(x) = f(x) - \alpha(x)$$

approaches zero more rapidly than $x - a$.

By Lemma 1.2, the problem of local linear approximation is equivalent to the problem of finding a linear function α and a function ε satisfying the following four conditions:

  (i) $\mathrm{dom}(\varepsilon) = \mathrm{dom}(f)$,
  (ii) $f(x) - \alpha(x) = \varepsilon(x)(x - a)$ for all $x \in \mathrm{dom}(f)$,
  (iii) $\varepsilon(a) = 0$, and
  (iv) $\lim\limits_{x \to a} \varepsilon(x) = 0$.

(See Problem 3.) The linear function α can be expressed in the form

$$\alpha(x) = mx + b$$

for all $x \in \mathbb{R}$, where m and b are fixed real numbers. Thus, Condition (ii) is equivalent to

$$f(x) - mx - b = \varepsilon(x)(x - a) \qquad \text{for all } x \in \mathrm{dom}(f).$$

By substituting $x = a$ in the last equation, we find that

$$f(a) - ma - b = 0,$$

so that

$$b = f(a) - ma.$$

Consequently, Condition (ii) can be rewritten in the equivalent form

$$f(x) - mx - (f(a) - ma) = \varepsilon(x)(x - a),$$

or

$$f(x) = f(a) + m(x - a) + \varepsilon(x)(x - a)$$

for all $x \in \text{dom}(f)$.

The considerations above lead us to our next definition.

1.3 DEFINITION

> **LOCAL LINEAR APPROXIMATION** If f is a function and $a \in \text{dom}(f)$, then the function α defined for all $x \in \mathbb{R}$ by
>
> $$\alpha(x) = f(a) + m(x - a)$$
>
> is said to be a **local linear approximation to f near a** if there is a function ε with $\text{dom}(\varepsilon) = \text{dom}(f)$, $\varepsilon(a) = 0$, and
>
> $$\lim_{x \to a} \varepsilon(x) = 0,$$
>
> such that, for all $x \in \text{dom}(f)$,
>
> $$f(x) = f(a) + m(x - a) + \varepsilon(x)(x - a).$$

Note that, if f has a local linear approximation α near a, then it is necessary that a be an accumulation point of $\text{dom}(f)$. [This follows from the fact that ε has a limit at a and $\text{dom}(\varepsilon) = \text{dom}(f)$.]

1.4 LEMMA

> **UNIQUENESS OF THE LOCAL LINEAR APPROXIMATION** If f is a function, $a \in \text{dom}(f)$, and α_1 and α_2 are local linear approximations to f near a, then $\alpha_1 = \alpha_2$.

PROOF Let α_1 and α_2 be given by

$$\alpha_1(x) = f(a) + m_1(x - a) \tag{1}$$

and

$$\alpha_2(x) = f(a) + m_2(x - a) \tag{2}$$

for all $x \in \mathbb{R}$. It will be sufficient to prove that $m_1 = m_2$. By hypothesis, there exist functions ε_1 and ε_2 with $\text{dom}(\varepsilon_1) = \text{dom}(\varepsilon_2) = \text{dom}(f)$ such that, for all $x \in \text{dom}(f)$,

$$f(x) = f(a) + m_1(x - a) + \varepsilon_1(x)(x - a), \tag{3}$$

$$f(x) = f(a) + m_2(x - a) + \varepsilon_2(x)(x - a), \tag{4}$$

and also

$$\lim_{x \to a} \varepsilon_1(x) = \lim_{x \to a} \varepsilon_2(x) = 0. \tag{5}$$

By subtracting Equation (4) from Equation (3), we find that

$$(m_1 - m_2)(x - a) = [\varepsilon_2(x) - \varepsilon_1(x)](x - a) \tag{6}$$

for all $x \in \text{dom}(f)$. Consequently,

$$m_1 - m_2 = \varepsilon_2(x) - \varepsilon_1(x) \tag{7}$$

holds for $x \in \text{dom}(f)$ with $x \neq a$. By Equations (7) and (5), we have

$$m_1 - m_2 = \lim_{x \to 0} \varepsilon_2(x) - \lim_{x \to a} \varepsilon_1(x) = 0, \tag{8}$$

and therefore $m_1 = m_2$. $\square$

1.5 DEFINITION

DIFFERENTIABILITY   If f is a function and $a \in \mathbb{R}$, we say that f is **differentiable at** a if $a \in \text{dom}(f)$ and f has a local linear approximation α near a. If $A \subseteq \text{dom}(f)$ and f is differentiable at every point $a \in A$, then we say that f is **differentiable on** A. If f is differentiable at every point $a \in \text{dom}(f)$, then we say that f is **differentiable**.

If f is differentiable at a and the local linear approximation α to f near a is defined by

$$\alpha(x) = f(a) + m(x - a)$$

for all $x \in \mathbb{R}$, then the graph of α is called the **tangent line to the graph of f at a**. Thus, the tangent line is the straight line through the point $(a, f(a))$ that best approximates the graph of f near $(a, f(a))$.

1.6 DEFINITION

THE DERIVATIVE OF A FUNCTION   Let f be a function and let D be the set of all points at which f is differentiable. If $D \neq \varnothing$, we define the **derivative** of f to be the function $f': D \to \mathbb{R}$ such that, for each $a \in D$,

$$f'(a) = m,$$

where the unique local linear approximation α to f near a is given by

$$\alpha(x) = f(a) + m(x - a)$$

for all $x \in \mathbb{R}$.

The notation f' for the derivative of f was introduced by the French mathematician Joseph Louis Lagrange (1736–1813). If f is differentiable at a, we sometimes refer to $f'(a)$ as the **derivative of f at a**, although it really should be called the *value* of the derivative of f at a. Note that the equation of the tangent line to the graph of f at a is

$$y = f(a) + f'(a)(x - a)$$

and that $f'(a)$ is the slope of this tangent line. The following lemma merely summarizes Definitions 1.3, 1.5, and 1.6 as well as Lemma 1.4.

1.7 LEMMA

Let f be a function and let $a \in \mathrm{dom}(f)$. Then f is differentiable at a if and only if there exists a function ε with $\mathrm{dom}(\varepsilon) = \mathrm{dom}(f)$ and $\lim_{x \to a} \varepsilon(x) = \varepsilon(a) = 0$ and there exists a real number m such that, for all $x \in \mathrm{dom}(f)$,

$$f(x) = f(a) + m(x - a) + \varepsilon(x)(x - a).$$

Furthermore, if such an m exists, then it is uniquely determined and $f'(a) = m$.

Intuitively, if a curve has a tangent line at a point, then the curve must be continuous at that point. We are now in a position to give a rigorous proof of this fact.

1.8 THEOREM

DIFFERENTIABILITY IMPLIES CONTINUITY If the function f is differentiable at the point a, then it is continuous at a.

PROOF Suppose f is differentiable at a. Then there exists a function ε satisfying the conditions of Lemma 1.7. In particular,

$$f(x) = f(a) + m(x - a) + \varepsilon(x)(x - a)$$

holds for all $x \in \mathrm{dom}(f)$. By taking the limit on both sides of this equation as x approaches a, we find that

$$\lim_{x \to a} f(x) = f(a). \qquad \square$$

The next theorem is usually paraphrased by the statement that the derivative of a sum is the sum of the derivatives.

1.9 THEOREM

SUM RULE FOR DERIVATIVES   Let f and g be functions, let $a \in$ $\mathrm{dom}(f) \cap \mathrm{dom}(g)$, and suppose that both f and g are differentiable at a. Then, if a is an accumulation point of $\mathrm{dom}(f) \cap \mathrm{dom}(g)$, it follows that $f + g$ is differentiable at a and that

$$(f + g)'(a) = f'(a) + g'(a).$$

PROOF   By hypothesis, there exist functions ε_1 and ε_2 and there exist real numbers $m_1 = f'(a)$ and $m_2 = g'(a)$ such that $\mathrm{dom}(\varepsilon_1) = \mathrm{dom}(f)$, $\mathrm{dom}(\varepsilon_2) = \mathrm{dom}(g)$,

$$\lim_{x \to a} \varepsilon_1(x) = \varepsilon_1(a) = 0 = \varepsilon_2(a) = \lim_{x \to a} \varepsilon_2(a),$$

$$f(x) = f(a) + m_1(x - a) + \varepsilon_1(x)(x - a), \qquad \forall x \in \mathrm{dom}(f),$$

and

$$g(x) = g(a) + m_2(x - a) + \varepsilon_2(x)(x - a), \qquad \forall x \in \mathrm{dom}(g).$$

Let $\varepsilon : \mathrm{dom}(f) \cap \mathrm{dom}(g) \to \mathbb{R}$ be defined by $\varepsilon = \varepsilon_1 + \varepsilon_2$. Then $\mathrm{dom}(\varepsilon) = \mathrm{dom}(f + g)$, $\varepsilon(a) = \varepsilon_1(a) + \varepsilon_2(a) = 0$, and

$$\lim_{x \to a} \varepsilon(x) = \lim_{x \to a} \left[\varepsilon_1(x) + \varepsilon_2(x) \right] = 0 + 0 = 0.$$

Furthermore,

$$(f + g)(x) = f(x) + g(x)$$
$$= f(a) + g(a) + (m_1 + m_2)(x - a) + \varepsilon(x)(x - a)$$
$$= (f + g)(a) + (m_1 + m_2)(x - a) + \varepsilon(x)(x - a).$$

Therefore, $f + g$ is differentiable at a and

$$(f + g)'(a) = m_1 + m_2 = f'(a) + g'(a). \qquad \square$$

By an argument similar to that in the proof of Theorem 1.9, it is not difficult to prove the theorem usually paraphrased by the statement that the derivative of a difference is the difference of the derivatives (Problem 12).

1.10 THEOREM

PRODUCT RULE FOR DERIVATIVES   Let f and g be functions, let $a \in$ $\mathrm{dom}(f) \cap \mathrm{dom}(g)$, and suppose that both f and g are differentiable at a. Then, if a is an accumulation point of $\mathrm{dom}(f) \cap \mathrm{dom}(g)$, it follows that $f \cdot g$ is differentiable at a and that

$$(f \cdot g)'(a) = f'(a)g(a) + f(a)g'(a).$$

PROOF Let ε_1, ε_2, m_1, and m_2 be as in the proof of Theorem 1.9. Let the function $\varepsilon:\mathrm{dom}(f) \cap \mathrm{dom}(g) \to \mathbb{R}$ be defined by

$$\varepsilon(x) = f(a)\varepsilon_2(x) + g(a)\varepsilon_1(x) + [m_1 + \varepsilon_1(x)][m_2 + \varepsilon_2(x)](x - a)$$

for all $x \in \mathrm{dom}(f) \cap \mathrm{dom}(g)$. Obviously $\varepsilon(a) = 0$ and $\lim_{x \to a} \varepsilon(x) = 0$. Also, by a direct calculation (Problem 13),

$$\begin{aligned}(f \cdot g)(x) &= f(x)g(x)\\ &= f(a)g(a) + [m_1 g(a) + f(a)m_2](x - a) + \varepsilon(x)(x - a)\\ &= (f \cdot g)(a) + [m_1 g(a) + f(a)m_2](x - a) + \varepsilon(x)(x - a)\end{aligned}$$

holds for every $x \in \mathrm{dom}(f \cdot g)$. Therefore, $f \cdot g$ is differentiable at a and

$$\begin{aligned}(f \cdot g)'(a) &= m_1 g(a) + f(a)m_2\\ &= f'(a)g(a) + f(a)g'(a).\end{aligned}$$ $\square$

1.11 LEMMA

THE DERIVATIVE OF $1/x$ Let $r(x) = 1/x$ for $x \neq 0$. Then r is differentiable and, for each $a \in \mathbb{R}$ with $a \neq 0$,

$$r'(a) = \frac{-1}{a^2}.$$

PROOF Define $\varepsilon:\mathbb{R} \setminus \{0\} \to \mathbb{R}$ by

$$\varepsilon(x) = \frac{1}{a^2} - \frac{1}{ax}$$

for all $x \neq 0$. Then $\varepsilon(a) = 0$, $\lim_{x \to a} \varepsilon(x) = 0$, and, by a direct calculation (Problem 19),

$$\frac{1}{x} = \frac{1}{a} + \frac{-1}{a^2}(x - a) + \varepsilon(x)(x - a)$$

holds for all $x \neq 0$. $\square$

1.12 THEOREM

THE CHAIN RULE Let f and g be functions, and let $a \in \mathrm{dom}(f \circ g)$. Suppose that g is differentiable at a and that f is differentiable at $g(a)$. Then, if a is an accumulation point of $\mathrm{dom}(f \circ g)$, it follows that $f \circ g$ is differentiable at a and that

$$(f \circ g)'(a) = f'(g(a))g'(a).$$

PROOF Since g is differentiable at a, there exists a function ε and a real number $m = g'(a)$ such that $\mathrm{dom}(\varepsilon) = \mathrm{dom}(g)$, $\lim_{x \to a} \varepsilon(x) = \varepsilon(a) = 0$, and

$$g(x) = g(a) + m(x - a) + \varepsilon(x)(x - a), \qquad \forall x \in \mathrm{dom}(g). \tag{1}$$

Let $b = g(a)$. Since f is differentiable at b, there exists a function γ and a real number $n = f'(b)$ such that $\mathrm{dom}(\gamma) = \mathrm{dom}(f)$, $\lim_{y \to b} \gamma(y) = \gamma(b) = 0$, and

$$f(y) = f(b) + n(y - b) + \gamma(y)(y - b), \qquad \forall y \in \mathrm{dom}(f). \tag{2}$$

Define $\lambda : \mathrm{dom}(f \circ g) \to \mathbb{R}$ by

$$\lambda(x) = \varepsilon(x)n + \gamma(g(x))[m + \varepsilon(x)] \tag{3}$$

for all $x \in \mathrm{dom}(f \circ g)$. Then, since $\varepsilon(a) = 0$ and $\gamma(g(a)) = \gamma(b) = 0$, we have

$$\lambda(a) = 0. \tag{4}$$

By Theorem 1.8, g is continuous at a. Also, in view of the fact that $\lim_{y \to b} \gamma(y) = \gamma(b)$, γ is continuous at $b = g(a)$; hence, by Theorem 3.10 in Chapter 4, $\gamma \circ g$ is continuous at a. Therefore,

$$\lim_{x \to a} \gamma(g(x)) = \lim_{x \to a} (\gamma \circ g)(x) = (\gamma \circ g)(a) \tag{5}$$
$$= \gamma(g(a)) = \gamma(b) = 0.$$

As a consequence of Equations (3) and (5), we have

$$\lim_{x \to a} \lambda(x) = 0 \cdot n + 0 \cdot (m + 0) = 0. \tag{6}$$

Now suppose that $x \in \mathrm{dom}(f \circ g)$ and let $y = g(x)$. Then, since $b = g(a)$, we can rewrite (1) in the form

$$y - b = m(x - a) + \varepsilon(x)(x - a). \tag{7}$$

As a consequence of (7), we have

$$n(y - b) = nm(x - a) + \varepsilon(x)n(x - a) \tag{8}$$

and

$$\gamma(y)(y - b) = \gamma(y)[m + \varepsilon(x)](x - a). \tag{9}$$

By combining (2), (3), (8), and (9), we find that

$$f(y) = f(b) + nm(x - a) + \lambda(x)(x - a). \tag{10}$$

Using the fact that $y = g(x)$ and $b = g(a)$, we can rewrite (10) as

$$(f \circ g)(x) = (f \circ g)(a) + nm(x - a) + \lambda(x)(x - a), \tag{11}$$

which, in view of (4) and (6), shows that $f \circ g$ is differentiable at a and that

$$(f \circ g)'(a) = nm = f'(b)g'(a) = f'(g(a))g'(a). \tag{12}$$

$\square$

1.13 THEOREM **RECIPROCAL RULE FOR DERIVATIVES** Let f be a function, let f be differentiable at a, and suppose that $f(a) \neq 0$. Also, let $M = \{x \in \text{dom}(f)\,|\, f(x) \neq 0\}$ and let $g:M \to \mathbb{R}$ be defined by $g(x) = 1/f(x)$ for all $x \in M$. Then g is differentiable at a and

$$g'(a) = \frac{-f'(a)}{[f(a)]^2}.$$

PROOF Let r be the function of Lemma 1.11. Then we have $g = r \circ f$, and the conclusion of the theorem follows from the chain rule (Theorem 1.12). $\square$

1.14 THEOREM **QUOTIENT RULE FOR DERIVATIVES** Let f and g be functions, let $a \in \text{dom}(f) \cap \text{dom}(g)$, and suppose that both f and g are differentiable at a. Then, if a is an accumulation point of $\text{dom}(f) \cap \text{dom}(g)$ and $g(a) \neq 0$, it follows that f/g is differentiable at a and that

$$\left(\frac{f}{g}\right)'(a) = \frac{g(a)f'(a) - f(a)g'(a)}{[g(a)]^2}.$$

PROOF Combine the product rule (Theorem 1.10) and the reciprocal rule (Theorem 1.13) (see Problem 23). $\square$

1.15 THEOREM **CONDITION FOR DIFFERENTIABILITY AT A POINT** Let f be a function and let $a \in \text{dom}(f)$. Then f is differentiable at a if and only if

$$\lim_{x \to a} \frac{f(x) - f(a)}{x - a}$$

exists. Furthermore, if this limit exists, then

$$f'(a) = \lim_{x \to a} \frac{f(x) - f(a)}{x - a}.$$

PROOF Define $q:\text{dom}(f) \setminus \{a\} \to \mathbb{R}$ by

$$q(x) = \frac{f(x) - f(a)}{x - a}.$$

Suppose that f is differentiable at a. Then there exists $\varepsilon:\text{dom}(f) \to \mathbb{R}$ such that $\varepsilon(a) = 0$, $\lim_{x \to a} \varepsilon(x) = 0$, and

$$f(x) = f(a) + f'(a)(x - a) + \varepsilon(x)(x - a).$$

By subtracting $f(a)$ from both sides of the last equation and dividing by $x - a$, we find that

$$q(x) = f'(a) + \varepsilon(x)$$

for all $x \in \text{dom}(f)$ with $x \neq a$. Therefore,

$$\lim_{x \to a} q(x) = f'(a) + \lim_{x \to a} \varepsilon(x) = f'(a) + 0 = f'(a).$$

Conversely, suppose that $\lim_{x \to a} q(x)$ exists and let

$$m = \lim_{x \to a} q(x).$$

Define $\varepsilon : \text{dom}(f) \to \mathbb{R}$ by

$$\varepsilon(x) = \begin{cases} q(x) - m, & \text{if } x \neq a \\ 0, & \text{if } x = a \end{cases}$$

for all $x \in \text{dom}(f)$. Then $\varepsilon(a) = 0$ and

$$\lim_{x \to a} \varepsilon(x) = \lim_{x \to a} [q(x) - m] = m - m = 0.$$

If $x \in \text{dom}(f)$ and $x \neq a$, then, from the definition of q,

$$f(x) - f(a) = q(x)(x - a)$$

and, from the definition of ε,

$$q(x) = m + \varepsilon(x).$$

Therefore, for $x \in \text{dom}(f)$ with $x \neq a$, we have

$$f(x) - f(a) = [m + \varepsilon(x)](x - a).$$

Note that the last equation holds even if $x = a$; hence,

$$f(x) = f(a) + m(x - a) + \varepsilon(x)(x - a)$$

holds for all $x \in \text{dom}(f)$, and it follows that f is differentiable at a.   □

1.16 Example   Let sinr be the function defined in Problem 10 of Problem Set 4.3 and let $f(x) = x \, \text{sinr}(x)$ for all $x \in \mathbb{R}$. Show that f is not differentiable at 0.

SOLUTION   By Theorem 1.13, f is differentiable at 0 if and only if

$$\lim_{x \to 0} \frac{f(x) - f(0)}{x - 0}$$

exists. Now, $f(0) = 0$ and, for $x \neq 0$, $f(x)/x = \text{sinr}(x)$, so it will be sufficient to show that sinr fails to have a limit at 0. Suppose, on the contrary, that

$$\lim_{x \to 0} \text{sinr}(x) = \lim_{x \to 0} \sin\left(\frac{1}{x}\right)$$

exists and equals L. Then there exists $\delta > 0$ such that

$$|x| < \delta \Rightarrow \left|\sin\left(\frac{1}{x}\right) - L\right| < 1.$$

By the Archimedean property of $\mathbb{R}$ (Theorem 4.11 in Chapter 3), there exists a natural number n such that

$$n(2\pi) > \frac{1}{\delta} - \frac{\pi}{2}.$$

Thus,

$$\frac{\pi}{2} + 2n\pi > \frac{1}{\delta}.$$

Let

$$u = \frac{1}{(\pi/2) + 2n\pi},$$

noting that

$$0 < u < \delta.$$

With $x = u$, we have $|x| = u < \delta$ and

$$f(x) = \sin\left(\frac{1}{x}\right) = \sin\left(\frac{\pi}{2} + 2n\pi\right) = 1,$$

so

$$|1 - L| < 1.$$

With $x = -u$, we have $|x| = u < \delta$ and

$$f(x) = \sin\left(\frac{1}{x}\right) = \sin\left(-\frac{\pi}{2} - 2n\pi\right) = -1,$$

so

$$|-1 - L| < 1.$$

Thus, by the triangle inequality,

$$\begin{aligned} 2 = |1 + 1| &= |(1 - L) + (1 + L)| \\ &\le |1 - L| + |1 + L| = |1 - L| + |-1 - L| \\ &< 1 + 1 = 2, \end{aligned}$$

a contradiction. $\qquad\square$

We leave the proof of the following corollary to Theorem 1.15 as an exercise (Problem 26).

1.17 COROLLARY

> **FORMULA FOR THE DERIVATIVE**   Let f be a function, let D be the set of all points in dom(f) at which f is differentiable, and suppose that $D \neq \emptyset$. Then, for all $x \in D$,
>
> $$f'(x) = \lim_{h \to 0} \frac{f(x + h) - f(x)}{h}.$$

The expression

$$\frac{f(x + h) - f(x)}{h}$$

in Corollary 1.17 is called a **difference quotient**. Although the derivative is often defined to be the limit of a difference quotient, our approach has the advantage that Definition 1.6 does not involve any division and hence it can be extended to vector-valued functions of a vector variable.[†] (Vectors can be added, subtracted, and multiplied by scalars but, in general, there is no meaning assigned to the quotient of two vectors.)

PROBLEM SET 5.1

1. Let w be a nonzero constant function on $\mathbb{R}$. If f is a function and a is an accumulation point of the domain of f, show that f approaches zero at a more rapidly than w does if and only if $\lim_{x \to a} f(x) = 0$.

2. Does $\sin x$ approach zero at 0 more rapidly than x does? Explain.

3. If a is an accumulation point of the domain of a function f, show that a linear function α is a local linear approximation to f at a if and only if there exists a function ε such that dom(ε) = dom(f), $\lim_{x \to a} \varepsilon(x) = \varepsilon(a) = 0$, and, for all $x \in$ dom(f), $f(x) - \alpha(x) = \varepsilon(x)(x - a)$.

4. Does $1 - \cos x$ approach zero at 0 more rapidly than x does? Explain.

5. Let $\mathscr{E}$ and w be functions and suppose that a is an accumulation point of dom($\mathscr{E}/w$). Let $|w|$:dom(w) $\to \mathbb{R}$ be defined by $|w|(x) = |w(x)|$ for all $x \in$ dom(w). Prove that $\mathscr{E}$ approaches zero at a more rapidly than w does if and only if $\mathscr{E}$ approaches zero at a more rapidly than $|w|$ does.

6. Let f be a polynomial function and let $a \in \mathbb{R}$. Let q be the quotient polynomial and R be the remainder obtained from the division algorithm when $f(x)$ is divided by $x - a$.
    (a) Prove that $f(x) - R - q(a)(x - a)$ approaches zero at a faster than $x - a$ does.

[†] See J. Marsden and A. Tromba, *Vector Calculus*, 3rd ed. (New York: Freeman, 1988).

(b) Show that f is differentiable at a and that $\alpha(x) = R + q(a)(x - a)$ is the local linear approximation to f near a.

7. If a is an accumulation point of the domain of a function f, show that the linear function $\alpha(x) = f(a) + m(x - a)$ is a local linear approximation to f near a if and only if there exists a function γ such that $\text{dom}(\gamma) = \text{dom}(f)$, $\lim_{x \to a} \gamma(x) = \gamma(a) = 0$, and, for all $x \in \text{dom}(f)$,

$$f(x) = f(a) + m(x - a) + \gamma(x)|x - a|.$$

8. Let f be the polynomial function $f(x) = x^3 + 3x^2 + 3x$ and let α be the local linear approximation to f near $a = 1$. If $\mathscr{E} = f - \alpha$, show that $\mathscr{E}$ approaches zero at 1 more rapidly than $(x - 1)^2$ does.

9. If f is a constant function, $a \in \text{dom}(f)$, and a is an accumulation point of $\text{dom}(f)$, prove that f is differentiable at a and that $f'(a) = 0$.

10. Let f and α be functions, let a be an accumulation point of the set $\text{dom}(f) \cap \text{dom}(\alpha)$, and suppose that α is continuous at a. If $f - \alpha$ approaches zero at a more rapidly than the function w does, and if there exists a neighborhood N of a and a constant M such that $|w(x)| \le M$ for all $x \in N \cap \text{dom}(f)$, prove that f is continuous at a.

11. Let f be the identity function, that is, $f(x) = x$ for all $x \in \mathbb{R}$. Prove that f is differentiable and that $f'(a) = 1$ for all $a \in \mathbb{R}$.

12. Give a careful statement and proof of the theorem usually paraphrased by the statement that the derivative of a difference is the difference of the derivatives.

13. Supply the details of the calculation in the proof of Theorem 1.10.

14. Let f be a function and let $a \in \text{dom}(f)$. If $M = [a, \infty)$, we say that f **has a derivative from the right at a** if the restricted function $f|_M$ has a derivative at a. If f has a derivative from the right at a, we define the **derivative from the right of f at a**, in symbols $f'(a^+)$, by $f'(a^+) = (f|_M)'(a)$. Provide a similar definition of $f'(a^-)$, the **derivative from the left of f at a**.

15. If n is a positive integer, prove that x^n is differentiable at every real number x and that its derivative is nx^{n-1}. [Hint: Use mathematical induction and Theorem 1.10.]

16. Suppose that the function f has derivatives from the right and from the left at a (Problem 14). Prove that f has a derivative at a and that $f'(a) = f'(a^+) = f'(a^-)$.

17. Prove that every polynomial function is differentiable.

18. If a is an interior point of $\text{dom}(f)$, show that f is differentiable at a if and only if f has derivatives from the right and from the left at a. (See Problems 14 and 16.)

19. Provide the direct calculation called for in the proof of Lemma 1.11.

20. If f is a function such that $\lim_{x \to 0} f(x) - 0$, and if $g : \text{dom}(f) \to \mathbb{R}$ is defined by $g(x) = x \cdot f(x)$ for all $x \in \text{dom}(f)$, prove that g is differentiable at 0 and that $g'(0) = 0$.

21. Let n be a positive integer, let f be a function that is differentiable at a, and define $g(x) = [f(x)]^n$ for all $x \in \text{dom}(f)$. Prove that g is differentiable at a and that $g'(a) = n[f(a)]^{n-1}f'(a)$. [Hint: Use the result of Problem 15 and the chain rule.]

22. Let sinr be the function defined in Problem 10 of Problem Set 4.3 and let $f(x) = x^2 \sin r(x^2)$ for all $x \in \mathbb{R}$.
     (a) Show that f is differentiable at every real number and find a formula for $f'(x)$.
     (b) Show that f' is unbounded on the interval $[-1, 1]$.
     (c) Show that f' is discontinuous at 0.

23. Supply the details for the proof of Theorem 1.14.

24. Let $f(x) = \sqrt{x}$ for all $x \geq 0$.
     (a) Prove by direct calculation of the limit of a difference quotient that $f'(x) = 1/(2\sqrt{x})$ holds for all $x > 0$.
     (b) Prove that f is not differentiable at 0.

25. If n is a negative integer and $f(x) = x^n$ for all $x \neq 0$, prove that f is differentiable and that $f'(x) = nx^{n-1}$ for all $x \neq 0$.

26. Prove Corollary 1.17.

27. Prove that every rational function is differentiable.

28. If n is a constant and $|x|$ is small, then $(1 + x)^n \approx 1 + nx$. Explain.

29. Show that, although the absolute value function $|x|$ is continuous at 0, it is not differentiable at 0.

30. If $|x|$ is small, then $\sin x \approx x$. Explain.

31. Let $f(x) = |x|$ for all $x \in \mathbb{R}$. If $x \neq 0$, prove that f is differentiable at x and that $f'(x) = x/|x|$.

32. If $|x|$ is small, then $e^x \approx 1 + x$. Explain.

33. If $f(x) = x|x|$ for all $x \in \mathbb{R}$, show that f is differentiable at each $x \in \mathbb{R}$ and find a formula for $f'(x)$.

34. Let sinr be the function defined in Problem 10 of Problem Set 4.3 and let $f(x) = x^3 \sin r(x)$ for all $x \in \mathbb{R}$.
     (a) Show that f is differentiable at every real number and find a formula for $f'(x)$.
     (b) Show that f' is continuous on $\mathbb{R}$.
     (c) Show that f' is differentiable on $\mathbb{R} \setminus \{0\}$.
     (d) Show that f' is not differentiable at 0.

35. Let X be a nonempty subset of $\mathbb{R}$ and assume that every point of X is an accumulation point of X. Let $\mathscr{D}$ be the set of all differentiable functions

$f : X \to \mathbb{R}$. Prove that $\mathscr{D}$ is a vector space of functions. (See Section 7 of Chapter 4, page 310.)

36. For which positive integers n is the function $f(x) = |x|^n$ differentiable at 0?

37. Let X be a nonempty subset of $\mathbb{R}$ and assume that every point of X is an accumulation point of X. Let $C^1(X)$ be the set of all differentiable functions $f : X \to \mathbb{R}$ such that $f' : X \to \mathbb{R}$ is continuous. Prove that $C^1(X)$ is a vector space of functions. (See Problem 35.)

2

THE MEAN VALUE AND INVERSE FUNCTION THEOREMS

Although the mean value theorem is one of the key theorems of differential calculus, its complete proof is rarely given in elementary courses because it depends on the extreme value theorem, the proof of which is too sophisticated to be presented at that level. Since we have proved the extreme value theorem (Theorem 5.14 in Chapter 4), we are now in a position to give a proof of the mean value theorem. We begin with some preliminary definitions and results, each of which is of some interest in its own right.

2.1 DEFINITION

> **MAXIMUM AND MINIMUM VALUES OF A FUNCTION** Let f be a function and let $X \subseteq \mathrm{dom}(f)$. If $q \in X$ and $f(x) \le f(q)$ for all $x \in X$, then we say that $f(q)$ is the **maximum value of f on X**. If $p \in X$ and $f(p) \le f(x)$ for all $x \in X$, then we say that $f(p)$ is the **minimum value of f on X**.

If $a \in X \subseteq \mathrm{dom}(f)$ and $f(a)$ is the maximum (or minimum) value of f on X, we often say that f **attains its maximum (or minimum) value on X at a**. Note that f attains its maximum (or minimum) value on X at a if and only if $a \in X$ and $f(a) = \sup[f(X)]$ (or $f(a) = \inf[f(X)]$). Maximum or minimum values of f on X, when they exist, are called **extreme values of f on X**, or **extrema of f on X**. The extreme value theorem guarantees that a continuous function takes on both a maximum and a minimum value on a compact, nonempty subset of its domain.

2.2 LEMMA

> Let f be a function, let $a \in \mathrm{dom}(f)$, let $c < a$, and let $X = (c, a] \cap \mathrm{dom}(f)$. Suppose that f attains a maximum value (or a minimum value) on X at a. Then, if a is an accumulation point of X, and if f is differentiable at a, it follows that $f'(a) \ge 0$ (or $f'(a) \le 0$).

PROOF   We give the proof for the case in which f attains a maximum value on X at a and leave the case in which it attains a minimum value on X at a as an exercise (Problem 1). Suppose, then, that $f(x) \leq f(a)$ holds for all $x \in X$. Note that $x \leq a$ also holds for all $x \in X$. Thus, for $x \in X$ with $x \neq a$, we have

$$\frac{f(x) - f(a)}{x - a} \geq 0. \tag{1}$$

We have to prove that $f'(a) \geq 0$. Suppose, on the contrary, that $f'(a) < 0$. Let

$$\varepsilon = -f'(a), \tag{2}$$

noting that $\varepsilon > 0$. By Theorem 1.15,

$$f'(a) = \lim_{x \to a} \frac{f(x) - f(a)}{x - a}. \tag{3}$$

Therefore, there exists $\delta > 0$ such that, for all $x \in \text{dom}(f)$,

$$0 < |x - a| < \delta \Rightarrow \left| \frac{f(x) - f(a)}{x - a} - f'(x) \right| < \varepsilon. \tag{4}$$

Let N be the neighborhood of a of radius δ. Since a is an accumulation point of X, there exists $x \in (N \setminus \{a\}) \cap X$. Thus, we have $x \in X$, $x \neq a$, and $0 < |x - a| < \delta$. Hence, by (4),

$$\left| \frac{f(x) - f(a)}{x - a} - f'(a) \right| < \varepsilon. \tag{5}$$

As a consequence of (5) and (2), we have

$$\frac{f(x) - f(a)}{x - a} < f'(a) + \varepsilon = f'(a) - f'(a) = 0, \tag{6}$$

in contradiction with (1).   □

An argument similar to that in the proof of Lemma 2.2 provides a proof of the following lemma (Problem 2).

2.3 LEMMA   Let f be a function, let $a \in \text{dom}(f)$, let $a < d$, and let $X = [a, d) \cap \text{dom}(f)$. Suppose that f attains a maximum value (or a minimum value) on X at a. Then, if a is an accumulation point of X, and if f is differentiable at a, it follows that $f'(a) \leq 0$ (or $f'(a) \geq 0$).

Lemmas 2.2 and 2.3 have illuminating graphical interpretations. For instance, according to Lemma 2.2, if f attains a maximum on a set

$(c, a] \cap \text{dom}(f)$ at a, then the tangent line to the graph of f at a cannot slope downward.

2.4 COROLLARY

Let the function f be defined on the closed bounded interval $[a, b]$ and let f be differentiable at the two endpoints a and b of $[a, b]$. Suppose that f attains a minimum value on $[a, b]$ at p and a maximum value on $[a, b]$ at q. Then:

(i) $f'(a) < 0 < f'(b) \Rightarrow a < p < b.$
(ii) $f'(b) < 0 < f'(a) \Rightarrow a < q < b.$

PROOF

We prove (i) and leave (ii) as an exercise (Problem 3). Thus, suppose that $f'(a) < 0 < f'(b)$. If $a = p$, then Lemma 2.3 with $d = b$ implies that $f'(a) \geq 0$, contradicting $f'(a) < 0$. Therefore, $a < p$. Likewise, if $p = b$, then Lemma 2.2 implies that $f'(b) \leq 0$, contradicting $f'(b) > 0$. Therefore, $p < b$. □

2.5 DEFINITION

RELATIVE MAXIMUM AND MINIMUM VALUES Let f be a function and let $a \in \text{dom}(f)$. We say that $f(a)$ is a **relative maximum** (or **relative minimum**) **value of** f if there is a neighborhood N of a such that $N \subseteq \text{dom}(f)$ and $f(a)$ is the maximum (or minimum) value of f on N.

If $f(a)$ is a relative maximum (or minimum) value of f, we often say that f **attains** (or, **has**) **a relative maximum** (or **minimum**) **value at** a. Relative maximum or minimum values of f, when they exist, are called **relative extreme values of** f, or **relative extrema**. Relative extrema are also called **local extrema**. Note that a relative extremum of f can be attained only at an interior point of the domain of f.

2.6 LEMMA

If f is a function, $a \in \text{dom}(f)$, f is differentiable at a, and f has a relative extremum at a, then $f'(a) = 0$.

PROOF

We assume that f has a relative maximum at a, leaving the opposite case as an exercise (Problem 6). By Definition 2.5, there exists $N = (c, d) \subseteq \text{dom}(f)$ with $c < a < d$ such that f attains a maximum value on N at a. Since $(c, a] \subseteq N \subseteq \text{dom}(f)$, we have $(c, a] \cap \text{dom}(f) = (c, a]$, and f attains a maximum value on $(c, a]$ at a. Since a is an accumulation point of $(c, a]$, it follows from Lemma 2.2 that $f'(a) \geq 0$. By a similar argument, this time using Lemma 2.3, we deduce that $f'(a) \leq 0$. Therefore, we have $f'(a) = 0$. □

2.7 LEMMA

> Let f be differentiable on the bounded interval $[a, b]$ and suppose that $f'(a)$ and $f'(b)$ have opposite algebraic signs. Then there exists $c \in (a, b)$ such that $f'(c) = 0$.

PROOF   We assume that $f'(a) < 0 < f'(b)$, leaving the remaining case as an exercise (Problem 8). By Theorem 1.8, f is continuous on $[a, b]$; hence, by the Heine–Borel theorem and the extreme value theorem (Theorems 5.7 and 5.14 in Chapter 4), f attains a minimum value at some point $c \in [a, b]$. By Part (i) of Corollary 2.4, $c \in (a, b)$; hence, f has a relative extremum at c. By Lemma 2.6, $f'(c) = 0$. □

In Theorem 4.17 of Chapter 4, we showed that continuous functions have the intermediate value property on connected sets, that is, they assume every value between any two of their values. Now we are going to show that derivatives, although they need not be continuous (see Problem 22 in Problem Set 5.1), also have the intermediate value property.

2.8 THEOREM

> **INTERMEDIATE VALUE THEOREM FOR DERIVATIVES**   Let f be differentiable on the bounded interval $[a, b]$ and suppose that k is any real number between $f'(a)$ and $f'(b)$. Then there exists $c \in [a, b]$ such that $f'(c) = k$.

PROOF   If $f'(a) = k$, we take $c = a$, and if $f'(b) = k$, we take $c = b$; hence, we can assume that k is strictly between $f'(a)$ and $f'(b)$. We define

$$F : [a, b] \to \mathbb{R}$$

by

$$F(x) = f(x) - kx$$

for all $x \in [a, b]$. Then F is differentiable on $[a, b]$ and

$$F'(x) = f'(x) - k$$

for all $x \in [a, b]$. Since k is strictly between $f'(a)$ and $f'(b)$, it follows that $F'(a)$ and $F'(b)$ have opposite algebraic signs. Therefore, by Lemma 2.7, there exists $c \in (a, b)$ such that $F'(c) = 0$. Thus, $f'(c) - k = 0$, and so $f'(c) = k$. □

The following theorem is named in honor of the French mathematician Michel Rolle (1652–1719), who formulated and proved a special case of the mean value theorem in 1690.

2.9 THEOREM

> **ROLLE'S THEOREM**   If the function f is continuous on the closed bounded interval $[a, b]$ and differentiable on the open interval (a, b), and if $f(a) = f(b)$, then there exists $c \in (a, b)$ such that $f'(c) = 0$.

PROOF   If f is constant on $[a, b]$, then (by Problem 9 in Problem Set 5.1), $f'(c) = 0$ will hold for any choice of $c \in (a, b)$. Therefore, we can assume that f is not constant on $[a, b]$. Thus, either there exists $d \in (a, b)$ such that $f(a) = f(b) < f(d)$, or there exists $d \in (a, b)$ such that $f(a) = f(b) > f(d)$. We assume that there exists $d \in (a, b)$ such that

$$f(a) = f(b) < f(d),$$

leaving the completion of the argument in the remaining case as an exercise (Problem 10). By the extreme value theorem, there exists $c \in [a, b]$ such that, for all $x \in [a, b]$,

$$f(x) \le f(c).$$

In particular, $f(d) \le f(c)$; hence,

$$f(a) = f(b) < f(d) \le f(c),$$

and it follows that $a \ne c$ and $b \ne c$. Therefore,

$$c \in (a, b),$$

and so f has a relative maximum at c. Thus, by Lemma 2.6,

$$f'(c) = 0. \qquad \square$$

The following theorem, which is a generalization of the mean value theorem of differential calculus, is attributed to the French mathematician Augustin Louis Cauchy (1789–1857), one of the great pioneers of mathematical analysis.

2.10 THEOREM

> **CAUCHY MEAN VALUE THEOREM**   If the functions f and g are continuous on the closed bounded interval $[a, b]$ and differentiable on the open interval (a, b), then there exists $c \in (a, b)$ such that
>
> $$f'(c)[g(b) - g(a)] = g'(c)[f(b) - f(a)].$$

PROOF   We introduce an auxilliary function

$$F:[a, b] \to \mathbb{R}$$

defined for $x \in [a, b]$ by

$$F(x) = f(x)[g(b) - g(a)] - g(x)[f(b) - f(a)].$$

Since f and g are continuous on $[a, b]$ and differentiable on (a, b), F inherits these properties as well. Furthermore, for all $x \in (a, b)$, we have

$$F'(x) = f'(x)[g(b) - g(a)] - g'(x)[f(b) - f(a)].$$

A direct calculation (Problem 12) shows that

$$F(a) = F(b);$$

hence, by Rolle's theorem, there exists $c \in (a, b)$ such that

$$F'(c) = 0.$$

Therefore,

$$f'(c)[g(b) - g(a)] - g'(c)[f(b) - f(a)] = 0,$$

from which the conclusion of the theorem follows.   □

The Cauchy mean value theorem has an interesting geometric interpretation. Consider the curve C in the xy plane given parametrically by the equations

$$x = f(t) \quad \text{and} \quad y = g(t) \quad \text{for } a \le t \le b.$$

If we assume that the initial point $(f(a), g(a))$ and the terminal point $(f(b), g(b))$ on C are distinct and that the straight line L containing these two points is nonvertical, then the slope of L is given by

$$\frac{g(b) - g(a)}{f(b) - f(a)}.$$

Furthermore, at a point $(f(t), g(t))$ on C at which the tangent line is nonvertical, the slope of the tangent line is given by

$$\frac{dy}{dx} = \frac{dy/dt}{dx/dt} = \frac{g'(t)}{f'(t)}.$$

Therefore, the conclusion of Theorem 2.10, rewritten in the form

$$\frac{g(b) - g(a)}{f(b) - f(a)} = \frac{g'(c)}{f'(c)},$$

says that, somewhere between the two endpoints of the curve C there is a point at which the tangent line is parallel to the straight line through the two endpoints.

The usual mean value theorem of differential calculus follows easily from the Cauchy mean value theorem.

2.11 THEOREM

> **THE MEAN VALUE THEOREM**   If the function f is continuous on the closed bounded interval $[a, b]$ and differentiable on the open interval (a, b), then there exists $c \in (a, b)$ such that
>
> $$f(b) - f(a) = f'(c)(b - a).$$

PROOF   Define $g:[a, b] \to \mathbb{R}$ by $g(x) = x$ for all $x \in [a, b]$, note that $g'(x) = 1$ for all $x \in (a, b)$ (Problem 11 in Problem Set 5.1), and apply the Cauchy mean value theorem to the two functions f and g. $\square$

2.12 Example   Prove that

$$\sqrt{1 + x} < 1 + \frac{x}{2} \qquad \text{for } x > 0.$$

SOLUTION   Let $f(t) = \sqrt{1 + t}$ for $t \in [0, x]$. By the mean value theorem, there is a $c \in (0, x)$ such that

$$f(x) - f(0) = f'(c)(x - 0),$$

that is,

$$\sqrt{1 + x} - 1 = f'(c) \cdot x.$$

Since $0 < c$, we have

$$1 < \sqrt{1 + c},$$

and it follows that

$$f'(c) = \frac{1}{2\sqrt{1 + c}} < \frac{1}{2}.$$

Therefore,

$$\sqrt{1 + x} - 1 = f'(c) \cdot x < \frac{x}{2},$$

and so

$$\sqrt{1 + x} < 1 + \frac{x}{2}. \qquad \square$$

The following lemma, the proof of which we leave as an exercise (Problem 14), will enable us to give a useful reformulation of the mean value theorem (Theorem 2.11). Recall that I^o denotes the interior of the set I (Definition 1.20 in Chapter 4).

2.13 LEMMA

> Let I be an interval (open, closed, half-open, bounded, or unbounded) and let $a, b \in I$ with $a < b$. Then $[a, b] \subseteq I$, I^o is an open interval, and $(a, b) \subseteq I^o \subseteq I$.

Note that a subset I of $\mathbb{R}$ is an interval if and only if it is connected and contains at least two distinct points (Problem 16).

2.14 THEOREM

> **MEAN VALUE THEOREM (ALTERNATIVE VERSION)**   Let f be a function, let I be an interval, and let $a \in I^o \subseteq I \subseteq \text{dom}(f)$. Suppose that f is continuous on I and differentiable on I^o. Then, for every $x \in I$, there exists a real number c between a and x such that
>
> $$f(x) = f(a) + f'(c)(x - a).$$
>
> Furthermore, if $x \neq a$, then c can be chosen to be strictly between a and x.

PROOF   We sketch the proof and leave the details as an exercise (Problem 18). If $x = a$, we can take $c = x = a$, so we can assume that $x \neq a$. By using Lemma 2.13 and applying the mean value theorem to the interval $[a, x]$ if $a < x$, or to the interval $[x, a]$ if $x < a$, we obtain the conclusion of the theorem.   □

The mean value theorem, as expressed in Theorem 2.14, gives us useful information about the behavior of a function f in terms of its value $f(a)$ at a single point a and the derivative f' of f.

2.15 THEOREM

> **FUNCTIONS WITH DERIVATIVE ZERO ARE CONSTANT**   Let I be an interval and suppose that f is a function that is continuous on I and differentiable on I^o. Then, if $f'(c) = 0$ for all $c \in I^o$, it follows that f is a constant function on I.

PROOF   Choose and fix $a \in I^o$ and let $x \in I$. By Theorem 2.14, there exists a number c between a and x such that

$$f(x) = f(a) + f'(c)(x - a).$$

By hypothesis, $f'(c) = 0$, so we have

$$f(x) = f(a)$$

for all $x \in I$.   □

2.16 COROLLARY

> **FUNCTIONS WITH THE SAME DERIVATIVE** Let I be an interval and suppose that f and g are functions such that $I \subseteq \operatorname{dom}(f) \cap \operatorname{dom}(g)$, both f and g are continuous on I, both f and g are differentiable on I^o, and
>
> $$f'(x) = g'(x)$$
>
> holds for all $x \in I^o$. Then there exists a constant real number C such that, for all $x \in I$,
>
> $$f(x) = g(x) + C.$$

PROOF Apply Theorem 2.15 to the function $f - g$.   □

Corollary 2.16, which plays a critical role in the theory of indefinite integration, is often paraphrased by the statement that functions with the same derivative differ by a constant. (Note, however, that the functions must be defined on an interval!)

2.17 THEOREM

> **DERIVATIVES AND MONOTONE FUNCTIONS** Let I be an interval, let f be a function such that $I \subseteq \operatorname{dom}(f)$, f is continuous on I, and f is differentiable on I^o. Then, if $f'(x) > 0$ (or $f'(x) < 0$) for all $x \in I^o$, it follows that f is strictly increasing (or strictly decreasing) on I.

PROOF We prove the theorem for the case in which $f'(x) > 0$ for all $x \in I^o$, and leave the remaining case as an exercise (Problem 26). Let $a, b \in I$ with $a < b$. We must prove that $f(a) < f(b)$. By Lemma 2.13, $[a, b] \subseteq I$ and $(a, b) \subseteq I^o$; hence, f is continuous on $[a, b]$, f is differentiable on (a, b), and $f'(x) > 0$ for all $x \in (a, b)$. By the mean value theorem (Theorem 2.11), there exists $c \in (a, b)$ such that

$$f(b) - f(a) = f'(c)(b - a).$$

Since $a < b$ and $f'(c) > 0$, it follows that

$$f(a) < f(b).$$ □

For the notation used in the next lemma, see Problem 27 in Problem Set 4.2.

2.18 LEMMA

Let J be an interval and let g be a function with $J \subseteq \mathrm{dom}(g)$. Then g is continuous on J if and only if the following two conditions are satisfied:

(i) If $b \in J$ and there exists a point $c \in J$ with $c < b$, then $\lim_{y \to b^-} g(y)$ exists and equals $g(b)$.

(ii) If $b \in J$ and there exists a point $d \in J$ with $b < d$, then $\lim_{y \to b^+} g(y)$ exists and equals $g(b)$.

The proof of Lemma 2.18 is not difficult and is left as an exercise (Problem 31).

2.19 LEMMA

Let I be an interval, let $f : I \to \mathbb{R}$ be a continuous and strictly increasing (or decreasing) function, and let $J = f(I)$. Then J is an interval, $f : I \to J$ is a bijection, and $f^{-1} : J \to I$ is a continuous and strictly increasing (or decreasing) function.

PROOF   We consider the case in which f is strictly increasing, leaving the opposite case as an exercise (Problem 32). Since f is strictly increasing, it is an injection, and therefore $f : I \to J$ is a bijection. Because f is continuous and I is a connected set, it follows from Theorem 4.16 in Chapter 4 that $J = f(I)$ is a connected set. Since I is an interval, it contains two distinct points, and therefore $J = f(I)$ contains two distinct points, so J is also an interval (Problem 16).

To show that f^{-1} is strictly increasing, suppose that $y_1, y_2 \in J$ with $y_1 < y_2$. Let $x_1 = f^{-1}(y_1)$ and $x_2 = f^{-1}(y_2)$. We must prove that $x_1 < x_2$. Note that $f(x_1) = y_1$ and $f(x_2) = y_2$. Since $y_1 \neq y_2$, we cannot have $x_1 = x_2$, and therefore either $x_1 < x_2$ or else $x_2 < x_1$. But, if $x_2 < x_1$, then $y_2 = f(x_2) < f(x_1) = y_1$, contradicting $y_1 < y_2$. Therefore, we must have $x_1 < x_2$.

Let $b \in J$ and suppose there exists $y_0 \in J$ with $y_0 < b$. Then, by Lemma 7.4 in Chapter 4,

$$\lim_{y \to b^-} f^{-1}(y) = \sup_{y < b} f^{-1}(y) \le f^{-1}(b)$$

(Problem 33). We propose to show that the last inequality is actually an equality. Suppose, on the contrary, that

$$\sup_{y < b} f^{-1}(y) < f^{-1}(b),$$

and let

$$a = \sup_{y < b} f^{-1}(y).$$

Because $y_0 \in J$ with $y_0 < b$, we have

$$f^{-1}(y_0) \leq \sup_{y<b} f^{-1}(y) = a < f^{-1}(b),$$

and, since both $f^{-1}(y_0)$ and $f^{-1}(b)$ belong to the interval I, it follows that $a \in I = \mathrm{dom}(f)$. Thus, we can apply the strictly increasing function f to both sides of the inequality $a < f^{-1}(b)$ to obtain the inequality

$$f(a) < b.$$

Now choose a real number y_1 such that

$$f(a) < y_1 < b.$$

Since both $f(a)$ and b belong to the interval J, it follows that $y_1 \in J$, and, consequently,

$$f^{-1}(y_1) \leq \sup_{y<b} f^{-1}(y) = a.$$

By applying the strictly increasing function f to both sides of the inequality $f^{-1}(y_1) \leq a$, we find that

$$y_1 \leq f(a),$$

contradicting the inequality $f(a) < y_1$.

The argument above shows that $f^{-1} : J \to \mathbb{R}$ satisfies Condition (i) in Lemma 2.18; a similar argument (Problem 34) shows that it satisfies Condition (ii) and therefore that f^{-1} is continuous.   □

2.20 THEOREM

> **INVERSE FUNCTION THEOREM**   Let I be an interval, let $f : I \to \mathbb{R}$ be a differentiable function such that $f'(x) \neq 0$ for all $x \in I$, and let $J = f(I)$. Then f is a strictly monotone function, J is an interval, and $f^{-1} : J \to \mathbb{R}$ is a differentiable function such that
>
> $$(f^{-1})'(y) = \frac{1}{f'(f^{-1}(y))}$$
>
> holds for all $y \in J$.

PROOF   Since f is differentiable, it is continuous by Theorem 1.8. As a consequence of Lemma 2.7 and the fact that $f'(x) \neq 0$ for all $x \in I$, $f'(x)$ must maintain a constant algebraic sign for all $x \in I$. We assume that $f'(x) > 0$ for all $x \in I$. [A similar argument applies if $f'(x) < 0$ for all $x \in I$.] By Lemma 2.19, J is an interval, $f : I \to J$ is a bijection, and $f^{-1} : J \to I$ is continuous and strictly increasing.

Choose and fix $b \in J$ and let $a = f^{-1}(b)$. We have to prove that $(f^{-1})'(b)$ exists and is equal to $1/f'(a)$. To this end, let $\varepsilon > 0$. By hypothesis, f is differentiable at a and $f'(a) \neq 0$, so, by Problem 23 in Problem Set 4.2,

$$\lim_{x \to a} \left[\frac{f(x) - f(a)}{x - a} \right]^{-1} = \frac{1}{f'(a)}. \tag{1}$$

By noting that $a = f^{-1}(b)$ and $f(a) = b$, we can rewrite (1) as

$$\lim_{x \to a} \frac{x - f^{-1}(b)}{f(x) - b} = \frac{1}{f'(a)}. \tag{2}$$

As a consequence of (2), there exists $\alpha > 0$ such that, for all $x \in I$,

$$0 < |x - a| < \alpha \Rightarrow \left| \frac{x - f^{-1}(b)}{f(x) - b} - \frac{1}{f'(a)} \right| < \varepsilon. \tag{3}$$

Because f^{-1} is continuous at b and $a = f^{-1}(b)$, there exists $\delta > 0$ such that

$$|y - b| < \delta \Rightarrow |f^{-1}(y) - a| < \alpha. \tag{4}$$

Now suppose that $y \in J$ and

$$0 < |y - b| < \delta. \tag{5}$$

Then $y \neq b$, so $f^{-1}(y) \neq f^{-1}(b) = a$, and it follows from (4) that

$$0 < |f^{-1}(y) - a| < \alpha. \tag{6}$$

Therefore, with $x = f^{-1}(y)$ in (3), we have $f(x) = y$ and

$$\left| \frac{f^{-1}(y) - f^{-1}(b)}{y - b} - \frac{1}{f'(a)} \right| < \varepsilon. \tag{7}$$

This shows that

$$\lim_{y \to b} \frac{f^{-1}(y) - f^{-1}(b)}{y - b} = \frac{1}{f'(a)}, \tag{8}$$

and the proof is complete.   □

In elementary calculus, the conclusion of Theorem 2.20 is often expressed in the form

$$\frac{dx}{dy} = \frac{1}{dy/dx}.$$

PROBLEM SET 5.2

1. Prove Lemma 2.2 for the case in which f attains a minimum value on X at a. [Hint: Either make a direct proof or apply what has already been proved to the function $-f$.]

2. Prove Lemma 2.3. [Hint: Either make a direct proof or apply Lemma 2.2 to the function $f \circ g$ where $g(x) = -x$ for all $x \in \mathbb{R}$.]

3. Prove Part (ii) of Corollary 2.4.

4. Give a graphical interpretation of each part of Corollary 2.4.

5. A point $a \in \mathrm{dom}(f)$ is called a **stationary point** of the domain of f if f is differentiable at a and $f'(a) = 0$. The point $a \in \mathrm{dom}(f)$ is called a **critical point** of the domain of f if either it is a stationary point or f is not differentiable at a. If f has a relative extremum at a, prove that a is a critical point of the domain of f.

6. Prove Lemma 2.6 for the case in which f has a relative minimum at a.

7. In elementary calculus, one finds the absolute extreme values of a continuous function f on a closed bounded interval $[a, b]$ by finding all of the critical points of f in (a, b) (see Problem 5), evaluating f at these critical points, evaluating f at the two endpoints a and b, and looking for the maximum and the minimum of the resulting function values. Justify this procedure.

8. Prove Lemma 2.7 for the case in which $f'(a) > 0 > f'(b)$.

9. In elementary calculus, a point $(a, f(a))$ is called a **point of inflection** of the graph of f if $a \in \mathrm{dom}(f)$ and there exists a neighborhood N of a such that $N \subseteq \mathrm{dom}(f)$, f and f' are differentiable on $(-\infty, a) \cap N$ and on $N \cap (a, \infty)$, and $f''(x)f''(y) < 0$ for all $x \in (-\infty, a) \cap N$ and all $y \in (a, \infty) \cap N$. If $(a, f(a))$ is a point of inflection of the graph of f and $f'(a)$ exists, show that a is a critical point of the domain of f'.

10. Complete the proof of Theorem 2.9 by taking care of the case in which there exists no $d \in (a, b)$ such that $f(a) = f(b) < f(d)$.

11. Use the Cauchy mean value theorem to prove the familiar theorem of elementary calculus known as **L'Hospital's rule**: Let f and g be continuous on $[a, b]$ and differentiable on (a, b), let $p \in (a, b)$, suppose that $g'(x) \neq 0$ for all $x \in (a, b)$ with $x \neq p$, let $f(p) = g(p) = 0$, and suppose that f'/g' has a limit at p. Then f/g has a limit at p and $\lim_{x \to p}[f(x)/g(x)] = \lim_{x \to p}[f'(x)/g'(x)]$.

12. Make the direct calculation called for in the proof of Theorem 2.10.

13. (a) Use L'Hospital's rule (Problem 11) to evaluate $\lim_{x \to 0}[(\sin x)/x]$.
    (b) Explain why the evaluation in Part (a) involves circular reasoning.

14. Prove Lemma 2.13.

15. Assume the relevant derivative formulas from elementary calculus and use L'Hospital's rule to evaluate each limit:

(a) $\lim\limits_{x \to 0} \dfrac{e^x - 1}{x}$ (b) $\lim\limits_{x \to 0} \dfrac{x^2}{1 - \cos 2x}$ (c) $\lim\limits_{x \to 7} \dfrac{\ln(x/7)}{7 - x}$

16. Prove that a subset I of $\mathbb{R}$ is an interval if and only if it is connected and contains at least two points.

17. State and prove the version of L'Hospital's rule that applies to limits as $x \to \pm\infty$.

18. Provide the details for the proof of Theorem 2.14.

19. Use the mean value theorem to prove **Bernoulli's inequality:**   If n is a real number, $n > 1$ and $x > 0$, then $(1 + x)^n > 1 + nx$. (You may assume the usual formula from elementary calculus for the derivative of x^n.)

20. Prove that Bernoulli's inequality (Problem 19) also holds if $-1 < x < 0$.

21. Use the mean value theorem to prove that $\ln x < x - 1$ holds for all $x > 1$. (You may assume the usual formula from elementary calculus for the derivative of $\ln x$.)

22. For $x > 1$, prove that $(x - 1)/x < \ln x$. [Hint: Apply the mean value theorem to $f(t) = t \ln t$.]

23. A function f is said to satisfy a **Lipschitz condition** on the interval I if $I \subseteq \text{dom}(f)$ and there exists a positive constant M such that $|f(x) - f(y)| \le M|x - y|$ holds for every $x, y \in I$.
(a) If f is differentiable on the interval I and if M is a positive constant such that $|f'(x)| \le M$ holds for all $x \in I$, prove that f satisfies a Lipschitz condition on I. [Hint: Use the mean value theorem.]
(b) If f satisfies a Lipschitz condition on the interval I, show that f is uniformly continuous on I.

24. Prove that $|\sin x - \sin y| \le |x - y|$ for all $x, y \in \mathbb{R}$. (You may assume the usual formula from elementary calculus for the derivative of $\sin x$.)

25. If $f:[a, b] \to \mathbb{R}$ is differentiable and $f':[a, b] \to \mathbb{R}$ is continuous, prove that f satisfies a Lipschitz condition on $[a, b]$. (See Problem 23.)

26. Complete the proof of Theorem 2.17 by taking care of the case in which $f'(x) < 0$ for all $x \in I^o$.

27. If f is differentiable on an interval I, prove that f is increasing (or decreasing) on I if and only if $f'(x) \ge 0$ (or $f'(x) \le 0$) holds for all $x \in I$.

28. Suppose that f is differentiable on an interval I, that $f'(x) \ge 0$ for all $x \in I$, and that, for every open interval $(a, b) \subseteq I$, there exists $x \in (a, b)$ such that $f'(x) > 0$. Prove that f is strictly increasing on I.

29. Give an example of a function that is differentiable and strictly increasing on an interval I, but whose derivative is not strictly greater than 0 on I.

30. Let f and g be continuous on an interval I and differentiable on I^o. Suppose there exists $a \in I^o$ such that $f(a) \le g(a)$ and that $f'(x) \le g'(x)$ holds for all $x \in I^o$. Prove that $f(x) \le g(x)$ holds for all $x \in I$.

31. Prove Lemma 2.18. [Hint: See Problem 27 in Problem Set 4.2.]

32. Complete the proof of Lemma 2.19 by taking care of the case in which f is strictly decreasing.

33. In the proof of Lemma 2.19, we used Lemma 7.4 in Chapter 4. Show that the hypotheses of this lemma are satisfied, that is, show that b is an accumulation point of $(-\infty, b) \cap \text{dom}(f^{-1})$.

34. Complete the proof of Lemma 2.19 by showing that f^{-1} satisfies Condition (ii) in Lemma 2.18.

35. Let n be a positive integer and let $f(x) = x^{1/n}$ be the principal nth root of x for $x > 0$. (See Theorem 4.19 in Chapter 4.) Prove that f is strictly increasing and differentiable and that $f'(x) = (1/n)x^{(1/n)-1}$ for all $x > 0$. [Hint: Use the inverse function theorem, Theorem 2.20.]

36. Prove the following "pointwise" version of the inverse function theorem: Suppose that $f:(c, d) \to \mathbb{R}$ is continuous and injective and that f is differentiable at the point $a \in (c, d)$ with $f'(a) \neq 0$. Let $b = f(a)$ and suppose that f^{-1} is continuous at b. Then f^{-1} is differentiable at b and $(f^{-1})'(b) = 1/f'(a)$.

37. Let m and n be integers, $n \neq 0$, and suppose that no integer $d > 1$ divides both m and n. Let $r = m/n$ and let D be the set of all real numbers that have a principal nth root. Define $x^r = (x^{1/n})^m$ for all $x \in D$. Study the function $f:D \to \mathbb{R}$ defined by $f(x) = x^r$ with regard to the properties of continuity, monotonicity, and differentiability. For which values of $x \in D$ is it true that $f'(x) = rx^{r-1}$?

3
_____

HIGHER-ORDER DERIVATIVES AND TAYLOR'S THEOREM

In this section we pursue our study of differential calculus by using the idea of approximation as a unifying concept. Here we consider the problem of approximating functions by polynomials and the problem of finding approximate solutions of equations. In both of these problems, higher-order derivatives play an important role.

Suppose that f is a function, A is the set of points in the domain of f at which f is differentiable, and $A \neq \varnothing$, so that $f':A \to \mathbb{R}$. If $a \in A$ and f' is differentiable at a, then we say that f is **twice differentiable** at the point a. If B is the subset of A consisting of the points at which f is twice differentiable, and if $B \neq \varnothing$, we define $f'': B \to \mathbb{R}$ by

$$f''(x) = (f')'(x) = \lim_{h \to 0} \frac{f'(x + h) - f'(x)}{h}$$

for all $x \in B$, and we call f'' the **second derivative** of f.

If B is the set of points at which f is twice differentiable, $a \in B$, and f'' is differentiable at a, then we say that f is **three times differentiable** at the point a. If C is the subset of B consisting of the points at which f is three times differentiable, and if $C \neq \varnothing$, we define $f''': C \to \mathbb{R}$ by

$$f'''(x) = (f'')'(x) = \lim_{h \to 0} \frac{f''(x + h) - f''(x)}{h}$$

for all $x \in C$, and we call f''' the **third derivative** of f.

By continuing in this manner, we define the **higher-order derivatives** of f, denoting the nth derivative of f (if it exists) by $f^{(n)}$. With this notation, the **first derivative** of f (if it exists) is the same as the derivative of f, so that

$$f^{(1)} = f',$$

and if n is an integer greater than 1, then the **nth derivative** of f (if it exists) is defined inductively by

$$f^{(n)} = (f^{(n-1)})'.$$

It is convenient to define the "zeroth-order derivative" of f to be f itself, so that, by definition,

$$f^{(0)} = f.$$

If $f^{(n)}$ exists and $a \in \mathrm{dom}(f^{(n)})$, we say that f is **n times differentiable** at the point a. Likewise, if $D \subseteq \mathrm{dom}(f^{(n)})$, we say that f is **n times differentiable** on the set D. Note that, if f is n times differentiable on D, then it is k times differentiable on D for every integer k with $0 \leq k \leq n$.

If p is a polynomial function of degree n, then there are coefficients $b_0, b_1, b_2, \ldots, b_n \in \mathbb{R}$ such that $b_n \neq 0$ and, for all $x \in \mathbb{R}$,

$$p(x) = b_0 + b_1 x + b_2 x^2 + \cdots + b_n x^n. \tag{1}$$

In what follows, it will be convenient to employ the usual summation notation to rewrite (1) in the more compact form:

$$p(x) = \sum_{k=0}^{n} b_k x^k, \tag{2}$$

it being understood that the equation holds for all $x \in \mathbb{R}$ and that the term corresponding to $k = 0$ is the constant term b_0 of the polynomial function p.

By taking the derivative on both sides of (2), we find that

$$p'(x) = \sum_{k=1}^{n} kb_k x^{k-1}. \tag{3}$$

Thus, p is differentiable and p' is a polynomial function of degree $n - 1$. Similarly, $p^{(2)} = (p')'$ is a polynomial function of degree $n - 2$, $p^{(3)} = (p^{(2)})'$ is a polynomial function of degree $n - 3$, and so forth. By continuing in this way, we find that, for each nonnegative integer $k \le n$, $p^{(k)}$ is a polynomial function of degree $n - k$. In particular, then, $p^{(n)}$ is a polynomial function of degree $n - n = 0$; in other words, $p^{(n)}$ is a constant function. Therefore, $p^{(n+1)}$ is the constant zero function:

$$p^{(n+1)} = 0. \tag{4}$$

As a consequence of (4), we have

$$p^{(k)} = 0 \qquad \text{for all integers } k > n. \tag{5}$$

3.1 LEMMA

Let p be a polynomial function of degree at most n and let $a \in \mathbb{R}$. Then there exist $c_0, c_1, c_2, \ldots, c_n \in \mathbb{R}$ such that

$$p(x) = \sum_{k=0}^{n} c_k (x - a)^k.$$

PROOF   Our proof is by induction on n. If $n = 0$, then p is a constant function, so that

$$p(x) = c$$

for all $x \in \mathbb{R}$. Thus, with $c_0 = c$, we have

$$p(x) = \sum_{k=0}^{0} c_k (x - a)^k.$$

Now assume that the statement to be proved holds for all polynomial functions of degree at most $n - 1$. Let p be a polynomial function of degree at most n. Then there exist coefficients $b_0, b_1, b_2, \ldots, b_n$ such that

$$p(x) = \sum_{k=0}^{n} b_k x^k.$$

Let q be the polynomial function defined for all $x \in \mathbb{R}$ by

$$q(x) = p(x) - b_n(x - a)^n,$$

noting that q has degree at most $n - 1$ (Problem 3). Therefore, by our induction hypothesis, there exist $c_0, c_1, c_2, \ldots, c_{n-1}$ such that

$$q(x) = \sum_{k=0}^{n-1} c_k(x - a)^k.$$

Consequently, if we let $c_n = b_n$, we have

$$
\begin{aligned}
p(x) &= q(x) + b_n(x - a)^n \\
&= \sum_{k=0}^{n-1} c_k(x - a)^k + c_n(x - a)^n \\
&= \sum_{k=0}^{n} c_k(x - a)^k.
\end{aligned}
$$

$\square$

The next lemma shows that the coefficients $c_0, c_1, c_2, \ldots, c_n$ in Lemma 3.1 are closely related to the successive derivatives of p evaluated at a. Recall that, if k is a positive integer, then, by definition,

$$k! = k(k - 1)(k - 2) \cdots 3 \cdot 2 \cdot 1.$$

Also, by definition,

$$0! = 1.$$

3.2 LEMMA

Let p be a polynomial function of degree at most n and let $a \in \mathbb{R}$. Then the coefficients $c_0, c_1, c_2, \ldots, c_n$ in Lemma 3.1 are given by

$$c_k = \frac{p^{(k)}(a)}{k!}$$

for $k = 0, 1, 2, \ldots, n$.

PROOF We have

$$p(x) = \sum_{k=0}^{n} c_k(x - a)^k. \tag{1}$$

Substituting $x = a$ in (1), we find that all terms in the sum vanish, except for the term $c_0(x - a)^0 = c_0$ corresponding to $k = 0$, and therefore,

$$c_0 = p(a) = \frac{p^{(0)}(a)}{0!}. \tag{2}$$

Differentiating both sides of (1), we obtain

$$p'(x) = \sum_{k=1}^{n} kc_k(x - a)^{k-1}. \tag{3}$$

Differentiating again, we find that

$$p''(x) = \sum_{k=2}^{n} k(k-1)c_k(x-a)^{k-2}. \tag{4}$$

Continuing in this way (Problem 5), we obtain the formula

$$p^{(m)}(x) = \sum_{k=m}^{n} k(k-1)(k-2)\cdots(k-m+1)c_k(x-a)^{k-m} \tag{5}$$

for $1 \le m \le n$. By substituting $x = a$ in (5), we again find that all terms in the sum vanish, except for the term corresponding to $k = m$, so that

$$p^{(m)}(a) = m(m-1)(m-2)\cdots(m-m+1)c_m(x-a)^{m-m}; \tag{6}$$

that is, for $1 \le m \le n$,

$$p^{(m)}(a) = m!c_m. \tag{7}$$

By (2) and (7), we have

$$c_k = \frac{p^{(k)}(a)}{k!} \tag{8}$$

for $0 \le k \le n$. $\square$

As a consequence of Lemmas 3.1 and 3.2, it is always possible to construct a polynomial function p whose successive derivatives $p^{(k)}$ for $k = 0, 1, 2, \ldots, n$ have preassigned values $p^{(k)}(a)$ at a given point a. In particular, we have the following important result:

3.3 COROLLARY

Let f be a function, let $a \in \text{dom}(f)$, let n be a nonnegative integer, and suppose that f is n times differentiable at a. Then there is a unique polynomial function p of degree at most n such that

$$f^{(k)}(a) = p^{(k)}(a)$$

for $k = 0, 1, 2, \ldots, n$.

PROOF Let p be the polynomial function defined by

$$p(x) = \sum_{k=0}^{n} \frac{f^{(n)}(a)}{k!}(x-a)^k.$$

Then, by Lemma 3.2,

$$\frac{f^{(k)}(a)}{k!} = \frac{p^{(k)}(a)}{k!}$$

holds for $k = 0, 1, 2, \ldots, n$. Therefore,

$$f^{(k)}(a) = p^{(k)}(a)$$

holds for $k = 0, 1, 2, \ldots, n$. This proves the existence of p. To prove the uniqueness, suppose that q is a polynomial function of degree at most n and that, for $k = 0, 1, 2, \ldots, n$,

$$f^{(k)}(a) = q^{(k)}(a).$$

By Lemma 3.1, there are coefficients $c_0, c_1, c_2, \ldots, c_n$ such that

$$q(x) = \sum_{k=0}^{n} c_k (x - a)^k,$$

and, by Lemma 3.2,

$$c_k = \frac{q^{(k)}(a)}{k!} = \frac{f^{(k)}(a)}{k!}$$

holds for $k = 0, 1, 2, \ldots, n$. Thus, the polynomial functions p and q have the same coefficients, and so $p = q$. $\quad\square$

3.4 DEFINITION

> **TAYLOR POLYNOMIAL** Let f be a function, let $a \in \mathrm{dom}(f)$, let n be a nonnegative integer, and suppose that f is n times differentiable at a. Then the polynomial function defined by
>
> $$p(x) = \sum_{k=0}^{n} \frac{f^{(k)}(a)}{k!} (x - a)^k$$
>
> is called the **nth Taylor polynomial** for f at the point a.

As is shown by the proof of Corollary 3.3, the nth Taylor polynomial for f at the point a is the unique polynomial function of degree at most n that agrees with f at a and whose successive derivatives, up to and including the nth derivative, also agree with the corresponding derivatives of f at a. Taylor polynomials were introduced by the English mathematician Brook Taylor (1685–1731) in 1715 in connection with the important formula now known as *Taylor's theorem*. The significance of Taylor's theorem was not appreciated until 1772, when the great French mathematician Joseph Louis Lagrange (1736–1813) declared it to be the "basic principle of the differential calculus." Among the different versions of Taylor's theorem is the following generalization of the mean value theorem.

3.5 THEOREM

> **EXTENDED MEAN VALUE THEOREM** Let n be a nonnegative integer, let f be n times differentiable on a closed bounded interval I with endpoints a and b, and let p be the nth Taylor polynomial for f at a. Then, if $f^{(n)}$ is continuous on I and differentiable on I^o, there exists a real number c strictly between a and b such that
>
> $$f(b) = p(b) + \frac{f^{(n+1)}(c)}{(n+1)!}(b-a)^{n+1}.$$

PROOF   If $n = 0$, the statement to be proved is equivalent to the mean value theorem (Problem 9); hence, we may assume that $n \geq 1$. We consider the case in which $a < b$, so that $I = [a, b]$ and $I^o = (a, b)$, and leave the opposite case as an exercise (Problem 10). Define $F : [a, b] \to \mathbb{R}$ and $G : [a, b] \to \mathbb{R}$ by

$$F(x) = f(b) - \sum_{k=0}^{n} \frac{f^{(k)}(x)}{k!}(b-x)^k \tag{1}$$

and

$$G(x) = (b-x)^{n+1} \tag{2}$$

for all $x \in [a, b]$. Since F and G satisfy the hypotheses of the Cauchy mean value theorem (Theorem 2.10), there exists $c \in (a, b)$ such that

$$F'(c)[G(b) - G(a)] = G'(c)[F(b) - F(a)]. \tag{3}$$

From (1), it follows that

$$F(b) - F(a) = p(b) - f(b) \tag{4}$$

(Problem 11) and, from (2), we have

$$G(b) - G(a) = -(b-a)^{n+1}. \tag{5}$$

Differentiating both sides of (1) for $x \in (a, b)$, simplifying, and substituting $x = c$ into the resulting equation, we find that

$$F'(c) = -\frac{f^{(n+1)}(c)}{n!}(b-c)^n \tag{6}$$

(Problem 15). By differentiating both sides of (2) and substituting $x = c$ into the resulting equation, we also find that

$$G'(c) = -(n+1)(b-c)^n. \tag{7}$$

Substituting (4), (5), (6), and (7) into (3) and canceling the factor $(b-c)^n$

from both sides of the resulting equation, we find that

$$\frac{f^{(n+1)}(c)}{n!}(b-a)^{n+1} = -(n+1)[p(b) - f(b)].\tag{8}$$

Dividing both sides of (8) by $n+1$, using the fact that $(n+1)! = n!(n+1)$, and rearranging the resulting equation, we obtain

$$f(b) = p(b) + \frac{f^{(n+1)}(c)}{(n+1)!}(b-a)^{n+1}.\tag{9}$$

$\square$

The following version of Taylor's theorem, attributed to Lagrange, is an immediate corollary of the extended mean value theorem.

3.6 THEOREM

> **TAYLOR'S THEOREM (LAGRANGE FORM)**   Let f be a function, let I be an interval, and let $a \in I^o \subseteq I \subseteq \mathrm{dom}(f)$. Let n be a nonnegative integer, suppose that f is n times differentiable on I, that $f^{(n)}$ is continuous on I, and that $f^{(n)}$ is differentiable on I^o. Let p be the nth Taylor polynomial for f at a. Then, for every $x \in I$, there exists a real number c between a and x such that
>
> $$f(x) = p(x) + \frac{f^{(n+1)}(c)}{(n+1)!}(x-a)^{n+1}.$$
>
> Furthermore, if $x \neq a$, then c can be chosen to be strictly between a and x.

PROOF   If $x = a$, let $c = a$; otherwise, apply Theorem 3.5 to the closed bounded interval with endpoints a and x. $\square$

In Section 1 we studied the problem of local linear approximation to a function f near a point $a \in \mathrm{dom}(f)$. Now we turn our attention to the analogous **problem of local approximation by a polynomial**. We are given a positive integer n, a function f, and an accumulation point a of $\mathrm{dom}(f)$ such that $a \in \mathrm{dom}(f)$. We seek a polynomial function p of degree at most n that approximates f near a in such a way that

$$f(a) = p(a)$$

and the error

$$\mathscr{E}(x) = f(x) - p(x)$$

approaches zero more rapidly than $(x-a)^n$ does.

Lemma 1.2 suggests the following reformulation of the notion of an approximating polynomial:

3.7 DEFINITION

LOCAL POLYNOMIAL APPROXIMATION   If n is a positive integer, f is a function, and $a \in \mathrm{dom}(f)$, then a polynomial function p is said to be a **local polynomial approximation of order n to f near a** if the degree of p does not exceed n and there is a function ε with $\mathrm{dom}(\varepsilon) = \mathrm{dom}(f)$ and $\lim_{x \to a} \varepsilon(x) = \varepsilon(a) = 0$, such that, for all $x \in \mathrm{dom}(f)$,

$$f(x) = p(x) + \varepsilon(x)(x - a)^n.$$

Note that a local polynomial approximation of order 1 to f near a is precisely the same thing as a local linear approximation to f near a.

3.8 LEMMA

Let n be a positive integer greater than 1 and let

$$p(x) = \sum_{k=0}^{n} c_k(x - a)^k$$

be a local polynomial approximation of order n to the function f near $a \in \mathrm{dom}(f)$. Then the function p^* defined for all $x \in \mathbb{R}$ by

$$p^*(x) = \sum_{k=0}^{n-1} c_k(x - a)^k$$

is a local polynomial approximation of order $n - 1$ to f near a.

PROOF   By Definition 3.7, there exists a function ε with $\mathrm{dom}(\varepsilon) = \mathrm{dom}(f)$ such that $\lim_{x \to a} \varepsilon(x) = \varepsilon(a) = 0$ and, for all $x \in \mathrm{dom}(f)$,

$$f(x) = p(x) + \varepsilon(x)(x - a)^n.$$

Thus, for all $x \in \mathrm{dom}(f)$,

$$
\begin{aligned}
f(x) &= p^*(x) + c_n(x - a)^n + \varepsilon(x)(x - a)^n \\
&= p^*(x) + \gamma(x)(x - a)^{n-1},
\end{aligned}
$$

where $\gamma : \mathrm{dom}(f) \to \mathbb{R}$ is defined by

$$\gamma(x) = [c_n + \varepsilon(x)](x - a)$$

for all $x \in \mathrm{dom}(f)$. Obviously,

$$\lim_{x \to a} \gamma(x) = \gamma(a) = 0,$$

so p^* is a local polynomial approximation of degree $n - 1$ to f near a.

$\square$

3.9 THEOREM

UNIQUENESS OF THE LOCAL POLYNOMIAL APPROXIMATION   If f is a function, n is a positive integer, $a \in \mathrm{dom}(f)$, and both p and q are local polynomial approximations of order n to f near a, then $p = q$.

PROOF   The proof is by induction on n. The case $n = 1$ is settled by Lemma 1.4. Thus, suppose $n > 1$, that the statement in question is true for $n - 1$, and that both p and q are local polynomial approximations of order n to f near a. Let

$$p(x) = p^*(x) + c_n(x - a)^n$$

and

$$q(x) = q^*(x) + d_n(x - a)^n$$

for all $x \in \mathbb{R}$, where p^* and q^* are polynomials of degree at most $n - 1$. By Lemma 3.8, both p^* and q^* are local polynomial approximations of order $n - 1$ to f near a; hence, by our induction hypothesis,

$$p^* = q^*.$$

Thus, to prove that $p = q$, it will be sufficient to prove that $c_n = d_n$.

By Definition 3.7, there exist functions ε_1 and ε_2 with $\mathrm{dom}(\varepsilon_1) = \mathrm{dom}(\varepsilon_2) = \mathrm{dom}(f)$ such that

$$\lim_{x \to a} \varepsilon_1(x) = \varepsilon_1(a) = \lim_{x \to a} \varepsilon_2(x) = \varepsilon_2(a) = 0,$$

and both of the equations

$$f(x) = p^*(x) + c_n(x - a)^n + \varepsilon_1(x)(x - a)^n$$

and

$$f(x) = q^*(x) + d_n(x - a)^n + \varepsilon_2(x)(x - a)^n$$

hold for all $x \in \mathrm{dom}(f)$. By subtracting the latter equation from the former, and using the fact that $p^* = q^*$, we find that

$$(c_n - d_n)(x - a)^n = [\varepsilon_2(x) - \varepsilon_1(x)](x - a)^n$$

for all $x \in \mathrm{dom}(f)$. Therefore, for $x \in \mathrm{dom}(f)$ with $x \neq a$, we have

$$c_n - d_n = \varepsilon_2(x) - \varepsilon_1(x).$$

Finally, by taking the limit as $x \to a$ on both sides of the last equation, we obtain $c_n - d_n = 0$, so $c_n = d_n$.   □

As the following theorem shows, if the function f and its derivatives are sufficiently well-behaved near a, then the nth Taylor polynomial for f at a is the unique local polynomial approximation of order n to f near a.

3.10 THEOREM

> **APPROXIMATION BY THE TAYLOR POLYNOMIAL**   Let $a \in \mathrm{dom}(f)$, let n be a positive integer, let N be a neighborhood of a, and suppose that f is $n - 1$ times differentiable on N and that $f^{(n-1)}$ is differentiable at a. Then the nth Taylor polynomial p for f at a is the unique local polynomial approximation of order n to f near a.

PROOF   By Definition 3.7 and Lemma 1.2, the statement to be proved is equivalent to the following:

If f is $n - 1$ times differentiable on N, $f^{(n-1)}$ is differentiable at a, and p is the nth Taylor polynomial for f at a, then

$$\lim_{x \to a} \frac{f(x) - p(x)}{(x - a)^n} = 0. \tag{*}$$

We prove (*) by induction on n.

For $n = 1$, the hypotheses of (*) are that f is defined on N, differentiable at a, and that p is given by

$$p(x) = f(a) + f'(a)(x - a).$$

Thus,

$$\lim_{x \to a} \frac{f(x) - p(x)}{(x - a)^1} = \lim_{x \to a} \left[\frac{f(x) - f(a)}{x - a} - f'(x) \right] = 0.$$

Now assume as the induction hypothesis that (*) holds for an arbitrary but fixed value of $n \geq 1$. To prove that (*) holds if n is replaced by $n + 1$, suppose that f is n times differentiable on N, that $f^{(n)}$ is differentiable at a, and that p is the $(n + 1)$st Taylor polynomial for f at a. Then f' is $n - 1$ times differentiable on N and $(f')^{(n-1)} = f^{(n)}$ is differentiable at a. Also, p' is the nth Taylor polynomial for f' at a (Problem 17). Therefore, by the induction hypothesis,

$$\lim_{x \to a} \frac{f'(x) - p'(x)}{(x - a)^n} = 0.$$

Since f is differentiable on N, it is continuous on N by Theorem 1.8. Hence, by applying L'Hospital's rule (Problem 11 in Problem Set 5.2), we find that

$$\lim_{x \to a} \frac{f(x) - p(x)}{(x - a)^{n+1}} = \lim_{x \to a} \frac{f'(x) - p'(x)}{(n + 1)(x - a)^n} = 0$$

(see Problem 19).   □

Taylor's theorem has important applications to the problem of finding approximate solutions to equations that cannot be solved exactly by algebraic means. If g and h are functions and $\text{dom}(g) \cap \text{dom}(h) \neq \varnothing$, then, by letting $f = g - h$, we can rewrite the equation

$$g(x) = h(x)$$

in the equivalent form

$$f(x) = 0.$$

A number $x \in \text{dom}(f)$ for which the last equation holds is called a **zero** of the function f. Thus, the problem of finding approximate solutions to the equation $g(x) = h(x)$ is equivalent to the problem of approximating the zeros of the function f.

In the seventeenth century, Sir Isaac Newton (1642–1727) discovered an elegant method for approximating zeros of differentiable functions. The idea behind **Newton's method** is illustrated in Figure 5-1. On the one hand, if we start at the point $(a, f(a))$ on the graph of f and follow the graph down to the x axis, we arrive at the zero r of f. On the other hand, if we start at the point $(a, f(a))$ and follow the *tangent line* down to the x axis, we arrive at an approximation b to r. In many cases, if a is a good approximation to r, then b is an even better approximation to r. If b is not a sufficiently good approximation to r for the purposes at hand, the method may be repeated, starting with b this time, to produce (one hopes!) an even better approximation. Iteration of the method in this way often produces a sequence of successively better and better approximations to r.

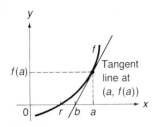

FIGURE 5-1

In Figure 5-1, the equation of the tangent line to the graph of f at $(a, f(a))$ is

$$y = f(a) + f'(a)(x - a).$$

Putting $y = 0$ and solving this equation for x, we find that the tangent line intersects the x axis at

$$b = a - \frac{f(a)}{f'(a)},$$

provided, of course, that $f'(a) \neq 0$. The critical question here is how much better an approximation to r is b than a? An answer is provided by the following theorem.

3.11 THEOREM

> **ERROR IN NEWTON'S METHOD**   Let f be twice differentiable on an interval I, suppose that there is a zero r of f in I^o, and that there are positive constants m and M such that $|f'(x)| \geq m$ and $|f''(x)| \leq M$ for all $x \in I$. Let $a \in I$ and let
>
> $$b = a - \frac{f(a)}{f'(a)}.$$
>
> Then
>
> $$|r - b| \leq \frac{M}{2m} |r - a|^2.$$

PROOF   By Taylor's theorem in Lagrange form (Theorem 3.6), there exists c between a and r such that

$$f(r) = f(a) + f'(a)(r - a) + \frac{f''(c)}{2}(r - a)^2. \tag{1}$$

Also, we have

$$b = a - \frac{f(a)}{f'(a)}. \tag{2}$$

Solving (2) for $f(a)$, we find that

$$f(a) = f'(a)(a - b). \tag{3}$$

Substituting (3) into (1) and using the fact that $f(r) = 0$, we obtain the equation

$$0 = f'(a)(a - b) + f'(a)(r - a) + \frac{f''(c)}{2}(r - a)^2, \tag{4}$$

that is,

$$-f'(a)(r - b) = \frac{f''(c)}{2}(r - a)^2. \tag{5}$$

Taking absolute values on both sides of (5), we find that

$$|f'(a)|\,|r - b| = \frac{|f''(c)|}{2} |r - a|^2. \tag{6}$$

Finally, in view of the fact that $|f'(a)| \geq m$ and that $|f''(c)| \leq M$, Equation (6) implies that

$$|r - b| \leq \frac{M}{2m} |r - a|^2. \tag{7}$$

$\square$

PROBLEM SET 5.3

1. If $p(x) = \sum_{k=0}^{n} c_k(x - a)^k$ and $c_n \neq 0$, prove that p is a polynomial function of degree n, that is, prove that there are coefficients $b_0, b_1, b_2, \ldots, b_n$ with $b_n \neq 0$ such that $p(x) = \sum_{k=0}^{n} b_k x^k$.

2. Let H be a finite nonempty subset of $\mathbb{R}$ with $\# H = n$, and let $h \in H$. Show that there exists a polynomial function p_h of degree $n - 1$ such that $p_h(h) = 1$ and $p_h(k) = 0$ for all $k \in H$ with $k \neq h$.

3. If $p(x) = \sum_{k=0}^{n} b_k x^k$ and $a \in \mathbb{R}$, prove that the function q defined by $q(x) = p(x) - b_n(x - a)^n$ for all $x \in \mathbb{R}$ is a polynomial function of degree at most $n - 1$.

4. With the notation of Problem 2, suppose that f is a function with $H \subseteq \text{dom}(f)$. Prove that there exists a polynomial function p of degree at most $n - 1$ that agrees with f on H in the sense that $p(h) = f(h)$ for all $h \in H$. [Hint: Let $p = \sum_{k \in H} f(k) p_k$.]

5. If $p(x) = \sum_{k=0}^{n} c_k(x - a)^k$ and m is an integer with $0 \leq m \leq n$, use mathematical induction on m to prove that

$$p^{(m)}(x) = \sum_{k=m}^{n} k(k - 1)(k - 2) \cdots (k - m + 1) c_k(x - a)^{k-m}.$$

6. Prove that the polynomial p in Problem 4 is uniquely determined by f and $H \subseteq \text{dom}(f)$. [Hint: The fact that the degree of p cannot exceed $n - 1$ is critical.]

7. With the notation of Problem 5, prove that

$$p^{(m)}(x) = \sum_{k=0}^{n-m} \frac{(m + k)!}{k!} c_{m+k}(x - a)^k.$$

[Hint: Begin by noting that $k(k - 1)(k - 2) \cdots (k - m + 1) = k!/(k - m)!$. Then change the summation index to $j = k - m$ as j runs from m to n. Finally, rewrite the result by renaming the summation index k rather than j.]

8. The development of Simpson's method in elementary calculus requires solving the following basic problem: Given a function f, a point $a \in \text{dom}(f)$, and a positive number h such that $[a - h, a + h] \subseteq \text{dom}(f)$, find constants A, B, and C such that the polynomial $Ax^2 + Bx + C$ agrees with f at the points $a - h$, a, and $a + h$. Solve this problem by finding formulas for A, B, and C in terms of f, a, and h. [Hint: See Problem 4.]

9. For $n = 0$, prove that the statement of the extended mean value theorem (Theorem 3.5) is equivalent to the mean value theorem (Theorem 2.11). [Hint: Note that the 0th Taylor polynomial p for f at a is the constant polynomial given by $p(x) = f(a)$ for all $x \in \mathbb{R}$.]

10. Give a proof of the extended mean value theorem (Theorem 3.5) for the case in which $b < a$. [Hint: Define F and G as in the given proof, but on the interval $[b, a]$, and note that you must interchange a and b in Equation (3) of the given proof.]

11. In the proof of Theorem 3.5, make the computation to show that $F(b) - F(a) = p(b) - f(b)$. [Hint: Show that $F(b) = 0$ and that $F(a) = f(b) - p(b)$.]

12. If f is a polynomial function of degree m and a is a real number, show that, for all integers $n \geq m$, f itself is the nth Taylor polynomial for f at a.

13. Let J be an interval, let $b \in \mathbb{R}$, suppose that f is $n + 1$ times differentiable on J, and define $\phi : J \to \mathbb{R}$ by

$$\phi(x) = \sum_{k=0}^{n} \frac{f^{(k)}(x)}{k!} (b - x)^k.$$

Prove that, for all $x \in J$,

$$\phi'(x) = \frac{f^{(n+1)}(x)}{n!} (b - x)^n.$$

[Hint: Compute the derivative using the sum, product, and power rules, then notice that all but one of the resulting terms cancel in pairs.]

14. Let $f(x) = \sqrt{1 + x}$ for $x > -1$. Find the nth Taylor polynomial for f at 0 for the indicated values of n:
   (a) $n = 1$ (b) $n = 2$ (c) $n = 3$ (d) $n = 4$

15. In the proof of Theorem 3.5, establish the result in Equation (6). [Hint: Use the result of Problem 13.]

16. Use the results of Problem 14 and Taylor's theorem in Lagrange form to prove that, for $x > -1$,

$$1 + \frac{x}{2} - \frac{x^2}{8} < \sqrt{1 + x} < 1 + \frac{x}{2} - \frac{x^2}{8} + \frac{x^3}{16}.$$

17. If f is n times differentiable at a, p is the nth Taylor polynomial for f at a, and k is an integer with $0 \leq k \leq n$, prove that $p^{(k)}$ is the $(n - k)$th Taylor polynomial for $f^{(k)}$ at a.

18. (a) If $x \neq 1$ and n is a nonnegative integer, prove that

$$\frac{1}{1 - x} = \sum_{k=0}^{n} x^k + \frac{x^{n+1}}{1 - x}.$$

   (b) Use the result of Part (a) to prove that, for the function f defined for $x \neq 1$ by $f(x) = 1/(1 - x)$, the unique local polynomial approximation of order n to f at 0 is given by $p(x) = \sum_{k=0}^{n} x^k$.

19. Where L'Hospital's rule is used in the proof of Theorem 3.10, show that the hypotheses of the rule are satisfied. (See Problem 11 in Problem Set 5.2.)

20. (a) Assume that the exponential function $f(x) = e^x$ has the properties developed in elementary calculus, and find the nth Taylor polynomial p for f at 0.

(b) Use the fact that $0 < e < 3$ to show that, for $x > 0$, the absolute value of the error in the approximation $e^x \approx p(x)$ does not exceed $3^x x^{n+1}/(n+1)!$.

(c) Find the smallest value of n for which the absolute value of the error in the approximation $e \approx p(1)$ does not exceed 0.0005.

21. Suppose that n is a positive integer, that $a \in \text{dom}(f)$, that f is $n-1$ times differentiable on a neighborhood of a, and that $f^{(n-1)}$ is differentiable at a. If p is a polynomial function whose degree does not exceed n, and if

$$\lim_{x \to a} \frac{f(x) - p(x)}{(x-a)^n} = 0,$$

prove that p is the nth Taylor polynomial for f at a.

22. Assume that the sine and cosine functions have the properties developed in elementary calculus, and find the nth Taylor polynomials p and q for sine and cosine, respectively, at 0.

23. Let f be n times differentiable on a neighborhood N of the point a, let p be the nth Taylor polynomial for f at a, let q be a polynomial function, and suppose that $b \in \mathbb{R}$ with $q(b) = a$. Prove that $p \circ q$ is the nth Taylor polynomial for $f \circ q$ at b.

24. (a) Show that the absolute value of the error involved in the approximation $\cos x \approx 1 - (x^2/2)$ does not exceed $x^4/24$.

(b) For $0 < x < \pi/2$, is $\cos x$ larger than or smaller than $1 - (x^2/2)$? Justify your answer. [Hint: Use the results of Problem 22.]

25. Suppose that f is twice differentiable on the closed bounded interval $[d, a]$, that f'' is continuous on $[d, a]$, and that $f(d) < 0$ and $f(a) > 0$. Suppose that m and M are positive constants such that $f'(x) \geq m$ and $0 < f''(x) \leq M$ for all $x \in [d, a]$.

(a) Prove that f has exactly one zero r in (d, a).

(b) If a is considered as an approximation to r, and b is the improved approximation to r obtained by Newton's method as in Theorem 3.11, prove that $r < b < a$.

26. It is desired to use Newton's method to find a solution of the equation $x^3 = \cos x$. A rough graph shows that a solution lies on the interval $[0.5, 1]$. Starting with $a = 1$ as an initial approximation to the solution, find the smallest number of iterations of Newton's method that will produce an approximate solution that is correct to 10 decimal places. [Hint: Make sure that the absolute value of the error does not exceed $10^{-10}/2$. (See Problem 7 in Problem Set 3.5.)]

27. The ancient Babylonians discovered the following rule for approximating the square root of a positive number c: Make a reasonable first approximation a to the square root of c. Then $b = (a + c/a)/2$ produces a better

approximation to the square root of c. Show that this **Babylonian method** is a special case of Newton's method.

28. Let $f(x) = (1 - 4x)/(1 + 4x)$ for $x \neq -\frac{1}{4}$. Then the only zero of f is $r = \frac{1}{4}$.
   (a) Try to use Newton's method with $a = 1$ as an initial approximation to r. What happens when you iterate the procedure?
   (b) Explain what goes wrong and why the difficulty does not contradict Theorem 3.11.

29. The equation $8x^3 - 6x - 1 = 0$ is used in the proof that it is not possible to trisect a $60°$ angle with straightedge and compass alone. One of the roots r of this equation lies in the interval $(0.5, 1)$.
   (a) Use a calculator and Newton's method to give an approximation of r that is correct to 8 decimal places.
   (b) Show that $r = \cos 20°$.

30. Let $f(x) = 2 + 3x^2 - x^4$.
   (a) Show that $r = \sqrt{(3 + \sqrt{17})/2}$ is the only positive zero of f.
   (b) Try to use Newton's method with $a = 1$ as an initial approximation to r. What happens when you iterate the procedure?
   (c) Explain what goes wrong and why the difficulty does not contradict Theorem 3.11.

31. Suppose that c is a positive constant.
   (a) If $a > 0$ is a first approximation to c^{-1}, show that Newton's method applied to the function $f(x) = x^{-1} - c$ leads to $b = 2a - ca^2$ as a second approximation.
   (b) If $0 < a < 2c^{-1}$, show that b is a better approximation to c^{-1} than a.

4
_____

THE DARBOUX INTEGRAL

In 1854, the German mathematician Georg Friedrich Bernhard Riemann (1826–1866) gave the formal analytic definition of the integral

$$\int_a^b f(x)\, dx$$

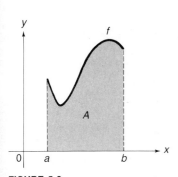

FIGURE 5-2

now found in most elementary calculus textbooks. If $f(x) \geq 0$ for $a \leq x \leq b$, the Riemann integral (when it exists) gives the area A of the region in the xy plane bounded by the x axis, the graph of the function f, and the vertical lines $x = a$ and $x = b$ (Figure 5-2). Riemann's version of the integral was an improvement of an earlier definition given by Cauchy. Both the Riemann and Cauchy definitions have historical antecedents extending to antiquity. For instance, about 367 B.C., Eudoxus of Cnidus derived exact formulas for areas of certain figures by finding successively better and better approximations using known areas of simpler figures. This procedure, which was further refined and exploited by Archimedes about a century later, came to be known by the mathematicians of the seventeenth century as the *method of exhaustion*, and it was the basis for the work of both Cauchy and Riemann.

In 1875, the French mathematician Jean Gaston Darboux (1842–1917) developed an alternative approach to the Riemann integral which has certain advantages, both conceptual and computational. In this section, we present Darboux's definition of the integral; in the next section, we prove that it is equivalent to Riemann's definition.

Suppose that $[a, b]$ is a closed bounded interval and that $f : [a, b] \to \mathbb{R}$ is a bounded function. In motivating and illustrating Darboux's approach to the integral, we usually think of f as a continuous function whose graph lies above the x axis as in Figure 5-2, but *we make no such requirements in our formal definitions*. With this in mind, consider the problem of finding an exact formula for the area A of the region between the graph of f and the x axis.

By assumption, f is bounded on $[a, b]$, that is, $f([a, b])$ is a bounded set; hence, since $f([a, b]) \neq \varnothing$, it follows from the completeness of $\mathbb{R}$ that both

$$m = \inf(f([a, b])) \qquad \text{and} \qquad M = \sup(f([a, b]))$$

exist. In what follows, we denote infima and suprema such as m and M by the simpler notation

$$m = \inf_{a \leq x \leq b} f(x) \qquad \text{and} \qquad M = \sup_{a \leq x \leq b} f(x).$$

Evidently, $m(b - a)$ is the area of a rectangle with base $[a, b]$ and lying below the graph of f (an **inscribed** rectangle) as in Figure 5-3, whereas $M(b - a)$ is the area of a rectangle with base $[a, b]$ and containing the graph of f (a **circumscribed** rectangle) as in Figure 5-4. Thus, $m(b - a)$ *underestimates* A, $M(b - a)$ *overestimates* A, and we have

$$m(b - a) \leq A \leq M(b - a).$$

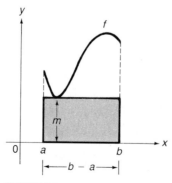

FIGURE 5-3

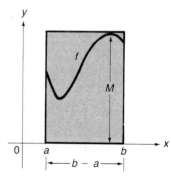

FIGURE 5-4

To obtain a better approximation for the area A, we divide the interval $[a, b]$ into subintervals and make similar estimates using inscribed and circumscribed rectangles on each subinterval. By forming the sum of the areas of the inscribed rectangles, we obtain an approximation of A from below (Figure 5-5) and, by forming the sum of the areas of the circumscribed rectangles, we obtain an approximation of A from above (Figure 5-6). The endpoints of the subintervals into which $[a, b]$ is divided form a *partition* of $[a, b]$ in accordance with Definition 4.1.

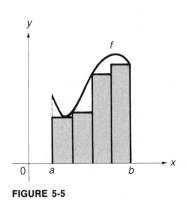

FIGURE 5-5

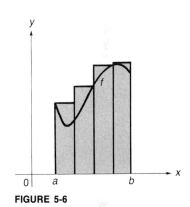

FIGURE 5-6

4.1 DEFINITION

PARTITION OF AN INTERVAL By a **partition** of the closed bounded interval $[a, b]$ we mean a finite set $P \subseteq \mathbb{R}$ such that $a, b \in P \subseteq [a, b]$.

If P is a partition of $[a, b]$ and $\#P = n + 1$, we use the subscript notation $x_0, x_1, x_2, \ldots, x_n$ to denote the elements P arranged in numerical order, so that

$$a = x_0 < x_1 < x_2 < \cdots < x_n = b. \tag{1}$$

Although the elements of the set P are not to be regarded as belonging to P in any particular order, it will be convenient in what follows to adopt the following convention: *Whenever we specify that*

$$P = \{x_0, x_1, x_2, \ldots, x_n\} \tag{2}$$

is a partition, we understand that the notation is so chosen that (1) *holds.* Note that, as a consequence,

$$[a, b] = [x_0, x_1] \cup [x_1, x_2] \cup \cdots \cup [x_{n-1}, x_n] \tag{3}$$

and that, for $i, j = 1, 2, \ldots, n$ with $i \neq j$, the intervals $[x_{i-1}, x_i]$ and $[x_{j-1}, x_j]$ are nonoverlapping in the sense that they have at most one point (an endpoint) in common. For $i = 1, 2, \ldots, n$, we refer to $[x_{i-1}, x_i]$

as the **ith subinterval** of $[a, b]$ corresponding to the partition P and we denote the length of the ith subinterval by

$$\Delta x_i = x_i - x_{i-1}. \tag{4}$$

Thus, the length of the interval $[a, b]$ is given by

$$b - a = \sum_{i=1}^{n} \Delta x_i \tag{5}$$

(see Problem 1).

As we have suggested, the *lower* and *upper Darboux sums* given in the next definition will provide estimates for the area A under the graph of the function f.

4.2 DEFINITION

> **LOWER AND UPPER DARBOUX SUMS**   Let $f : [a, b] \rightarrow \mathbb{R}$ be a bounded function and let $P = \{x_0, x_1, x_2, \ldots, x_n\}$ be a partition of $[a, b]$. For each $i = 1, 2, \ldots, n$, define
>
> $$m_i = \inf_{x_{i-1} \le x \le x_i} f(x) \quad \text{and} \quad M_i = \sup_{x_{i-1} \le x \le x_i} f(x).$$
>
> Then the **lower** and **upper Darboux sums** $L(f, P)$ and $U(f, P)$, respectively, are defined by
>
> $$L(f, P) = \sum_{i=1}^{n} m_i \Delta x_i \quad \text{and} \quad U(f, P) = \sum_{i=1}^{n} M_i \Delta x_i.$$

In Definition 4.2, note that, since f is bounded on $[a, b]$, it is bounded on the subinterval $[x_{i-1}, x_i] \subseteq [a, b]$, and, therefore, m_i and M_i exist by the completeness of $\mathbb{R}$. Intuitively, we have

$$L(f, P) \le A \le U(f, P),$$

and both $L(f, P)$ and $U(f, P)$ can be considered as approximations to A with a maximum error given by

$$U(f, P) - L(f, P).$$

Our intuition also suggests that, by using finer and finer partitions P, and thus dividing $[a, b]$ into more and more subintervals, we can make the error $U(f, P) - L(f, P)$ as small as we please. Let's see just what does happen.

4.3 DEFINITION

> **REFINEMENT OF A PARTITION**   Let P and Q be partitions of $[a, b]$. If $P \subseteq Q$, we say that Q is a **refinement** of P.

The following lemma and its corollary show that, as we refine the partition P of $[a, b]$, the lower Darboux sums $L(f, P)$ can only increase and the upper Darboux sums $U(f, P)$ can only decrease.

4.4 LEMMA

Let $f:[a, b] \to \mathbb{R}$ be bounded, let P be a partition of $[a, b]$, and suppose that $x^* \in (a, b)$. Then, if

$$Q = P \cup \{x^*\},$$

it follows that

$$L(f, P) \le L(f, Q) \quad \text{and} \quad U(f, P) \ge U(f, Q).$$

PROOF

We prove here that $L(f, P) \le L(f, Q)$ and leave the similar proof that $U(f, P) \ge U(f, Q)$ as an exercise (Problem 4). If $x^* \in P$, then $Q = P$, and there is nothing to prove. Therefore, we can assume that $x^* \notin P$. Let

$$P = \{x_0, x_1, x_2, \ldots, x_n\}.$$

Then there exists $k = 1, 2, \ldots, n$ such that

$$x_{k-1} < x^* < x_k$$

and the terms in the sum $L(f, P)$ match the corresponding terms in the sum $L(f, P^*)$ except for those that involve the interval $[x_{k-1}, x_k]$ or the intervals $[x_{k-1}, x^*]$ and $[x^*, x_k]$.
Let

$$m_k = \inf_{x_{k-1} \le x \le x_k} f(x),$$

noting that the term in the sum $L(f, P)$ involving the interval $[x_{k-1}, x_k]$ is

$$m_k \Delta x_k = m_k(x_k - x_{k-1}).$$

Similarly, let

$$m^* = \inf_{x_{k-1} \le x \le x^*} f(x) \quad \text{and} \quad m^{**} = \inf_{x^* \le x \le x_k} f(x),$$

noting that the terms in the sum $L(f, Q)$ involving the intervals $[x_{k-1}, x^*]$ and $[x^*, x_k]$ are

$$m^*(x^* - x_{k-1}) \quad \text{and} \quad m^{**}(x_k - x^*).$$

Thus, it will be sufficient to prove that

$$m_k(x_k - x_{k-1}) \le m^*(x^* - x_{k-1}) + m^{**}(x_k - x^*).$$

Since $[x_{k-1}, x^*] \subseteq [x_{k-1}, x_k]$, it follows that

$$m_k \leq m^*$$

(Problem 3c). Likewise, since $[x^*, x_k] \subseteq [x_{k-1}, x_k]$, we have

$$m_k \leq m^{**}.$$

Therefore, since $x^* - x_{k-1}$ and $x_k - x^*$ are positive,

$$
\begin{aligned}
m_k(x_k - x_{k-1}) = m_k[(x_k - x^*) + (x^* - x_{k-1})] \\
= m_k(x_k - x_{k-1}) + m_k(x^* - x_{k-1}) \\
\leq m^*(x^* - x_{k-1}) + m^{**}(x_k - x^*).
\end{aligned}
$$

$\square$

If Q is a refinement of the partition P, then Q can be obtained from P by successively adjoining points, one by one, as in Lemma 4.4. Therefore, by using mathematical induction (Problem 6), we obtain the following corollary.

4.5 COROLLARY

> Let $f:[a, b] \to \mathbb{R}$ be bounded, let P and Q be partitions of $[a, b]$, and suppose that Q is a refinement of P. Then
>
> $$L(f, P) \leq L(f, Q) \qquad \text{and} \qquad U(f, P) \geq U(f, Q).$$

4.6 LEMMA

> Let $f:[a, b] \to \mathbb{R}$ be bounded, suppose that P is a partition of $[a, b]$, and let
>
> $$m = \inf_{a \leq x \leq b} f(x) \qquad \text{and} \qquad M = \sup_{a \leq x \leq b} f(x).$$
>
> Then
>
> $$m(b - a) \leq L(f, P) \leq U(f, P) \leq M(b - a).$$

PROOF   Note that $\{a, b\}$ is a partition of $[a, b]$ and that P is a refinement of $\{a, b\}$. Furthermore,

$$L(f, \{a, b\}) = m(b - a) \qquad \text{and} \qquad U(f, \{a, b\}) = M(b - a).$$

Therefore, by Corollary 4.5,

$$m(b - a) \leq L(f, P) \qquad \text{and} \qquad U(f, P) \leq M(b - a).$$

It remains to prove that

$$L(f, P) \leq U(f, P).$$

If $P = \{x_0, x_1, x_2, \ldots, x_n\}$, then, for each $i = 1, 2, \ldots, n$,

$$m_i = \inf_{x_{i-1} \le x \le x_i} f(x) \le \sup_{x_{i-1} \le x \le x_i} f(x) = M_i,$$

and, since each $\Delta_i = x_i - x_{i-1}$ is positive, it follows that

$$L(f, P) = \sum_{i=1}^{n} m_i \Delta x_i \le \sum_{i=1}^{n} M_i \Delta x_i = U(f, P). \qquad \square$$

4.7 COROLLARY

Let $f : [a, b] \to \mathbb{R}$ be bounded and suppose that P and Q are partitions of $[a, b]$. Then, with the notation of Lemma 4.6,

$$m(b - a) \le L(f, P) \le U(f, Q) \le M(b - a).$$

PROOF   Lemma 4.6 implies that

$$m(b - a) \le L(f, P) \qquad \text{and} \qquad U(f, Q) \le M(b - a),$$

and it remains to prove that

$$L(f, P) \le U(f, Q).$$

Let $R = P \cup Q$, noting that R is a partition of $[a, b]$ and that R refines both P and Q. Thus, by Corollary 4.5 and Lemma 4.6,

$$L(f, P) \le L(f, R) \le U(f, R) \le U(f, Q). \qquad \square$$

Let $\mathscr{P}$ be the set of all partitions of $[a, b]$. By Lemma 4.6, the nonempty set

$$\{L(f, P) \mid P \in \mathscr{P}\}$$

is bounded above by $M(b - a)$, and it follows that

$$\sup(\{L(f, P) \mid P \in \mathscr{P}\})$$

exists. In what follows, we denote this supremum simply as

$$\sup_{P} L(f, P).$$

Likewise, by Lemma 4.6, the nonempty set

$$\{U(f, P) \mid P \in \mathscr{P}\}$$

is bounded below by $m(b - a)$, and it follows that

$$\inf(\{U(f, P) \mid P \in \mathscr{P}\})$$

exists. In what follows, we denote this infimum simply as

$$\inf_{P} U(f, P).$$

4.8 DEFINITION

> **UPPER AND LOWER DARBOUX INTEGRALS**   Let $f:[a, b] \to \mathbb{R}$ be a bounded function. We define the **lower** and **upper Darboux integrals** of f, in symbols $\underline{\int} f$ and $\overline{\int} f$, respectively, by
>
> $$\underline{\int} f = \sup_{P} L(f, P) \qquad \text{and} \qquad \overline{\int} f = \inf_{P} U(f, P).$$

4.9 LEMMA

> Let $f:[a, b] \to \mathbb{R}$ be a bounded function. Then
>
> $$m(b - a) \le \underline{\int} f \le \overline{\int} f \le M(b - a).$$

PROOF   That

$$m(b - a) \le \underline{\int} f \qquad \text{and} \qquad \overline{\int} f \le M(b - a)$$

is an immediate consequence of Lemma 4.6 (see Problem 7a and b). Choose and (temporarily) fix a partition Q of $[a, b]$. Then, by Corollary 4.7,

$$L(f, P) \le U(f, Q)$$

holds for all $P \in \mathscr{P}$, and it follows that

$$\underline{\int} f = \sup_{P} L(f, P) \le U(f, Q).$$

But Q was an arbitrary partition of $[a, b]$, and therefore

$$\underline{\int} f \le U(f, P)$$

holds for all $P \in \mathscr{P}$. Consequently,

$$\underline{\int} f \le \inf_{P} U(f, P) = \overline{\int} f. \qquad \square$$

As we have observed on intuitive grounds, if A is the area under the graph of f, then we expect

$$L(f, P) \le A \le U(f, P)$$

to hold for all $P \in \mathscr{P}$. If this is so, then it follows that

$$\sup_{P} L(f, P) \le A \le \inf_{P} U(f, P),$$

so that

$$\underline{\int} f \le A \le \overline{\int} f.$$

In particular, if the lower and upper Darboux integrals happen to be equal, then their common value must be A. This suggests our next definition.

4.10 DEFINITION

> **THE DARBOUX INTEGRAL** Let $f:[a, b] \to \mathbb{R}$. If f is bounded and the lower and upper Darboux integrals of f are equal, we say that f is **Darboux integrable**, and we define the **Darboux integral** of f, in symbols $\int f$, to be their common value, so that
>
> $$\int f = \underline{\int} f = \overline{\int} f.$$

4.11 Example Let f be a constant function on $[a, b]$, say, $f(x) = k$ for all $x \in [a, b]$. Prove that f is Darboux integrable and that $\int f = k(b - a)$.

SOLUTION Let $P = \{x_0, x_1, x_2, \ldots, x_n\}$ be a partition of $[a, b]$. For each $i = 1, 2, \ldots, n$, we have

$$m_i = \inf_{x_{i-1} \le x \le x_i} f(x) = k \quad \text{and} \quad M_i = \sup_{x_{i-1} \le x \le x_i} f(x) = k;$$

hence,

$$L(f, P) = \sum_{i=1}^{n} m_i \Delta x_i = \sum_{i=1}^{n} k \Delta x_i = k \sum_{i=1}^{n} \Delta x_i = k(b - a).$$

Since $L(f, P) = k(b - a)$ holds for every partition P of $[a, b]$, it follows that

$$\underline{\int} f = \sup_P L(f, P) = k(b - a).$$

A similar argument (Problem 8) shows that

$$\overline{\int} f = k(b - a),$$

and it follows that f is Darboux integrable and that

$$\int f = k(b - a). \qquad \square$$

4.12 Example Let $f:[a, b] \to \mathbb{R}$ be defined by

$$f(x) = \begin{cases} 1, & \text{if } x \in \mathbb{Q} \\ 0, & \text{if } x \notin \mathbb{Q} \end{cases}$$

for all $x \in [a, b]$. Prove that f is not Darboux integrable.

SOLUTION Let $P = \{x_0, x_1, x_2, \ldots, x_n\}$ be a partition of $[a, b]$. As a consequence of Theorems 5.14 and 5.15 in Chapter 3, every subinterval $[x_{i-1}, x_i]$ of $[0, 1]$

contains both a rational number and an irrational number. Hence, for every $i = 1, 2, \ldots, n$,

$$m_i = \inf_{x_{i-1} \leq x \leq x_i} f(x) = 0 \qquad \text{and} \qquad M_i = \sup_{x_{i-1} \leq x \leq x_i} f(x) = 1.$$

Therefore,

$$L(f, P) = \sum_{i=1}^{n} m_i \Delta x_i = \sum_{i=1}^{n} 0 \cdot \Delta x_i = 0.$$

Since $L(f, P) = 0$ holds for every partition P of $[a, b]$, it follows that

$$\underline{\int} f = \sup_{P} L(f, P) = 0.$$

A similar argument (Problem 10) shows that

$$\overline{\int} f = b - a.$$

Therefore, the lower and upper Darboux integrals are not equal, and it follows that f is not Darboux integrable. □

The elegance and simplicity of the following theorem make Darboux's approach to the integral especially attractive.

4.13 THEOREM

> **DARBOUX'S INTEGRABILITY CONDITION**
> Let $f : [a, b] \to \mathbb{R}$ be bounded. Then f is Darboux integrable if and only if, for every $\varepsilon > 0$, there exists a partition P of $[a, b]$ such that
>
> $$U(f, P) - L(f, P) < \varepsilon.$$

PROOF   First, suppose that f is Darboux integrable so that the lower and upper Darboux integrals are equal, and let $\varepsilon > 0$. By Part (i) of Lemma 3.23 in Chapter 3 and the fact that

$$\int f - \frac{\varepsilon}{2} = \underline{\int} f - \frac{\varepsilon}{2} < \underline{\int} f = \sup_{P} L(f, P),$$

there exists a partition P_1 of $[a, b]$ such that

$$\int f - \frac{\varepsilon}{2} < L(f, P_1).$$

Similarly, by Part (ii) of the same lemma, there exists a partition P_2 of $[a, b]$ such that

$$U(f, P_2) < \int f + \frac{\varepsilon}{2}.$$

Let $P = P_1 \cup P_2$ and note that P refines both P_1 and P_2. Therefore, by Corollary 4.5,

$$U(f, P) - L(f, P) \leq U(f, P_2) - L(f, P_1)$$

$$< \left(\int f + \frac{\varepsilon}{2} \right) - \left(\int f - \frac{\varepsilon}{2} \right) = \varepsilon.$$

Conversely, suppose that, for every $\varepsilon > 0$, there exists a partition P of $[a, b]$ such that

$$U(f, P) < L(f, P) + \varepsilon.$$

Since the upper Darboux integral is the infimum of all such numbers $U(f, P)$, it follows that, for every $\varepsilon > 0$, there exists a partition P of $[a, b]$ such that

$$\overline{\int} f < L(f, P) + \varepsilon.$$

Since the lower Darboux integral is the supremum of all such numbers $L(f, P)$, it follows that, for every $\varepsilon > 0$,

$$\overline{\int} f < \underline{\int} f + \varepsilon.$$

Because ε is an arbitrary positive number, we can conclude that

$$\overline{\int} f \leq \underline{\int} f$$

(see Problem 15). By Lemma 4.9, the opposite inequality always holds, and therefore f is Darboux integrable. □

Note how the idea of compactness developed in Section 5 of Chapter 4 figures in the proof of the following important theorem.

4.14 THEOREM | **INTEGRABILITY OF A CONTINUOUS FUNCTION** If $f:[a, b] \to \mathbb{R}$ is continuous, then it is Darboux integrable.

PROOF Let $f:[a, b] \to \mathbb{R}$ be continuous. By the Heine–Borel theorem (Theorem 5.7 in Chapter 4), $[a, b]$ is compact; hence, by the extreme value theorem (Theorem 5.14 in Chapter 4), f is bounded. Let $\varepsilon > 0$. In view of Theorem 4.13, it will be sufficient to find a partition P of $[a, b]$ such that $U(f, P) - L(f, P) < \varepsilon$. By Theorem 5.18 in Chapter 4, f is uniformly continuous on the compact set $[a, b]$. Therefore, there exists $\delta > 0$ such that, for all $p, q \in [a, b]$,

$$|p - q| < \delta \Rightarrow |f(p) - f(q)| < \frac{\varepsilon}{b - a}.$$

By Corollary 4.12 in Chapter 3, there exists a positive integer n such that

$$\frac{b - a}{\delta} < n.$$

For each $i = 0, 1, 2, \ldots, n$, define

$$x_i = a + \frac{i(b - a)}{n}.$$

Then $P = \{x_0, x_1, x_2, \ldots, x_n\}$ is a partition of $[a, b]$ and, for each $i = 1, 2, \ldots, n$, the ith subinterval $[x_{i-1}, x_i]$ has length

$$\Delta x_i = x_i - x_{i-1} = \frac{b - a}{n} < \delta$$

(Problem 17). Also, as a consequence of the extreme value theorem, there exist points $p_i, q_i \in [x_{i-1}, x_i]$ such that

$$m_i = \inf_{x_{i-1} \le x \le x_i} f(x) = f(q_i) \qquad \text{and} \qquad M_i = \sup_{x_{i-1} \le x \le x_i} f(x) = f(p_i).$$

Therefore,

$$M_i - m_i = f(p_i) - f(q_i).$$

Since $p_i, q_i \in [x_{i-1}, x_i]$ and $\Delta x_i < \delta$, it follows that

$$M_i - m_i = |M_i - m_i| = |f(p_i) - f(q_i)|$$

$$< \frac{\varepsilon}{b - a}.$$

Consequently,

$$U(f, P) - L(f, P) = \sum_{i=1}^{n} M_i \Delta x_i - \sum_{i=1}^{n} m_i \Delta x_i$$

$$= \sum_{i=1}^{n} (M_i - m_i) \Delta x_i$$

$$< \frac{\varepsilon}{b - a} \sum_{i=1}^{n} \Delta x_i = \frac{\varepsilon}{b - a} (b - a)$$

$$= \varepsilon. \qquad \qquad \square$$

4.15 THEOREM | **INTEGRABILITY OF A MONOTONE FUNCTION** Let $f:[a, b] \to \mathbb{R}$ be monotone. Then f is Darboux integrable.

PROOF   We assume that f is increasing, leaving the opposite case as an exercise (Problem 18). Then, for each $x \in [a, b]$, we have $f(a) \le f(x) \le f(b)$, and therefore f is bounded. Let $\varepsilon > 0$. We must find a partition P such that $U(f, P) - L(f, P) < \varepsilon$.

By Corollary 4.12 in Chapter 3, there exists a positive integer n such that

$$(b - a)\frac{f(b) - f(a)}{\varepsilon} < n.$$

For each $i = 0, 1, 2, \ldots, n$, define

$$x_i = a + \frac{i(b - a)}{n}.$$

Then $P = \{x_0, x_1, x_2, \ldots, x_n\}$ is a partition of $[a, b]$ and, for each $i = 1, 2, \ldots, n$, the ith subinterval $[x_{i-1}, x_i]$ has length

$$\Delta x_i = x_i - x_{i-1}$$
$$= \frac{b - a}{n} < \frac{\varepsilon}{f(b) - f(a)}.$$

(Problem 19). Since f is increasing, we have

$$m_i = \inf_{x_{i-1} \le x \le x_i} f(x) = f(x_{i-1}) \qquad \text{and} \qquad M_i = \sup_{x_{i-1} \le x \le x_i} f(x) = f(x_i).$$

Consequently,

$$U(f, P) - L(f, P) = \sum_{i=1}^{n} M_i \Delta x_i - \sum_{i=1}^{n} m_i \Delta x_i$$
$$= \sum_{i=1}^{n} (M_i - m_i) \Delta x_i$$
$$< \left[\sum_{i=1}^{n} (M_i - m_i) \right] \frac{\varepsilon}{f(b) - f(a)}.$$

However,

$$\sum_{i=1}^{n} (M_i - m_i) = \sum_{i=1}^{n} [f(x_i) - f(x_{i-1})]$$
$$= f(x_n) - f(x_0)$$
$$= f(b) - f(a)$$

(Problem 1c), and it follows that $U(f, P) - L(f, P) < \varepsilon$.   □

4.16 LEMMA

> Let $f:[a, c] \to \mathbb{R}$ be bounded and let $b \in (a, c)$. Let $g:[a, b] \to \mathbb{R}$ and $h:[b, c] \to \mathbb{R}$ be the restrictions $g = f|_{[a, b]}$ and $h = f|_{[b, c]}$ of f to $[a, b]$ and $[b, c]$, respectively. Then
>
> $$\overline{\int} f = \overline{\int} g + \overline{\int} h.$$

PROOF

We sketch the proof and leave the details as an exercise (Problem 25). Let ε be an arbitrary positive number. Then (as in the proof of Theorem 4.13) there exist partitions P_1 of $[a, b]$ and P_2 of $[b, c]$ such that

$$U(g, P_1) < \overline{\int} g + \frac{\varepsilon}{2} \quad \text{and} \quad U(h, P_2) < \overline{\int} h + \frac{\varepsilon}{2}.$$

Let $P = P_1 \cup P_2$, noting that P is a partition of $[a, c]$ and that

$$\overline{\int} f \le U(f, P) = U(g, P_1) + U(h, P_2) < \overline{\int} g + \overline{\int} h + \varepsilon.$$

Since ε is an arbitrary positive number, it follows that

$$\overline{\int} f \le \overline{\int} g + \overline{\int} h.$$

To prove that equality holds, assume the contrary, so that

$$\inf_P U(f, P) = \overline{\int} f < \overline{\int} g + \overline{\int} h.$$

Then there exists a partition R of $[a, c]$ such that

$$U(f, R) < \overline{\int} g + \overline{\int} h.$$

Let $Q = R \cup \{b\}$, noting that Q is a refinement of R so that, by Lemma 4.4,

$$U(f, Q) \le U(f, R) < \overline{\int} g + \overline{\int} h.$$

Now let

$$P_3 = Q \cap [a, b] \quad \text{and} \quad P_4 = Q \cap [b, c],$$

noting that P_3 is a partition of $[a, b]$, P_4 is a partition of $[b, c]$, and

$$U(g, P_3) + U(h, P_4) = U(f, Q) < \overline{\int} g + \overline{\int} h.$$

But

$$\overline{\int} g = \inf_P U(g, P) \le U(g, P_3) \quad \text{and} \quad \overline{\int} h = \inf_P U(h, P) \le U(h, P_4),$$

and so we obtain the self-contradictory inequality

$$U(g, P_3) + U(h, P_4) < U(g, P_3) + U(h, P_4). \qquad \square$$

In Problem 27, we ask you to show that the conclusion of Lemma 4.16 also holds for the lower Darboux integral.

PROBLEM SET 5.4

1. Let $P = \{x_0, x_1, x_2, \ldots, x_n\}$ be a partition of $[a, b]$.
   (a) Prove that $[a, b] = \bigcup_{i=1}^{n}[x_{i-1}, x_i]$.
   (b) If i and j are integers between 1 and n and $i \neq j$, prove that the intersection $[x_{i-1}, x_i] \cap [x_{j-1}, x_j]$ contains at most one point.
   (c) If $y_0, y_1, y_2, \ldots, y_n \in \mathbb{R}$, prove that $\sum_{i=1}^{n}(y_i - y_{i-1}) = y_n - y_0$. (This is called a **telescoping sum**.)
   (d) If $\Delta x_i = x_i - x_{i-1}$ for $i = 1, 2, \ldots, n$, use the result in Part (c) to prove that $b - a = \sum_{i=1}^{n} \Delta x_i$.

2. If $a < b$ and $\delta > 0$, prove that there exists a partition of $[a, b]$ given by $P = \{x_0, x_1, x_2, \ldots, x_n\}$ such that $x_i - x_{i-1} < \delta$ for all $i = 1, 2, \ldots, n$.

3. Let f be bounded on $[p, q]$ and suppose that $[r, s] \subseteq [p, q]$.
   (a) Prove that f is bounded on $[r, s]$.
   (b) Prove that $\sup_{r \leq x \leq s} f(x) \leq \sup_{p \leq x \leq q} f(x)$.
   (c) Prove that $\inf_{r \leq x \leq s} f(x) \geq \inf_{p \leq x \leq q} f(x)$.

4. Complete the proof of Lemma 4.4 by showing that $U(f, P) \geq U(f, Q)$.

5. Let P, Q, and R be partitions of $[a, b]$.
   (a) Prove that $P \cup Q$ is a partition of $[a, b]$ and that it is a refinement of both P and Q.
   (b) Prove that R is a refinement of both P and Q if and only if R is a refinement of $P \cup Q$.
   (c) If P is a refinement of Q and Q is a refinement of R, prove that P is a refinement of R.

6. Prove Corollary 4.5.

7. Let $f:[a, b] \rightarrow \mathbb{R}$ be bounded, $m = \inf_{a \leq x \leq b} f(x)$, and $M = \sup_{a \leq x \leq b} f(x)$.
   (a) Prove that $m(b - a) \leq \underline{\int} f$.
   (b) Prove that $\overline{\int} f \leq M(b - a)$.
   (c) If P is a partition of $[a, b]$, prove that $L(f, P) \leq \underline{\int} f \leq \overline{\int} f \leq U(f, P)$.

8. Complete the solution of Example 4.11 by showing that $\overline{\int} f = k(b - a)$.

9. Let $f:[a, b] \rightarrow \mathbb{R}$ be bounded, let P be a partition of $[a, b]$, and define $-f:[a, b] \rightarrow \mathbb{R}$ by $(-f)(x) = -f(x)$ for all $x \in [a, b]$.
   (a) Prove that $-f$ is bounded.
   (b) Prove that $L(-f, P) = -U(f, P)$. [Hint: See Lemma 3.21, Chapter 3.]
   (c) Prove that $U(-f, P) = -L(f, P)$.
   (d) Prove that $\overline{\int}(-f) = -\underline{\int} f$.
   (e) Prove that $\underline{\int}(-f) = -\overline{\int} f$.

10. Complete the solution of Example 4.12 by showing that $\overline{\int} f = b - a$.

11. Let $f:[a, b] \to \mathbb{R}$ be Darboux integrable. Use the notation and results of Problem 9 to show that $-f$ is Darboux integrable and that $\int(-f) = -\int f$.

12. Let $a = 0$, $b = 1$, and let $f:[a, b] \to \mathbb{R}$ be the function of Example 4.12. Define $g:[0, 1] \to \mathbb{R}$ by $g(x) = xf(x)$ for all $x \in [0, 1]$. Prove that g is not Darboux integrable.

13. Let $f:[a, b] \to \mathbb{R}$ be bounded, let P be a partition of $[a, b]$, and let k be a positive constant. Define $kf:[a, b] \to \mathbb{R}$ by $(kf)(x) = kf(x)$ for all $x \in [a, b]$.
    (a) Prove that kf is bounded.
    (b) Prove that $L(kf, P) = kL(f, P)$.
    (c) Prove that $U(kf, P) = kU(f, P)$.
    (d) Prove that $\overline{\int}(kf) = k \overline{\int} f$.
    (e) Prove that $\underline{\int}(kf) = k \underline{\int} f$.

14. Let $f:[a, b] \to \mathbb{R}$ be defined by $f(x) = x$ for all $x \in [a, b]$. Let n be a positive integer, and let P be the partition of $[a, b]$ consisting of n subintervals all of the same length.
    (a) Find $L(f, P)$ in closed form (that is, not involving a summation symbol).
    (b) Find $U(f, P)$ in closed form.
    (c) Find $\int f$.

15. If u and v are real numbers such that $u < v + \varepsilon$ holds for every $\varepsilon > 0$, prove that $u \le v$.

16. Work Problem 14 with f replaced by $f(x) = x^2$ for all $x \in [a, b]$.

17. In the proof of Theorem 4.14, show that $\Delta x_i < \delta$.

18. Prove Theorem 4.15 for the case in which f is a decreasing function. [Hint: Either make a direct proof, or apply what has already been proved to the function $-f$ and use the result of Problem 11.]

19. In the proof of Theorem 4.15, show that $\Delta x_i < \varepsilon/[f(b) - f(a)]$.

20. Let $f:[a, b] \to \mathbb{R}$ be bounded and suppose that, for every $n \in \mathbb{N}$, there exists a partition P of $[a, b]$ such that $U(f, P) < L(f, P) + 1/n$. Prove that f is Darboux integrable.

21. Suppose that $f:[a, b] \to \mathbb{R}$ is Darboux integrable. Let $[c, d] \subseteq [a, b]$ and let $g = f|_{[c, d]}$ be the restriction of f to the subinterval $[c, d]$ of $[a, b]$. Prove that g is Darboux integrable.

22. Define $h:[0, 1] \to \mathbb{R}$ by $h(x) = 1$ if $1/x \in \mathbb{N}$ and $h(x) = 0$ otherwise. Prove that h is Darboux integrable and that $\int h = 0$.

23. If $f:[a, b] \to \mathbb{R}$ is bounded above and $0 \le f(x)$ for all $x \in [a, b]$, prove that $0 \le \overline{\int} f$.

24. If $f, g:[a, b] \to \mathbb{R}$ are bounded functions and $f(x) \le g(x)$ holds for all $x \in [a, b]$, prove that $\underline{\int} f \le \underline{\int} g$ and that $\overline{\int} f \le \overline{\int} g$.

25. Supply the details for the proof of Lemma 4.16.

26. Give an example of a bounded function $f:[a, b] \to \mathbb{R}$ such that $|f|$ is Darboux integrable, but f is not Darboux integrable.

27. Under the hypotheses of Lemma 4.16, prove that $\underline{\int} f = \underline{\int} g + \underline{\int} h$. [Hint: Either make a direct proof patterned after the proof of Lemma 4.16, or apply Lemma 4.16 to $-f$ and use the results of Problem 9.]

28. Let $f, g:[a, b] \to \mathbb{R}$ be bounded functions such that, for every subinterval $[c, d] \subseteq [a, b]$,

$$\sup_{c \le x \le d} g(x) - \inf_{c \le x \le d} g(x) \le \sup_{c \le x \le d} f(x) - \inf_{c \le x \le d} f(x).$$

If f is Darboux integrable, prove that g is Darboux integrable.

29. Let $f:[a, c] \to \mathbb{R}$ and let $b \in (a, c)$. Let $g = f|_{[a, b]}$ and let $h = f|_{[b, c]}$ as in Lemma 4.16. If g and h are Darboux integrable, prove that f is Darboux integrable and that $\int f = \int g + \int h$. [Hint: Use Lemma 4.16 and Problem 27.] This is called the **interval additivity property** of the Darboux integral.

30. Let $f:[a, b] \to \mathbb{R}$ be Darboux integrable.
(a) Prove that $|f|$ is Darboux integrable. [Hint: Use Problem 28 with $g = |f|$.]
(b) Prove that $\left| \int f \right| \le \int |f|$. [Hint: Use Problem 11, Problem 24, and the fact that $-|f(x)| \le f(x) \le |f(x)|$.]

31. Let $f:[a, c] \to \mathbb{R}$ and let $P = \{x_0, x_1, x_2, \ldots, x_n\}$ be a partition of $[a, c]$. For each $i = 1, 2, \ldots, n$, define $f_i:[x_{i-1}, x_i] \to \mathbb{R}$ by $f_i(x) = f(x)$ for all $x \in [x_{i-1}, x_i]$.
(a) Prove that f is Darboux integrable if and only if f_i is Darboux integrable for $i = 1, 2, \ldots, n$.
(b) If f is Darboux integrable, prove that $\int f = \sum_{i=1}^{n} \int f_i$. [Hint: Use Problem 21, Problem 29, and mathematical induction.]

32. Let $f:[a, b] \to \mathbb{R}$ be Darboux integrable. Suppose $f(x) \ge 0$ for all $x \in [a, b]$ and that there exists a point $c \in (a, b)$ such that f is continuous at c and $f(c) > 0$. Prove that $\int f > 0$. [Hint: Find $\delta > 0$ such that $[c - \delta, c + \delta] \subseteq [a, b]$ and $f(x) \ge f(c)/2$ for all $x \in [c - \delta, c + \delta]$. Use Problem 31 to break $\int f$ into the sum of three integrals, one over $[a, c - \delta]$, one over $[c - \delta, c + \delta]$, and one over $[c + \delta, b]$. Use Problem 23 to show that the first and third of these integrals are nonnegative, and prove that the second is no smaller than $\delta f(c)$.]

5

THE RIEMANN INTEGRAL

In this section, we give the familiar definition of the Riemann integral as a limit of Riemann sums, and we show that the Riemann and Darboux integrals are the same.

5.1 DEFINITION

> **NORM OF A PARTITION**   Let $P = \{x_0, x_1, x_2, \ldots, x_n\}$ be a partition of the closed bounded interval $[a, b]$ and, for $i = 1, 2, \ldots, n$, let $\Delta x_i = x_i - x_{i-1}$ denote the length of the ith subinterval corresponding to P. Then the **norm** of P, in symbols $\|P\|$, is defined by
>
> $$\|P\| = \max\{\Delta x_1, \Delta x_2, \ldots, \Delta x_n\}.$$

According to the following lemma, as a partition is refined, its norm can only decrease.

5.2 LEMMA

> If P and Q are partitions of $[a, b]$ and if P is a refinement of Q, then $\|P\| \leq \|Q\|$.

PROOF   We outline the proof and leave the details as an exercise (Problem 1). If P is obtained by appending a single point to Q, then exactly one of the subintervals corresponding to Q is broken into two shorter subintervals by the appended point; the remaining subintervals are unaffected, and therefore $\|P\| \leq \|Q\|$.

More generally, if P is a refinement of Q, then P is obtained by starting with Q and appending new points, one by one, in this fashion. As the points are appended, the norms of the resulting partitions cannot increase.   □

5.3 LEMMA

> If $[a, b]$ is a bounded closed interval and $\delta > 0$, there exists a partition P of $[a, b]$ such that $\|P\| < \delta$ and all of the subintervals of $[a, b]$ corresponding to P have the same length $\Delta x < \delta$.

PROOF   Choose a positive integer n such that

$$\frac{b - a}{\delta} < n$$

and let

$$\Delta x = \frac{b - a}{n}.$$

Define

$$x_i = a + i \cdot \Delta x$$

for $i = 0, 1, 2, \ldots, n$ and let

$$P = \{x_0, x_1, x_2, \ldots, x_n\}.$$

Then each subinterval $[x_{i-1}, x_i]$ has length $\Delta x < \delta$, and it follows that $\|P\| < \delta$.   □

A partition P for which all of the corresponding subintervals have the same length Δx, as in Lemma 5.3, is called a **regular** partition.

5.4 DEFINITION

> **SAMPLE FOR A PARTITION**   Let $P = \{x_0, x_1, x_2, \ldots, x_n\}$ be a partition of $[a, b]$. By a **sample** for P, we mean a function $x^*: \{1, 2, \ldots, n\} \to [a, b]$ such that $x^*(i) \in [x_{i-1}, x_i]$ for each $i = 1, 2, \ldots, n$.

Note that a sample x^* for the partition P has the effect of choosing a point from each subinterval corresponding to P. We usually write the point chosen from the ith subinterval in the alternative subscript form $x_i^* = x^*(i)$. Thus,

$$x_{i-1} \leq x_i^* \leq x_i$$

holds for $i = 1, 2, \ldots, n$.

5.5 DEFINITION

> **RIEMANN SUM**   Let $f:[a, b] \to \mathbb{R}$, let $P = \{x_0, x_1, x_2, \ldots, x_n\}$ be a partition of $[a, b]$, and let x^* be a sample for P. For $i = 1, 2, \ldots, n$, let $\Delta x_i = x_i - x_{i-1}$. We define the **Riemann sum** for f corresponding to P and x^* by
>
> $$S(f, P, x^*) = \sum_{i=1}^{n} f(x_i^*)\Delta x_i.$$

Because it is difficult to write an effective computer program for calculating general infima and suprema, machine calculation of upper and lower Darboux sums can be impractical. By contrast, it is relatively easy to write a computer program for calculating Riemann sums. An obvious relationship between Darboux and Riemann sums is brought forth by the following lemma, the proof of which we leave as an exercise (Problem 3).

5.6 LEMMA

> Let $f:[a, b] \to \mathbb{R}$ be bounded, let P be a partition of $[a, b]$, and let x^* be a sample for P. Then
>
> $$L(f, P) \leq S(f, P, x^*) \leq U(f, P).$$

5.7 DEFINITION

> **LIMIT OF RIEMANN SUMS**   Let $f:[a, b] \to \mathbb{R}$. We say that the real number λ is a **limit** of the Riemann sums $S(f, P, x^*)$ of f as $\|P\| \to 0$ if, for every $\varepsilon > 0$, there exists $\delta > 0$ such that, for every partition P of $[a, b]$ and every sample x^* for P,
>
> $$\|P\| < \delta \Rightarrow |S(f, P, x^*) - \lambda| < \varepsilon.$$

Although Definition 5.7 is somewhat analogous to the definition of a limit of a function, it differs from Definition 2.1 in Chapter 4 in several obvious ways. (For instance, S is not a function of a real variable.) However, the analogy is close enough so that some of the arguments made for limits of functions can be adapted to limits of Riemann sums. Thus, the proof of the following theorem is quite similar to the proof of Theorem 2.5 in Chapter 4 and can safely be left as an exercise (Problem 5).

5.8 THEOREM

> **UNIQUENESS OF LIMITS OF RIEMANN SUMS**   Let $f:[a, b] \to \mathbb{R}$ and suppose that λ and λ' are limits of the Riemann sums $S(f, P, x^*)$ of f as $\|P\| \to 0$. Then $\lambda = \lambda'$.

Having secured the uniqueness of limits of Riemann sums, we are entitled to establish suitable notation for these limits.

5.9 DEFINITION

> **NOTATION FOR LIMITS OF RIEMANN SUMS**   If $f:[a, b] \to \mathbb{R}$ and if λ is the limit of the Riemann sums $S(f, P, x^*)$ of f as $\|P\| \to 0$, then we write
>
> $$\lambda = \lim_{\|P\| \to 0} S(f, P, x^*).$$

With the understanding that $P = \{x_0, x_1, x_2, \ldots, x_n\}$, $\Delta x_i = x_i - x_{i-1}$, and $x_{i-1} \le x_i^* \le x_i$ for $i = 1, 2, \ldots, n$, the limit in Definition 5.9 is often written in the alternative form

$$\lambda = \lim_{\|P\| \to 0} \sum_{i=1}^{n} f(x_i^*)\Delta x_i.$$

5.10 DEFINITION

> **THE RIEMANN INTEGRAL**   If $f:[a, b] \to \mathbb{R}$ and if the limit of the Riemann sums $S(f, P, x^*)$ exists, then we say that f is **Riemann integrable** and we define the **Riemann integral** of f by
>
> $$\int_a^b f(x)\, dx = \lim_{\|P\| \to 0} S(f, P, x^*).$$

The notation used for the Riemann integral serves as a reminder that

$$\int_a^b f(x)\,dx = \lim_{\|P\| \to 0} \sum_{i=1}^n f(x_i^*)\Delta x_i.$$

Roughly speaking, in the limit as $\|P\| \to 0$, $\sum$ becomes $\int$, $f(x_i^*)$ becomes $f(x)$, and Δx_i becomes dx.

5.11 LEMMA | If $f:[a, b] \to \mathbb{R}$ is Riemann integrable, then it is bounded.

PROOF   We prove here that the values of f are bounded above. A similar argument (Problem 6) will show that they are bounded below. Thus, let

$$\lambda = \lim_{\|P\| \to 0} S(f, P, x^*).$$

Then there exists $\delta > 0$ such that, for every partition P with $\|P\| < \delta$,

$$|S(f, P, x^*) - \lambda| < 1.$$

Choose and fix a regular partition $P = \{x_0, x_1, x_2, \ldots, x_n\}$, as in Lemma 5.3, for which all of the corresponding subintervals have the same length $\Delta x < \delta$. Then $\|P\| < \delta$, and therefore, if x^* is any sample for P,

$$\lambda - 1 < S(f, P, x^*) < \lambda + 1.$$

If r^* is any other sample for P, we also have

$$\lambda - 1 < S(f, P, r^*) < \lambda + 1.$$

In particular,

$$S(f, P, x^*) < \lambda + 1 \qquad \text{and} \qquad -S(f, P, r^*) < 1 - \lambda.$$

Adding the last two inequalities, we find that

$$S(f, P, x^*) - S(f, P, r^*) < 2.$$

Therefore, if x^* and r^* are any two samples for the partition P, we have

$$S(f, P, x^*) < 2 + S(f, P, r^*);$$

that is,

$$\sum_{i=1}^n f(x_i^*)\Delta x < 2 + \sum_{i=1}^n f(r_i^*)\Delta x.$$

Let r^* be the sample for P that selects the right endpoint of each subinterval, so that, for $i = 1, 2, \ldots, n$,

$$r_i^* = x_i.$$

Thus, if x^* is any sample for P, we have

$$\sum_{i=1}^{n} f(x_i^*)\Delta x < 2 + \sum_{i=1}^{n} f(x_i)\Delta x.$$

Now suppose that c is an arbitrary point in $[a, b]$ and let

$$K = \max\{f(x_i)|i = 1, 2, \ldots, n\}.$$

Since $c \in [a, b]$, there exists $k \in \{1, 2, \ldots, n\}$ such that $c \in [x_{k-1}, x_k]$. Define $x^*:\{1, 2, \ldots, n\} \to \mathbb{R}$ by

$$x_i^* = \begin{cases} x_i, & \text{if } i \neq k \\ c, & \text{if } i = k \end{cases}$$

for $i = 1, 2, \ldots, n$. Then all terms in the sums on both sides of the inequality

$$\sum_{i=1}^{n} f(x_i^*)\Delta x < 2 + \sum_{i=1}^{n} f(x_i)\Delta x$$

will cancel, except for the terms corresponding to $i = k$, and we have

$$f(c)\Delta x < 2 + f(x_k)\Delta x.$$

Therefore,

$$f(c) < \frac{2}{\Delta x} + f(x_k) \leq \frac{2}{\Delta x} + K.$$

Since c was an arbitrary point in $[a, b]$, it follows that

$$f(x) < \frac{2}{\Delta x} + K$$

for all $x \in [a, b]$, that is, the values of the function f on $[a, b]$ are bounded above.   □

5.12 LEMMA

> Let $f:[a, b] \to \mathbb{R}$ be Riemann integrable. Then f is Darboux integrable and
>
> $$\int_a^b f(x)\, dx = \int f.$$

PROOF   Suppose that f is Riemann integrable and let

$$\lambda = \int_a^b f(x)\, dx = \lim_{||P|| \to 0} S(f, P, x^*).$$

By Lemma 5.11, f is bounded. Let $\varepsilon > 0$. Then there exists $\delta > 0$ such that, for every partition P with $\|P\| < \delta$ and for every sample x^* for P,

$$\lambda - \frac{\varepsilon}{2} < S(f, P, x^*) < \lambda + \frac{\varepsilon}{2}.$$

Choose and fix a partition

$$P = \{x_0, x_1, x_2, \ldots, x_n\}$$

of $[a, b]$ with $\|P\| < \delta$. For $i = 1, 2, \ldots, n$, let

$$m_i = \inf_{x_{i-1} \leq x \leq x_i} f(x) \qquad \text{and} \qquad M_i = \sup_{x_{i-1} \leq x \leq x_i} f(x).$$

By Lemma 3.23 in Chapter 3, for each $i = 1, 2, \ldots, n$, there exist points $p_i, q_i \in [x_{i-1}, x_i]$ such that

$$f(p_i) < m_i + \frac{\varepsilon}{2(b-a)} \qquad \text{and} \qquad M_i - \frac{\varepsilon}{2(b-a)} < f(q_i).$$

Define samples p^* and q^* for the partition P by

$$p_i^* = p_i \qquad \text{and} \qquad q_i^* = q_i$$

for $i = 1, 2, \ldots, n$. Then we have

$$S(f, P, p^*) = \sum_{i=1}^n f(p_i)\Delta x_i < \sum_{i=1}^n \left[m_i + \frac{\varepsilon}{2(b-a)} \right]\Delta x_i$$

$$= \sum_{i=1}^n m_i\Delta x_i + \frac{\varepsilon}{2(b-a)} \sum_{i=1}^n \Delta x_i$$

$$= L(f, P) + \frac{\varepsilon}{2}.$$

Likewise,

$$S(f, P, q^*) = \sum_{i=1}^n f(q_i)\Delta x_i > \sum_{i=1}^n \left[M_i - \frac{\varepsilon}{2(b-a)} \right]\Delta x_i$$

$$= \sum_{i=1}^n M_i\Delta x_i - \frac{\varepsilon}{2(b-a)} \sum_{i=1}^n \Delta x_i$$

$$= U(f, P) - \frac{\varepsilon}{2}.$$

Therefore,

$$\underline{\int} f \geq L(f, P) > S(f, P, p^*) - \frac{\varepsilon}{2} > \left(\lambda - \frac{\varepsilon}{2} \right) - \frac{\varepsilon}{2} = \lambda - \varepsilon$$

and

$$\overline{\int} f \leq U(f, P) < S(f, P, q^*) + \frac{\varepsilon}{2} < \left(\lambda + \frac{\varepsilon}{2} \right) + \frac{\varepsilon}{2} = \lambda + \varepsilon.$$

Because ε is arbitrary, it follows that

$$\underline{\int} f \geq \lambda \qquad \text{and} \qquad \overline{\int} f \leq \lambda.$$

Consequently,

$$\lambda \leq \underline{\int} f \leq \overline{\int} f \leq \lambda,$$

and therefore f is Darboux integrable with

$$\int_a^b f(x)\, dx = \lambda = \underline{\int} f = \overline{\int} f = \int f. \qquad \square$$

5.13 LEMMA

Let $f : [a, b] \to \mathbb{R}$ and suppose that $|f(x)| \leq B$ for all $x \in [a, b]$. Let P and Q be partitions of $[a, b]$ and suppose that $\#Q = m + 1$. Then

$$U(f, P) \leq U(f, Q) + 2B(m - 1)\|P\|.$$

PROOF   Let

$$P = \{x_0, x_1, x_2, \ldots, x_n\} \qquad \text{and} \qquad Q = \{y_0, y_1, y_2, \ldots, y_m\}.$$

For each $j = 1, 2, \ldots, m$, define

$$C_j = \{i \mid i = 1, 2, \ldots, n \text{ and } [x_{i-1}, x_i] \subseteq [y_{j-1}, y_j]\}.$$

For simplicity, we refer to a subinterval corresponding to P (or Q) as a P-subinterval (or a Q-subinterval). Thus, C_j is the set of all i such that the ith P-subinterval is contained in the jth Q-subinterval. Note that the sets $C_1, C_2, \ldots, C_m$ are pairwise disjoint (Problem 7). Define

$$C = C_1 \cup C_2 \cup \cdots \cup C_m.$$

Thus, C is the set of all i such that the ith P-subinterval is contained in some Q-subinterval.

For each $j = 1, 2, \ldots, m$, define

$$\mu_j = \sum_{i \in C_j} \Delta x_i$$

so that μ_j is the sum of the lengths of all of the P-subintervals that are contained in the jth Q-subinterval. Consequently, $\mu_j \leq \Delta y_j$, and so

$$\Delta y_j - \mu_j \geq 0.$$

Let

$$\mu = \sum_{j=1}^m \mu_j,$$

noting that μ is the sum of the lengths of all the P-subintervals that are contained in some Q-subinterval. Define

$$C' = \{1, 2, \ldots, n\} \setminus C,$$

so that $i \in C'$ if and only if the ith P-subinterval is contained in no Q-subinterval. Consequently,

$$i \in C' \Rightarrow \exists j = 1, 2, \ldots, m - 1, \qquad y_j \in (x_{i-1}, x_i),$$

(Problem 9), and it follows that

$$\#C' \le m - 1$$

(Problem 11). Therefore, since $(b - a) - \mu$ is the sum of the lengths of the P-subintervals corresponding to indices $i \in C'$, and none of these sub-intervals have length greater than $\|P\|$, we have

$$(b - a) - \mu \le (m - 1)\|P\|.$$

For $i = 1, 2, \ldots, n$ and $j = 1, 2, \ldots, m$, define

$$M_i = \sup_{x_{i-1} \le x \le x_i} f(x) \quad \text{and} \quad \hat{M}_j = \sup_{y_{j-1} \le y \le y_j} f(y).$$

Note that, for $i = 1, 2, \ldots, n$,

$$M_i = \sup_{x_{i-1} \le x \le x_i} f(x) \le \sup_{x_{i-1} \le x \le x_i} |f(x)| \le B,$$

and that, for $j = 1, 2, \ldots, m$,

$$-\hat{M}_j = \inf_{y_{j-1} \le y \le y_j} (-f(y)) \le \inf_{y_{j-1} \le y \le y_j} |f(y)| \le B.$$

Also, if $i \in C_j$, then $[x_{i-1}, x_i] \subseteq [y_{j-1}, y_j]$, and it follows that

$$i \in C_j \Rightarrow M_i \le \hat{M}_j.$$

Therefore,

$$\sum_{i \in C} M_i \Delta x_i = \sum_{j=1}^{m} \left(\sum_{i \in C_j} M_i \Delta x_i \right) \le \sum_{j=1}^{m} \left(\sum_{i \in C_j} \hat{M}_j \Delta x_i \right)$$

$$= \sum_{j=1}^{m} \hat{M}_j \left(\sum_{i \in C_j} \Delta x_i \right) = \sum_{j=1}^{m} \hat{M}_j \mu_j$$

$$= U(f, Q) - U(f, Q) + \sum_{j=1}^{m} \hat{M}_j \mu_j$$

$$= U(f, Q) - \sum_{j=1}^{m} \hat{M}_j \Delta y_j + \sum_{j=1}^{m} \hat{M}_j \mu_j$$

$$= U(f, Q) + \sum_{j=1}^{m} [-\hat{M}_j(\Delta y_j - \mu_j)].$$

Therefore, because $-\hat{M}_j \le B$ and $\Delta y_j - \mu_j \ge 0$,

$$\sum_{i \in C} M_i \Delta x_i \le U(f, Q) + B \sum_{j=1}^{m} (\Delta y_j - \mu_j)$$
$$= U(f, Q) + B\left(\sum_{j=1}^{m} \Delta y_j - \sum_{j=1}^{m} \mu_j \right)$$
$$= U(f, Q) + B[(b - a) - \mu]$$
$$\le U(f, Q) + B(m - 1)\|P\|.$$

Furthermore, because $\#C' \le m - 1$ and because, for each $i = 1, 2, \ldots, n$, we have $\Delta x_i \le \|P\|$, it follows that

$$\sum_{i \in C'} M_i \Delta x_i \le B \sum_{i \in C'} \Delta x_i \le B(m - 1)\|P\|.$$

Therefore,

$$U(f, P) = \sum_{i \in C} M_i \Delta x_i + \sum_{i \in C'} M_i \Delta x_i$$
$$\le U(f, Q) + B(m - 1)\|P\| + B(m - 1)\|P\|$$
$$= U(f, Q) + 2B(m - 1)\|P\|. \qquad \square$$

5.14 COROLLARY

> Let $f : [a, b] \to \mathbb{R}$ and suppose that $|f(x)| \le B$ for all $x \in [a, b]$. Let P and Q be partitions of $[a, b]$, let $\#Q = m + 1$, and let $k = 2B(m - 1)\|P\|$. Then
>
> $$L(f, Q) - k \le L(f, P) \le U(f, P) \le U(f, Q) + k.$$

PROOF   By Lemma 5.13,

$$U(f, P) \le U(f, Q) + k.$$

By applying Lemma 5.13 to $-f$ and using the result of Part (c) of Problem 9 in Problem Set 5.4, we find that

$$-L(f, P) \le -L(f, Q) + k,$$

so that

$$L(f, Q) - k \le L(f, P). \qquad \square$$

5.15 THEOREM

> **EQUIVALENCE OF RIEMANN AND DARBOUX INTEGRALS**
> Let $f : [a, b] \to \mathbb{R}$. Then f is Riemann integrable if and only if it is Darboux integrable. Furthermore, if f is either Riemann or Darboux integrable, then
>
> $$\int_a^b f(x)\, dx = \int f.$$

PROOF   By Lemma 5.12, if f is Riemann integrable, then it is Darboux integrable and the two integrals coincide. Conversely, suppose that f is Darboux integrable. We have to show that f is Riemann integrable. Thus, let $\varepsilon > 0$. Since

$$\int f = \underline{\int} f = \sup_P L(f, P) \quad \text{and} \quad \int f = \overline{\int} f = \inf_P U(f, P),$$

it follows from Lemma 3.23 in Chapter 3 that there exist partitions Q_1 and Q_2 of $[a, b]$ such that

$$\int f - \frac{\varepsilon}{2} < L(f, Q_1) \quad \text{and} \quad U(f, Q_2) < \int f + \frac{\varepsilon}{2}.$$

Let $Q = Q_1 \cup Q_2$, noting that Q is a refinement of both Q_1 and Q_2, so that, by Corollary 4.5,

$$\int f - \frac{\varepsilon}{2} < L(f, Q) \quad \text{and} \quad U(f, Q) < \int f + \frac{\varepsilon}{2}.$$

Since f is Darboux integrable, it is bounded. Let

$$B = \sup_{a \le x \le b} |f(x)|$$

and let $\#Q = m + 1$. If $B = 0$, let $\delta = 1$, and if $B \ne 0$, let

$$\delta = \frac{\varepsilon}{4B(m - 1)}.$$

Now suppose that P is any partition of $[a, b]$ such that

$$\|P\| < \delta.$$

Then

$$2B(m - 1)\|P\| < \frac{\varepsilon}{2},$$

and it follows from Corollary 5.14 that

$$L(f, Q) < L(f, P) + \frac{\varepsilon}{2} \quad \text{and} \quad U(f, P) - \frac{\varepsilon}{2} < U(f, Q).$$

Therefore,

$$\int f - \frac{\varepsilon}{2} < L(f, P) + \frac{\varepsilon}{2} \quad \text{and} \quad U(f, P) - \frac{\varepsilon}{2} < \int f + \frac{\varepsilon}{2},$$

so that

$$\int f - \varepsilon < L(f, P) \quad \text{and} \quad U(f, P) < \int f + \varepsilon.$$

Therefore, by Lemma 5.6, if x^* is any sample for P, we have

$$\int f - \varepsilon < S(f, P, x^*) < \int f + \varepsilon,$$

that is,

$$\left| S(f, P, x^*) - \int f \right| < \varepsilon.$$

Consequently, f is Riemann integrable and

$$\int_a^b f(x)\, dx = \int f. \qquad \qquad \square$$

5.16 LEMMA

> Let $f, g : [a, b] \to \mathbb{R}$, let $c \in [a, b]$ and suppose that $f(x) = g(x)$ for all $x \in [a, b]$ except possibly for $x = c$. Then, if f is Riemann integrable, so is g, and
>
> $$\int_a^b f(x)\, dx = \int_a^b g(x)\, dx.$$

PROOF   If $f(c) = g(c)$, there is nothing to prove, so we assume that $f(c) \neq g(c)$. Let $\varepsilon > 0$, and choose δ_1 such that, for every partition P of $[a, b]$ and every sample x^* for P,

$$\|P\| < \delta_1 \Rightarrow \left| S(f, P, x^*) - \int_a^b f(x)\, dx \right| < \frac{\varepsilon}{2}. \tag{1}$$

Let

$$\delta = \min \left\{ \delta_1, \frac{\varepsilon}{4|f(c) - g(c)|} \right\}, \tag{2}$$

suppose that $P = \{x_0, x_1, x_2, \ldots, x_n\}$ is a partition of $[a, b]$, that x^* is a sample for P, and that

$$\|P\| < \delta. \tag{3}$$

Then $\delta \leq \delta_1$, so, in view of (1), we have

$$\left| S(f, P, x^*) - \int_a^b f(x)\, dx \right| < \frac{\varepsilon}{2}. \tag{4}$$

Also, for each $i = 1, 2, \ldots, n$, we have $\Delta x_i \leq \|P\| < \delta$, and so

$$|f(c)\Delta x_i - g(c)\Delta x_i| = |f(c) - g(c)|\Delta x_i < \frac{\varepsilon}{4}. \tag{5}$$

Note that all terms in the difference $S(f, P, x^*) - S(g, P, x^*)$ will cancel in pairs, with the exception of at most two, each of which has the form

$f(c)\Delta x_i - g(c)\Delta x_i$ (Problem 13), and it follows from (5) that

$$|S(f, P, x^*) - S(g, P, x^*)| < \frac{\varepsilon}{4} + \frac{\varepsilon}{4} = \frac{\varepsilon}{2}. \tag{6}$$

By combining (4) and (6), we conclude that

$$\left| S(g, P, x^*) - \int_a^b f(x)\,dx \right| < \varepsilon. \tag{7}$$

Since P was an arbitrary partition with $\|P\| < \delta$, we conclude that g is Riemann integrable and that

$$\int_a^b g(x)\,dx = \int_a^b f(x)\,dx. \tag{8}$$

□

5.17 THEOREM

> **FUNCTIONS DIFFERING AT FINITELY MANY POINTS**
> Let $f, g:[a, b] \to \mathbb{R}$ and suppose that $f(x) = g(x)$ for all but at most finitely many values of $x \in [a, b]$. Then, if f is Riemann integrable, so is g, and
> $$\int_a^b f(x)\,dx = \int_a^b g(x)\,dx.$$

PROOF   Use Lemma 5.16 and mathematical induction on the number of values of x for which $f(x) \neq g(x)$ (Problem 17).   □

5.18 DEFINITION

> **EXTENSIONS OF THE NOTION OF THE RIEMANN INTEGRAL**   Let f be a function, let $[a, b] \subseteq \mathrm{dom}(f)$, and let $f|_{[a, b]}$ denote the restriction of f to $[a, b]$.
>
> (i) If $f|_{[a, b]}$ is Riemann integrable, we say that f is **Riemann integrable on** $[a, b]$ and we define
> $$\int_a^b f(x)\,dx = \int_a^b (f|_{[a, b]})(x)\,dx.$$
>
> (ii) If f is Riemann integrable on $[a, b]$, we define
> $$\int_b^a f(x)\,dx = -\int_a^b f(x)\,dx.$$
>
> (iii) If $c \in \mathrm{dom}(f)$, we define
> $$\int_c^c f(x)\,dx = 0.$$

A proof of the following theorem can be based on Problems 21 and 29 in Problem Set 5.4, the fact that Darboux and Riemann integrals coincide, and Definition 5.18 (Problem 19).

5.19 THEOREM

INTERVAL ADDITIVITY OF THE RIEMANN INTEGRAL Let f be a function, suppose that f is Riemann integrable on a closed bounded interval $I \subseteq \mathrm{dom}(f)$, and let a, b, $c \in I$. Then

$$\int_a^c f(x)\,dx = \int_a^b f(x)\,dx + \int_b^c f(x)\,dx.$$

PROBLEM SET 5.5

1. Supply the details of the proof of Lemma 5.2. [Hint: Use mathematical induction on $k = \#P - \#Q$.]

2. In the proof of Lemma 5.3, how do we know that there is a positive integer n such that $(b - a)/\delta < n$?

3. Prove Lemma 5.6.

4. Suppose that $f:[a, b] \to \mathbb{R}$, $\lambda \in \mathbb{R}$, and, for every $n \in \mathbb{N}$, there exists a $\delta_n > 0$ such that, for every partition P of $[a, b]$ and every sample x^* for P, $\|P\| < \delta_n$ implies that $|S(f, P, x^*) - \lambda| < \varepsilon$. Prove that f is Riemann integrable and that $\int_a^b f(x)\,dx = \lambda$.

5. Prove Theorem 5.8.

6. Complete the proof of Lemma 5.11 by showing that the values of f are bounded below.

7. In the proof of Lemma 5.13, show that the sets C_j are pairwise disjoint for $j = 1, 2, \ldots, m$. [Hint: Two different Q-subintervals can have at most one point in common, and a P-subinterval contains infinitely many points.]

8. In the proof of Lemma 5.13, exactly where is use made of the fact that the sets C_j are pairwise disjoint for $j = 1, 2, \ldots, m$?

9. In the proof of Lemma 5.13, show that, if $i \in C'$, then there must exist an integer j such that $0 < j < m$ and $y_j \in (x_{i-1}, x_i)$. [Hint: You might try proving the contrapositive.]

10. Let $f:[a, b] \to \mathbb{R}$ be a bounded function and let $\varepsilon > 0$. Prove that there exists $\delta > 0$ such that, for every partition P of $[a, b]$ with $\|P\| < \delta$, $\int f - \varepsilon < L(f, P) \le U(f, P) < \overline{\int} f + \varepsilon$. [Hints: Choose partitions Q_1 and Q_2 of $[a, b]$ such that $\int f - (\varepsilon/2) < L(f, Q_1)$ and $U(f, Q_2) < \overline{\int} f + (\varepsilon/2)$. Let $Q = Q_1 \cup Q_2$, and apply Corollary 5.14 much as in the proof of Theorem 5.15.]

11. In the proof of Lemma 5.13, explain why $\# C' \leq m - 1$.

12. True or false: For every partition P, there is a regular partition Q (one for which all of the subintervals have the same length) that is a refinement of P. If true, prove it; if false, give a counterexample.

13. In the proof of Lemma 5.16, provide the details in the argument that inequality (6) holds.

14. Let $f:[a, b] \to \mathbb{R}$ be Darboux integrable and let $\varepsilon > 0$. Prove that there is a partition Q of $[a, b]$ such that, for every partition P and every sample x^* for P, if P is a refinement of Q, then $|S(f, P, x^*) - \int f| < \varepsilon$.

15. Let $f:[a, b] \to \mathbb{R}$ be bounded, let $P = \{x_0, x_1, x_2, \ldots, x_n\}$ be a partition of $[a, b]$, and, for each $i = 1, 2, \ldots, n$, let

$$m_i = \inf_{x_{i-1} \leq x \leq x_i} f(x) \quad \text{and} \quad M_i = \sup_{x_{i-1} \leq x \leq x_i} f(x).$$

Suppose that $\varepsilon > 0$.
(a) Show that there exist samples p^* and q^* for P such that, for each $i = 1, 2, \ldots, n$,

$$f(p_i^*) < m_i + \frac{\varepsilon}{2(b - a)} \quad \text{and} \quad M_i - \frac{\varepsilon}{2(b - a)} < f(q_i^*).$$

(b) Show that $U(f, P) - L(f, P) < S(f, P, q^*) - S(f, P, p^*) + \varepsilon$.

16. Let $f:[a, b] \to \mathbb{R}$ be bounded. Suppose that $\lambda \in \mathbb{R}$ has the property that, for every $\varepsilon > 0$, there exists a partition Q of $[a, b]$ such that, for every partition P and every sample x^* for P, if P is a refinement of Q, then $|S(f, P, x^*) - \lambda| < \varepsilon$. Prove that f is Darboux integrable and that $\int f = \lambda$. [Hint: Use the results of Problem 15.]

17. Supply the details for the proof of Theorem 5.17.

18. If the function f is Riemann integrable on a closed bounded interval $I \subseteq \text{dom}(f)$, and if $a, b \in I$, prove that $\int_a^b f(x)\,dx = -\int_b^a f(x)\,dx$. [Note that there is no requirement here that $a < b$ as there is in Part (ii) of Definition 5.18.]

19. Prove Theorem 5.19. [Hint: There are several possible cases according to whether some of the points a, b, c coincide and according to the order in which a, b, c occur on the interval I.]

20. Suppose that f is Riemann integrable on $[a, b]$. Given $\varepsilon > 0$, prove that there exists a positive integer n such that

$$\left| \frac{1}{n} \sum_{i=1}^{n} f\left(a + \frac{i(b - a)}{n}\right) - \int_a^b f(x)\,dx \right| < \varepsilon.$$

21. Suppose that $f:[a, b] \to \mathbb{R}$ is bounded and that f is continuous at each point in $[a, b]$ except for a possible discontinuity at a single point $d \in [a, b]$. Prove that f is Riemann integrable.

22. Let sinr be the function defined in Problem 10 of Problem Set 4.3. Prove that sinr is integrable over every bounded closed interval $[a, b]$.

23. Suppose that $f:[a, b] \to \mathbb{R}$ is bounded and that D is the set of points in $[a, b]$ at which f is discontinuous. If D is a finite set, prove that f is Riemann integrable. [Hint: Use the result of Problem 21 and mathematical induction.]

24. Suppose that $f:[a, b] \to \mathbb{R}$ is bounded and that, for each ε with $0 < \varepsilon < b - a$, f is Riemann integrable on $[a + \varepsilon, b]$. Prove that f is Riemann integrable and that

$$\int_a^b f(x)\,dx = \lim_{\varepsilon \to 0} \int_{a+\varepsilon}^b f(x)\,dx.$$

25. If $f:[a, b] \to \mathbb{R}$ is bounded and piecewise continuous (see page 309), prove that f is Riemann integrable.

26. Let $f:[a, b] \to \mathbb{R}$ be Riemann integrable and for each positive integer n, let $\Delta x = (b - a)/n$ and define

$$T_n = \frac{f(a) + f(b)}{2} + \sum_{i=1}^{n-1} f(a + i \cdot \Delta x)\Delta x.$$

Prove that, for every $\varepsilon > 0$, there exists a positive integer n such that $\left| T_n - \int_a^b f(x)\,dx \right| < \varepsilon$. (This is the **trapezoidal rule** of elementary calculus.)

6

PROPERTIES OF THE INTEGRAL

By Theorem 5.15, the Darboux and Riemann integrals coincide. Thus, for convenience in what follows, we refer to a Darboux, or Riemann, integral simply as an *integral*. Also, if we say that a function f is *integrable* on a bounded closed interval $[a, b]$, we mean that it is Riemann integrable on $[a, b]$ in the sense of Part (i) of Definition 5.18. In establishing the properties of the integral, we can use either the Darboux definition (Definition 4.10) or the Riemann definition (Definition 5.10), whichever is more advantageous.

Several important properties of the integral have been proved already in Sections 4 and 5. Our purpose in the present section is to develop some additional properties.

6.1 THEOREM

ADDITIVITY OF THE INTEGRAL   If f and g are integrable functions on $[a, b]$, then $f + g$ is integrable on $[a, b]$ and

$$\int_a^b (f(x) + g(x))\,dx = \int_a^b f(x)\,dx + \int_a^b g(x)\,dx.$$

PROOF Let ε be a positive number. By hypothesis and Definition 5.10, there exist positive numbers δ_1 and δ_2 such that, for every partition P of $[a, b]$ and every sample x^* for P,

$$\|P\| < \delta_1 \Rightarrow \left| S(f, P, x^*) - \int_a^b f(x)\,dx \right| < \frac{\varepsilon}{2}$$

and

$$\|P\| < \delta_2 \Rightarrow \left| S(g, P, x^*) - \int_a^b g(x)\,dx \right| < \frac{\varepsilon}{2}.$$

If $P = \{x_0, x_1, x_2, \ldots, x_n\}$ is a partition of $[a, b]$ and x^* is a sample for P, then

$$S(f + g, P, x^*) = \sum_{i=1}^n (f + g)(x_i^*)\Delta x_i = \sum_{i=1}^n [f(x_i^*) + g(x_i^*)]\Delta x_i$$

$$= \sum_{i=1}^n f(x_i^*)\Delta x_i + \sum_{i=1}^n g(x_i^*)\Delta x_i$$

$$= S(f, P, x^*) + S(g, P, x^*).$$

Now let $\delta = \min(\delta_1, \delta_2)$. Suppose that P is a partition of $[a, b]$, that x^* is a sample for P, and that $\|P\| < \delta$. Then $\|P\| < \delta_1$, $\|P\| < \delta_2$, and, by the triangle inequality, we have

$$\left| S(f + g, P, x^*) - \left(\int_a^b f(x)\,dx + \int_a^b g(x)\,dx \right) \right|$$

$$= \left| S(f, P, x^*) - \int_a^b f(x)\,dx + S(g, P, x^*) - \int_a^b g(x)\,dx \right|$$

$$\leq \left| S(f, P, x^*) - \int_a^b f(x)\,dx \right| + \left| S(g, P, x^*) - \int_a^b g(x)\,dx \right|$$

$$< \frac{\varepsilon}{2} + \frac{\varepsilon}{2} = \varepsilon,$$

and it follows from Definition 5.10 that $f + g$ is integrable with

$$\int_a^b (f + g)(x)\,dx = \int_a^b f(x)\,dx + \int_a^b g(x)\,dx.$$

Since $(f + g)(x) = f(x) + g(x)$ holds for all $x \in [a, b]$, we may rewrite the last equation as

$$\int_a^b (f(x) + g(x))\,dx = \int_a^b f(x)\,dx + \int_a^b g(x)\,dx. \qquad \Box$$

The following theorem is a consequence of the results of Problems 11 and 13 in Problem Set 5.4. We leave the proof as an exercise (Problem 1).

6.2 THEOREM

HOMOGENEITY OF THE INTEGRAL   Let f be an integrable function on the bounded closed interval $[a, b]$ and let k be a constant. Then kf is an integrable function on $[a, b]$ and

$$\int_a^b kf(x)\, dx = k \int_a^b f(x)\, dx.$$

By combining Theorems 6.1 and 6.2, we may conclude that the integral *preserves linear combinations* in the following sense: If f, g are integrable on $[a, b]$ and if A and B are constants, then the linear combination $Af + Bg$ is integrable on $[a, b]$ and

$$\int_a^b (Af(x) + Bg(x))\, dx = A \int_a^b f(x)\, dx + B \int_a^b f(x)\, dx.$$

This is called the **linearity property** of the integral. In particular, by putting $A = 1$ and $B = -1$, we find that the integral *preserves subtraction* in the sense that

$$\int_a^b [f(x) - g(x)]\, dx = \int_a^b f(x)\, dx - \int_a^b g(x)\, dx.$$

6.3 THEOREM

MONOTONICITY OF THE INTEGRAL   Suppose that f and g are integrable functions on the bounded closed interval $[a, b]$ and that $g(x) \le f(x)$ holds for all $x \in [a, b]$. Then

$$\int_a^b g(x)\, dx \le \int_a^b f(x)\, dx.$$

PROOF   By Problem 23 in Problem Set 5.4,

$$0 \le \int_a^b [f(x) - g(x)]\, dx = \int_a^b f(x)\, dx - \int_a^b g(x)\, dx. \qquad \square$$

6.4 COROLLARY

Suppose that f is an integrable function on the bounded closed interval $[a, b]$ and that m and M are constants such that, for every $x \in [a, b]$,

$$m \le f(x) \le M.$$

Then

$$m(b - a) \le \int_a^b f(x)\, dx \le M(b - a).$$

PROOF   Let $g, h:[a, b] \to \mathbb{R}$ be the constant functions defined by $g(x) = m$ and $h(x) = M$ for all $x \in [a, b]$. As a consequence of Example 4.11, g and h are integrable and we have

$$\int_a^b g(x)\, dx = m(b - a) \qquad \text{and} \qquad \int_a^b h(x)\, dx = M(b - a).$$

Also, for all $x \in [a, b]$,

$$g(x) \leq f(x) \leq h(x),$$

and it follows from Theorem 6.3 that

$$\int_a^b g(x)\, dx \leq \int_a^b f(x)\, dx \leq \int_a^b h(x)\, dx.$$

Therefore,

$$m(b - a) \leq \int_a^b f(x)\, dx \leq M(b - a). \qquad \square$$

6.5 THEOREM

> **ABSOLUTE-VALUE PROPERTY**   If f is an integrable function on the bounded closed interval $[a, b]$, then $|f|$ is integrable on $[a, b]$ and
>
> $$\left| \int_a^b f(x)\, dx \right| \leq \int_a^b |f(x)|\, dx.$$

PROOF   See Problem 30 in Problem Set 5.4. $\qquad \square$

Recall that the **arithmetic mean**, or **average**, of the real numbers $y_1, y_2, y_3, \dots, y_n$ is defined to be the real number

$$\bar{y} = \frac{1}{n} \sum_{i=1}^n y_i.$$

How shall we define the mean, or average, value of a function f on an interval $[a, b] \subseteq \text{dom}(f)$? One possibility is as follows: For each positive integer n, we let

$$\Delta x = \frac{b - a}{n}$$

and form the regular partition $P = \{x_0, x_1, x_2, \dots, x_n\}$ of $[a, b]$ given by

$$x_i = a + i \cdot \Delta x$$

for $i = 0, 1, 2, \dots, n$. Consider the sample x^* for this partition given by $x_i^* = x_i$, compute the corresponding values $y_i = f(x_i)$ of f for $i = 1, 2, \dots, n$, and form the average $\bar{y}$ of these values. If $\bar{y}$ tends to a limiting value A as n grows larger and larger (so that $\Delta x \to 0$), it seems reasonable to regard A as the *mean value* of f on $[a, b]$.

Proceeding heuristically, we now suppose that f is integrable on $[a, b]$. Then, with the notation in the paragraph above,

$$S(f, P, x^*) = \sum_{i=1}^{n} f(x_i^*)\Delta x = \sum_{i=1}^{n} y_i \left(\frac{b-a}{n}\right) = \left(\frac{b-a}{n}\right) \sum_{i=1}^{n} y_i.$$

Therefore,

$$S(f, P, x^*) = (b - a)\bar{y}.$$

By taking the limit on both sides of the last equation as n grows larger and larger (so that $\Delta x \to 0$), we obtain

$$\int_a^b f(x)\, dx = (b - a)A,$$

so that

$$A = \frac{1}{b-a} \int_a^b f(x)\, dx.$$

These considerations motivate the following definition.

6.6 DEFINITION

> **MEAN VALUE OF A FUNCTION ON AN INTERVAL**   If the function f is integrable on the bounded interval $[a, b]$, then the **mean value of f on $[a, b]$** is defined to be
>
> $$\frac{1}{b-a} \int_a^b f(x)\, dx.$$

If f is integrable on $[a, b]$, then f takes on its own mean value at a point $c \in [a, b]$ if and only if

$$f(c) = \frac{1}{b-a} \int_a^b f(x)\, dx.$$

This equation may be rewritten in the form

$$\int_a^b f(x)\, dx = f(c)(b - a).$$

Thus, in words, the next theorem states that, *on a bounded closed interval a continuous function takes on its own mean value.*

6.7 THEOREM

> **MEAN VALUE THEOREM FOR INTEGRALS**   If the function f is continuous on the bounded closed interval $[a, b]$, then there is at least one point $c \in [a, b]$ such that
>
> $$\int_a^b f(x)\, dx = f(c)(b - a).$$

PROOF Assume that f is continuous on $[a, b]$. By Theorem 4.14, f is integrable on $[a, b]$. By the Heine–Borel and extreme value theorems (Theorems 5.7 and 5.14 in Chapter 4), there are points $p, q \in [a, b]$ such that, for all $x \in [a, b]$,

$$f(p) \le f(x) \le f(q).$$

Therefore, by Corollary 6.4, we have

$$f(p)(b - a) \le \int_a^b f(x)\, dx \le f(q)(b - a),$$

and so

$$f(p) \le \frac{1}{b - a} \int_a^b f(x)\, dx \le f(q).$$

Consequently, by the intermediate value theorem (Theorem 4.17 in Chapter 4), there exists a point c between p and q such that

$$f(c) = \frac{1}{b - a} \int_a^b f(x)\, dx,$$

that is,

$$\int_a^b f(x)\, dx = f(c)(b - a). \qquad \square$$

It is often useful to form a *weighted average* of numbers $y_1, y_2, y_3, \ldots, y_n$ in which some of the numbers are counted more heavily than others. For instance, in averaging grades, exam scores may count more heavily than quiz scores. Such a weighted average is formed by selecting **weighting factors** $g_1, g_2, g_3, \ldots, g_n$ such that $g_i \ge 0$ for all i and at least one g_i is nonzero. The corresponding **weighted average** A is then defined by

$$A = \frac{\sum_{i=1}^n g_i y_i}{\sum_{i=1}^n g_i}.$$

(For instance, if y_2 is to be counted three times as heavily as y_1, we would make sure that $g_3 = 3g_1$.) An analogous concept is available for functions.

6.8 DEFINITION

WEIGHTED MEAN VALUE OF A FUNCTION Let the function g be defined and integrable on the bounded closed interval $[a, b]$. Suppose that $g(x) \ge 0$ for all $x \in [a, b]$ and that $\int_a^b g(x)\, dx \ne 0$. If f is a function defined on $[a, b]$, and if fg is integrable on $[a, b]$, then we define the **g-weighted mean value of f on $[a, b]$** to be

$$\frac{\int_a^b g(x) f(x)\, dx}{\int_a^b g(x)\, dx}.$$

Note that, if g is the constant function defined by $g(x) = 1$ for all $x \in [a, b]$, then the g-weighted mean value of f on $[a, b]$ is the mean value of f on $[a, b]$ as in Definition 6.6. As a consequence of the following theorem, *if f is continuous on $[a, b]$ and if g satisfies the conditions in Definition 6.8, then f takes on its own g-weighted mean value at some point in $[a, b]$.* The proof, which is quite similar to the proof of Theorem 6.7, is left as an exercise (Problem 9).

6.9 THEOREM

GENERALIZED MEAN VALUE THEOREM FOR INTEGRALS
Let the function g be integrable on the bounded closed interval $[a, b]$ and suppose that $g(x) \geq 0$ for all $x \in [a, b]$. Then, if the function f is continuous on $[a, b]$, there exists $c \in [a, b]$ such that

$$\int_a^b g(x) f(x)\, dx = f(c) \int_a^b g(x)\, dx.$$

According to Theorem 3.10 in Chapter 4, the composition of continuous functions is again a continuous function. Likewise, by the chain rule (Theorem 1.12), the composition of differentiable functions is again a differentiable function. An obvious question now arises: Is the composition of integrable functions again an integrable function? In Example 6.11 on page 411, we show that the answer is no! To establish this negative result, we are going to use the so-called *Dirichlet function*, named after Peter Gustav Lejeune Dirichlet (1805—1859), a German mathematician who made a number of important discoveries in the fields of analysis, number theory, and mechanics.

6.10 Example   Let $g: [0, 1] \to [0, 1] \subseteq \mathbb{R}$ be the **Dirichlet function** defined by

$$g(x) = \left\{ \begin{array}{ll} 1/n, & \text{if } x = m/n, \text{ a rational number in reduced form}^\dagger \\ 0, & \text{if } x \text{ is not a rational number} \end{array} \right\}.$$

Prove that g is integrable on $[0, 1]$.

SOLUTION   Since $0 \leq g(x) \leq 1$ holds for every $x \in [0, 1]$, it follows that g is bounded. We show that g is integrable by applying Darboux's integrability condition (Theorem 4.13). Because every interval of real numbers contains an irrational number (Theorem 5.15 in Chapter 3), it follows that $L(g, P) = 0$ for every partition P of $[0, 1]$. Therefore, it will be sufficient to show that, for every $\varepsilon > 0$, there exists a partition P of $[0, 1]$ such that $U(g, P) < \varepsilon$.

---

† See Section 3.5, page 210.

Let $\varepsilon > 0$. If $\varepsilon > 1$, let $P = \{0, 1\}$ and we have $U(g, P) = 1 < \varepsilon$ (Problem 15a). Thus, we may assume that $0 < \varepsilon \leq 1$. Let

$$A = \left\{ x \in [0, 1] \,\middle|\, g(x) \geq \frac{\varepsilon}{2} \right\}.$$

In other words, A is the set of all rational numbers that can be written in reduced form m/n with

$$0 \leq m \leq n \leq \frac{2}{\varepsilon}.$$

Since there are only finitely many positive integers n that are less than or equal to $2/\varepsilon$, it follows that A is a finite set (Problem 15b). Let

$$N = \#A,$$

noting that, because $\varepsilon \leq 1$, we have $0 = \frac{0}{1} \in A$ and $1 = \frac{1}{1} \in A$, so $N \geq 2$.

Now choose a prime number p with

$$\frac{2N}{\varepsilon} < p$$

(Problem 17b). Since $N \geq 2 > 1$, we have

$$\frac{2}{\varepsilon} < \frac{2N}{\varepsilon} < p,$$

and therefore

$$\frac{1}{p} < \frac{\varepsilon}{2N} < \frac{\varepsilon}{2}.$$

Let $P = \{x_0, x_1, x_2, \ldots, x_p\}$ be the partition of $[0, 1]$ defined by

$$x_i = \frac{i}{p}$$

for $i = 0, 1, 2, \ldots, p$. Thus, P is a regular partition with

$$\Delta x_i = x_i - x_{i-1} = \frac{1}{p} < \frac{\varepsilon}{2N}$$

for $i = 1, 2, \ldots, p$. Also,

$$\sum_{i=1}^{p} \Delta x_i = 1.$$

Let

$$M_i = \sup_{x_{i-1} \leq x \leq x_i} g(x),$$

noting that $M_i \leq 1$ for $i = 1, 2, \ldots, p$.

Since p is a prime, it follows that $x_i = i/p$ is a rational number in reduced form unless $i = 0$ or $i = p$. Therefore, for $i = 1, 2, \ldots, p - 1$, we have

$$g(x_i) = \frac{1}{p} < \frac{\varepsilon}{2},$$

and it follows from the definition of A that

$$P \cap A = \{0, 1\}.$$

Now let

$$I = \{i \in \mathbb{N} \,|\, 1 \le i \le p, \, [x_{i-1}, x_i] \cap A = \varnothing\}$$

and let

$$J = \{j \in \mathbb{N} \,|\, 1 \le j \le p, \, [x_{j-1}, x_j] \cap A \ne \varnothing\}.$$

If $i \in I$ and $x \in [x_{i-1}, x_i]$, then $x \notin A$, so that

$$g(x) < \frac{\varepsilon}{2}.$$

Consequently, for $i \in I$, we have

$$M_i = \sup_{x_{i-1} \le x \le x_i} g(x) \le \frac{\varepsilon}{2},$$

and therefore

$$\sum_{i \in I} M_i \Delta x_i \le \sum_{i \in I} \frac{\varepsilon}{2} \Delta x_i = \frac{\varepsilon}{2} \sum_{i \in I} \Delta x_i \le \frac{\varepsilon}{2} \sum_{i=1}^{p} \Delta x_i = \frac{\varepsilon}{2}.$$

For each $j \in J$, choose a point $a_j \in [x_{j-1}, x_j] \cap A$. Define $\phi : J \to A$ by $\phi(j) = a_j$ for all $j \in J$. Because $P \cap A = \{0, 1\}$, it follows that $\phi : J \to A$ is an injection (Problem 19), and therefore

$$\#J \le \#A = N.$$

Hence, in view of the fact that $M_j \le 1$ for all $j \in J$, we have

$$\sum_{j \in J} M_j \Delta x_j \le \sum_{j \in J} \Delta x_j \le (\#J)\frac{1}{p} < N\frac{\varepsilon}{2N} = \frac{\varepsilon}{2}.$$

Consequently,

$$U(g, P) = \sum_{i=1}^{p} M_i \Delta x_i = \sum_{i \in I} M_i \Delta x_j + \sum_{j \in J} M_j \Delta x_j$$

$$< \frac{\varepsilon}{2} + \frac{\varepsilon}{2} = \varepsilon. \qquad \square$$

6.11 Example   Show that the composition of integrable functions need not be integrable.

SOLUTION   Let $g:[0, 1] \to [0, 1]$ be the Dirichlet function (Example 6.10) and let $h:[0, 1] \to \mathbb{R}$ be defined by

$$h(x) = \left\{\begin{array}{ll} 1, & \text{if } x \neq 0 \text{ and } 1/x \in \mathbb{N} \\ 0, & \text{otherwise} \end{array}\right\}$$

for all $x \in [0, 1]$. By Example 6.10, g is integrable on $[0, 1]$ and, by Problem 22 in Problem Set 5.4, h is integrable on $[0, 1]$. Let $f:[0, 1] \to \mathbb{R}$ be defined by

$$f = h \circ g.$$

Then

$$f(x) = \left\{\begin{array}{ll} 1, & \text{if } x \in \mathbb{Q} \\ 0, & \text{if } x \notin \mathbb{Q} \end{array}\right\}$$

for all $x \in [0, 1]$. By Example 4.12, f is not integrable on $[0, 1]$.   □

Although, in general, the composition of integrable functions need not be integrable, we do have the following important result:

6.12 THEOREM

> **INTEGRABILITY OF A COMPOSITION OF FUNCTIONS**   Suppose that g is an integrable function on the bounded closed interval $[a, b]$, that $[c, d]$ is a bounded closed interval containing $g([a, b])$, and that the function f is continuous on $[c, d]$. Then $f \circ g$ is integrable on $[a, b]$.

PROOF   Since f is continuous on the compact set $[c, d]$, it follows there exists $K > 0$ such that

$$y \in [c, d] \Rightarrow |f(y)| < K. \tag{1}$$

If $x \in [a, b]$, then (by hypothesis), $g(x) \in [c, d]$, and therefore

$$x \in [a, b] \Rightarrow |(f \circ g)(x)| = |f(g(x))| < K. \tag{2}$$

Therefore, the composite function $f \circ g$ is bounded on $[a, b]$. To prove that it is integrable on $[a, b]$, we propose to use Darboux's integrability condition (Theorem 4.13).

Let $\varepsilon > 0$ and let

$$\eta = \frac{\varepsilon}{b - a + 2K}. \tag{3}$$

Since f is continuous on $[c, d]$, it is uniformly continuous on $[c, d]$ (Theorem 5.18 in Chapter 4), and it follows that there exists $\delta > 0$ such that, for all y, $y' \in [c, d]$,

$$|y - y'| < \delta \Rightarrow |f(y) - f(y')| < \eta. \tag{4}$$

By replacing δ by $\min(\delta, \eta)$ if necessary, we can assume without loss of generality that

$$\delta \le \eta. \tag{5}$$

Because g is integrable on $[a, b]$, Darboux's integrability condition implies the existence of a partition $P = \{x_0, x_1, x_2, \ldots, x_n\}$ of $[a, b]$ such that

$$U(g, P) - L(g, P) < \delta^2. \tag{6}$$

For $i = 1, 2, \ldots, n$, define

$$m_i = \inf_{x_{i-1} \le x \le x_i} g(x), \qquad M_i = \sup_{x_{i-1} \le x \le x_i} g(x), \tag{7}$$

and define

$$m_i^* = \inf_{x_{i-1} \le x \le x_i} (f \circ g)(x), \qquad M_i^* = \sup_{x_{i-1} \le x \le x_i} (f \circ g)(x). \tag{8}$$

Note that, as a consequence of (2) and (8), we have

$$M_i^* - m_i^* \le 2K \tag{9}$$

(see Problem 21). Let

$$I = \{i \in \mathbb{N} \,|\, 1 \le i \le n, \quad M_i - m_i < \delta\} \tag{10}$$

and

$$J = \{j \in \mathbb{N} \,|\, 1 \le j \le n, \, M_j - m_j \ge \delta\}. \tag{11}$$

Suppose that

$$i \in I \quad \text{and} \quad x, x' \in [x_{i-1}, x_i]. \tag{12}$$

Then, by (7),

$$m_i \le g(x), \qquad g(x') \le M_i, \tag{13}$$

and it follows from (10) that

$$|g(x) - g(x')| \le M_i - m_i < \delta \tag{14}$$

(Problem 23). Therefore, by (4), we have

$$|(f \circ g)(x) - (f \circ g)(x')| = |f(g(x)) - f(g(x'))| < \eta. \tag{15}$$

As a consequence of (8) and (15), we have

$$i \in I \Rightarrow M_i^* - m_i^* \le \eta \tag{16}$$

(Problem 25). Therefore,

$$\sum_{i \in I} (M_i^* - m_i^*)\Delta x_i \le \eta \sum_{i \in I} \Delta x_i \le \eta \sum_{i=1}^{n} \Delta x_i = \eta(b - a). \tag{17}$$

If $j \in J$, then, by (11), $1 \le (M_j - m_j)/\delta$, and it follows that

$$\sum_{j \in J} \Delta x_j \le \sum_{j \in J} \frac{M_j - m_j}{\delta} \Delta x_j \le \frac{1}{\delta} \sum_{j=1}^{n} (M_j - m_j)\Delta x_j$$

$$= \frac{1}{\delta} [U(g, P) - L(g, P)]. \tag{18}$$

By combining (18), (6), and (5), we find that

$$\sum_{j \in J} \Delta x_j < \delta \le \eta. \tag{19}$$

As a consequence of (9) and (19), we have

$$\sum_{j \in J} (M_j^* - m_j^*)\Delta x_j \le 2K \sum_{j \in J} \Delta x_j < 2K\eta. \tag{20}$$

Thus, by (17), (20), and (3),

$$U(f \circ g, P) - L(f \circ g, P) = \sum_{i=1}^{n} (M_i^* - m_i^*)\Delta x_i$$

$$= \sum_{i \in I} (M_i^* - m_i^*)\Delta x_i + \sum_{j \in J} (M_j^* - m_j^*)\Delta x_j$$

$$< \eta(b - a) + 2K\eta = \eta(b - a + 2K) = \varepsilon. \tag{21}$$

$\square$

If $n \in \mathbb{N}$ and g is a function, we define $g^n : \mathrm{dom}(g) \to \mathbb{R}$ by

$$g^n(x) = (g(x))^n$$

for all $x \in \mathrm{dom}(g)$. Note that $g^n = f \circ g$, where $f : \mathbb{R} \to \mathbb{R}$ is defined by $f(y) = y^n$ for all $y \in \mathbb{R}$. Since f is continuous, we have the following corollary of Theorem 6.12.

6.13 COROLLARY

If g is an integrable function on the bounded closed interval $[a, b]$, and if $n \in \mathbb{N}$, then g^n is integrable on $[a, b]$.

6.14 THEOREM

THE PRODUCT OF INTEGRABLE FUNCTIONS IS INTEGRABLE Suppose that the functions f and g are integrable on the bounded closed interval $[a, b]$. Then the product function fg is integrable on $[a, b]$.

PROOF   We have

$$(f + g)^2 = f^2 + 2fg + g^2.$$

Therefore,

$$fg = \tfrac{1}{2}[(f + g)^2 - (f^2 + g^2)].$$

By Corollary 6.13, $(f + g)^2$, f^2, and g^2 are integrable on $[a, b]$. By Theorem 6.1, $f^2 + g^2$ is integrable on $[a, b]$; hence, by Theorem 6.2 (with $k = -1$), $-(f^2 + g^2)$ is integrable on $[a, b]$. Therefore, by Theorem 6.1 again,

$$(f + g)^2 - (f^2 + g^2) = (f + g)^2 + [-(f^2 + g^2)]$$

is integrable on $[a, b]$. Finally, by Theorem 6.2 (with $k = \tfrac{1}{2}$ this time), fg is integrable on $[a, b]$.   □

PROBLEM SET 5.6

1. Prove Theorem 6.2.

2. If $\mathscr{R}[a, b]$ denotes the set of all functions that are Riemann integrable on the bounded closed interval $[a, b]$, show that $\mathscr{R}[a, b]$ forms a vector space of functions. (See Section 7 of Chapter 4.)

3. Suppose that f is integrable on the bounded closed interval $[a, b]$ and that $f(x) \geq 0$ for all $x \in [a, b]$. Then the area under the graph of f between $x = a$ and $x = b$ is the same as the area of the rectangle bounded by $x = a$, $x = b$, $y = 0$, and $y = A$, where A is the mean value of f on $[a, b]$. Explain why.

4. Suppose that f and g are integrable functions on the bounded closed interval $[a, b]$ and that ε is a positive number such that $|f(x) - g(x)| \leq \varepsilon$ holds for all $x \in [a, b]$. Prove that

$$\left| \int_a^b f(x)\, dx - \int_a^b g(x)\, dx \right| \leq \varepsilon(b - a).$$

5. Suppose that f and g are integrable functions on the bounded closed interval $[a, b]$. Prove that

$$\left| \int_a^b [f(x) + g(x)]\, dx \right| \leq \int_a^b |f(x)|\, dx + \int_a^b |g(x)|\, dx.$$

6. If f is a function, define $f^+ : \mathrm{dom}(f) \to \mathbb{R}$ by $f^+(x) = \max(f(x), 0)$ for all $x \in \mathrm{dom}(f)$ and define f^- by $f^- = (-f)^+$.
   (a) Prove that $f^+ = \tfrac{1}{2}(f + |f|)$.
   (b) Prove that $|f| = f^+ + f^-$.
   (c) Prove that $f = f^+ - f^-$.
   (d) Prove that f is integrable on the bounded closed interval $[a, b]$ if and only if both f^+ and f^- are integrable on $[a, b]$.

7. Prove the following more general version of Theorem 6.7: If f is continuous on an interval I and if $a, b \in I$, then there is a point c between a and b such that $\int_a^b f(x) \, dx = f(c)(b - a)$. (Note that there is no restriction here that $a \le b$.)

8. The force F required to stretch a perfectly elastic spring x units from its relaxed position is given by $F = kx$, where k is a constant depending on the stiffness of the spring.
   (a) Assume the usual integration formulas developed in elementary calculus to find the mean value of F if the spring is stretched from $x = a$ to $x = b$.
   (b) Find a number c between a and b such that the value of F when $x = c$ is the same as this mean value.

9. Prove Theorem 6.9. [Hint: With the notation in the proof of Theorem 6.7, we have $f(p) \le f(x) \le f(q)$ for all $x \in [a, b]$. Since $g(x) \ge 0$ for all $x \in [a, b]$, it follows that $g(x)f(p) \le g(x)f(x) \le g(x)f(q)$ for all $x \in [a, b]$. Now use Theorem 6.3.]

10. Let f be an integrable function on the closed bounded interval $[a, c]$ and let $b \in (a, c)$. Let $A_{[a, b]}$, $A_{[b, c]}$, and $A_{[a, c]}$ be the mean values of f on $[a, b]$, $[b, c]$, and $[a, c]$, respectively. Prove that there is a number $t \in [0, 1]$ such that $A_{[a, c]} = tA_{[a, b]} + (1 - t)A_{[b, c]}$.

11. Prove that the conclusion of Theorem 6.9 still holds if the condition that $g(x) \ge 0$ for all $x \in [a, b]$ is replaced by the condition that $g(x) \le 0$ for all $x \in [a, b]$.

12. If $f : [a, b] \to \mathbb{R}$ is monotone, prove that there is a point $c \in [a, b]$ such that $\int_a^b f(x) \, dx = f(a)(c - a) + f(b)(b - c)$. [Hint: Define the function $g(x) = f(a)(x - a) + f(b)(b - x)$ for $x \in [a, b]$, note that g is continuous, and show that $\int_a^b f(x) \, dx$ lies between $g(a)$ and $g(b)$.]

13. Prove the following alternative version of Theorem 6.9: Let f and g be continuous functions on an interval I and suppose that $g(x) \ne 0$ for all $x \in I$. Then, if $a, b \in I$, there exists a number c between a and b such that $\int_a^b g(x)f(x) \, dx = f(c) \int_a^b g(x) \, dx$.

14. Let g be the Dirichlet function (Example 6.10).
   (a) Prove that g is discontinuous at each rational number $a \in [0, 1]$.
   (b) Prove that g is continuous at each irrational number $a \in [0, 1]$. [Hint: Given $\varepsilon > 0$, make use of the fact that there are only finitely many rational numbers in $[0, 1]$ whose denominators (in reduced form) are less than $1/\varepsilon$.]

15. In the solution of Example 6.10:
   (a) If P is the partition of $[0, 1]$ given by $P = \{0, 1\}$, show that $U(g, P) = 1$.
   (b) Explain in detail why the set A is finite.

16. If g is the Dirichlet function (Example 6.10), find the numerical value of $\int_0^1 g(x) \, dx$.

17. (a) Prove that there are infinitely many prime numbers. [Hint: Suppose that there are only finitely many prime numbers, say, $p_1, p_2, p_3, \ldots, p_k$. Let $n = 1 + p_1 p_2 p_3 \cdots p_k$ and argue that n must have a prime factor different from $p_1, p_2, p_3, \ldots, p_k$.]

(b) Use the result of Part (a) to justify the reasoning in the solution of Example 6.10 in which a prime $p > 2N/\varepsilon$ is chosen. How do we know that such a prime exists?

18. Let $\mathscr{R}[a, b]$ denote the set of all functions that are Riemann integrable on the bounded closed interval $[a, b]$. Show that $\mathscr{R}[a, b]$ is an algebra of functions. (See Section 7 of Chapter 4.)

19. In the solution of Example 6.10, fill in the details of the proof that $\phi : J \to A$ is an injection.

20. With the notation of Problem 18, define the **inner product** of the functions $f, g \in \mathscr{R}[a, b]$, in symbols $\langle f, g \rangle$, by $\langle f, g \rangle = \int_a^b f(x)g(x)\,dx$. If $f, g, h \in \mathscr{R}[a, b]$ and $k \in \mathbb{R}$, prove the following:

(a) $\langle f + g, h \rangle = \langle f, h \rangle + \langle g, h \rangle$ (b) $\langle kf, g \rangle = k \langle f, g \rangle$
(c) $\langle f, g + h \rangle = \langle f, g \rangle + \langle f, h \rangle$ (d) $\langle f, kg \rangle = k \langle f, g \rangle$
(e) $\langle f, g \rangle = \langle g, f \rangle$ (f) $\langle f, f \rangle \geq 0$

21. (a) Suppose that $K > 0$ and that S is a nonempty set with $S \subseteq [-K, K]$. Prove that $\sup(S) - \inf(S) \leq 2K$.

(b) Use the result of Part (a) to establish Inequality (9) in the proof of Theorem 6.12. [Hint: Let $S = \{(f \circ g)(x) \mid x_{i-1} \leq x \leq x_i\}$.]

22. With the notation of Problem 20, define the **norm** of a function $f \in \mathscr{R}[a, b]$, in symbols $\|f\|$, by $\|f\| = (\langle f, f \rangle)^{1/2}$. Prove the **Schwarz inequality:**   For $f, g \in \mathscr{R}[a, b]$, $|\langle f, g \rangle| \leq \|f\| \|g\|$. [Hint: Let $\alpha = \langle g, g \rangle$, $\beta = -\langle f, g \rangle$, and use the fact that $\langle \alpha f + \beta g, \alpha f + \beta g \rangle \geq 0$.]

23. (a) If $m \leq y$, $y' \leq M$, prove that $|y - y'| \leq M - m$.

(b) Use the result of Part (a) to establish the first part of Inequality (14) in the proof of Theorem 6.12.

24. With the notation of Problems 20 and 22, prove the **triangle inequality for norms:**   For $f, g \in \mathscr{R}[a, b]$, $\|f + g\| \leq \|f\| + \|g\|$. [Hint: Prove that $\|f + g\|^2 \leq (\|f\| + \|g\|)^2$, and then take the square root on both sides.]

25. (a) Suppose that S is a nonempty bounded subset of $\mathbb{R}$, $m = \inf(S)$, and $M = \sup(S)$. If $\eta > 0$ and $|s - s'| \leq \eta$ holds for all $s, s' \in S$, prove that $M - m \leq \eta$.

(b) Use the result of Part (a) to establish Implication (16) in the proof of Theorem 6.12.

26. Let the function f be defined on the bounded closed interval $[a, b]$ and assume that $m \leq f(x) \leq M$ for all $x \in [a, b]$. Suppose that, for every positive number ε there exists a positive number $\lambda < \varepsilon$, a partition $P = \{x_0, x_1, x_2, \ldots, x_n\}$ of $[a, b]$, and a decomposition of $\{1, 2, \ldots, n\}$ into two disjoint subsets I and J such that,

(i) for $i \in I$, $M_i - m_i < \dfrac{\lambda}{b-a}$ and   (ii) $\displaystyle\sum_{j \in J} \Delta x_j < \dfrac{\varepsilon - \lambda}{M - m}$.

(Here we use the usual notation $m_i = \inf_{x_{i-1} \leq x \leq x_i} f(x)$, $M_i = \sup_{x_{i-1} \leq x \leq x_i} f(x)$, and $\Delta x_i = x_i - x_{i-1}$ for $i = 1, 2, \ldots, n$.) Prove that f is integrable on $[a, b]$.

27. Suppose that f is integrable on $[a, b]$ and that $f(x) \geq 0$ for all $x \in \mathrm{dom}(f)$. Define $\sqrt{f} : \mathrm{dom}(f) \to \mathbb{R}$ by $\sqrt{f}(x) = \sqrt{f(x)}$ for all $x \in \mathrm{dom}(f)$. Prove that $\sqrt{f}$ is integrable on $[a, b]$.

28. The result of Problem 26 can be paraphrased *informally* as follows: If there are partitions P of $[a, b]$ such that the variation of f on some of the subintervals of P is sufficiently small and the sum of the lengths of the remaining subintervals is also sufficiently small, then f is integrable. Explain how this helps to motivate the solution of Example 6.10 and the proof of Theorem 6.12.

29. Suppose that f is integrable on $[a, b]$, that $f(x) \neq 0$ for all $x \in [a, b]$, and that the function $1/f$ is bounded on $[a, b]$. Prove that $1/f$ is integrable on $[a, b]$.

7

THE FUNDAMENTAL THEOREM OF CALCULUS

Geometrically, the derivative and the integral may be regarded as tools to solve the apparently unrelated problems of finding the tangent line to a curve and determining the area under a curve. Several special cases of the tangent problem had been solved prior to the work of Newton and Leibniz; in fact, Isaac Barrow (1630–1677), one of Newton's teachers, used a technique for finding tangent lines that was strongly suggestive of the concept of the derivative. Also, as we have mentioned, the ancient Greeks were able to solve particular cases of the area problem by a technique (the method of exhaustion) that portends the Cauchy/Riemann/Darboux definition of the integral. Thus, both the derivative and the integral were "in the air" even before Newton and Leibniz began their great work.

Newton and Leibniz are regarded as the architects of analysis not because they single-handedly discovered the derivative and the integral but because they realized and exploited the *connection* between these two concepts. This connection, which was also known to Barrow, is called the *Fundamental Theorem of Calculus*, abbreviated FTC. Roughly speaking, the FTC states that *integration and differentiation are inverse processes*. In this section, we shall present two different, but related, versions of the FTC.

Before we proceed, it may be helpful to clear up a few minor matters of notation and terminology. In the notation $\int_a^b f(x)\,dx$, the symbol $\int$ is called the **integral sign**, a and b are referred to as the lower and upper **limits of integration**, x is called the **variable of integration**, $f(x)$ is referred to as the **integrand**, and f is called the function **under the integral sign**.

Notice that $\int_a^b f(x)\,dx$ is a real number depending only on the function f and the bounded closed interval $[a, b]$ over which f is integrated. The integral has nothing to do with x, or, for that matter, with dx, and it could just as well be written as $\int_a^b f$. The more elaborate notation merely serves to remind us that the integral is a limit of Riemann sums. In particular, the variable of integration is a dummy variable, and there is no special reason to use x for this variable—any other letter (apart from a, b, f, or d) would do as well. For instance,

$$\int_a^b f(x)\,dx = \int_a^b f(t)\,dt = \int_a^b f(u)\,du = \int_a^b f(v)\,dv,$$

and so on.

In the following definition, it will be convenient to use the letter x for the upper limit of integration. Accordingly, we choose to use the letter t for the (dummy) variable of integration.

7.1 DEFINITION   **INDEFINITE INTEGRAL**   Let f be a function, let I be an interval contained in $\mathrm{dom}(f)$, let $a \in I$, and suppose that, for every $x \in I$ with $x \neq a$, f is integrable on the bounded closed interval between a and x. Then the function $F: I \to \mathbb{R}$ defined by

$$F(x) = \int_a^x f(t)\,dt$$

for every $x \in I$ is called an **indefinite integral of f on I**.

Note that the integral $\int_a^b f(x)\,dx$ is a *real number*, whereas an indefinite F of f is a *function*. To emphasize the distinction, $\int_a^b f(x)\,dx$ is often referred to as a **definite integral**.

As a consequence of Problems 21 and 29 in Problem Set 5.4, the condition in Definition 7.1 that f be integrable on the bounded closed interval between a and x for every $x \in I$ with $x \neq a$ is equivalent to the requirement that f be integrable on every bounded closed interval $[p, q]$ contained in I (Problem 1). If either of these equivalent conditions holds, we say that f is **indefinitely integrable on I**. For instance, if f is either continuous on I or monotone on I, then it is indefinitely integrable on I. (See Theorems 4.14 and 4.15.)

If f is indefinitely integrable on I, then, in general, there are infinitely many different indefinite integrals F of f on I, depending on the choice of the lower limit of integration a in the formula

$$F(x) = \int_a^x f(t)\, dt.$$

However, as we prove in the next lemma, *any two indefinite integrals of f on I differ by a constant.*

7.2 LEMMA

Let the function f be indefinitely integrable on the interval $I \subseteq \mathbb{R}$ and suppose that F and G are indefinite integrals of f on I. Then there exists a constant C such that

$$F(x) - G(x) = C$$

for all $x \in I$.

PROOF   Since F and G are indefinite integrals of f on I, there are points $a, a' \in I$ such that

$$F(x) = \int_a^x f(t)\, dt \qquad \text{and} \qquad G(x) = \int_{a'}^x f(t)\, dt$$

hold for every $x \in I$. Consequently, by Definition 5.18 and Theorem 5.19, we have

$$F(x) - G(x) = \int_a^x f(t)\, dt - \int_{a'}^x f(t)\, dt = \int_a^x f(t)\, dt + \int_x^{a'} f(t)\, dt$$

$$= \int_a^{a'} f(t)\, dt$$

for all $x \in I$. Thus, if we define the constant $C \in \mathbb{R}$ by

$$C = \int_a^{a'} f(t)\, dt,$$

then we have

$$F(x) - G(x) = C$$

for all $x \in I$.   □

7.3 LEMMA

If f is indefinitely integrable on the interval I and if F is an indefinite integral of f on I, then, for $p, q \in I$, we have

$$F(p) - F(q) = \int_q^p f(t)\, dt.$$

PROOF   By definition, there is a point $a \in I$ such that

$$F(x) = \int_a^x f(t)\, dt$$

for every $x \in I$. Consequently, by Definition 5.18 and Theorem 5.19, we have

$$F(p) - F(q) = \int_a^p f(t)\, dt - \int_a^q f(t)\, dt = \int_a^p f(t)\, dt + \int_q^a f(t)\, dt$$

$$= \int_q^a f(t)\, dt + \int_a^p f(t)\, dt = \int_q^p f(t)\, dt. \qquad \square$$

7.4 LEMMA

Let f be indefinitely integrable on the interval I, let $p, q \in I$, and suppose that $|f(t)| \le K$ holds for all t between p and q. Then

$$\left| \int_q^p f(t)\, dt \right| \le K|p - q|.$$

PROOF   If $p = q$, the inequality in question holds because both sides are equal to 0. If $q < p$, then, by Theorem 6.5, and Corollary 6.4,

$$\left| \int_q^p f(t)\, dt \right| \le \int_q^p |f(t)|\, dt \le K(p - q) = K|p - q|.$$

In the only remaining case, we have $p < q$, so that

$$\left| \int_q^p f(t)\, dt \right| = \left| -\int_p^q f(t)\, dt \right| = \left| \int_p^q f(t)\, dt \right| \le \int_p^q |f(t)|\, dt \le K(q - p)$$

$$= K|p - q|. \qquad \square$$

7.5 COROLLARY

Let f be indefinitely integrable on the interval I, let F be an indefinite integral of f on I, and let $p, q \in I$ be such that $|f(t)| \le K$ for all t between p and q. Then

$$|F(p) - F(q)| \le K|p - q|.$$

PROOF   Combine Lemmas 7.3 and 7.4. $\qquad \square$

7.6 THEOREM

CONTINUITY OF AN INDEFINITE INTEGRAL   If f is indefinitely integrable on the interval I and if F is an indefinite integral of f on I, then F is continuous.

PROOF   We must prove that F is continuous at each point $b \in I$. We consider the case in which b is an interior point of the interval I, leaving the argument for the case in which b is an endpoint of I as an exercise (Problem 3). Thus, we may assume that there exists $\eta > 0$ such that

$$[b - \eta, b + \eta] \subseteq I.$$

Since f is integrable on $[b - \eta, b + \eta]$, it is bounded on $[b - \eta, b + \eta]$ by Lemma 5.11, so there exists $K > 0$ such that

$$t \in [b - \eta, b + \eta] \Rightarrow |f(t)| \leq K.$$

Let $\varepsilon > 0$, let

$$\delta = \min\left(\frac{\varepsilon}{K}, \eta\right),$$

and suppose that $x \in I$ with

$$|x - b| < \delta.$$

Then $|x - b| < \eta$, so

$$x \in (b - \eta, b + \eta).$$

Hence, if t is between b and x, then t belongs to the interval $[b - \eta, b + \eta]$, and it follows that $|f(t)| \leq K$. Therefore, by Corollary 7.5,

$$|F(x) - F(b)| \leq K|x - b| < K\delta \leq K\frac{\varepsilon}{K} = \varepsilon. \qquad \square$$

A function f may have discontinuities on an interval I and still be indefinitely integrable on I; however, according to Theorem 7.6, an indefinite integral F of f on I is always continuous. Thus, indefinite integration is a "smoothing operation" in the sense that it can take a discontinuous function f and convert it into a continuous function F. The following theorem, which is one form of the Fundamental Theorem of Calculus (FTC), reinforces this idea by showing that, if f is continuous at c, then not only is F continuous at c, it is actually *differentiable* at c.

7.7 THEOREM

FUNDAMENTAL THEOREM OF CALCULUS (FIRST FORM) Let f be indefinitely integrable on the interval I and let F be an indefinite integral of f on I. Suppose that f is continuous at the point $c \in I$. Then F is differentiable at c and

$$F'(c) = f(c).$$

PROOF We have to show that

$$\lim_{h \to 0} \frac{F(c + h) - F(c)}{h} = f(c). \qquad (*)$$

Let $\varepsilon > 0$. Thus, we have to find a $\delta > 0$ such that, for all h with $c + h \in I$,

$$0 < |h| < \delta \Rightarrow \left| \frac{F(c + h) - F(c)}{h} - f(c) \right| < \varepsilon. \tag{**}$$

Since f is continuous at c, there exists $\delta > 0$ such that, for every $t \in I$,

$$|t - c| < \delta \Rightarrow |f(t) - f(c)| < \frac{\varepsilon}{2}. \tag{1}$$

We propose to show that, for this δ, Condition (**) holds.

Suppose that $h \neq 0$ and $c + h \in I$. Then, by Lemma 7.3, we have

$$\frac{F(c + h) - F(c)}{h} = \frac{1}{h} \int_c^{c+h} f(t)\, dt. \tag{2}$$

Also, since $f(c)$ is a constant, Example 4.11 implies that

$$\int_c^{c+h} f(c)\, dt = f(c)[(c + h) - c] = f(c)h, \tag{3}$$

and therefore

$$f(c) = \frac{1}{h} \int_c^{c+h} f(c)\, dt. \tag{4}$$

By combining (2) and (4), we obtain

$$\frac{F(c + h) - F(c)}{h} - f(c) = \frac{1}{h} \int_c^{c+h} f(t)\, dt - \frac{1}{h} \int_c^{c+h} f(c)\, dt$$

$$= \frac{1}{h} \int_c^{c+h} \left[f(t) - f(c) \right] dt. \tag{5}$$

Now, to prove (**), suppose that $c + h \in I$ and that

$$0 < |h| < \delta. \tag{6}$$

Then, if t is between c and $c + h$, it follows that $|t - c| < \delta$ (Problem 5), and hence, by (1), we have $|f(t) - f(c)| < \varepsilon/2$. Therefore, by (5) and Lemma 7.4,

$$\left| \frac{F(c + h) - F(c)}{h} - f(c) \right| = \frac{1}{|h|} \left| \int_c^{c+h} \left[f(t) - f(c) \right] dt \right|$$

$$\leq \frac{1}{|h|} \cdot \frac{\varepsilon}{2} |(c + h) - c|$$

$$= \frac{\varepsilon}{2} < \varepsilon. \tag{7}$$

$\square$

The following result is an immediate consequence of Theorem 7.7.

7.8 COROLLARY

Let $f:I \to \mathbb{R}$ be a continuous function and let $F:I \to \mathbb{R}$ be an indefinite integral of f on the interval $I \subseteq \mathbb{R}$. Then F is differentiable and

$$F' = f.$$

In Corollary 7.8 we again see that indefinite integration is a smoothing operation in the sense that it can convert a continuous, but perhaps non-differentiable, function f into a function F that is not only continuous but differentiable as well. The conclusion of Corollary 7.8 may be rewritten in the form

$$\frac{d}{dx} \int_a^x f(t)\, dt = f(x),$$

which clearly shows the sense in which (indefinite) integration and differentiation are inverse operations—*if you form the indefinite integral of a continuous function and then take the derivative of the result, you obtain the original function.*

7.9 Example Many elementary calculus textbooks define the natural logarithm function $\ln:(0, \infty) \to \mathbb{R}$ by

$$\ln(x) = \int_1^x \frac{1}{t}\, dt$$

for $x > 0$. Find $(d/dx)\ln(x)$ for $x > 0$.

SOLUTION By Corollary 7.8,

$$\frac{d}{dx} \ln(x) = \frac{d}{dx} \int_1^x \frac{1}{t}\, dt = \frac{1}{x}$$

for all $x > 0$. □

7.10 DEFINITION

ANTIDERIVATIVE Let $I \subseteq \mathbb{R}$ be an interval and let $f:I \to \mathbb{R}$. A function $F:I \to \mathbb{R}$ is said to be an **antiderivative** of f on I if F is differentiable and

$$F' = f.$$

By using the notion of an antiderivative, we may paraphrase Corollary 7.8 as follows:

> *If f is defined and continuous on an interval I, then f has an antiderivative on I. In fact, any indefinite integral of f on I is an antiderivative of f on I.*

Suppose that I is an interval and that $f : I \to \mathbb{R}$ has an antiderivative $F : I \to \mathbb{R}$ on I. Then, since F is differentiable on I, it follows from Theorem 1.8 that F is continuous on I. In short:

> *An antiderivative is necessarily a continuous function.*

Furthermore, although f need not be continuous, Theorem 2.8 and the fact that $f = F'$ imply that f has the intermediate value property on I. Thus:

> *If a function has an antiderivative on an interval, then it has the intermediate value property on that interval.*

7.11 THEOREM

> **ANTIDERIVATIVES OF A FUNCTION DIFFER BY A CONSTANT**   Let f be continuous on the interval I and suppose that F and G are antiderivatives of f on I. Then there exists a constant real number C such that
>
> $$G(x) = F(x) + C$$
>
> holds for all $x \in I$.

PROOF   We have $F'(x) = G'(x) = f(x)$ for all $x \in I$, and so the conclusion of the theorem follows from Corollary 2.16.   □

If f is a continuous function, $\operatorname{dom}(f) = I$ is an interval, F is an antiderivative of f on I, C is a constant real number, and $G : I \to \mathbb{R}$ is defined by

$$G(x) = F(x) + C$$

for all $x \in \mathbb{R}$, then, for all $x \in \mathbb{R}$,

$$G'(x) = F'(x) + 0 = f(x),$$

and it follows that G is also an antiderivative of f on I. Conversely, by Theorem 7.11, every antiderivative G of f on I can be obtained in this way by assigning a suitable value to the constant C. It is customary to use the notation $\int f(x)\,dx$ to represent an arbitrary antiderivative of f, so that

$$\int f(x)\,dx = F(x) + C,$$

where it is understood that C is an arbitrary constant, called the **constant of integration**. Although this notation often proves to be handy (for instance, in tables of integrals), it must be used with care because $\int f(x)\,dx$ actually represents a whole class of functions (all of the antiderivatives of f on I) rather than a single function. Also, it is customary to refer to $\int f(x)\,dx$ as an *indefinite integral*; however this usage is different from ours as set forth in Definition 7.1. (One can usually tell from the context which usage is intended.)

In the next example, we use Corollary 7.8 to derive a version of the FTC that is undoubtedly familiar from elementary calculus.

7.12 Example   If f is a continuous function on the bounded closed interval $[a, b]$, and if F is an antiderivative of f on $[a, b]$, show that

$$\int_a^b f(x)\,dx = F(b) - F(a).$$

SOLUTION   Define $G:[a, b] \to \mathbb{R}$ by

$$G(x) = \int_a^x f(t)\,dt. \tag{1}$$

By Corollary 7.8, G is an antiderivative of f on $[a, b]$. Therefore, by Theorem 7.11, there is a constant $C \in \mathbb{R}$ such that

$$\int_a^x f(t)\,dt = F(x) + C \tag{2}$$

for all $x \in [a, b]$. Letting $x = a$ in (2), we find that $0 = F(a) + C$, and so

$$C = -F(a). \tag{3}$$

Combining (2) and (3), we obtain

$$\int_a^x f(t)\,dt = F(x) - F(a) \tag{4}$$

for all $x \in [a, b]$. Now, by letting $x = b$ in (4), we find that

$$\int_a^b f(t)\,dt = F(b) - F(a). \tag{5}$$

Finally, in view of the fact that the variable of integration is a dummy variable, we may rewrite (5) in the form

$$\int_a^b f(x)\,dx = F(b) - F(a). \tag{6}$$

$\square$

The formula $\int_a^b f(x)\, dx = F(b) - F(a)$ in Example 7.12 is so effective, and it is used so routinely in elementary calculus, that many students come to believe it is the very *definition* of the definite integral. On the contrary, it is really a deep and powerful *theorem*—one of the versions of the Fundamental Theorem of Calculus!

7.13 Example   If $n \in \mathbb{N}$ and $a, b \in \mathbb{R}$, prove that

$$\int_a^b x^n\, dx = \frac{b^{n+1} - a^{n+1}}{n+1}.$$

SOLUTION   Let $f : [a, b] \to \mathbb{R}$ be defined by $f(x) = x^n$ for all $x \in [a, b]$. If $F : [a, b] \to \mathbb{R}$ is defined by

$$F(x) = \frac{x^{n+1}}{n+1}$$

for all $x \in [a, b]$, then $F'(x) = x^n = f(x)$ holds for all $x \in [a, b]$. Since f is continuous on $[a, b]$, the result in Example 7.12 shows that

$$\int_a^b x^n\, dx = \int_a^b f(x)\, dx = F(b) - F(a) = \frac{b^{n+1} - a^{n+1}}{n+1}. \qquad \square$$

In our next theorem, which is a second form of the FTC, we show that the hypothesis in Example 7.12 that f be continuous can be relaxed somewhat. All that is really required is that f be *integrable* on $[a, b]$.

7.14 THEOREM

> **FUNDAMENTAL THEOREM OF CALCULUS (SECOND FORM)**   If f is an integrable function on the bounded closed interval $[a, b]$, and if F is an antiderivative of f on $[a, b]$, then
>
> $$\int_a^b f(x)\, dx = F(b) - F(a).$$

PROOF   Let ε be a positive number. Since f is integrable on $[a, b]$, there exists $\delta > 0$ such that, for every partition P of $[a, b]$ with $\|P\| < \delta$, and every sample x^* for P,

$$\left| S(f, P, x^*) - \int_a^b f(x)\, dx \right| < \varepsilon.$$

Let $P = \{x_0, x_1, x_2, \ldots, x_n\}$ be a partition of $[a, b]$ with $\|P\| < \delta$ (Lemma 5.3). For each $i = 1, 2, \ldots, n$, we apply the mean value theorem (Theorem 2.11) to the function F on the interval $[x_{i-1}, x_i]$ and conclude that there exists $x_i^* \in (x_{i-1}, x_i)$ such that

$$F(x_i) - F(x_{i-1}) = F'(x_i^*)(x_i - x_{i-1}),$$

that is,

$$F(x_i) - F(x_{i-1}) = f(x_i^*)\Delta x_i.$$

Therefore,

$$S(f, P, x^*) = \sum_{i=1}^{n} f(x_i^*)\Delta x_i = \sum_{i=1}^{n} [F(x_i) - F(x_{i-1})].$$

The last sum "telescopes," and we have

$$S(f, P, x^*) = F(x_n) - F(x_0) = F(b) - F(a).$$

Consequently,

$$\left| F(b) - F(a) - \int_a^b f(x)\,dx \right| < \varepsilon$$

holds for every positive number ε, and it follows that

$$\int_a^b f(x)\,dx = F(b) - F(a). \qquad \square$$

Theorem 7.14 can be restated in the following form (Problem 11):

> *If f is a differentiable function on the bounded closed interval $[a, b]$, and if f' is integrable on $[a, b]$, then*
>
> $$\int_a^b f'(x)\,dx = f(b) - f(a).$$

7.15 THEOREM

INTEGRATION BY PARTS   Let f and g be differentiable on $[a, b]$ and suppose that f' and g' are integrable on $[a, b]$. Then fg' and $f'g$ are integrable on $[a, b]$ and

$$\int_a^b f(x)g'(x)\,dx = f(b)g(b) - f(a)g(a) - \int_a^b f'(x)g(x)\,dx.$$

PROOF   Since f and g are differentiable on $[a, b]$, they are continuous on $[a, b]$, and therefore they are integrable on $[a, b]$. By hypothesis, f' and g' are integrable on $[a, b]$, and it follows from Theorem 6.14 that fg' and $f'g$ are integrable on $[a, b]$. By the product rule for differentiation, fg is differentiable on $[a, b]$ and

$$(fg)' = f'g + fg'. \qquad (1)$$

Since both $f'g$ and fg' are integrable on $[a, b]$, it follows from (1) and Theorem 6.1 that $(fg)'$ is integrable on $[a, b]$ and that

$$\int_a^b (fg)'(x)\, dx = \int_a^b f'(x)g(x)\, dx + \int_a^b f(x)g'(x)\, dx. \tag{2}$$

But, by Theorem 7.14,

$$\int_a^b (fg)'(x)\, dx = (fg)(b) - (fg)(a) = f(b)g(b) - f(a)g(a). \tag{3}$$

By combining (2) and (3), and rearranging the resulting terms, we obtain the formula to be proved.   □

Our work in the present section raises three basic questions, all of which are answered in more advanced courses in real analysis:[†]

1. Which functions $f:[a, b] \to \mathbb{R}$ have antiderivatives?

2. What are the most general conditions under which

$$\frac{d}{dx} \int_a^x f(t)\, dt = f(x)?$$

3. What are the most general conditions under which

$$\int_a^b f'(x)\, dx = f(b) - f(a)?$$

To answer these three questions completely, it is necessary to employ a theory of integration more general than the Riemann/Darboux theory. The appropriate extended theory of integration was formulated in the early 1900s by the French mathematician Henri Léon Lebesgue (1875–1941), and it is the Lebesgue integral, not the Riemann/Darboux integral, which prevails in modern real analysis.

PROBLEM SET 5.7

1. Let I be an interval contained in the domain of the function f and let $a \in I$. Prove that f is integrable on the bounded closed interval between a and x for every $x \in I$ with $x \neq a$ if and only if f is integrable on every bounded closed interval $[p, q] \subseteq I$.

2. Let $f:\mathbb{R} \to \mathbb{R}$ be defined by $f(x) = 0$ for $x \leq 0$ and $f(x) = 1$ for $x > 0$. Note that f is discontinuous at 0.
   (a) Prove that f is indefinitely integrable on $\mathbb{R}$.
   (b) Prove that the continuous function F defined by $F(x) = (x + |x|)/2$ for all $x \in \mathbb{R}$ is an indefinite integral of f on $\mathbb{R}$.
   (c) Is F an antiderivative of f on $\mathbb{R}$? Why or why not?

[†] See Chapter 5 of H. L. Royden's *Real Analysis*, 3rd ed. (New York: Macmillan, 1988).

3. Complete the proof of Theorem 7.6 by taking care of the case in which b is a left or right endpoint of the interval I.

4. Let $f:\mathbb{R} \to \mathbb{R}$ be defined by $f(x) = |x|$ for all $x \in \mathbb{R}$.
    (a) Prove that f is indefinitely integrable on $\mathbb{R}$.
    (b) Prove that the continuous function F defined by $F(x) = x|x|/2$ for all $x \in \mathbb{R}$ is an indefinite integral of f on $\mathbb{R}$.
    (c) Is F an antiderivative of f on $\mathbb{R}$? Why or why not?

5. Supply the following detail in the proof of Theorem 7.7: Suppose that $0 < |h| < \delta$ and that t is between c and $c + h$, and prove that $|t - c| < \delta$.

6. By Corollary 7.8, an indefinite integral F of a continuous function f on an interval I is an antiderivative of f on I. Is it true that an antiderivative F of a continuous function f on an interval I is an indefinite integral of f? If so, prove it; if not, give a counterexample.

7. Let sinr be the function defined in Problem 10 of Problem Set 4.3 and let $F:\mathbb{R} \to \mathbb{R}$ be defined by $F(x) = x^2 \operatorname{sinr}(x)$ for all $x \in \mathbb{R}$.
    (a) Show that F is differentiable on $\mathbb{R}$ and let $f = F'$.
    (b) Prove that f is discontinuous at 0.
    (c) Prove that f is indefinitely integrable on $\mathbb{R}$. [Hint: Use the result of Problem 21 in Problem Set 5.5.]

8. Give an example of a function $f:\mathbb{R} \to \mathbb{R}$ that has an antiderivative on $\mathbb{R}$, but is not indefinitely integrable on $\mathbb{R}$. [Hint: Use the results in Problem 22 of Problem Set 5.1.]

9. For the natural logarithm function defined in Example 7.9, prove that:
    (a) ln is a strictly increasing function on $(0, \infty)$.
    (b) $\ln(1) = 0$.

10. For the natural logarithm function defined in Example 7.9, prove that $\ln(ab) = \ln(a) + \ln(b)$ for all $a, b \in (0, \infty)$. [Hint: Define $f:(0, \infty) \to \mathbb{R}$ by $f(x) = \ln(xb) - \ln(x)$. Prove that $f'(x) = 0$ for all $x \in (0, \infty)$, conclude that there is a constant C such that $\ln(xb) - \ln(x) = C$ for all $x \in (0, \infty)$, determine the value of C, and then let $x = a$.]

11. Show that Theorem 7.14 can be restated as indicated on page 427.

12. Since ln is a strictly increasing function on $(0, \infty)$, it has an inverse. The **exponential function** exp is defined by $\exp = \ln^{-1}$. Show that exp is differentiable and that it is its own derivative.

13. Criticize the following: If $f(x) = -1/x$, then $f'(x) = 1/x^2$, and so
    $$\int_{-1}^{1} f'(x)\, dx = f(1) - f(-1) = -2.$$

14. Let I and J be intervals, let $f:I \to \mathbb{R}$ be continuous, and let $\phi:J \to \mathbb{R}$ be a differentiable function such that $\phi(J) \subseteq I$. Let $a \in I$ and define $G:J \to \mathbb{R}$ by $G(x) = \int_a^{\phi(x)} f(t)\, dt$ for all $x \in J$. Prove that G is differentiable on J and that $G'(x) = (f \circ \phi)(x)\phi'(x)$ for all $x \in J$. [Hint: Use the FTC and the chain rule.]

15. By using an appropriate version of the FTC, we can see why Theorem 2.11 is called the *mean value theorem*: Supposing that f is differentiable on the bounded closed interval $[a, b]$, that f' is integrable on $[a, b]$, and that A is the mean value of f' on $[a, b]$ (see Definition 6.6), show that Theorem 2.11 implies that there is a point $c \in (a, b)$ at which f' takes on the value A.

16. Let I and J be intervals, let $f:I \to \mathbb{R}$ be continuous and suppose that $\phi:J \to \mathbb{R}$ and $\psi:J \to \mathbb{R}$ are differentiable functions such that $\phi(J) \subseteq I$ and $\psi(J) \subseteq I$. Define $G:J \to \mathbb{R}$ by $G(x) = \int_{\psi(x)}^{\phi(x)} f(t)\, dt$ for all $x \in J$. Prove that G is differentiable on J, and obtain a formula for $G'(x)$.

17. If f is continuous on the interval I and $b \in I$, prove that $(d/dx)\int_x^b f(t)\, dt = -f(x)$ for all $x \in I$.

18. If $y = \int_{x^4}^{x^-x^2} \sqrt{t^3 + 1}\, dt$, find dy/dx. [Hint: Use the result of Problem 16].

19. If $n \in \mathbb{N}$, find an antiderivative on $\mathbb{R}$ of the function $f(x) = |x|^n$.

20. Let $a > 0$ and suppose that f is a continuous function on the interval $[-a, a]$. Define $G:[-a, a]$ by $G(x) = \int_0^{-x} f(t)\, dt + \int_0^x f(-t)\, dt$ for all $x \in [-a, a]$.
(a) Prove that $G'(x) = 0$ for all $x \in [-a, a]$.
(b) Use Part (a) to prove that $G(x) = 0$ for all $x \in [-a, a]$.
(c) Conclude that $\int_{-x}^0 f(t)\, dt = \int_0^x f(-t)\, dt$.

21. Prove the **change of variable theorem**: Let ϕ be differentiable on the closed bounded interval $[a, b]$ and let ϕ' be integrable on $[a, b]$. Suppose that f is continuous on $\phi([a, b])$. Then $\int_a^b f(\phi(t))\phi'(t)\, dt = \int_{\phi(a)}^{\phi(b)} f(x)\, dx$. [Hint: Note that ϕ is continuous on $[a, b]$, so that $\phi([a, b])$ is an interval. Define $F(x) = \int_{\phi(a)}^x f(u)\, du$ for all $x \in \phi([a, b])$. Define $g(t) = F(\phi(t))$ for $t \in [a, b]$, noting that, by the chain rule and appropriate versions of the FTC, $g'(t) = f(\phi(t))\phi'(t)$ and $\int_a^b g'(t)\, dt = g(b) - g(a)$.]

22. Let $a > 0$ and suppose that f is a continuous function on $[-a, a]$.
(a) If f is an even function, that is, if $f(-x) = f(x)$ for all $x \in [-a, a]$, prove that $\int_{-a}^a f(x)\, dx = 2 \int_0^a f(x)\, dx$.
(b) If f is an odd function, that is, if $f(-x) = -f(x)$ for all $x \in [-a, a]$, prove that $\int_{-a}^a f(x)\, dx = 0$.

23. The technique of integration known as ***u*-substitution** is employed in almost all elementary calculus textbooks. The technique works as follows: Suppose that f is an integrable function on the interval $[a, b]$ and it is desired to evaluate $\int_a^b f(x)\, dx$. We choose a differentiable function $\phi:[a, b] \to \mathbb{R}$ such that ϕ' is continuous, and we write $u = \phi(x)$. [Usually, u is taken to be an appropriate portion of the integrand $f(x)$.] We differentiate both sides of the equation $u = \phi(x)$ and rewrite the result in differential form $du = \phi'(x)\, dx$. Now, using the two equations $u = \phi(x)$ and $du = \phi'(x)\, dx$, we rewrite the entire expression $f(x)\, dx$ in terms of u and du only, so that, say, $f(x)\, dx = h(u)\, du$ and h is defined on $\phi([a, b])$. Then, if h is continuous, we conclude that $\int_a^b f(x)\,dx = \int_{\phi(a)}^{\phi(b)} h(u)\,du$. In many cases the integral on the right of the last equation is easier to evaluate than the original integral. Justify this procedure. [Hint: Use the change of variable theorem (Problem 21) with f replaced by h, noting that the equation $f(x)\,dx = h(u)\,du$ is shorthand for the condition that $f(x) = h(u)(du/dx)$, that is, $f(x) = h(\phi(x))\phi'(x)$ for all $x \in [a, b]$.]

24. If I is an interval contained in the domain of the function f, justify the following statements:

(a) It is possible for f to be indefinitely integrable on I and yet not have an antiderivative on I.

(b) It is possible for f to have an antiderivative on I and yet not to be indefinitely integrable on I.

(c) However, if f is indefinitely integrable on I, and if f has an antiderivative on I, then every indefinite integral of f on I is an antiderivative of f on I. [Hint: Let F be an antiderivative of f on I, let $a \in I$, use Theorem 7.14 to obtain the equation $\int_a^x f(t)\,dt = F(x) - F(a)$ for every $x \in I$, and use this equation to draw the desired conclusion.]

25. Evaluate $\int_2^6 x^2\sqrt{2x-3}\,dx$ by using the u-substitution $u = 2x - 3$ (Problem 23).

26. Prove **Taylor's theorem with integral remainder**: Let n be a nonnegative integer and suppose that f and its first $n + 1$ derivatives are continuous on an interval I and that $a \in I$. Then, for all $x \in I$,

$$f(x) = \sum_{k=0}^{n} \frac{f^{(k)}(a)}{k!}(x - a)^k + R_n(x),$$

where the remainder $R_n(x)$ is given by $R_n(x) = (1/n!)\int_a^x (x - t)^n f^{(n+1)}(t)\,dt$. [Hint: Use mathematical induction on n (starting from $n = 0$) and integration by parts (Theorem 7.15).]

27. Evaluate $\int_0^3 (x/\sqrt{x+1})\,dx$ using the indicated u-substitution (Problem 23).

(a) $u = x + 1$ (b) $u = \sqrt{x+1}$

28. Show that Taylor's theorem in the Lagrange form (Theorem 3.6) can be obtained for the case in which f and its first $n + 1$ derivatives are continuous on the interval I by combining Taylor's theorem with integral remainder (Problem 26) with the generalized mean value theorem for integrals (Theorem 6.9).

HISTORICAL NOTES

From the title of Leibniz's first work on differential calculus, *Nova Methodus Pro Maximis et Minimis*, it is clear what one of the motivating factors was in the development of calculus. The remarkable thing about this paper is the fact that, aside from it's being in Latin, a minor inconvenience, it is easy to read and looks amazingly like a modern text in calculus. One quickly recognizes the sum, product, and quotient rules for differentiation of functions. The reason is simple. Leibniz came upon notation that is almost perfect for "doing" calculus, and it has survived almost intact for over 300 years. Unfortunately, Newton was not nearly so fortunate in his choice of notation, the single dot over the variable, and this notation only survives in physics courses where parametrically expressed functions are differentiated with respect to time.

The Newton–Leibniz controversy over priority probably had a damaging effect on British science, since Newton enlisted some of his protégés in the Royal Society (which he headed) to participate in the dispute. But the controversary does not seem to have had much effect on the European continent. The Leibniz notation was clearly superior for day-to-day use of calculus. (And, of course, it is also possible that in the period following the discovery of calculus, the better mathematicians were on the continent.) During the eighteenth century, mathematics was largely dominated by the French school, represented by such towering figures as Lagrange, Pierre Simon de Laplace (1749–1827), and Adrien Marie Legendre (1752–1833). With the arrival of Gauss on the scene, the center shifted to Germany, and by the late nineteenth century, Paris (which had Poincaré) and the University of Göttingen were the centers of mathematical research.

The nineteenth century was a period not only of the movement to make calculus more rigorous, it was also a period of startling disclosures. Things are seldom what they seem, it turns out. Bolzano had as early as 1834 discovered a continuous function over an interval that failed to have a derivative at any point in that interval. Unfortunately, this did not become widely known, and it was not until Weierstrass gave an example of such a function in classroom lectures in 1861 (and in a paper to the Berlin Academy in 1872) that the mathematical world became aware of this startling and rather disconcerting result. The discovery of such nonintuitive results made a more rigorous foundation for analysis all the more necessary.

There are other surprising curves that defy easy analysis. Jean Gaston Darboux (1842–1917) showed that there is a function that takes on all the intermediate values between two given values in passing from $x = a$ to $x = b$ but is not continuous. Thus, he demonstrated that a basic property of continuous functions is not sufficient to ensure continuity! Giuseppe Peano (1858–1932) discovered a curve that covers the plane, and Hilbert, George Pólya (1887–1985), and others found similar curves with the same property. In 1904, Helge von Koch (1870–1924) described a continuous closed curve that has no tangent anywhere and, furthermore, showed that the arc length between any two points on the curve is infinite!

It is clear from examples like these that when dealing with the infinite, as one is when taking limits, intuition breaks down, and to save ourselves from unwarranted conclusions, we must be very careful to have accurate definitions and to avoid using intuitive geometric or physical arguments.

Newton had emphasized that integration is the inverse of differentiation, while Leibniz thought of integrals as limits of sums. Cauchy favored the Leibniz view. Cauchy, in fact, defined areas, arc lengths, volumes, and such as the appropriate integrals, rather than beginning with the geometric figure and assuming that the integral would calculate some geometric quantity. But in defining the arc length, for example, as an integral, since the integrand involves the derivative of the original function, one ends up assuming that the function is differentiable. In order to take into account functions of a less restricted character, more general definitions of integration were needed. One motivating factor was the need for a better understanding of Fourier series. These are series of trigonometric functions developed by Jean Baptiste Joseph Fourier (1768–1830) to solve problems in the theory of heat flow. Unfortunately, though his series worked, Fourier's derivations involved a number of dubious steps that made mathematicians uneasy about the results.

Georg Friedrich Bernhard Riemann (1826–1866) was one of the giants of late nineteenth century mathematics. Like two famous mathematicians earlier in the century, Niels Henrik Abel (1802–1829) and Évariste Galois, he died young, but only after having a profound influence on a variety of mathematical fields. His contributions to non-Euclidean geometry were important to Einstein in the development of the general theory of relativity. Elsewhere, he is known for his fundamental work in complex analysis (the Cauchy–Riemann equations are a reminder), with the Riemann ζ-function fundamental to many questions in number theory, among others.

Riemann was able to demonstrate a function $f(x)$ that is discontinuous at infinitely many points in an interval, but has an integral that exists and defines a continuous function $F(x)$ that, for the infinity of points described, fails to have a derivative. To handle such functions, it was necessary to go beyond the Cauchy definition of the integral and move on to what is now called the Riemann integral, the integral seen in essentially all elementary texts in analysis.

ANNOTATED BIBLIOGRAPHY

Boas, Ralph P., Jr. *A Primer of Real Functions*, 3rd ed. (Washington, D.C.: Mathematical Association of America), 1982.

> *This is a classic monograph on real functions, beautifully written and accessible, and full of surprises. This is as entertaining and rewarding a book as one will find on this subject.*

Hall, A. Rupert. *Philosophers at War: The Quarrel between Newton and Leibniz* (Cambridge: Cambridge University Press), 1980.

> *A modern appraisal of the famous Newton–Leibniz controversy over priority in the discovery of calculus. Over the years, there have been so many conflicting reports by both sides, often biased due to nationalistic prejudices, it is good to find something recent that tries to weed out the nonsense from the real evidence.*

Manuel, Frank E. *A Portrait of Isaac Newton.* (Cambridge, Mass.: Harvard University Press), 1968.

> *A "psychobiography" of Newton by a respected historian. Sometimes the Freudian tone grates, but the book is full of interesting material on Newton and his contemporaries. Newton does not always appear as god-like as earlier biographies, particularly the British, portrayed him. Definitely worth reading.*

Palter, Robert, ed. *The Annus Mirabilis of Sir Isaac Newton 1666–1966* (Cambridge, Mass.: MIT Press), 1970.

> *If you really want to know what Newton did in that famous year, 1666, you will find out from this set of essays by experts on Newton. It may be more than anyone really wanted to know!*

Smith, David Eugene. *A Source Book in Mathematics* (New York: Dover), 1959.

> *If you want to read the whole essay on fluxions by Bishop Berkeley, or see just what the first printed version of differential calculus looks like, this is a good book to browse through. For example, one finds in the Leibniz piece the following formula: $dx^a = ax^{a-1}\,dx$, something familiar to every high school student of calculus today.*

6

INFINITE SERIES

Although infinite sequences and infinite series are usually introduced in elementary calculus courses, it is not possible at that level to pay much attention to careful definitions and proofs. In this final chapter, we briefly review these topics, this time with more emphasis on mathematical rigor.

1

SEQUENCES AND SERIES

In Section 6 of Chapter 4, we defined a sequence in the set X to be a mapping $s: \mathbb{N} \to X$. In this chapter, we consider only sequences in $\mathbb{R}$. If $s: \mathbb{N} \to \mathbb{R}$ is such a sequence, and if $n \in \mathbb{N}$, we define $s_n \in \mathbb{R}$ by

$$s_n = s(n),$$

and we refer to s_n as the **nth term** of the sequence. For purposes of intuition, we imagine that the terms of the sequence are arranged in an endless progression $s_1, s_2, s_3, \ldots$.

The expression

$$(s_n)_{n \in \mathbb{N}}$$

is often used as alternative notation for the sequence $s: \mathbb{N} \to \mathbb{R}$. For instance,

$$(n^2)_{n \in \mathbb{N}}$$

denotes the sequence 1, 4, 9, 16, 25, ... of squares of the positive integers. Of course, different letters of the alphabet may be used for different sequences, for example:

$$(a_n)_{n \in \mathbb{N}}, \qquad (b_i)_{i \in \mathbb{N}}, \qquad (c_j)_{j \in \mathbb{N}},$$

and so on.

The definition of a limit of a sequence is similar to the definition of a limit of a function (Definition 2.1 in Chapter 4).

1.1 DEFINITION

LIMIT OF A SEQUENCE   A number $L \in \mathbb{R}$ is said to be a **limit** of the sequence $(s_n)_{n \in \mathbb{N}}$ if, for every $\varepsilon > 0$, there exists $N \in \mathbb{N}$ such that, for every integer $n \in \mathbb{N}$,

$$n \geq N \Rightarrow |s_n - L| < \varepsilon.$$

The condition that a sequence has L as a limit is often expressed by saying that the sequence **converges to L**. This is in accord with the terminology in the following definition.

1.2 DEFINITION

CONVERGENT AND DIVERGENT SEQUENCES   A sequence that has a limit is said to **converge**, or to be **convergent**. A sequence that has no limit is said to **diverge**, or to be **divergent**.

The idea in the following definition will enable us to recast Definition 1.1 in the language of neighborhoods.

1.3 DEFINITION

SEQUENCES EVENTUALLY IN A SET   We say that the sequence $(s_n)_{n \in \mathbb{N}}$ is **eventually in the set** $A \subseteq \mathbb{R}$ if there exists $N \in \mathbb{N}$ such that, for every $n \in \mathbb{N}$,

$$n \geq N \Rightarrow s_n \in A.$$

Note that a sequence is eventually in a set if all but possibly finitely many of its terms are in that set (Problem 1). A simple argument combining Definitions 1.1 and 1.3 proves the following lemma (Problem 3).

1.4 LEMMA

A real number L is a limit of a sequence if and only if the sequence is eventually in each neighborhood of L.

Now we are in a position to prove the analogue for sequences of the uniqueness theorem for limits of functions (Theorem 2.5 in Chapter 4).

1.5 THEOREM

> **UNIQUENESS OF LIMITS OF SEQUENCES** If L and L' are limits of the sequence $(s_n)_{n \in \mathbb{N}}$, then $L = L'$.

PROOF Suppose that L and L' are limits of $(s_n)_{n \in \mathbb{N}}$ and that $L \neq L'$. By Lemma 2.4 in Chapter 4 (the Hausdorff property of $\mathbb{R}$), there exist neighborhoods V and V' of L and L', respectively, such that

$$V \cap V' = \emptyset.$$

By Lemma 1.4, $(s_n)_{n \in \mathbb{N}}$ is eventually in both V and V'. Therefore, there exist $N, N' \in \mathbb{N}$ such that, for all $n \in \mathbb{N}$,

$$n \geq N \Rightarrow s_n \in V$$

and

$$n \geq N' \Rightarrow s_n \in V'.$$

Thus, choosing $n \in \mathbb{N}$ with $n \geq N, N'$, we obtain the contradiction

$$s_n \in V \cap V' = \emptyset. \qquad \square$$

1.6 DEFINITION

> **NOTATION FOR THE LIMIT OF A SEQUENCE** The notation
>
> $$s_n \to L \qquad \text{as } n \to \infty,$$
>
> or,
>
> $$\lim_{n \to \infty} s_n = L$$
>
> is used to indicate that $(s_n)_{n \in \mathbb{N}}$ is a convergent sequence and that its limit is L.

1.7 Example Show that $\lim_{n \to \infty}(1/n) = 0$.

SOLUTION Let $\varepsilon > 0$. By the Archimedian property of $\mathbb{R}$, there exists $N \in \mathbb{N}$ such that

$$\frac{1}{\varepsilon} < N.$$

Thus, for $n \in \mathbb{N}$ with $n \geq N$, we have

$$\left| \frac{1}{n} - 0 \right| = \frac{1}{n} \leq \frac{1}{N} < \varepsilon. \qquad \square$$

The next lemma is often useful in computing with sequences. We leave the simple proof as an exercise (Problem 5).

1.8 LEMMA

Let $(a_n)_{n \in \mathbb{N}}$ be a sequence, let $k \in \mathbb{N}$, and let the sequence $(b_n)_{n \in \mathbb{N}}$ be defined by $b_n = a_{n+k}$ for all $n \in \mathbb{N}$. Then if either of the sequences $(a_n)_{n \in \mathbb{N}}$ or $(b_n)_{n \in \mathbb{N}}$ converges, they both converge and

$$\lim_{n \to \infty} a_n = \lim_{n \to \infty} b_n.$$

Note that the conclusion of Lemma 1.8 may be rewritten in the form

$$\lim_{n \to \infty} a_n = \lim_{n \to \infty} a_{n+k}.$$

In addition to the concept of limit, several other ideas that we have used in connection with functions can be adapted to sequences. For instance, a sequence $(s_n)_{n \in \mathbb{N}}$ is said to be **increasing** (or **decreasing**) if $n \leq m$ implies that $s_n \leq s_m$ (or $s_n \geq s_m$), and a sequence that is either increasing or decreasing is said to be **monotone**. We note that $(s_n)_{n \in \mathbb{N}}$ is increasing (or decreasing) if and only if $s_n \leq s_{n+1}$ (or if $s_n \geq s_{n+1}$) holds for all $n \in \mathbb{N}$ (Problem 7).

If $s_n \leq K$ (or if $K \leq s_n$) holds for all $n \in \mathbb{N}$, we say that K is an **upper bound** (or a **lower bound**) for the sequence $(s_n)_{n \in \mathbb{N}}$. A sequence that has an upper (or a lower) bound is said to be **bounded above** (or **below**), and a sequence that is bounded both above and below is said to be **bounded**. Note that $(s_n)_{n \in \mathbb{N}}$ is bounded if and only if there is a positive number K such that $|s_n| \leq K$ for all $n \in \mathbb{N}$.

In the following theorem we establish the important fact that bounded monotone sequences are convergent.

1.9 THEOREM

LIMITS OF BOUNDED MONOTONIC SEQUENCES If $(s_n)_{n \in \mathbb{N}}$ is an increasing (or decreasing) sequence that is bounded above (or below), and if

$$S = \{s_n | n \in \mathbb{N}\},$$

then $L = \sup(S)$ (or $L = \inf(S)$) exists and

$$\lim_{n \to \infty} s_n = L.$$

PROOF   We prove the theorem for the case in which $(s_n)_{n \in \mathbb{N}}$ is an increasing sequence that is bounded above, and leave the proof in the remaining case as an exercise (Problem 9). By hypothesis, the set S is bounded above, and, since it is nonempty, S has a supremum in $\mathbb{R}$. Let

$$L = \sup(S).$$

To show that L is the limit of the sequence, let $\varepsilon > 0$. Then, by Part (i) of Lemma 3.23 in Chapter 3, there exists $N \in \mathbb{N}$ such that

$$L - \varepsilon < s_N.$$

Since $(s_n)_{n \in \mathbb{N}}$ is an increasing sequence with L as an upper bound, $n \geq N$ implies that

$$L - \varepsilon < s_N \leq s_n \leq L < L + \varepsilon,$$

and therefore

$$|s_n - L| < \varepsilon. \qquad \square$$

1.10 Example   If $r \in \mathbb{R}$ with $0 \leq r < 1$, prove that $\lim_{n \to \infty} r^n = 0$.

SOLUTION   Since $0 \leq r < 1$, it follows that $(r^n)_{n \in \mathbb{N}}$ is a decreasing sequence with 0 as a lower bound (Problem 11). Therefore, by Theorem 1.9,

$$L = \inf_{n \in \mathbb{N}} r^n = \lim_{n \to \infty} r^n.$$

Since $0 \leq r^n$ for all $n \in \mathbb{N}$, we have

$$0 \leq L.$$

Let $\varepsilon > 0$. Then, by Part (ii) of Lemma 3.23 in Chapter 3, there exists $k \in \mathbb{N}$ such that

$$r^k < L + \varepsilon.$$

Multiplying both sides of the last inequality by the nonnegative number r, we find that

$$r^{k+1} \leq r(L + \varepsilon) = rL + r\varepsilon.$$

Because L is a lower bound for $(r^n)_{n \in \mathbb{N}}$, we also have

$$L \leq r^{k+1},$$

and it follows that

$$L \leq rL + r\varepsilon.$$

Therefore,

$$(1 - r)L \leq r\varepsilon,$$

and, since $0 < 1 - r$, we infer that

$$0 \leq L \leq \frac{r}{1 - r}\varepsilon.$$

Finally, since ε is an arbitrarily small positive number, we conclude that

$$L = 0. \qquad \square$$

We say that a sequence $(s_n)_{n \in \mathbb{N}}$ is **constant** if there is a fixed real number c such that $s_n = c$ for every $n \in \mathbb{N}$. Such a sequence is eventually in (indeed, it is *always* in) each neighborhood of c, and it follows that

$$c = \lim_{n \to \infty} s_n.$$

The last equation is often written in the equivalent form

$$c = \lim_{n \to \infty} c.$$

Sequences may be added, subtracted, multiplied, and divided (provided that zeros do not appear in denominators) in much the same way as functions. For instance, if $(a_n)_{n \in \mathbb{N}}$ and $(b_n)_{n \in \mathbb{N}}$ are sequences, and if we define $(s_n)_{n \in \mathbb{N}}$ by

$$s_n = a_n + b_n$$

for all $n \in \mathbb{N}$, then we say that $(s_n)_{n \in \mathbb{N}}$ is the **sum** of the sequences $(a_n)_{n \in \mathbb{N}}$ and $(b_n)_{n \in \mathbb{N}}$, and we write $(s_n)_{n \in \mathbb{N}}$ as

$$(s_n)_{n \in \mathbb{N}} = (a_n + b_n)_{n \in \mathbb{N}}.$$

Similar definitions and notation apply to differences, products, and quotients of sequences. The arguments given in Section 2 of Chapter 4 pertaining to limits of sums, products, differences, and quotients of functions are easily adapted to sequences, and so we can leave the proofs of the various parts of the next theorem as exercises (Problems 13, 15, and 17).

1.11 THEOREM   **LIMITS OF ALGEBRAIC COMBINATIONS OF SEQUENCES**   Let $(a_n)_{n \in \mathbb{N}}$ and $(b_n)_{n \in \mathbb{N}}$ be sequences, suppose that

$$\lim_{n \to \infty} a_n = A \qquad \text{and} \qquad \lim_{n \to \infty} b_n = B,$$

and let $c \in \mathbb{R}$. Then:

(i) $\lim_{n \to \infty} ca_n = cA,$

(ii) $\lim_{n \to \infty} (a_n \pm b_n) = A \pm B,$

(iii) $\lim_{n \to \infty} (a_n b_n) = AB,$

and, provided that zeros do not appear in any denominators,

(iv) $\lim_{n \to \infty} \dfrac{a_n}{b_n} = \dfrac{A}{B}.$

Our next definition introduces a condition on a sequence that is considerably weaker than being eventually in a set.

1.12 DEFINITION

> **SEQUENCE FREQUENTLY IN A SET**   We say that a sequence $(s_n)_{n \in \mathbb{N}}$ is **frequently in the set** $A \subseteq \mathbb{R}$ if, for every $k \in \mathbb{N}$, there exists $n \in \mathbb{N}$ such that $n \geq k$ and $s_n \in A$.

In other words, a sequence is frequently in a set if infinitely many of its terms belong to the set (Problem 21). If a sequence is eventually in a set, then it is frequently in that set (Problem 23). Note the similarity between Lemma 1.4 and our next definition.

1.13 DEFINITION

> **CLUSTER POINT**   A real number c is said to be a **cluster point** of a sequence if and only if the sequence is frequently in each neighborhood of c.

Clearly, the limit of a convergent sequence is a cluster point of the sequence. The sequence $(a_n)_{n \in \mathbb{N}}$ defined for all $n \in \mathbb{N}$ by $a_n = 0$ if n is even and $a_n = 1$ if n is odd has no limit, but it has both 0 and 1 as cluster points. The sequence $(b_n)_{n \in \mathbb{N}}$ defined for all $n \in \mathbb{N}$ by $b_n = n$ has no cluster point. The following theorem, which can be regarded as a version of the Bolzano–Weierstrass theorem for sequences, guarantees that bounded sequences have cluster points.

1.14 THEOREM

> **BOLZANO–WEIERSTRASS THEOREM FOR SEQUENCES**   A bounded sequence has at least one cluster point.

PROOF   Suppose, on the contrary, that $(s_n)_{n \in \mathbb{N}}$ is a sequence with no cluster point and that there exist $a, b \in \mathbb{R}$ such that $s_n \in [a, b]$ for every $n \in \mathbb{N}$. If $x \in [a, b]$, then x is not a cluster point of $(s_n)_{n \in \mathbb{N}}$, and therefore there exists a neighborhood V_x of x such that $s_n \in V_x$ holds only for finitely many values of $n \in \mathbb{N}$. Since

$$\{V_x \mid x \in [a, b]\}$$

is an open cover of the compact set $[a, b]$, it follows that there exists a finite subset F of $[a, b]$ such that

$$[a, b] \subseteq \bigcup_{x \in F} V_x.$$

Hence, since F is a finite set, there can be only finitely many values of $n \in \mathbb{N}$ for which $s_n \in [a, b]$, contradicting the fact that $s_n \in [a, b]$ for every integer $n \in \mathbb{N}$.   □

Intuitively, the terms of a convergent sequence come closer and closer to the limit of the sequence, and therefore these terms must be coming closer and closer to each other. This idea gives rise to the following definition.

1.15 DEFINITION

CAUCHY SEQUENCE   A sequence $(s_n)_{n \in \mathbb{N}}$ is said to be a **Cauchy sequence** if, for every $\varepsilon > 0$, there exists $N \in \mathbb{N}$ such that, for all $n, m \in \mathbb{N}$,

$$n, m \geq N \Rightarrow |s_n - s_m| < \varepsilon.$$

1.16 Example   Show that every convergent sequence is a Cauchy sequence.

SOLUTION   Suppose that $(s_n)_{n \in \mathbb{N}}$ is a convergent sequence, let

$$\lim_{n \to \infty} s_n = L,$$

and let $\varepsilon > 0$. Then, by Definition 1.1, there exists $N \in \mathbb{N}$ such that, for every $n \in \mathbb{N}$,

$$n \geq N \Rightarrow |s_n - L| < \frac{\varepsilon}{2}.$$

Consequently, if $n, m \in \mathbb{N}$ and $n, m \geq N$, we have

$$|s_n - s_m| = |s_n - L + L - s_m| \leq |s_n - L| + |L - s_m|$$

$$< \frac{\varepsilon}{2} + \frac{\varepsilon}{2} = \varepsilon.$$ □

1.17 LEMMA   Every Cauchy sequence is bounded.

PROOF   Let $(s_n)_{n \in \mathbb{N}}$ be a Cauchy sequence. Then there exists $N \in \mathbb{N}$ such that, for all $n, m \in \mathbb{N}$,

$$n, m \geq N \Rightarrow |s_n - s_m| < 1.$$

In particular, if $n \in \mathbb{N}$ and $n \geq N$, then

$$|s_n| = |s_n - s_N + s_N| \leq |s_n - s_N| + |s_N| < 1 + |s_N|.$$

Let B denote the maximum of the numbers

$$|s_1|, |s_2|, |s_3|, \ldots, |s_N|,$$

so that, for every integer n with $1 \leq n \leq N$,

$$|s_n| \leq B.$$

Consequently, for every integer $n \in \mathbb{N}$,

$$|s_n| \leq \max(B, 1 + |s_N|). \qquad \square$$

1.18 THEOREM

> **CAUCHY CRITERION FOR CONVERGENCE**   A sequence converges if and only if it is a Cauchy sequence.

PROOF   Example 1.16 shows that a convergent sequence is a Cauchy sequence. Conversely, suppose that $(s_n)_{n \in \mathbb{N}}$ is a Cauchy sequence. By Lemma 1.17, the sequence $(s_n)_{n \in \mathbb{N}}$ is bounded; hence, by Theorem 1.14, it has a cluster point L. We are going to show that L is, in fact, the limit of the sequence.

Let $\varepsilon > 0$. Since $(s_n)_{n \in \mathbb{N}}$ is a Cauchy sequence, there exists $N' \in \mathbb{N}$ such that, for all $n, m \in \mathbb{N}$,

$$n, m \geq N' \Rightarrow |s_n - s_m| < \frac{\varepsilon}{2}.$$

Since L is a cluster point of $(s_n)_{n \in \mathbb{N}}$, there exists $N \in \mathbb{N}$ such that

$$N \geq N'$$

and

$$s_N \in \left(L - \frac{\varepsilon}{2}, L + \frac{\varepsilon}{2} \right),$$

that is,

$$|s_N - L| < \frac{\varepsilon}{2}.$$

Now suppose that $n \in \mathbb{N}$ and $n \geq N$. Then $n, N \geq N'$, and therefore

$$|s_n - s_N| < \frac{\varepsilon}{2}.$$

Consequently,

$$|s_n - L| = |s_n - s_N + s_N - L| \leq |s_n - s_N| + |s_N - L|$$
$$< \frac{\varepsilon}{2} + \frac{\varepsilon}{2} = \varepsilon. \qquad \square$$

The indicated sum

$$\sum_{k=1}^{\infty} a_k$$

of all the terms of a sequence $(a_k)_{k \in \mathbb{N}}$ is called an **infinite series** (or, for short, simply a **series**) and, for $k \in \mathbb{N}$, a_k is called the **kth term** of the series. If $n \in \mathbb{N}$, then the finite sum

$$s_n = a_1 + a_2 + a_3 + \cdots + a_n = \sum_{k=1}^{n} a_k$$

of the first n terms of the series is referred to as the **nth partial sum** of the series. [Many authors *define* the infinite series to be the sequence $(s_n)_{n \in \mathbb{N}}$ of partial sums, but we prefer to be not quite so fussy here.]

As an aid to intuition, we often write a series

$$\sum_{k=1}^{\infty} a_k$$

in the alternative form

$$a_1 + a_2 + a_3 + \cdots + a_k + \cdots,$$

or, if the kth term is understood, simply as

$$a_1 + a_2 + a_3 + \cdots.$$

1.19 DEFINITION

> **CONVERGENCE OF A SERIES**   If the sequence $(s_n)_{n \in \mathbb{N}}$ of partial sums of the series
>
> $$\sum_{k=1}^{\infty} a_k$$
>
> has a limit S, we say that the series **converges**, we refer to S as the **sum** of the series, and we write
>
> $$\sum_{k=1}^{\infty} a_k = S;$$
>
> otherwise, we say that the series **diverges**.

Thus, by definition,

$$\sum_{k=1}^{\infty} a_k = \lim_{n \to \infty} \sum_{k=1}^{n} a_k,$$

provided that the limit on the right exists; otherwise we do not assign any numerical value to the symbol

$$\sum_{k=1}^{\infty} a_k.$$

If $(b_n)_{n \in \mathbb{N}}$ is a sequence, then a series of the form

$$\sum_{n=1}^{\infty} (b_k - b_{k+1})$$

is called a **telescoping series**. Note that the nth partial sum of such a series is given by

$$s_n = (b_1 - b_2) + (b_2 - b_3) + \cdots + (b_{n-1} - b_n) + (b_n - b_{n+1}),$$

and, since all terms b_k for $1 < k < n + 1$ cancel in pairs on the right, we have

$$s_n = b_1 - b_{n+1}.$$

Consequently, the telescoping series converges if and only if the sequence $(b_n)_{n \in \mathbb{N}}$ converges; and, if

$$B = \lim_{n \to \infty} b_n = \lim_{n \to \infty} b_{n+1}$$

(see Lemma 1.8), then

$$\sum_{k=1}^{\infty} (b_k - b_{k+1}) = b_1 - B.$$

1.20 Example   Show that the series

$$\sum_{k=1}^{\infty} \frac{1}{k(k+1)}$$

converges and find its sum.

SOLUTION   Because

$$\frac{1}{k(k+1)} = \frac{1}{k} - \frac{1}{k+1},$$

the series telescopes, and we have

$$\sum_{k=1}^{\infty} \frac{1}{k(k+1)} = \sum_{k=1}^{\infty} \left(\frac{1}{k} - \frac{1}{k+1} \right) = \lim_{n \to \infty} \sum_{k=1}^{n} \left(\frac{1}{k} - \frac{1}{k+1} \right)$$

$$= \lim_{n \to \infty} \left(1 - \frac{1}{n+1} \right)$$

$$= 1 - \lim_{n \to \infty} \frac{1}{n+1} = 1 - 0 = 1. \qquad \square$$

1.21 DEFINITION

> **GEOMETRIC SERIES**   A series
>
> $$\sum_{k=1}^{\infty} a_n$$
>
> is said to be **geometric** with **ratio** r if r is a fixed real number and
>
> $$a_{n+1} = a_n r$$
>
> holds for all $n \in \mathbb{N}$.

In other words, in a geometric series, successive terms have a constant ratio r. Let

$$\sum_{k=1}^{\infty} a_n \qquad (1)$$

be a geometric series with ratio r and let $a = a_1$ be its initial term. Then

$$a_2 = ar, \qquad a_3 = a_2 r = ar^2, \qquad a_4 = a_3 r = ar^3, \qquad (2)$$

and so forth. By mathematical induction, we have

$$a_k = ar^{k-1} \qquad (3)$$

for all $k \in \mathbb{N}$. Therefore, the geometric series (1) with ratio r and initial term a may be rewritten in the form (Problem 37)

$$\sum_{k=1}^{\infty} ar^{k-1}. \qquad (4)$$

Let

$$s_n = a + ar + ar^2 + \cdots + ar^{n-1} \qquad (5)$$

be the nth partial sum of the geometric series (4). By multiplying both sides of (5) by r, we obtain

$$s_n r = ar + ar^2 + \cdots + ar^{n-1} + ar^n. \qquad (6)$$

By subtracting (6) from (5), we find that

$$s_n - s_n r = a - ar^n, \qquad (7)$$

and therefore

$$s_n(1 - r) = a(1 - r^n). \qquad (8)$$

Consequently, if $r \neq 1$, we have

$$s_n = a\left(\frac{1 - r^n}{1 - r}\right). \qquad (9)$$

1.22 THEOREM

CONVERGENCE OF A GEOMETRIC SERIES A geometric series

$$\sum_{k=1}^{\infty} ar^{k-1}$$

with ratio r and initial term $a \neq 0$ converges if and only if $|r| < 1$, in which case

$$\sum_{k=1}^{\infty} ar^{k-1} = \frac{a}{1-r}.$$

PROOF We consider the case in which $|r| < 1$, leaving the case in which $|r| \geq 1$ as an exercise (Problem 39b). Since $|r| < 1$, we have

$$\lim_{n \to \infty} |r|^n = 0$$

(see Example 1.10). Therefore, if $\varepsilon > 0$, there exists $N \in \mathbb{N}$ such that, for all $n \in \mathbb{N}$,

$$n \geq N \Rightarrow |r^n - 0| = |r^n| = |r|^n < \varepsilon.$$

Consequently,

$$\lim_{n \to \infty} r^n = 0.$$

The nth partial sum of the geometric series is given by

$$s_n = a\left(\frac{1 - r^n}{1 - r}\right),$$

and it follows that

$$\lim_{n \to \infty} s_n = a\left(\frac{1 - 0}{1 - r}\right) = \frac{a}{1-r}. \qquad \square$$

1.23 Example Show that the geometric series

$$-1 + \frac{2}{3} - \frac{4}{9} + \frac{8}{27} - \frac{16}{81} + \cdots$$

is convergent and find its sum.

SOLUTION The initial term is $a = -1$ and the ratio is $r = -\frac{2}{3}$, so, by Theorem 1.22, the series converges and we have

$$-1 + \frac{2}{3} - \frac{4}{9} + \frac{8}{27} - \frac{16}{81} + \cdots = \frac{a}{1-r} = \frac{-1}{1 + \frac{2}{3}} = -\frac{3}{5}. \qquad \square$$

PROBLEM SET 6.1

1. Prove that a sequence $(s_n)_{n \in \mathbb{N}}$ is eventually in the set $A \subseteq \mathbb{R}$ if and only if $\{n \in \mathbb{N} \mid s_n \notin A\}$ is a finite set.

2. If $\mathbb{N}$ is contained in the domain of the function f, and if the sequence $(s_n)_{n \in \mathbb{N}}$ is defined by $s_n = f(n)$ for all $n \in \mathbb{N}$, prove that $\lim_{n \to \infty} s_n = L$ if and only if $\lim_{x \to \infty} f(x) = L$. (See Problem 31 in Problem Set 4.2.)

3. Prove Lemma 1.4.

4. Prove that *continuous functions preserve limits of convergent sequences* in the following sense: If $(s_n)_{n \in \mathbb{N}}$ is a convergent sequence, $\{s_n \mid n \in \mathbb{N}\}$ is contained in the domain of the function f, and f is continuous at $L = \lim_{n \to \infty} s_n$, then the sequence $(f(s_n))_{n \in \mathbb{N}}$ is convergent and $\lim_{n \to \infty} f(s_n) = f(\lim_{n \to \infty} s_n)$.

5. Prove Lemma 1.8.

6. Prove the converse of the result in Problem 4: If f is a function, L is a point of accumulation of $\mathrm{dom}(f)$, and if, for every sequence $(s_n)_{n \in \mathbb{N}}$ such that $\{s_n \mid n \in \mathbb{N}\} \subseteq \mathrm{dom}(f)$ and $L = \lim_{n \to \infty} s_n$, the condition $\lim_{n \to \infty} f(s_n) = f(L)$ holds, then f is continuous at L.

7. Let $(s_n)_{n \in \mathbb{N}}$ be a sequence. Prove:
   (a) If $s_n \leq s_{n+1}$ holds for all $n \in \mathbb{N}$, then $(s_n)_{n \in \mathbb{N}}$ is increasing.
   (b) If $s_n \geq s_{n+1}$ holds for all $n \in \mathbb{N}$, then $(s_n)_{n \in \mathbb{N}}$ is decreasing.

8. Let $(s_n)_{n \in \mathbb{N}}$ be a sequence. Prove:
   (a) If $s_n < s_{n+1}$ holds for all $n \in \mathbb{N}$, then $(s_n)_{n \in \mathbb{N}}$ is strictly increasing.
   (b) If $s_n > s_{n+1}$ holds for all $n \in \mathbb{N}$, then $(s_n)_{n \in \mathbb{N}}$ is strictly decreasing.

9. Complete the proof of Theorem 1.9 by considering the case in which $(s_n)_{n \in \mathbb{N}}$ is a decreasing sequence that is bounded below.

10. A sequence $(s_n)_{n \in \mathbb{N}}$ is said to have a property **eventually** if there is an integer $k \geq 0$ such that the sequence $(s_{n+k})_{n \in \mathbb{N}}$ has the property. Prove that a bounded eventually monotone sequence converges.

11. If $0 \leq r < 1$, prove that the sequence $(r^n)_{n \in \mathbb{N}}$ is decreasing.

12. If $0 < a < 2$, prove that the sequence $(a^n/n!)_{n \in \mathbb{N}}$ is strictly decreasing.

13. (a) Prove Part (i) of Theorem 1.11.
    (b) Prove Part (ii) of Theorem 1.11.

14. Prove that $\lim_{n \to \infty} n^{1/n} = 1$.

15. Prove Part (iii) of Theorem 1.11. [Hint: Use Lemma 1.4 together with Lemma 2.13 in Chapter 4.]

16. Let c be a real number. Prove:
    (a) There exists a sequence of rational numbers with c as a limit.
    (b) There exists a sequence of irrational numbers with c as a limit.

17. Prove Part (iv) of Theorem 1.11. [Hint: Begin by proving that if $(b_n)_{n \in \mathbb{N}}$ is a convergent sequence of nonzero numbers and if $B = \lim_{n \to \infty} b_n \neq 0$, then $\lim_{n \to \infty} (1/b_n) = 1/B$. To do this, use Lemma 1.4 together with Lemma

2.15 in Chapter 4. Then note that $a_n/b_n = a_n(1/b_n)$ and use the result of Problem 15.]

18. If $M \subseteq \mathbb{R}$ and $p \in \mathbb{R}$, prove that p is an accumulation point of M if and only if there exists a sequence of numbers in $M \setminus \{p\}$ that converges to p.

19. If $(a_n)_{n \in \mathbb{N}}$ is a bounded sequence and if $(b_n)_{n \in \mathbb{N}}$ is a sequence that converges to 0, prove that $(a_n b_n)_{n \in \mathbb{N}}$ converges to 0.

20. If $M \subseteq \mathbb{R}$ and $p \in \mathbb{R}$, prove that $p \in \bar{M}$ if and only if there is a sequence of numbers in M that converges to p. [Hint: See Problem 18.]

21. Prove that a sequence $(s_n)_{n \in \mathbb{N}}$ is frequently in the set $A \subseteq \mathbb{R}$ if and only if $\{n \in \mathbb{N} \mid s_n \in A\}$ is an infinite set.

22. Prove that a sequence $(s_n)_{n \in \mathbb{N}}$ is frequently in a set $A \subseteq \mathbb{R}$ if and only if it is not eventually in $\mathbb{R} \setminus A$.

23. If a sequence is eventually in a set, prove that it is frequently in that set.

24. If $(n_k)_{k \in \mathbb{N}}$ is a strictly increasing sequence of positive integers, then the sequence $(s_{n_k})_{k \in \mathbb{N}}$ is called a **subsequence** of the sequence $(s_n)_{n \in \mathbb{N}}$. If $(s_n)_{n \in \mathbb{N}}$ converges to the limit L, prove that every subsequence of $(s_n)_{n \in \mathbb{N}}$ converges to L.

25. If $(a_n)_{n \in \mathbb{N}}$ and $(b_n)_{n \in \mathbb{N}}$ are convergent sequences, and if $a_n \leq b_n$ holds for all $n \in \mathbb{N}$, prove that $\lim_{n \to \infty} a_n \leq \lim_{n \to \infty} b_n$.

26. Prove that a point is a cluster point of a sequence if and only if there is a subsequence of the sequence that converges to the point. (See Problem 24.)

27. Show that the idea of a limit of a sequence is topological. [Hint: Show that the words *neighborhood of L* in Lemma 1.4 can be replaced by the words *open set containing L*.]

28. Prove that the set of all cluster points of a sequence is a closed set.

29. Prove that a sequence $(s_n)_{n \in \mathbb{N}}$ is a Cauchy sequence if and only if it satisfies the following condition: For every $\varepsilon > 0$ there exists $N \in \mathbb{N}$ such that, for every integer $n \geq 0$, $|s_{N+n} - s_N| \leq \varepsilon$.

30. Let $(s_n)_{n \in \mathbb{N}}$ be a bounded sequence.
(a) The **limit superior** of $(s_n)_{n \in \mathbb{N}}$ is defined to be $\inf_{k \in \mathbb{N}} \sup_{n \geq k} s_n$. Prove that the limit superior exists and that it is the largest cluster point of the sequence.
(b) The **limit inferior** of $(s_n)_{n \in \mathbb{N}}$ is defined to be $\sup_{k \in \mathbb{N}} \inf_{n \geq k} s_n$. Prove that the limit inferior exists and that it is the smallest cluster point of the sequence.

31. Prove that every convergent sequence is bounded.

32. Prove that a set $S \subseteq \mathbb{R}$ is compact if and only if every sequence in S has a cluster point.

33. If $(s_n)_{n \in \mathbb{N}}$ is a sequence, prove that there exists a series $\sum_{k=1}^{\infty} a_k$ with $(s_n)_{n \in \mathbb{N}}$ as its sequence of partial sums.

34. If $c_0 \in \mathbb{R}$ and $(c_n)_{n \in \mathbb{N}}$ is a sequence, prove that the series $\sum_{k=1}^{\infty}(c_n - c_{n-1})$ telescopes and that, for each $n \in \mathbb{N}$, its nth partial sum is $c_n - c_0$.

35. Prove that a series $\sum_{k=1}^{\infty} a_k$ can always be rewritten as a telescoping series $\sum_{k=1}^{\infty}(b_k - b_{k+1})$. [Hint: Let s_n be the nth partial sum of the series and let $b_k = -s_k$ for $k \in \mathbb{N}$.]

36. If $(a_n)_{n \in \mathbb{N}}$ is a sequence of positive numbers, find a formula for the nth partial sum s_n of the series $\sum_{k=1}^{\infty} \ln(a_n/a_{n+1})$.

37. Prove that a geometric series with initial term a and ratio r can be written in the form $\sum_{k=1}^{\infty} ar^{k-1}$.

38. If $r \neq 1$, then the nth partial sum of the geometric series with initial term a and ratio r is $a(1 - r^n)/(1 - r)$. Find a formula for the nth partial sum for the case in which $r = 1$.

39. (a) If $|r| > 1$, prove that the sequence $(r^n)_{n \in \mathbb{N}}$ diverges.
   (b) Complete the proof of Theorem 1.22 by proving that a geometric series with initial term $a \neq 0$ and ratio r diverges if $|r| \geq 1$.

40. A beaker originally contains a grams of salt dissolved in V milliliters of water. Let $0 < W < V$. The following procedure is performed repeatedly: Exactly W milliliters of saltwater is poured out of the beaker and replaced by $V - W$ milliliters of pure water, and then the mixture is thoroughly stirred.
   (a) After this procedure is repeated n times, how many grams of salt remain in the beaker?
   (b) If the procedure is repeated "infinitely often," how much salt will remain in the beaker?

2

PROPERTIES OF SERIES

In Section 1, we were able to find formulas for the partial sums of telescoping and geometric series. However, apart from these special cases, it is rarely possible to find tractable formulas for partial sums, so it is important to develop general methods for testing a series for convergence or divergence, and for dealing with its sum if it does converge. We devote the present section to a brief study of some of these techniques.

2.1 THEOREM

CAUCHY CRITERION FOR CONVERGENCE OF A SERIES
A series $\sum_{k=1}^{\infty} a_k$ converges if and only if it satisfies the following condition: For every $\varepsilon > 0$, there exists $M \in \mathbb{N}$ such that, for every integer $q \geq 0$, $\left|\sum_{k=M}^{M+q} a_k\right| < \varepsilon$.

PROOF   For $n \in \mathbb{N}$, let $s_n = \sum_{k=1}^{n} a_k$ be the nth partial sum of the series $\sum_{k=1}^{\infty} a_k$. By Theorem 1.18, $\sum_{k=1}^{\infty} a_k$ converges if and only if $(s_n)_{n \in \mathbb{N}}$ is a Cauchy sequence.

Suppose that $(s_n)_{n \in \mathbb{N}}$ is a Cauchy sequence and let $\varepsilon > 0$. Then there exists $N \in \mathbb{N}$ such that, for $m, n \in \mathbb{N}$,

$$m, n \geq N \Rightarrow |s_n - s_m| < \varepsilon.$$

Let $M = N + 1$ and let q be an integer with $q \geq 0$. Then $N + 1 + q, N \geq N$, and it follows that

$$\left| \sum_{k=M}^{M+q} a_k \right| = \left| \sum_{k=1}^{M+q} a_k - \sum_{k=1}^{M-1} a_k \right| = \left| s_{M+q} - s_{M-1} \right|$$

$$= \left| s_{N+1+q} - s_N \right| < \varepsilon.$$

Conversely, suppose that the given condition holds and let $\varepsilon > 0$. Then there exists $M \in \mathbb{N}$ such that, for all integers q with $q \geq 0$,

$$\left| \sum_{k=M}^{M+q} a_k \right| < \frac{\varepsilon}{4}.$$

To prove that $(s_n)_{n \in \mathbb{N}}$ is a Cauchy sequence, it will be sufficient to show that, if $n, m \geq M$, then $|s_n - s_m| < \varepsilon$. We have

$$|-a_M| = |a_M| = \left| \sum_{k=M}^{M+0} a_k \right| < \frac{\varepsilon}{4},$$

and therefore, if q is an integer and $q \geq 1$, then

$$\left| \sum_{k=M+1}^{M+q} a_k \right| = \left| \sum_{k=M}^{M+q} a_k - a_M \right| \leq \left| \sum_{k=M}^{M+q} a_k \right| + |-a_M|$$

$$< \frac{\varepsilon}{4} + \frac{\varepsilon}{4} = \frac{\varepsilon}{2}.$$

Let $n, m \in \mathbb{N}$ with $n, m \geq M$. If $n > M$, then, with $q = n - M$ in the inequality obtained above, we have

$$|s_n - s_M| = |s_{M+q} - s_M| = \left| \sum_{k=M+1}^{M+q} a_k \right| < \frac{\varepsilon}{2}.$$

If $n = M$, then $|s_n - s_M| = 0$, so, in any case, the fact that $n \geq M$ implies that

$$|s_n - s_M| < \frac{\varepsilon}{2}.$$

Similarly, the fact that $m \geq M$ implies that

$$|s_M - s_m| = |s_m - s_M| < \frac{\varepsilon}{2},$$

and consequently,

$$|s_n - s_m| = |s_n - s_M + s_M - s_m| \leq |s_n - s_M| + |s_M - s_m|$$

$$< \frac{\varepsilon}{2} + \frac{\varepsilon}{2} = \varepsilon. \qquad \square$$

2.2 THEOREM

A NECESSARY CONDITION FOR CONVERGENCE
If the series $\sum_{k=1}^{\infty} a_k$ converges, then $\lim_{n \to \infty} a_n = 0$.

PROOF Let $(s_n)_{n \in \mathbb{N}}$ be the sequence of partial sums of the series and let S be the sum of the series. Then, for every $n \in \mathbb{N}$,

$$a_{n+1} = s_{n+1} - s_n,$$

and it follows from by Lemma 1.8 that

$$\lim_{n \to \infty} a_n = \lim_{n \to \infty} a_{n+1} = \lim_{n \to \infty} (s_{n+1} - s_n) = S - S = 0. \qquad \square$$

2.3 Example Show that the converse of Theorem 2.2 is false by exhibiting a divergent series whose terms converge to 0.

SOLUTION Recall from elementary calculus that the series

$$\sum_{k=1}^{\infty} \frac{1}{k}$$

diverges in spite of the fact that

$$\lim_{n \to \infty} \frac{1}{n} = 0.$$

Another interesting example is the series

$$\sum_{k=1}^{\infty} \ln \frac{k}{k+1}.$$

Here we assume the usual properties of the natural logarithm function ln as developed in elementary calculus. Because

$$\ln \frac{k}{k+1} = \ln k - \ln(k+1),$$

this series telescopes, and its nth partial sum is

$$s_n = \ln 1 - \ln(n+1) = -\ln(n+1).$$

However, $-\ln(n + 1)$ has no limit as $n \to \infty$, so the series diverges in spite of the fact that

$$\lim_{n \to \infty} \left(\ln \frac{n}{n + 1} \right) = \ln 1 = 0.$$ □

As an immediate consequence of Theorem 2.2 we have the following useful test for the divergence of a series:

2.4 COROLLARY

TEST FOR DIVERGENCE If the sequence $(a_n)_{n \in \mathbb{N}}$ diverges, or if it converges to a number other than 0, then the series $\sum_{k=1}^{\infty} a_k$ diverges.

2.5 Example Show that the series $\sum_{k=1}^{\infty} \sec(\pi/3k)$ diverges.

SOLUTION

$$\lim_{n \to \infty} \left(\sec \frac{\pi}{3n} \right) = \sec 0 = 1 \neq 0.$$ □

The proof of the next lemma, which follows easily from Theorem 1.11, is left as an exercise (Problem 7).

2.6 LEMMA

Let $\sum_{k=1}^{\infty} a_n$ and $\sum_{k=1}^{\infty} b_n$ be convergent series and let c be a real number. Then:

(i) $\displaystyle\sum_{k=1}^{\infty} (a_n \pm b_n) = \sum_{k=1}^{\infty} a_n \pm \sum_{k=1}^{\infty} b_n.$

(ii) $\displaystyle\sum_{k=1}^{\infty} ca_n = c \sum_{k=1}^{\infty} a_n.$

2.7 Example If c is nonzero constant and $\sum_{k=1}^{\infty} a_n$ diverges, prove that $\sum_{k=1}^{\infty} ca_n$ diverges.

SOLUTION Suppose, on the contrary, that $\sum_{k=1}^{\infty} ca_n$ converges. Then, by Lemma 2.6,

$$\sum_{k=1}^{\infty} a_n = \sum_{k=1}^{\infty} \frac{1}{c} ca_n$$

converges, contrary to the hypothesis. □

If none of the terms of a series is negative, then the sequence of partial sums of the series is increasing. Therefore, the following lemma is a consequence of Theorem 1.9 (Problem 11).

2.8 LEMMA

Suppose that $\sum_{k=1}^{\infty} a_k$ is a series of nonnegative terms.

(i) If there exists $B \geq 0$ such that $\sum_{k=1}^{n} a_k \leq B$ holds for all $n \in \mathbb{N}$, then the series $\sum_{k=1}^{\infty} a_n$ converges.

(ii) If the series $\sum_{k=1}^{\infty} a_n$ converges, then, for all $n \in \mathbb{N}$,
$$\sum_{k=1}^{n} a_k \leq \sum_{k=1}^{\infty} a_k.$$

In the notation

$$\sum_{k=1}^{\infty} a_k$$

we refer to k as the **summation index**. Note that the summation index is a dummy variable, so that

$$\sum_{k=1}^{\infty} a_k = \sum_{n=1}^{\infty} a_n = \sum_{j=1}^{\infty} a_j = \sum_{i=1}^{\infty} a_i,$$

and so forth.

In computing with series, we often perform simple algebraic manipulations involving the summation indices. For instance, it may be convenient to start a summation with an integer other than $k = 1$. Thus, if q is an integer (positive, negative, or zero), and if $a_n \in \mathbb{R}$ for every $n \in \mathbb{Z}$ with $n \geq q$, we understand that

$$\sum_{n=q}^{\infty} a_n = \sum_{k=1}^{\infty} a_{q+k-1} = a_q + a_{q+1} + a_{q+2} + \cdots.$$

For instance, a geometric series may be written as

$$\sum_{n=0}^{\infty} ar^n = \sum_{k=1}^{\infty} ar^{k-1} = a + ar + ar^2 + ar^3 + \cdots.$$

Our next lemma shows that the summation index may be changed in a series in much the same way that the variable of integration may be changed in an integral. We leave the proof as an exercise (Problem 13).

2.9 LEMMA

CHANGE OF SUMMATION INDEX   Let p and q be integers and suppose that $a_n \in \mathbb{R}$ for every integer n with $n \geq q$. Then, if we put $n = p + k$, we have

$$\sum_{n=q}^{\infty} a_n = \sum_{k=q-p}^{\infty} a_{p+k}.$$

2.10 Example   Rewrite the series

$$\sum_{k=2}^{\infty} \frac{1}{k(k-1)}$$

by changing the summation index from k to $j = k - 1$.

SOLUTION   If $j = k - 1$, then $j = 1$ when $k = 2$. Also, $k = j + 1$, and so

$$\sum_{k=2}^{\infty} \frac{1}{k(k-1)} = \sum_{j=1}^{\infty} \frac{1}{(j+1)j} = \sum_{j=1}^{\infty} \frac{1}{j(j+1)}.$$

Since the summation index j in the series on the right is a dummy variable, it may be replaced by k if desired, so that

$$\sum_{k=2}^{\infty} \frac{1}{k(k-1)} = \sum_{k=1}^{\infty} \frac{1}{k(k+1)}. \qquad \square$$

The following theorem shows that the first M terms of a series have no effect whatsoever on its convergence or divergence—it is only the "tail end" of a series that determines whether it converges or diverges.

2.11 THEOREM

> **REMOVING THE FIRST M TERMS FROM A SERIES**   Let M be a positive integer. Then the series $\sum_{k=1}^{\infty} a_k$ converges if and only if the series $\sum_{k=M+1}^{\infty} a_k$ converges and, if these series converge, then
>
> $$\sum_{k=1}^{\infty} a_k = \sum_{k=1}^{M} a_k + \sum_{k=M+1}^{\infty} a_k.$$

PROOF   For any integer $n > M$,

$$\sum_{k=1}^{n} a_k = \sum_{k=1}^{M} a_k + \sum_{k=M+1}^{n} a_k$$

and therefore $\lim_{n \to \infty} \sum_{k=1}^{n} a_k$ exists if and only if $\lim_{n \to \infty} \sum_{k=M+1}^{n} a_k$ exists. Furthermore, if these limits exist, then

$$\sum_{k=1}^{\infty} a_k = \lim_{n \to \infty} \sum_{k=1}^{n} a_k = \sum_{k=1}^{M} a_k + \lim_{n \to \infty} \sum_{k=M+1}^{n} a_k$$

$$= \sum_{k=1}^{M} a_k + \sum_{k=M+1}^{\infty} a_k. \qquad \square$$

2.12 DEFINITION

> **EVENTUAL DOMINATION**   We say that the series $\sum_{k=1}^{\infty} b_k$ **eventually dominates** the series $\sum_{k=1}^{\infty} a_k$ if there is a positive integer M such that
>
> $$0 \le a_k \le b_k$$
>
> holds for every $k \ge M$.

If $M = 1$ in Definition 2.12, so that

$$0 \le a_k \le b_k$$

holds for all $k \in \mathbb{N}$, we say that the series $\sum_{k=1}^{\infty} b_k$ **dominates** the series $\sum_{k=1}^{\infty} a_k$.

In our next theorem, we establish the fundamental **comparison test** for series. This theorem provides the basis for many of the standard tests for convergence or divergence of series.

2.13 THEOREM

COMPARISON TEST   Suppose that the series $\sum_{k=1}^{\infty} b_k$ eventually dominates the series $\sum_{k=1}^{\infty} a_k$. Then:

(i) If $\sum_{k=1}^{\infty} b_k$ converges, then $\sum_{k=1}^{\infty} a_k$ converges.

(ii) If $\sum_{k=1}^{\infty} a_k$ diverges, then $\sum_{k=1}^{\infty} b_k$ diverges.

PROOF   We prove Part (i) and leave the proof of Part (ii) as an exercise (Problem 17). Suppose that $\sum_{k=1}^{\infty} b_k$ converges. By hypothesis, there exists a positive integer M such that, for every integer $k \ge M$,

$$0 \le a_k \le b_k.$$

Consequently, for every integer $n > M$, we have

$$\sum_{k=M+1}^{n} a_k \le \sum_{k=M+1}^{n} b_k.$$

By Theorem 2.11, $\sum_{k=M+1}^{\infty} b_k$ converges. Therefore, by Part (ii) of Lemma 2.8,

$$\sum_{k=M+1}^{n} a_k \le \sum_{k=M+1}^{n} b_k \le \sum_{k=M+1}^{\infty} b_k$$

holds for every integer $n > M$, and it follows from Part (i) of Lemma 2.8 that $\sum_{k=M+1}^{\infty} a_k$ converges. Hence, $\sum_{k=1}^{\infty} a_k$ converges by Theorem 2.11. $\square$

In practice, the following corollary of the comparison test is often easier to apply than the comparison test itself.

2.14 COROLLARY

LIMIT COMPARISON TEST   Let $\sum_{k=1}^{\infty} a_k$ be a series with nonnegative terms, let $\sum_{k=1}^{\infty} b_k$ be a series with positive terms, and suppose that $\lim_{n \to \infty} (a_n / b_n) = L > 0$. Then either both series converge or else both series diverge.

PROOF   Let $\varepsilon = L/2$. Then, by hypothesis, there exists a positive integer M such that, for every integer $n \geq M$,

$$\frac{L}{2} = L - \varepsilon < \frac{a_n}{b_n} < L + \varepsilon = \frac{3L}{2},$$

and therefore

$$\frac{L}{2} b_n < a_n < \frac{3L}{2} b_n.$$

Consequently, the series $\sum_{k=1}^{\infty} a_k$ eventually dominates the series $\sum_{k=1}^{\infty} (L/2)b_k$, and the series $\sum_{k=1}^{\infty} (3L/2)b_k$ eventually dominates the series $\sum_{k=1}^{\infty} a_k$. Thus, the statements to be proved follow from Part (ii) of Lemma 2.6 and Theorem 2.13.   □

2.15 Example   Prove that the series $\sum_{k=1}^{\infty} (1/k^2)$ converges.

SOLUTION   In Example 1.20, we showed that the telescoping series $\sum_{k=1}^{\infty} [1/k(k+1)]$ converges. Let $a_n = 1/n^2$ and $b_n = 1/n(n+1)$ for each $n \in \mathbb{N}$. By applying the limit comparison test (Corollary 2.14), we find that

$$\lim_{n \to \infty} \frac{a_n}{b_n} = \lim_{n \to \infty} \frac{n(n+1)}{n^2} = \lim_{n \to \infty} \frac{n+1}{n} = \lim_{n \to \infty} \left(1 + \frac{1}{n}\right) = 1 > 0,$$

from which we may conclude that $\sum_{k=1}^{\infty} (1/k^2)$ converges.   □

Although the argument in Example 2.15 shows that the series $\sum_{k=1}^{\infty} (1/k^2)$ converges, it gives no clue as to its numerical value. In 1734 the Swiss mathematician Leonhard Euler[†] (1707–1783), probably the most prolific mathematician of all time, was able to show that

$$\sum_{k=1}^{\infty} \frac{1}{k^2} = \frac{\pi^2}{6}.$$

For an elegant proof of this remarkable result (who would have suspected that the ratio between the circumference and diameter of a circle is related to the reciprocals of the squares of the positive integers?), see Ioannis Papadimitriou's Classroom Note, "A Simple Proof of the Formula $\sum_{k=1}^{\infty} k^{-2} = \pi^2/6$," *The American Mathematical Monthly* **80**(4), April 1973, 424–425.

[†] For a fascinating account of Euler's work on infinite series, see Raymond Ayoub, "Euler and the Zeta Function," *The American Mathematical Monthly* **81**(10), December 1974, 1067–1086.

2.16 THEOREM

COMPARISON OF INTEGRALS AND SUMS   Let $n \in \mathbb{N}$ and suppose that the function f is decreasing on the interval $[1, n+1]$. Then

$$\int_1^{n+1} f(x)\, dx \le \sum_{k=1}^n f(k) \le f(1) + \int_1^n f(x)\, dx.$$

PROOF   We assume that $n \ge 2$, leaving it as an exercise to complete the proof for the case in which $n = 1$ (Problem 25). By Theorem 4.15 in Chapter 5, f is integrable on $[1, n+1]$. Let k be an integer with $1 \le k \le n$. Then

$$f(k+1) \le f(x) \le f(k)$$

holds for all $x \in [k, k+1]$, and it follows from Corollary 6.4 in Chapter 5 that

$$f(k+1) \le \int_k^{k+1} f(x)\, dx \le f(k).$$

Consequently, by the interval additivity of the integral (Theorem 5.19 in Chapter 5),

$$\sum_{k=1}^n f(k) = f(1) + \sum_{k=2}^n f(k)$$
$$= f(1) + \sum_{k=1}^{n-1} f(k+1)$$
$$\le f(1) + \sum_{k=1}^{n-1} \int_k^{k+1} f(x)\, dx = f(1) + \int_1^n f(x)\, dx$$

and

$$\int_1^{n+1} f(x)\, dx = \sum_{k=1}^n \int_k^{k+1} f(x)\, dx \le \sum_{k=1}^n f(k). \qquad \square$$

By combining Theorem 2.16 and Lemma 2.8, we obtain an important corollary, known as the **integral test**, which was first proved by Cauchy in 1837. We leave the proof as an exercise (Problem 27).

2.17 COROLLARY

THE INTEGRAL TEST   Let f be a function that has only positive values and is decreasing on the interval $[1, \infty]$. Then the series $\sum_{k=1}^{\infty} f(k)$ converges if and only if $L = \lim_{n \to \infty} \int_1^n f(x)\, dx$ exists. Moreover, if this limit exists, then

$$L \le \sum_{k=1}^{\infty} f(k) \le f(1) + L.$$

2.18 Example   Show that the **harmonic series** $\sum_{k=1}^{\infty}(1/k)$ diverges.

SOLUTION   By applying the integral test (Corollary 2.17) with $f(x) = 1/x$, we find that

$$\int_1^n \frac{1}{x}\,dx = \ln n - \ln 1 = \ln n,$$

and, since $\ln n$ has no upper bound as $n \to \infty$, $\lim_{n \to \infty} \int_1^n (1/x)\,dx$ fails to exist.   □

The harmonic series $\sum_{k=1}^{\infty}(1/k)$ in Example 2.18 and the series $\sum_{k=1}^{\infty}(1/k^2)$ in Example 2.15 are special cases, for $p = 1$ and $p = 2$, respectively, of the *p-series* defined as follows:

2.19 DEFINITION   *p*-**SERIES**   If $p \in \mathbb{R}$, then the series

$$\sum_{k=1}^{\infty} \frac{1}{k^p}$$

is called the *p*-**series**.

In working with *p*-series, we assume familiarity with the properties of (not necessarily rational) exponents as developed in elementary calculus courses.

2.20 Example   Show that the *p*-series $\sum_{k=1}^{\infty}(1/k^p)$ diverges if $p \le 1$ and converges if $p > 1$.

SOLUTION   For $p \le 1$ and $k \in \mathbb{N}$, we have $k^p \le k$, and so $1/k \le 1/k^p$. Therefore, $\sum_{k=1}^{\infty}(1/k^p)$ diverges by comparison with the divergent harmonic series $\sum_{k=1}^{\infty}(1/k)$ (Theorem 2.13).

Now suppose that $p > 1$. We are going to apply the integral test to the series $\sum_{k=1}^{\infty}(1/k^p)$ using the function $f:[1, \infty) \to \mathbb{R}$ defined by

$$f(x) = \frac{1}{x^p} = x^{-p}$$

for $x \ge 1$. We have

$$\int_1^n x^{-p}\,dx = \frac{n^{1-p} - 1}{1 - p}.$$

Since $p > 1$, it follows that $\lim_{n \to \infty} n^{1-p} = 0$, and so

$$\lim_{n \to \infty} \int_1^n x^{-p}\,dx = \lim_{n \to \infty} \frac{n^{1-p} - 1}{1 - p} = \frac{1}{p - 1}.$$

Therefore, $\sum_{k=1}^{\infty}(1/k^p)$ converges by the integral test.   □

The inequalities in Theorem 2.16 suggest that definite integrals might be used to approximate partial sums of series. The following lemma and theorem show that, under certain conditions, it is possible to write partial sums in terms of definite integrals.

2.21 LEMMA

Let $n \in \mathbb{N}$ and suppose that the function f has a continuous first derivative on the interval $[1, n]$. Then

$$\sum_{k=1}^{n} f(k) = nf(n) - \int_{1}^{n} [\![x]\!] f'(x)\, dx.$$

PROOF   If k is an integer, then $[\![x]\!] = k$ holds for all $x \in [k, k+1)$, and therefore

$$\int_{1}^{n} [\![x]\!] f'(x)\, dx = \sum_{k=1}^{n-1} \int_{k}^{k+1} [\![x]\!] f'(x)\, dx = \sum_{k=1}^{n-1} \int_{k}^{k+1} k f'(x)\, dx$$

$$= \sum_{k=1}^{n-1} k \int_{k}^{k+1} f'(x)\, dx = \sum_{k=1}^{n-1} k[f(k+1) - f(k)]$$

$$= \sum_{k=1}^{n-1} k f(k+1) - \sum_{k=1}^{n-1} k f(k).$$

By changing the summation index, we find that

$$\sum_{k=1}^{n-1} k f(k+1) = \sum_{k=2}^{n} (k-1) f(k) = \sum_{k=1}^{n} (k-1) f(k)$$

$$= \sum_{k=1}^{n} k f(k) - \sum_{k=1}^{n} f(k)$$

$$= \sum_{k=1}^{n-1} k f(k) + nf(n) - \sum_{k=1}^{n} f(k),$$

and it follows that

$$\int_{1}^{n} [\![x]\!] f'(x)\, dx = nf(n) - \sum_{k=1}^{n} f(k). \qquad \square$$

In Theorem 2.16, we showed that, for a decreasing function f,

$$\sum_{k=1}^{n} f(k) \leq f(1) + \int_{1}^{n} f(x)\, dx.$$

The following theorem, which is a version of **Euler's summation formula**, gives an expression for the error $\mathscr{E}$ involved in the approximation

$$\sum_{k=1}^{n} f(k) \approx f(1) + \int_{1}^{n} f(x)\, dx.$$

2.22 THEOREM

ERROR IN APPROXIMATING A SUM BY AN INTEGRAL
Let $n \in \mathbb{N}$ and suppose that the function f has a continuous first derivative on the interval $[1, n]$. Then

$$\sum_{k=1}^{n} f(k) = f(1) + \int_{1}^{n} f(x) \, dx + \mathscr{E},$$

where

$$\mathscr{E} = \int_{1}^{n} (x - [\![x]\!]) f'(x) \, dx.$$

PROOF An integration by parts shows that

$$\int_{1}^{n} x f'(x) \, dx = n f(n) - f(1) - \int_{1}^{n} f(x) \, dx.$$

Therefore,

$$\mathscr{E} = \int_{1}^{n} (x - [\![x]\!]) f'(x) \, dx = \int_{1}^{n} x f'(x) \, dx - \int_{1}^{n} [\![x]\!] f'(x) \, dx$$

$$= n f(n) - f(1) - \int_{1}^{n} f(x) \, dx - \int_{1}^{n} [\![x]\!] f'(x) \, dx,$$

and it follows that

$$n f(n) - \int_{1}^{n} [\![x]\!] f'(x) \, dx = f(1) + \int_{1}^{n} f(x) \, dx + \mathscr{E}.$$

Hence, by Lemma 2.21,

$$\sum_{k=1}^{n} f(k) = f(1) + \int_{1}^{n} f(x) \, dx + \mathscr{E}. \qquad \square$$

PROBLEM SET 6.2

1. Give an alternative proof of Theorem 2.2 by showing that it is a corollary of the Cauchy criterion for convergence of a series (Theorem 2.1). [Hint: Begin by choosing $M \in \mathbb{N}$ such that, for every integer $q \geq 0$, $\left| \sum_{k=M}^{M+q} a_k \right| < \varepsilon/2$.]

2. Prove the following alternative version of the Cauchy criterion for convergence of a series: The series $\sum_{k=1}^{\infty} a_k$ converges if and only if, for every $\varepsilon > 0$, there exists $N \in \mathbb{N}$ such that, for all $n, m \in \mathbb{N}$ with $n \geq m \geq N$, $\left| \sum_{k=m}^{n} a_k \right| < \varepsilon$.

3. Show that the series $\sum_{k=1}^{\infty} \cos(\pi/k)$ diverges.

4. Use the Cauchy criterion for convergence to prove the following without invoking Theorem 2.11: If M is a positive integer, then the series $\sum_{k=1}^{\infty} a_k$ converges if and only if the series $\sum_{k=M}^{\infty} a_k$ converges.

5. Criticize the following argument: Since

$$\lim_{n \to \infty} \frac{1}{n \ln(n + 1)} = 0,$$

the series

$$\sum_{k=1}^{\infty} \frac{1}{k \ln(k + 1)}$$

must converge.

6. Show that each of the following series is divergent:

(a) $\displaystyle\sum_{k=1}^{\infty} \frac{k}{3k - 2}$ (b) $\displaystyle\sum_{k=1}^{\infty} \sin \frac{\pi k}{4}$ (c) $\displaystyle\sum_{k=0}^{\infty} (-1)^k$ (d) $\displaystyle\sum_{k=1}^{\infty} \frac{k!}{2^k}$

7. (a) Prove Part (i) of Lemma 2.6.
    (b) Prove Part (ii) of Lemma 2.6.

8. Find the sum of the series

$$\sum_{k=1}^{\infty} \left[\frac{1}{k(k + 1)} - \frac{2}{3^k} \right].$$

9. Prove the following: If the series $\sum_{k=1}^{\infty} a_k$ converges and the series $\sum_{k=1}^{\infty} b_k$ diverges, then the series $\sum_{k=1}^{\infty}(a_k + b_k)$ diverges. [Hint: Make an indirect proof by using Lemma 2.6.]

10. Prove that the convergence or divergence of a series is unaffected by changing, deleting, or appending a finite number of terms.

11. Prove Lemma 2.8.

12. Criticize the following argument: For a telescoping series, we have

$$\sum_{k=1}^{\infty} (b_k - b_{k+1}) = \sum_{k=1}^{\infty} b_k - \sum_{k=1}^{\infty} b_{k+1} = \sum_{k=1}^{\infty} b_k - \sum_{k=2}^{\infty} b_k = b_1.$$

13. Prove Lemma 2.9.

14. Under what conditions (if any) is the argument in Problem 12 valid?

15. By changing the summation index, rewrite the series

$$\sum_{k=2}^{\infty} \frac{1}{\ln k}$$

as a series of the form $\sum_{k=1}^{\infty} a_k$.

16. Show that the series

$$\sum_{k=0}^{\infty} \frac{1}{4k^2 + 8k + 3}$$

converges, and find its sum. [Hint: One approach is to begin by changing the summation index so that the series has the form $\sum_{k=1}^{\infty} a_k$.]

17. Prove Part (ii) of Theorem 2.13. [Hint: What is the contrapositive of Part (i)?]

18. Let $(\alpha_n)_{n \in \mathbb{N}}$ be a sequence of integers such that $0 \le \alpha_n \le 9$ for each $n \in \mathbb{N}$. When we write the real number $x \in [0, 1]$ in decimal form as $x = 0.\alpha_1\alpha_2\alpha_3\ldots$, we mean that $x = \sum_{k=1}^{\infty}(\alpha_k/10^k)$.
    (a) Prove that the series $\sum_{k=1}^{\infty}(\alpha_k/10^k)$ converges by comparing it with the convergent geometric series $\sum_{k=1}^{\infty}(1/10^{k-1})$.
    (b) If $\alpha_n = 9$ for all $n \in \mathbb{N}$, what is the value of x?

19. Show that the series $\sum_{k=1}^{\infty} \sin(1/k)$ diverges by using the limit comparison test (Corollary 2.14) and the fact that the harmonic series $\sum_{k=1}^{\infty}(1/k)$ diverges. [Hint: Recall from elementary calculus that $\lim_{x \to 0}[(\sin x)/x] = 1$.]

20. Let $x \in (0, 1)$. Prove that x has a decimal expansion (see Problem 18) as follows: Define a sequence $(\alpha_n)_{n \in \mathbb{N}}$ by $\alpha_n = [\![10^n x]\!] - 10[\![10^{n-1}x]\!]$ for each $n \in \mathbb{N}$.
    (a) Prove that α_n is an integer and $0 \le \alpha_n \le 9$ for each $n \in \mathbb{N}$.
    (b) Prove that $0 \le x - \sum_{k=1}^{n}(\alpha_k/10^k) < 10^{-n}$ for each $n \in \mathbb{N}$.
    (c) Conclude that $x = \sum_{k=1}^{\infty}(\alpha_k/10^k)$.

21. If $\sum_{k=1}^{\infty} a_k$ is a series with nonnegative terms and $\sum_{k=1}^{\infty} b_k$ is a convergent series with positive terms such that the sequence $(a_n/b_n)_{n \in \mathbb{N}}$ is bounded above, prove that $\sum_{k=1}^{\infty} a_k$ converges.

22. Prove the following modified version of the limit comparison test: Let $\sum_{k=1}^{\infty} a_k$ be a series of nonnegative terms and let $\sum_{k=1}^{\infty} b_k$ be a convergent series of positive terms. Then $\lim_{n \to \infty}(a_n/b_n) = 0$ implies that $\sum_{k=1}^{\infty} a_k$ converges.

23. Assume that f is continuous, decreasing, and takes only positive values on the interval $[1, n + 1]$, and draw a sketch to illustrate the inequality $\int_1^{n+1} f(x)\, dx \le \sum_{k=1}^{n} f(k)$ in Theorem 2.16. [Hint: Interpret the integral as an area and interpret the sum as a sum of areas of circumscribed rectangles.]

24. Prove the following modified version of the limit comparison test: Let $\sum_{k=1}^{\infty} a_k$ be a series of nonnegative terms and let $\sum_{k=1}^{\infty} b_k$ be a divergent series of positive terms. Then $\lim_{n \to \infty}(a_n/b_n) = \infty$ implies that $\sum_{k=1}^{\infty} a_k$ diverges.

25. Complete the proof of Theorem 2.16 by considering the case in which $n = 1$.

26. If n is an integer, $n \ge 2$, and if the function f is decreasing on the interval $[1, n]$, prove that

$$\sum_{k=2}^{n} f(k) \le \int_1^n f(x)\, dx \le \sum_{k=1}^{n-1} f(k).$$

27. Prove Corollary 2.17.

28. If $n, q \in \mathbb{N}$ with $q \leq n$, and if the function f is decreasing on the interval $[1, n]$, prove that

$$\sum_{k=1}^{n} f(k) \leq \sum_{k=1}^{q} f(k) + \int_{q}^{n} f(x)\, dx.$$

29. If n is an integer, $n \geq 2$, and if the function f is increasing on the interval $[1, n]$, prove that

$$\sum_{k=1}^{n-1} f(k) \leq \int_{1}^{n} f(x)\, dx \leq \sum_{k=2}^{n} f(k).$$

[Hint: Note that $-f$ is decreasing on the interval $[1, n]$ and use Theorem 2.16 with n replaced by $n - 1$.]

30. (a) If $n, q \in \mathbb{N}$ with $q \leq n$, use the result of Problem 28 to prove that

$$\sum_{k=1}^{n} \frac{1}{k} \leq \sum_{k=1}^{q} \frac{1}{k} + \ln\left(\frac{n}{q}\right).$$

(b) Use a calculator and the inequality in Part (a) with $q = 10$ and $n = 10^{6}$ to prove that the sum of the first million terms of the harmonic series does not exceed 15.

(c) Show that the sum of the first billion terms of the harmonic series does not exceed 22.

31. Use the result of Problem 29 with $f(x) = \ln x$ to prove that $n^{n}e^{1-n} \leq n! \leq (n + 1)^{n+1}e^{-n}$ holds for all integers $n \geq 2$.

32. Let $n \in \mathbb{N}$ and suppose that the function f has a continuous first derivative on the interval $[1, n]$. Prove that the error $\mathscr{E}$ in the approximation $\sum_{k=1}^{n} f(k) \approx f(1) + \int_{1}^{n} f(x)\, dx$ does not exceed $K(n - 1)/2$, where $K = \sup_{1 \leq x \leq n} |f'(x)|$.

33. Show that the series below converges if $p > 1$ and diverges if $p \leq 1$.

$$\sum_{k=2}^{\infty} \frac{1}{k(\ln k)^{p}}$$

34. Derive Euler's summation formula in the following alternative form: If $n \in \mathbb{N}$ and the function f has a continuous first derivative on the interval $[1, n]$, then

$$\sum_{k=1}^{n} f(k) = \int_{1}^{n} f(x)\, dx + \int_{1}^{n} \left(x - [\![x]\!] - \frac{1}{2}\right) f'(x)\, dx + \frac{f(1) + f(n)}{2}.$$

3 _____

ABSOLUTE AND CONDITIONAL CONVERGENCE

The tests for convergence developed in Section 2 on the basis of Lemma 2.8 pertain to series whose terms are eventually positive. In this section, we develop methods to deal with series whose terms continually change sign.

3.1 DEFINITION

ALTERNATING SERIES   If $(a_n)_{n \in \mathbb{N}}$ is a sequence of nonnegative numbers, then a series having either of the forms

$$\sum_{k=1}^{\infty} (-1)^{k+1} a_k \qquad \text{or} \qquad \sum_{k=1}^{\infty} (-1)^k a_k$$

is called an **alternating series**.

Since the second series in Definition 3.1 can be obtained from the first by multiplication by -1, we concentrate our attention on alternating series of the form

$$\sum_{k=1}^{\infty} (-1)^{k+1} a_k = a_1 - a_2 + a_3 - a_4 + \cdots.$$

3.2 LEMMA

Let $(a_n)_{n \in \mathbb{N}}$ be a decreasing sequence of nonnegative numbers and let s_n denote the nth partial sum of the alternating series $\sum_{k=1}^{\infty} (-1)^{k+1} a_k$. Then:

(i) $(s_{2n})_{n \in \mathbb{N}}$ is an increasing sequence of nonnegative numbers.
(ii) $(s_{2n-1})_{n \in \mathbb{N}}$ is a decreasing sequence of nonnegative numbers.
(iii) For all $n, m \in \mathbb{N}$,

$$0 \le s_{2n} \le \lim_{k \to \infty} s_{2k} \le \lim_{k \to \infty} s_{2k-1} \le s_{2m-1} \le s_1.$$

PROOF   Grouping terms in pairs, we find that

$$s_{2n} = (a_1 - a_2) + (a_3 - a_4) + \cdots + (a_{2n-1} - a_{2n}). \tag{1}$$

Since $(a_n)_{n \in \mathbb{N}}$ is a decreasing sequence, each expression in parentheses is nonnegative, and it follows that

$$0 \le s_{2n} \tag{2}$$

holds for all $n \in \mathbb{N}$. Also, for all $n \in \mathbb{N}$,

$$s_{2n} + (a_{2n+1} - a_{2n+2}) = s_{2n+2}, \tag{3}$$

and again the expression in parentheses is nonnegative, so

$$s_{2n} \le s_{2n+2}. \tag{4}$$

Combining (2) and (4), we have

$$0 \le s_2 \le s_4 \le s_6 \le \cdots, \tag{5}$$

and Part (i) is proved.

Note that, for all $n \in \mathbb{N}$,

$$s_{2n+1} = s_{2n-1} - a_{2n} + a_{2n+1} = s_{2n-1} - (a_{2n} - a_{2n-1}), \tag{6}$$

and therefore

$$s_{2n+1} \leq s_{2n-1}. \tag{7}$$

Consequently,

$$\cdots \leq s_7 \leq s_5 \leq s_3 \leq s_1. \tag{8}$$

Also, for any $n \in \mathbb{N}$,

$$s_{2n} + a_{2n+1} = s_{2n+1}, \tag{9}$$

and therefore, in view of (2) and the fact that $a_{2n+1} \geq 0$,

$$0 \leq s_{2n} \leq s_{2n+1}. \tag{10}$$

Combining (8) and (10), we have

$$0 \leq \cdots \leq s_7 \leq s_5 \leq s_3 \leq s_1, \tag{11}$$

and Part (ii) is proved.

Let $n, m \in \mathbb{N}$. If $q = \max(m, n)$, then $n, m \leq q$, and it follows from (5) and (8) that

$$s_{2n} \leq s_{2q} \quad \text{and} \quad s_{2q+1} \leq s_{2m-1}. \tag{12}$$

But, according to (10) with n replaced by q, $s_{2q} \leq s_{2q+1}$, and so (12) implies that

$$s_{2n} \leq s_{2m-1} \tag{13}$$

for all $n, m \in \mathbb{N}$. By combining (5), (11), and (13), we now have

$$0 \leq s_2 \leq s_4 \leq \cdots \leq s_{2n} \leq s_{2m-1} \leq \cdots \leq s_5 \leq s_3 \leq s_1. \tag{14}$$

Let $m \in \mathbb{N}$. By (14), the sequence $(s_{2n})_{n \in \mathbb{N}}$ is increasing and bounded above by s_{2m-1}. By Theorem 1.9, $\lim_{k \to \infty} s_{2k}$ exists and, for every $n \in \mathbb{N}$,

$$0 \leq s_{2n} \leq \lim_{k \to \infty} s_{2k} = \sup_{k \in \mathbb{N}} s_{2k} \leq s_{2m-1}. \tag{15}$$

By (11) and (15), the sequence $(s_{2m-1})_{m \in \mathbb{N}}$ is decreasing and bounded below by $\lim_{k \to \infty} s_{2k}$. By Theorem 1.9 again, $\lim_{k \to \infty} s_{2k-1}$ exists and, for every $m \in \mathbb{N}$,

$$\lim_{k \to \infty} s_{2k} \leq \inf_{k \in \mathbb{N}} s_{2k-1} = \lim_{k \to \infty} s_{2k-1} \leq s_{2m-1} \leq s_1. \tag{16}$$

Part (iii) now follows from (15) and (16). $\qquad\square$

If the sequence $(a_n)_{n \in \mathbb{N}}$ in Lemma 3.2 is *strictly* decreasing and its terms are all *positive*, then it is not difficult to see that $(s_{2n})_{n \in \mathbb{N}}$ is a strictly increasing sequence and that $(s_{2n-1})_{n \in \mathbb{N}}$ is a strictly decreasing sequence (Problem 1).

The following theorem, which was discovered by Leibniz in 1705, is a consequence of the results in Lemma 3.2.

3.3 THEOREM

> **LEIBNIZ'S ALTERNATING SERIES TEST**   If $(a_n)_{n \in \mathbb{N}}$ is a decreasing sequence of nonnegative terms and if $\lim_{n \to \infty} a_n = 0$, then the alternating series $\sum_{k=1}^{\infty} (-1)^{k+1} a_k$ converges. Moreover, if S is its sum and s_n is its nth partial sum, then
>
> $$0 \le (-1)^n (S - s_n) \le a_{n+1}.$$

PROOF

Let $\varepsilon > 0$ and choose $n \in \mathbb{N}$ large enough so that $a_{2n+1} < \varepsilon$. By Part (iii) of Lemma 3.2 with $m = n + 1$, we have

$$s_{2n} \le \lim_{k \to \infty} s_{2k} \le \lim_{k \to \infty} s_{2k-1} \le s_{2n+1},$$

and it follows that

$$0 \le \lim_{k \to \infty} s_{2k-1} - \lim_{k \to \infty} s_{2k} \le s_{2n+1} - s_{2n} = a_{2n+1} < \varepsilon.$$

Since ε is an arbitrary positive number, we may infer that

$$\lim_{k \to \infty} s_{2k} = \lim_{k \to \infty} s_{2k-1}.$$

Thus, if we let

$$S = \lim_{k \to \infty} s_{2k} = \lim_{k \to \infty} s_{2k-1},$$

we have

$$S = \lim_{k \to \infty} s_n = \sum_{k=1}^{n} (-1)^{k+1} a_k$$

(Problem 5).

Let $n \in \mathbb{N}$. If n is even, then $n + 1$ is odd, and it follows from Part (iii) of Lemma 3.2 that

$$s_n \le S \le s_{n+1},$$

and so

$$0 \le S - s_n \le s_{n+1} - s_n = (-1)^{(n+1)+1} a_{n+1}$$
$$= (-1)^n a_{n+1} = a_{n+1}.$$

Thus, if n is even, we have

$$0 \le (-1)^n (S - s_n) \le a_{n+1}.$$

If n is odd, then $n + 1$ is even, and Part (iii) of Lemma 3.2 shows that

$$s_{n+1} \leq S \leq s_n,$$

and so

$$0 \leq s_n - S \leq s_n - s_{n+1} = -(s_{n+1} - s_n)$$
$$= -(-1)^{(n+1)+1}a_{n+1} = -(-1)^n a_{n+1} = a_{n+1}.$$

Thus, if n is odd, we also have

$$0 \leq (-1)^n (S - s_n) \leq a_{n+1}. \qquad \square$$

3.4 Example   Show that the **alternating harmonic series**

$$\sum_{k=1}^{\infty} (-1)^{k+1} \frac{1}{k} = 1 - \frac{1}{2} + \frac{1}{3} - \frac{1}{4} + \cdots$$

converges.

SOLUTION   The sequence $(1/n)_{n \in \mathbb{N}}$ is decreasing and $\lim_{n \to \infty}(1/n) = 0$, so the series converges by the alternating series test (Theorem 3.3). $\qquad \square$

If the hypotheses of Theorem 3.3 are satisfied, then we may consider the nth partial sum s_n as an approximation to the sum

$$S = \sum_{k=1}^{\infty} (-1)^k a_k$$

of the series. If n is even, then

$$0 \leq S - s_n \leq a_{n+1},$$

and therefore s_n *underestimates* S with an error of at most a_{n+1}. If n is odd, then

$$0 \leq s_n - S \leq a_{n+1},$$

and therefore s_n *overestimates* S with an error of at most a_{n+1}. Thus, the alternating series test may be paraphrased as follows:

> *If the terms of an alternating series are decreasing in absolute value and approaching zero as a limit, then the series converges and the absolute value of the error in approximating the sum of the series by a partial sum does not exceed the absolute value of the first omitted term.*

If the sequence $(a_n)_{n \in \mathbb{N}}$ in Theorem 3.3 is *strictly* decreasing and all of its terms are *positive*, then it is easy to see that

$$0 < (-1)^n (S - s_n) < a_{n+1}$$

holds for all $n \in \mathbb{N}$ (Problem 3); hence, under these conditions, the absolute value of the error in approximating S by a partial sum is strictly less than the first omitted term.

In making decimal approximations, we adhere to the following **rounding-off rule**:

> *If an approximation in decimal form is to be rounded off to d decimal places, then the error involved in this approximation should not exceed* 0.5×10^{-d}.

The rationale for the rounding-off rule can be found in Problem 7 of Problem Set 3.5.

3.5 Example   Find an approximation, rounded off to three decimal places, for

$$S = \sum_{k=0}^{\infty} \frac{(-1)^k}{k!} = 1 - 1 + \frac{1}{2} - \frac{1}{6} + \frac{1}{24} - \frac{1}{120} + \cdots.$$

SOLUTION   We propose to approximate S by the partial sum

$$s_n = \sum_{k=0}^{n} \frac{(-1)^k}{k!}.$$

The error $\mathscr{E}$ involved in the approximation $S \approx s_n$ will be less than the absolute value of the first omitted term, that is,

$$\mathscr{E} < \left| \frac{(-1)^{n+1}}{(n+1)!} \right| = \frac{1}{(n+1)!}.$$

According to the rounding-off rule, we need

$$\mathscr{E} \le 0.5 \times 10^{-3},$$

and we can guarantee this by making sure that

$$\frac{1}{(n+1)!} \le 0.5 \times 10^{-3} = \frac{1}{2 \times 10^3} = \frac{1}{2000}.$$

Thus, $n \in \mathbb{N}$ must be large enough so that

$$2000 < (n+1)!.$$

Since $6! = 720$ and $7! = 5040$, the smallest value of n for which the desired inequality will hold is $n = 6$. Now

$$s_6 = 1 - 1 + \frac{1}{2} - \frac{1}{6} + \frac{1}{24} - \frac{1}{120} + \frac{1}{720} = \frac{53}{144} \approx 0.368,$$

where we have rounded off to three decimal places. So $S \approx 0.368$.   □

Although the harmonic series diverges (Example 2.18), the alternating harmonic series converges (Example 3.4). Thus, we have a series that converges only because some of its terms are negative—if all of the negative signs are removed and replaced by plus signs, the convergence is destroyed. To study this phenomenon more generally, we introduce the following definition.

3.6 DEFINITION

ABSOLUTE AND CONDITIONAL CONVERGENCE We say that the series $\sum_{k=1}^{\infty} a_k$ is **absolutely convergent** if the series $\sum_{k=1}^{\infty} |a_k|$ is convergent. We say that a convergent series $\sum_{k=1}^{\infty} a_k$ is **conditionally convergent** if the series $\sum_{k=1}^{\infty} |a_k|$ is divergent.

The alternating harmonic series provides an example of a conditionally convergent series. As a consequence of the following theorem, we can produce examples of absolutely convergent series by changing the algebraic signs of some or all of the terms of a convergent series of positive terms.

3.7 THEOREM

ABSOLUTE CONVERGENCE IMPLIES CONVERGENCE If a series is absolutely convergent, then it is convergent.

PROOF   Suppose that $\sum_{k=1}^{\infty} |a_k|$ is convergent and let

$$b_n = |a_n| - a_n$$

for all $n \in \mathbb{N}$. Then

$$0 \le b_n \le 2|a_n|$$

holds for all $n \in \mathbb{N}$, so the series $\sum_{k=1}^{\infty} b_k$ converges by comparison with the series $\sum_{k=1}^{\infty} 2|a_k|$ (Theorem 2.13). Therefore, by Part (i) of Lemma 2.6,

$$\sum_{k=1}^{\infty} a_k = \sum_{k=1}^{\infty} (|a_k| - b_k)$$

converges.   □

3.8 Example   Show that the series

$$\sum_{k=1}^{\infty} \frac{\sin k}{k^2}$$

converges.

SOLUTION   Note that the given series contains both positive and negative terms but is not an alternating series. However,

$$\left|\frac{\sin k}{k^2}\right| = \frac{|\sin k|}{k^2} \leq \frac{1}{k^2},$$

and so the series

$$\sum_{k=1}^{\infty} \left|\frac{\sin k}{k^2}\right|$$

converges by comparison with the convergent p-series $\sum_{k=1}^{\infty}(1/k^2)$. Thus,

$$\sum_{k=1}^{\infty} \left|\frac{\sin k}{k^2}\right|$$

is absolutely convergent, and so it converges by Theorem 3.7.   □

The following simple lemma is the basis for two of the most practical tests for absolute convergence, the *root test* and the *ratio test*, both of which were discovered by Cauchy.

3.9 LEMMA

If there exist $c, r \in \mathbb{R}$ and $N \in \mathbb{N}$ such that $0 \leq c, 0 \leq r < 1$, and, for all integers $n \geq N$,

$$|a_n| \leq cr^n,$$

then the series $\sum_{k=1}^{\infty} a_k$ is absolutely convergent.

PROOF   If the given condition holds, then the series $\sum_{k=1}^{\infty} |a_k|$ is eventually dominated by the convergent geometric series $\sum_{k=1}^{\infty} cr^k$.   □

3.10 COROLLARY

THE ROOT TEST   Let $\sum_{k=1}^{\infty} a_k$ be a series and suppose that $L = \lim_{n \to \infty} |a_n|^{1/n}$ exists. Then:

   (i) If $L < 1$, the series converges absolutely.
   (ii) If $L > 1$, the series diverges.
   (iii) If $L = 1$, the test is inconclusive.

PROOF   Suppose that $L < 1$ and choose r such that $L < r < 1$. Then there exists $N \in \mathbb{N}$ such that, for all integers $n \geq N$,

$$|a_n|^{1/n} < r$$

(Problem 13a). Consequently,

$$|a_n| < r^n$$

holds for all integers $n \geq N$, and it follows from Lemma 3.9, with $c = 1$, that the series converges absolutely. This proves Part (i).

To prove Part (ii), suppose that $L > 1$. Then there exists $N \in \mathbb{N}$ such that, for all integers $n \geq N$,

$$1 < |a_n|^{1/n}$$

(Problem 13b). Consequently,

$$1 < |a_n|$$

holds for all integers $n \geq N$, and therefore it cannot be true that $a_n \to 0$ as $n \to \infty$. Hence, the series diverges by the divergence test (Corollary 2.4). We ask you to confirm Part (iii) in Problem 15.   □

3.11 Example   Test the series

$$\sum_{k=1}^{\infty} \left(\frac{-k}{2k+1} \right)^k$$

for convergence.

SOLUTION   Since

$$\lim_{n \to \infty} \left| \left(\frac{-n}{2n+1} \right)^n \right|^{1/n} = \lim_{n \to \infty} \frac{n}{2n+1} = \lim_{n \to \infty} \frac{1}{2 + (1/n)} = \frac{1}{2} < 1,$$

the series converges absolutely by the root test.   □

3.12 THEOREM

> **THE RATIO TEST**   Let $\sum_{k=1}^{\infty} a_k$ be a series of nonzero terms and suppose that
>
> $$L = \lim_{n \to \infty} \left| \frac{a_{n+1}}{a_n} \right|.$$
>
> Then:
>
> (i) If $L < 1$, the series converges.
> (ii) If $L > 1$, the series diverges.
> (iii) If $L = 1$, the test is inconclusive.

PROOF   Suppose that $L < 1$ and choose r such that $L < r < 1$. Then there exists N such that, for all $n \in \mathbb{N}$,

$$n \geq N \Rightarrow \left| \frac{a_{n+1}}{a_n} \right| < r.$$

Therefore,

$$n \geq N \Rightarrow |a_{n+1}| < |a_n| r.$$

Consequently,

$$|a_{N+1}| < |a_N| r,$$

and so

$$|a_{N+2}| < |a_{N+1}| r < (|a_N| r) r = |a_N| r^2.$$

Continuing in this way by induction, we find that

$$|a_{N+j}| < |a_N| r^j$$

for all $j \in \mathbb{N}$ (Problem 17). Therefore, for all integers $n \geq N$,

$$|a_n| < |a_N| r^{-N} r^n,$$

and it follows from Lemma 3.9, with $c = |a_N| r^{-N}$, that the series converges absolutely. This proves Part (i).

The proof of Part (ii) is similar to the proof of Part (ii) of Corollary 3.10 and is left as an exercise (Problem 19). The same examples that confirmed Part (iii) of Corollary 3.10 are effective for Part (iii) of the present theorem. □

Since the root and ratio tests are consequences of Lemma 3.9, they are merely special cases of the comparison test in disguise. Although the root test is more powerful than the ratio test (the root test sometimes works when the ratio test fails), the ratio test is often easier to apply.

3.13 Example If c is a positive constant, test the series $\sum_{k=0}^{\infty} (k^c/k!)$ for convergence.

SOLUTION The series has the form $\sum_{k=0}^{\infty} a_k$ with $a_k = k^c/k!$ for all $k \in \mathbb{N}$. Thus,

$$\lim_{n \to \infty} \frac{a_{n+1}}{a_n} = \lim_{n \to \infty} \frac{(n+1)^c/(n+1)!}{n^c/n!} = \lim_{n \to \infty} \left(\frac{n+1}{n} \right)^c \frac{n!}{(n+1)!}$$

$$= \lim_{n \to \infty} \left(1 + \frac{1}{n} \right)^c \frac{1}{n+1} = 1 \cdot 0 = 0.$$

Therefore, the series converges by the ratio test. □

The summation formula in the next lemma, which is attributed to the Norwegian mathematician Niels Henrik Abel (1802–1829), is analogous to integration by parts.

3.14 LEMMA

SUMMATION BY PARTS For each $n \in \mathbb{N}$, let s_n denote the nth partial sum of the series $\sum_{k=1}^{\infty} a_k$. Then, if $(b_n)_{n \in \mathbb{N}}$ is a sequence,

$$\sum_{k=1}^{n} a_k b_k = s_n b_{n+1} - \sum_{k=1}^{n} s_k (b_{k+1} - b_k).$$

PROOF Let $s_0 = 0$. Then

$$\sum_{k=1}^{n} a_k b_k = \sum_{k=1}^{n} (s_k - s_{k-1}) b_k = \sum_{k=1}^{n} s_k b_k - \sum_{k=1}^{n} s_{k-1} b_k.$$

By a change of summation index, we have

$$\sum_{k=1}^{n} s_{k-1} b_k = \sum_{k=0}^{n-1} s_k b_{k+1} = s_0 b_1 + \sum_{k=1}^{n-1} s_k b_{k+1}$$

$$= \sum_{k=1}^{n-1} s_k b_{k+1} = \sum_{k=1}^{n} s_k b_{k+1} - s_n b_{n+1}.$$

Therefore,

$$\sum_{k=1}^{n} a_k b_k = \sum_{k=1}^{n} s_k b_k - \sum_{k=1}^{n} s_k b_{k+1} + s_n b_{n+1}$$

$$= s_n b_{n+1} - \sum_{k=1}^{n} s_k (b_{k+1} - b_k). \qquad \square$$

3.15 COROLLARY

ABEL'S TEST FOR CONVERGENCE Let $\sum_{k=1}^{\infty} a_k$ be a convergent series and let $(b_n)_{n \in \mathbb{N}}$ be a convergent monotone sequence with $B = \lim_{n \to \infty} b_n$. Let s_n denote the nth partial sum of $\sum_{k=1}^{\infty} a_k$. Then both of the series $\sum_{k=1}^{\infty} a_k b_k$ and $\sum_{k=1}^{\infty} s_k (b_{k+1} - b_k)$ are convergent and

$$\sum_{k=1}^{\infty} a_k b_k = \left(\sum_{k=1}^{\infty} a_k \right) B - \sum_{k=1}^{\infty} s_k (b_{k+1} - b_k).$$

PROOF Since the sequence $(s_n)_{n \in \mathbb{N}}$ of partial sums has a limit, it is bounded, so there exists $M > 0$ such that, for all $n \in \mathbb{N}$,

$$|s_n| \le M.$$

We have

$$\lim_{n \to \infty} s_n b_{n+1} = \left(\lim_{n \to \infty} s_n \right) \left(\lim_{n \to \infty} b_{n+1} \right) = \left(\sum_{k=1}^{\infty} a_k \right) B,$$

so, by Lemma 3.14, it will be sufficient to show that the series

$$\sum_{k=1}^{\infty} s_k (b_{k+1} - b_k)$$

is convergent. We assume that $(b_n)_{n \in \mathbb{N}}$ is a decreasing sequence, leaving the case in which it is increasing as an exercise (Problem 27). Now

$$|s_n (b_{n+1} - b_n)| = |s_n| |b_{n+1} - b_n| \le M |b_{n+1} - b_n|$$
$$= M (b_n - b_{n+1}),$$

so the series $\sum_{k=1}^{\infty} |s_k(b_{k+1} - b_k)|$ is dominated by the convergent telescoping series $\sum_{k=1}^{\infty} M(b_k - b_{k+1})$. Therefore, by the comparison test, not only does the series $\sum_{k=1}^{\infty} s_k(b_{k+1} - b_k)$ converge, it converges absolutely. $\qquad\square$

PROBLEM SET 6.3

1. In Lemma 3.2, suppose that the sequence $(a_n)_{n \in \mathbb{N}}$ is strictly decreasing and that its terms are all positive. Prove that $(s_{2n})_{n \in \mathbb{N}}$ is a strictly increasing sequence and that $(s_{2n-1})_{n \in \mathbb{N}}$ is a strictly decreasing sequence.

2. In the statement of Theorem 3.3, replace the condition that $(a_n)_{n \in \mathbb{N}}$ is a decreasing sequence by the condition that it is *eventually* decreasing in the sense that there exists $N \in \mathbb{N}$ such that $a_n \geq a_{n+1}$ holds for all $n \in \mathbb{N}$ with $n \geq N$.
   (a) Prove that the alternating series still converges to a sum S.
   (b) Prove that $0 \leq (-1)^n(S - s_n) \leq a_{n+1}$ still holds provided that $n \geq N$.

3. In Theorem 3.3, suppose that the sequence $(a_n)_{n \in \mathbb{N}}$ is strictly decreasing and that its terms are all positive. Prove that $0 < (-1)^n(S - s_n) < a_{n+1}$ holds for all $n \in \mathbb{N}$.

4. Let $(a_n)_{n \in \mathbb{N}}$ and $(b_n)_{n \in \mathbb{N}}$ be sequences of nonnegative numbers such that $\lim_{n \to \infty} a_n = \lim_{n \to \infty} b_n = 0$. Let $(c_n)_{n \in \mathbb{N}}$ be the sequence $a_1, b_1, a_2, b_2, a_3, b_3, \ldots$ obtained by alternately choosing terms from $(a_n)_{n \in \mathbb{N}}$ and $(b_n)_{n \in \mathbb{N}}$ so that, if n is odd, $c_n = a_{(n+1)/2}$, and, if n is even, $c_n = b_{n/2}$. Suppose there exists $N \in \mathbb{N}$ such that $a_n \geq b_n \geq a_{n+1}$ holds for all $n \in \mathbb{N}$ with $n \geq N$.
   (a) Prove that the alternating series $\sum_{k=1}^{\infty} (-1)^{k+1} c_k$ converges.
   (b) If S is the sum of the alternating series in Part (a), prove that

$$S = \lim_{n \to \infty} \left(\sum_{k=1}^{n} a_k - \sum_{k=1}^{n-1} b_k \right).$$

   [Hint: Look at $\lim_{n \to \infty} s_{2n-1}$.]

5. Suppose that $(s_n)_{n \in \mathbb{N}}$ is a sequence such that both of the sequences $(s_{2n})_{n \in \mathbb{N}}$ and $(s_{2n-1})_{n \in \mathbb{N}}$ converge to the same limit L. Prove that $(s_n)_{n \in \mathbb{N}}$ converges to L.

6. With the notation of Problem 4, let $a_n = 1/n$ and let $b_n = 2^{-n}$ for all $n \in \mathbb{N}$.
   (a) Prove that the alternating series $\sum_{k=1}^{\infty} (-1)^{k+1} c_k$ diverges.
   (b) Explain why the result of Part (a) does not contradict Theorem 3.3.

7. Find an approximation, rounded off to four decimal places, for

$$\sum_{k=1}^{\infty} \frac{(-1)^{k+1}}{(2k)!}.$$

8. With the notation of Problem 4, let $a_n = 1/n$ and let $b_n = \int_n^{n+1} (1/x)\, dx$ for all $n \in \mathbb{N}$.

(a) Prove that $a_n \le b_n \le a_{n+1}$ holds for every integer $n \ge 2$. [Hint: Use Corollary 6.4 in Chapter 5.]

(b) Prove that $\lim_{n \to \infty} a_n = \lim_{n \to \infty} b_n = 0$.

(c) Use the result of Problem 4 to prove that there exists a positive number γ such that

$$\gamma = \lim_{n \to \infty} \left(\sum_{k=1}^{n} \frac{1}{k} - \ln n \right).$$

The number γ is called **Euler's constant**.

9. Give an alternative proof of Theorem 3.7 by using the Cauchy criterion for convergence of a series (Theorem 2.1) and the triangle inequality.

10. Let $\sum_{k=1}^{\infty} a_k$ be a series. Define $a_k^+ = \max(a_k, 0)$ and $a_k^- = \max(-a_k, 0)$ for all $k \in \mathbb{N}$. If $\sum_{k=1}^{\infty} a_k$ converges absolutely, prove that $\sum_{k=1}^{\infty} a_k^+$ and $\sum_{k=1}^{\infty} a_k^-$ both converge and that $\sum_{k=1}^{\infty} a_k = \sum_{k=1}^{\infty} a_k^+ - \sum_{k=1}^{\infty} a_k^-$. [Hint: $\max(a, 0) = (|a| + a)/2$.]

11. Prove that the series $\sum_{k=1}^{\infty} \sin^3(1/k)$ converges absolutely. [Hint: Begin by using the mean value theorem to prove that $|\sin x| \le |x|$ holds for all $x \in \mathbb{R}$.]

12. With the notation of Problem 10, prove that, if $\sum_{k=1}^{\infty} a_k$ is conditionally convergent, then both $\sum_{k=1}^{\infty} a_k^+$ and $\sum_{k=1}^{\infty} a_k^-$ diverge. [Hint: $\sum_{k=1}^{\infty} \frac{1}{2} a_k$ converges, but $\sum_{k=1}^{\infty} \frac{1}{2} |a_k|$ diverges.]

13. To clarify the proof of Corollary 3.10, prove the following:
(a) If $\lim_{n \to \infty} b_n < r$, then there exists $N \in \mathbb{N}$ such that, for all integers $n \ge N$, $b_n < r$.
(b) If $r < \lim_{n \to \infty} b_n$, then there exists $N \in \mathbb{N}$ such that, for all integers $n \ge N$, $r < b_n$.

14. If A and B are positive constants, C is a constant such that $|C| < 1/B$, and $(a_n)_{n \in \mathbb{N}}$ is a sequence such that $|a_n| \le AB^n$ for all positive integers n, show that the series $\sum_{k=1}^{\infty} a_k C^k$ converges absolutely.

15. Although the p-series $\sum_{k=1}^{\infty}(1/k^2)$ converges, and the harmonic series $\sum_{k=1}^{\infty}(1/k)$ diverges, show that both the root and the ratio tests are inconclusive for these series.

16. If $\sum_{k=1}^{\infty} a_k$ converges absolutely, show that $\sum_{k=1}^{\infty} a_k^2$ converges.

17. Fill in the following detail in the proof of Theorem 3.12: If r is a positive number, $N \in \mathbb{N}$, and, for every $n \in \mathbb{N}$, $n \ge N \Rightarrow |a_{n+1}| < |a_n|r$, then prove that $|a_{N+j}| < |a_N|r^j$ holds for every $j \in \mathbb{N}$.

18. Give an example of a convergent series $\sum_{k=1}^{\infty} a_k$ of positive terms such that $\lim_{n \to \infty}(a_{n+1}/a_n)$ does not exist.

19. Complete the proof of Theorem 3.12 by proving Part (ii).

20. If the series $\sum_{k=1}^{\infty} a_k$ converges absolutely, show that $\left| \sum_{k=1}^{\infty} a_k \right| \le \sum_{k=1}^{\infty} |a_k|$.

21. Test each series for absolute convergence, conditional convergence, or divergence:

(a) $\sum_{k=1}^{\infty} (-1)^k \left(\dfrac{k^2}{1+k^2} \right)$ 　(b) $\sum_{k=1}^{\infty} (-1)^k \left(\dfrac{\ln k}{k^2} \right)$

(c) $\sum_{k=1}^{\infty} \left(\dfrac{-1}{2} - \dfrac{1}{k} \right)^k$ 　(d) $\sum_{k=2}^{\infty} \dfrac{(-1)^k}{k \ln k}$

22. (a) If $(a_n)_{n \in \mathbb{N}}$ is a sequence of positive terms such that $L = \lim_{n \to \infty} (a_{n+1}/a_n)$ exists, prove that $\lim_{n \to \infty} (a_n)^{1/n}$ exists and equals L.
    (b) Explain the sense in which the result in Part (a) shows that the root test is at least as "powerful" as the ratio test.

23. Test each series for convergence or divergence:

(a) $\sum_{k=1}^{\infty} \dfrac{2 \cdot 4 \cdot 6 \cdots (2k)}{1 \cdot 4 \cdot 7 \cdots (3k-2)}$ 　(b) $\sum_{k=1}^{\infty} \dfrac{k!}{k^k}$ 　(c) $\sum_{k=1}^{\infty} \dfrac{k!}{1 \cdot 3 \cdot 5 \cdots (2k-1)}$

24. Let $0 < r < 1$ and, for all $n \in \mathbb{N}$, define $a_n = r^{(n+3)/2}$ if n is odd and $a_n = r^{n/2}$ if n is even.
    (a) Show that $\lim_{n \to \infty} (a_{n+1}/a_n)$ does not exist, so that the ratio test is not applicable to the series $\sum_{k=1}^{\infty} a_k$.
    (b) Show that the series $\sum_{k=1}^{\infty} a_k$ converges by applying the root test.
    (c) Explain the sense in which the root test is more "powerful" than the ratio test.

25. Prove the following alternative version of the root test: If there is a real number r such that $0 \le r < 1$ and a positive integer N such that $|a_n|^{1/n} \le r$ holds for every integer $n \ge N$, then the series $\sum_{k=1}^{\infty} a_k$ converges absolutely.

26. State and prove an alternative version of the ratio test analogous to the alternative version of the root test in Problem 25.

27. Complete the proof of Corollary 3.15 by taking care of the case in which $(b_n)_{n \in \mathbb{N}}$ is an increasing sequence.

28. Let $(a_n)_{n \in \mathbb{N}}$ be a sequence of nonzero terms and suppose that L is the limit superior of the sequence $(|a_{n+1}/a_n|)_{n \in \mathbb{N}}$ (see Problem 30 in Problem Set 6.1). If $L < 1$, prove that the series $\sum_{k=1}^{\infty} a_k$ converges absolutely.

29. If $|a_k| \le k$ for all $k \in \mathbb{N}$ and $|r| < 1$, prove that the series $\sum_{k=0}^{\infty} a_k r^k$ converges absolutely.

30. Prove the **Cauchy condensation test**: If $(|a_n|)_{n \in \mathbb{N}}$ is a decreasing sequence, then the series $\sum_{k=1}^{\infty} a_k$ converges absolutely if and only if the series $\sum_{k=1}^{\infty} 2^k a_{2^k}$ converges absolutely. [Hint: Let $s_n = \sum_{k=1}^{n} |a_k|$ and $t_n = \sum_{k=1}^{n} 2^k |a_{2^k}|$. If $n < 2^m$, prove that $s_n \le t_m$, and, if $n > 2^m$, prove that $s_n \ge t_m/2$.]

31. If $\phi : \mathbb{N} \to \mathbb{N}$ is a bijection, then the series $\sum_{k=1}^{\infty} a_{\phi(k)}$ is called a **rearrangement** of the series $\sum_{k=1}^{\infty} a_k$. If $\sum_{k=1}^{\infty} a_k$ is a convergent series of positive terms with sum S, prove that every rearrangement of the series converges

and has sum S. [Hint: Show that all of the partial sums of the rearranged series are bounded, conclude that the rearranged series converges, and then show that the partial sums of the series and its rearrangement must have the same limit S.]

32. Prove that, if a series is conditionally convergent, and C is a real number, then there exists a rearrangement of the series (see Problem 31) that converges and has sum C. [Hint: First take care of the case in which $C \geq 0$. Using the result of Problem 12, begin by taking just enough positive terms of the series so that the sum of these terms is greater than or equal to C. Then take just enough negative terms to reduce the sum to a number less than or equal to C. Continue in this way by induction.]

33. Prove that every rearrangement of an absolutely convergent series converges and has the same sum as the original series. [Hint: Use the result of Problem 31.]

34. A series is said to be **unconditionally convergent** if all of its rearrangements converge (see Problems 31, 32, and 33). Prove that a series is unconditionally convergent if and only if it is absolutely convergent.

35. Prove **Dirichlet's test**: If the partial sums of the series $\sum_{k=1}^{\infty} a_k$ form a bounded sequence, and if $(b_n)_{n \in \mathbb{N}}$ is a decreasing sequence that converges to zero, then the series $\sum_{k=1}^{\infty} a_k b_k$ converges. [Hint: The proof is quite similar to the proof of Corollary 3.15.]

4

SEQUENCES AND SERIES OF FUNCTIONS

If a function f_n is defined for each $n \in \mathbb{N}$, we may consider the **sequence of functions** $(f_n)_{n \in \mathbb{N}}$ and inquire in what sense we might be willing to say that a function f is a *limit* of this sequence. Perhaps the most obvious answer to this question is provided by the following definition.

4.1 DEFINITION

> **POINTWISE CONVERGENCE** Let $(f_n)_{n \in \mathbb{N}}$ be a sequence of functions and suppose that S is a set such that $S \subseteq \mathrm{dom}(f_n)$ for every $n \in \mathbb{N}$. We say that a function f is the **pointwise limit of** $(f_n)_{n \in \mathbb{N}}$ **on** S if $S \subseteq \mathrm{dom}(f)$ and, for each point $x \in S$, the sequence $(f_n(x))_{n \in \mathbb{N}}$ converges to $f(x)$.

In other words, the sequence $(f_n)_{n \in \mathbb{N}}$ converges pointwise to f on S if, for each point $x \in S$,

$$\lim_{n \to \infty} f_n(x) = f(x).$$

Note that each $x \in S$ is held fixed while the limit of $f_n(x)$ is calculated as $n \to \infty$ to obtain $f(x)$.

4.2 Example   For each $n \in \mathbb{N}$ let $f_n:[0, 1] \to \mathbb{N}$ be defined by $f_n(x) = x^n$ for all $x \in [0, 1]$. Find the pointwise limit of the sequence $(f_n)_{n \in \mathbb{N}}$ on $[0, 1]$.

SOLUTION   If $x \in [0, 1]$ and $x \neq 1$, then, by Example 1.10,

$$\lim_{n \to \infty} f_n(x) = \lim_{n \to \infty} x^n = 0.$$

On the other hand, if $x = 1$, then

$$\lim_{n \to \infty} f_n(x) = \lim_{n \to \infty} 1^n = 1.$$

Therefore, the pointwise limit of the sequence $(f_n)_{n \in \mathbb{N}}$ on $[0, 1]$ is the function f given by

$$f(x) = \begin{cases} 0, & \text{if } x \neq 1 \\ 1, & \text{if } x = 1 \end{cases}$$

for all $x \in [0, 1]$.   □

As Example 4.2 illustrates, ill-behaved functions can arise as pointwise limits of well-behaved functions. Indeed, each function f_n in this example is both continuous and differentiable on the interval $[0, 1]$, and yet the pointwise limit f is neither continuous nor differentiable at $x = 1$. This phenomenon is illustrated in Figure 6-1.

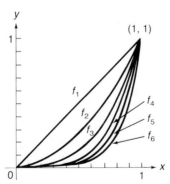

FIGURE 6-1

Suppose that f is the pointwise limit on the set S of a sequence of functions $(f_n)_{n \in \mathbb{N}}$, so that

$$f(x) = \lim_{n \to \infty} f_n(x)$$

for all $x \in S$. If $a \in S$, and if a is a point of accumulation of S, then the condition that f be continuous at a is that

$$\lim_{x \to a} f(x) = f(a).$$

In other words, f is continuous at a if and only if

$$\lim_{x \to a} \left[\lim_{n \to \infty} f_n(x) \right] = \lim_{n \to \infty} f_n(a).$$

If we assume that each function f_n is continuous at a, then we have

$$f_n(a) = \lim_{x \to a} f_n(x),$$

and the condition that f be continuous at a may be rewritten as

$$\lim_{x \to a} \left[\lim_{n \to \infty} f_n(x) \right] = \lim_{n \to \infty} \left[\lim_{x \to a} f_n(x) \right].$$

Thus, the continuity of the pointwise limit hinges on the question of whether the limit operations

$$\lim_{x \to a} \quad \text{and} \quad \lim_{n \to \infty}$$

can be interchanged. Example 4.2 shows that, in general, they cannot!

A number of questions in mathematical analysis revolve around the issue of whether various limit operations *commute*, that is, whether the order in which the operations are performed is of any consequence. In general, unless suitable precautions are taken, limit operations do not commute; hence, theorems stating that limit operations commute under certain conditions are especially noteworthy. These conditions often involve the notion of *uniform convergence*, introduced in the next definition.

4.3 DEFINITION

UNIFORM CONVERGENCE   Let $(f_n)_{n \in \mathbb{N}}$ be a sequence of functions and suppose that S is a set such that $S \subseteq \mathrm{dom}(f_n)$ for every $n \in \mathbb{N}$. We say that the sequence $(f_n)_{n \in \mathbb{N}}$ **converges uniformly to the function** f **on** S if $S \subseteq \mathrm{dom}(f)$ and, for every $\varepsilon > 0$, there is a positive integer N such that, for all $n \in \mathbb{N}$ and for all $x \in S$,

$$n \geq N \Rightarrow |f_n(x) - f(x)| < \varepsilon.$$

The condition $|f_n(x) - f(x)| < \varepsilon$ in Definition 4.3 may be rewritten in the form

$$f(x) - \varepsilon < f_n(x) < f(x) + \varepsilon$$

and visualized geometrically as the requirement that the graph of f_n lies in a band of height 2ε centered about the graph of f (Figure 6-2). The following theorem shows that uniform convergence is a stronger condition than pointwise convergence.

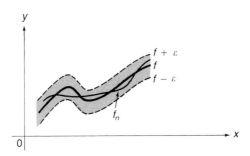

FIGURE 6-2

4.4 THEOREM

UNIFORM CONVERGENCE IMPLIES POINTWISE CONVERGENCE If a sequence of functions $(f_n)_{n \in \mathbb{N}}$ converges uniformly to a function f on the set S, then it converges pointwise to f on S.

PROOF Suppose that $(f_n)_{n \in \mathbb{N}}$ converges uniformly to f on S and let a be an arbitrary point in S. We must prove that

$$\lim_{n \to \infty} f_n(a) = f(a).$$

Let $\varepsilon > 0$. By Definition 4.3, there exists $N \in \mathbb{N}$ such that, for all $n \in \mathbb{N}$ and all $x \in S$,

$$n \geq N \Rightarrow |f_n(x) - f(x)| < \varepsilon.$$

Therefore, in particular,

$$n \geq N \Rightarrow |f_n(a) - f(a)| < \varepsilon,$$

and so $f_n(a) \to f(a)$ as $n \to \infty$. □

We write

$$f_n \to f \quad \text{uniformly on } S,$$

or

$$f_n(x) \to f(x) \quad \text{uniformly for } x \in S,$$

or

$$\lim_{n \to \infty} f_n(x) = f(x) \quad \text{uniformly for } x \in S$$

to mean that the sequence of functions $(f_n)_{n \in \mathbb{N}}$ converges uniformly to the function f on the set S.

It is instructive to compare the definitions of pointwise and uniform convergence. If

$$\lim_{n \to \infty} f_n(x) = f(x),$$

pointwise for $x \in S$, then, for every $\varepsilon > 0$ and every point $x \in S$, there exists $N \in \mathbb{N}$, *possibly depending on both ε and x*, such that, for every $n \in \mathbb{N}$,

$$n \geq N \Rightarrow |f_n(x) - f(x)| < \varepsilon.$$

But, if

$$\lim_{n \to \infty} f_n(x) = f(x) \qquad \text{uniformly for } x \in S,$$

then, for every $\varepsilon > 0$, there exists $N \in \mathbb{N}$, *possibly depending on ε*, such that, for all $n \in \mathbb{N}$ and *every point $x \in S$*,

$$n \geq N \Rightarrow |f_n(x) - f(x)| < \varepsilon.$$

For pointwise convergence, we might have to use different values of N for different points $x \in S$; for uniform convergence, a single value of N (depending only on ε) will do. Thus, the relationship between pointwise and uniform convergence is quite analogous to the relationship between continuity and uniform continuity—both involve only an interchange of the order in which quantifiers appear. (See Definition 5.17 in Chapter 4.)

4.5 Example   In Example 4.2, show that the sequence of functions $(f_n)_{n \in \mathbb{N}}$ does not converge uniformly to the function f.

SOLUTION   This can be seen geometrically by observing that for $0 < \varepsilon \leq \frac{1}{2}$, none of the functions f_n have graphs that remain in the band between $f - \varepsilon$ and $f + \varepsilon$. For a more analytic proof, let $\varepsilon = \frac{1}{2}$ and suppose that $N \in \mathbb{N}$ is such that, for all $n \in \mathbb{N}$ and all $x \in [0, 1]$,

$$n \geq N \Rightarrow |f_n(x) - f(x)| < \frac{1}{2}.$$

For $x \in [0, 1]$ with $x \neq 1$, we have $f_n(x) = x^n$ and $f(x) = 0$, and so

$$n \geq N \Rightarrow |x^n - 0| = x^n < \frac{1}{2}.$$

In particular, then,

$$x < \frac{1}{2^N}$$

holds for all x such that $0 \leq x < 1$. But $0 \leq 1/2^N < 1$, so, with $x = 1/2^N$, we arrive at the contradiction

$$\frac{1}{2^N} < \frac{1}{2^N}.$$

□

4.6 THEOREM

CONTINUITY OF UNIFORM LIMITS   Suppose $f:S \to \mathbb{R}$, that $a \in S$, and that

$$f_n \to f \qquad \text{uniformly on the set } S.$$

Then, if f_n is continuous at the point a for each $n \in \mathbb{N}$, it follows that f is continuous at a.

PROOF   If a is not an accumulation point of S, then f is automatically continuous at a. Therefore, we may assume that a is an accumulation point of S. Let $\varepsilon > 0$. Then there exists $N \in \mathbb{N}$ such that, for all $n \in \mathbb{R}$ and all $x \in S$,

$$n \geq N \Rightarrow |f_n(x) - f(x)| < \frac{\varepsilon}{3}.$$

In particular, for all $x \in S$,

$$|f(x) - f_N(x)| = |f_N(x) - f(x)| < \frac{\varepsilon}{3},$$

and, since $a \in S$,

$$|f_N(a) - f(a)| < \frac{\varepsilon}{3}.$$

Because f_N is continuous at a, there exists $\delta > 0$ such that, for every $x \in S$,

$$|x - a| < \delta \Rightarrow |f_N(x) - f_N(a)| < \frac{\varepsilon}{3}.$$

Now assume that $x \in S$ and that $|x - a| < \delta$. Then, by the triangle inequality (Problem 9),

$$|f(x) - f(a)| \leq |f(x) - f_N(x)| + |f_N(x) - f_N(a)| + |f_N(a) - f(a)|$$
$$< \frac{\varepsilon}{3} + \frac{\varepsilon}{3} + \frac{\varepsilon}{3} = \varepsilon,$$

and it follows that f is continuous at a.

□

4.7 COROLLARY

Suppose that f_n is continuous on S for each $n \in \mathbb{N}$ and that

$$f_n \to f \qquad \text{uniformly on the set } S.$$

Then f is continuous on S.

Corollary 4.7 is extremely important and widely used in mathematical analysis. The converse of Corollary 4.7 is false, that is, merely because a sequence of continuous functions converges pointwise to a continuous function is no guarantee that the convergence is uniform (Problem 6). However, we do have the following theorem.

4.8 THEOREM

DINI'S THEOREM   Let S be a compact set and let $(f_n)_{n \in \mathbb{N}}$ be a sequence of continuous functions that converges pointwise to the continuous function f on S. Suppose, moreover, that $(f_n)_{n \in \mathbb{N}}$ is an increasing sequence on S in the sense that, for all $n \in \mathbb{N}$ and for all $x \in S$,

$$f_n(x) \leq f_{n+1}(x).$$

Then $f_n \to f$ uniformly on S.

PROOF   Let $g_n(x) = f(x) - f_n(x)$ for all $n \in \mathbb{N}$ and all $x \in S$. Then $(g_n)_{n \in \mathbb{N}}$ is a sequence of continuous functions that converges pointwise to the constant function 0 on S. Moreover, for all $n \in \mathbb{N}$ and for all $x \in S$,

$$g_n(x) \geq g_{n+1}(x) \geq 0. \tag{1}$$

It will be sufficient to prove that $g_n \to 0$ uniformly on S (Problem 11).

Let $\varepsilon > 0$. If $y \in S$, then, owing to the fact that the sequence $(g_n(y))_{n \in \mathbb{N}}$ converges to the limit 0, there exists $N_y \in \mathbb{N}$ such that

$$0 \leq g_{N_y}(y) < \frac{\varepsilon}{2}. \tag{2}$$

Also, since g_{N_y} is continuous at y, there exists a neighborhood W_y of y such that, for all $x \in S$,

$$x \in W_y \Rightarrow \left| g_{N_y}(x) - g_{N_y}(y) \right| < \frac{\varepsilon}{2}. \tag{3}$$

In particular, then, for all $x \in S$,

$$x \in W_y \Rightarrow g_{N_y}(x) < g_{N_y}(y) + \frac{\varepsilon}{2}. \tag{4}$$

Now, by (1), (4), and (2), if $x \in W_y$ and $n \in \mathbb{N}$ with $n \geq N_y$, then

$$0 \leq g_n(x) \leq g_{N_y}(x) < g_{N_y}(y) + \frac{\varepsilon}{2} < \frac{\varepsilon}{2} + \frac{\varepsilon}{2} = \varepsilon. \tag{5}$$

Because S is compact, there are finitely many points $y_1, y_2, \ldots, y_k$ in S such that

$$S \subseteq W_{y_1} \cup W_{y_2} \cup \cdots \cup W_{y_k}. \tag{6}$$

Let

$$N = \max(N_{y_1}, N_{y_2}, \ldots, N_{y_k}). \tag{7}$$

As a consequence of (6), (7), and (5), if $n \in \mathbb{N}$ and $x \in S$,

$$n \geq N \Rightarrow 0 \leq g_n(x) \leq \varepsilon. \tag{8}$$

Hence, $g_n \to 0$ uniformly on S. □

4.9 THEOREM

> **UNIFORM CONVERGENCE AND INTEGRATION** If $(f_n)_{n \in \mathbb{N}}$ is a sequence of functions each of which is Riemann integrable on the bounded closed interval $[a, b]$, and if $f_n \to f$ uniformly on $[a, b]$, then f is Riemann integrable on $[a, b]$ and
>
> $$\lim_{n \to \infty} \int_a^b f_n(x)\, dx = \int_a^b f(x)\, dx.$$

PROOF Let $\varepsilon > 0$. Choose $N \in \mathbb{N}$ large enough so that

$$\left| f_N(x) - f(x) \right| < \frac{\varepsilon}{3(b - a)} \tag{1}$$

for all $x \in [a, b]$. By Darboux's integrability condition (Theorem 4.13 in Chapter 5), there exists a partition $P = \{x_0, x_1, x_2, \ldots, x_n\}$ of $[a, b]$ such that

$$U(f_N, P) - L(f_N, P) < \frac{\varepsilon}{3}. \tag{2}$$

For $i = 1, 2, \ldots, n$, let

$$M_i = \sup_{x_{i-1} \leq x \leq x_i} f(x) \quad \text{and} \quad L_i = \sup_{x_{i-1} \leq x \leq x_i} f_N(x). \tag{3}$$

According to (1), we have

$$f(x) < f_N(x) + \frac{\varepsilon}{3(b - a)} \tag{4}$$

for all $x \in [a, b]$, and therefore

$$M_i \leq L_i + \frac{\varepsilon}{3(b - a)} \tag{5}$$

holds for $i = 1, 2, \ldots, n$. Consequently,

$$U(f, P) = \sum_{i=1}^n M_i \Delta x_i \leq \sum_{i=1}^n L_i \Delta x_i + \sum_{i=1}^n \frac{\varepsilon}{3(b - a)} \Delta x_i$$

$$= U(f_N, P) + \frac{\varepsilon}{3}. \tag{6}$$

A similar argument shows that

$$L(f, P) \geq L(f_N, P) - \frac{\varepsilon}{3} \tag{7}$$

(Problem 15). By combining (2), (6), and (7), we find that

$$U(f, P) - L(f, P) < \frac{\varepsilon}{3} + \frac{\varepsilon}{3} + \frac{\varepsilon}{3} = \varepsilon, \tag{8}$$

and it follows from Darboux's integrability condition that f is integrable.

To complete the argument, we again let $\varepsilon > 0$, but this time we choose $N \in \mathbb{N}$ so that, for all $n \in \mathbb{N}$ with $n \geq N$, and for all $x \in [a, b]$,

$$|f_n(x) - f(x)| < \frac{\varepsilon}{2(b - a)}. \tag{9}$$

Then, by Theorem 6.5 in Chapter 5, for $n \geq N$, we have

$$\left| \int_a^b f_n(x)\, dx - \int_a^b f(x)\, dx \right| = \left| \int_a^b [f_n(x) - f(x)]\, dx \right|$$

$$\leq \int_a^b |f_n(x) - f(x)|\, dx \leq \frac{\varepsilon}{2(b - a)} (b - a) < \varepsilon.$$

$\square$

Note that the conclusion of Theorem 4.9 may be written in the form

$$\lim_{n \to \infty} \int_a^b f_n(x)\, dx = \int_a^b \left[\lim_{n \to \infty} f_n(x) \right] dx,$$

and, since the integral is a limit (of Riemann sums), we again have a situation in which—under suitable conditions—limit operations commute.

As we have seen, continuity and integrability are preserved under uniform limits. Alas, differentiability does not fare so well! Not only are there examples to show that the uniform limit of a sequence of differentiable functions need not be differentiable (Problem 12), but it turns out that, even if it is differentiable, the limit of the derivatives need not be the derivative of the limits (Problem 14). However, as the following corollary to Theorem 4.9 shows, we can make the derivative behave itself if we impose sufficiently strong conditions.

4.10 COROLLARY

UNIFORM CONVERGENCE AND DIFFERENTIATION   Let $(g_n)_{n \in \mathbb{N}}$ be a sequence of functions each of which has a continuous derivative on the closed bounded interval $[a, b]$. Suppose that $g_n \to g$ pointwise on $[a, b]$ and also that $g_n' \to f$ uniformly on $[a, b]$. Then g is differentiable on $[a, b]$ and $g' = f$.

PROOF Let $f_n = g'_n$ for each $n \in \mathbb{N}$. By hypothesis, each f_n is continuous on $[a, b]$ and $f_n \to f$ uniformly on $[a, b]$; hence, by Corollary 4.7, f is continuous on $[a, b]$. Also, by the second form of the Fundamental Theorem of Calculus (Theorem 7.14 in Chapter 5),

$$\int_a^x f_n(t)\, dt = g_n(x) - g_n(a)$$

holds for every $x \in [a, b]$. Furthermore, since $g_n(x)$ converges to $g(x)$ as $n \to \infty$, we have

$$\lim_{n \to \infty} \int_a^x f_n(t)\, dt = \lim_{n \to \infty} [g_n(x) - g_n(a)] = g(x) - g(a)$$

for every $x \in [a, b]$. The fact that $f_n \to f$ uniformly on $[a, b]$ implies that

$$f_n \to f \qquad \text{uniformly on } [a, x]$$

for all $x \in [a, b]$. Moreover, since f_n is continuous on $[a, x]$, it is Riemann integrable on $[a, x]$, and thus, by Theorem 4.9,

$$\lim_{n \to \infty} \int_a^x f_n(t)\, dt = \int_a^x f(t)\, dt.$$

Therefore, for every $x \in [a, b]$,

$$\int_a^x f(t)\, dt = g(x) - g(a),$$

so that

$$g(x) = g(a) + \int_a^x f(t)\, dt.$$

Consequently, by the first form of the Fundamental Theorem of Calculus (Theorem 7.7 in Chapter 5), we have

$$g'(x) = \frac{d}{dx}\left[g(a) + \int_a^x f(t)\, dt \right] = \frac{d}{dx} \int_a^x f(t)\, dt = f(x)$$

for every $x \in [a, b]$. □

4.11 THEOREM

CAUCHY CRITERION FOR UNIFORM CONVERGENCE Let $(f_n)_{n \in \mathbb{N}}$ be a sequence of functions and suppose that $S \subseteq \text{dom}(f_n)$ for every $n \in \mathbb{N}$. Then there exists a function f with domain S such that $f_n \to f$ uniformly on S if and only if the following condition is satisfied: For every $\varepsilon > 0$, there exists $N \in \mathbb{N}$ such that, for all $n, m \in \mathbb{N}$ and all $x \in S$,

$$n, m \geq N \Rightarrow |f_n(x) - f_m(x)| < \varepsilon.$$

PROOF   Suppose that

$$f_n \to f \qquad \text{uniformly on } S,$$

and let $\varepsilon > 0$. Then there exists $N \in \mathbb{N}$ such that, for all $n \in \mathbb{N}$ and all $x \in S$,

$$n \geq N \Rightarrow |f_n(x) - f(x)| < \frac{\varepsilon}{2}.$$

Therefore, for all $n, m \geq N$ and all $x \in S$,

$$|f_n(x) - f_m(x)| \leq |f_n(x) - f(x)| + |f(x) - f_m(x)|$$
$$< \frac{\varepsilon}{2} + \frac{\varepsilon}{2} = \varepsilon.$$

Conversely, suppose that the given condition is satisfied. Then, for each $x \in S$, $(f_n(x))_{n \in \mathbb{N}}$ is a Cauchy sequence of real numbers (Problem 17), and it follows from Theorem 1.18 that it converges to a limit (which may well depend on x). Define $f : S \to \mathbb{R}$ by

$$f(x) = \lim_{n \to \infty} f_n(x).$$

Thus, the sequence of functions $(f_n)_{n \in \mathbb{N}}$ converges *pointwise* to the function f on S. We have to prove that the convergence is uniform.

Let $\varepsilon > 0$. By hypothesis, there exists $N \in \mathbb{N}$ such that, for all $n, m \in \mathbb{N}$ and all $x \in S$,

$$n, m \geq N \Rightarrow |f_n(x) - f_m(x)| < \frac{\varepsilon}{2}.$$

We claim that, for all $n \in \mathbb{N}$ and for all $x \in S$,

$$n \geq N \Rightarrow |f_n(x) - f(x)| < \varepsilon,$$

and therefore that $f_n \to f$ uniformly on S. To see that this is so, let $x \in S$ and suppose that $n \geq N$. Since the sequence $(f_n(x))_{n \in \mathbb{N}}$ converges to $f(x)$ as a limit, there exists $M \in \mathbb{N}$ such that, for all $m \in \mathbb{N}$,

$$m \geq M \Rightarrow |f_m(x) - f(x)| < \frac{\varepsilon}{2}.$$

Choose $m \geq \max(M, N)$. Then, since $n, m \geq N$, we have

$$|f_n(x) - f_m(x)| < \frac{\varepsilon}{2},$$

and, since $m \geq M$, we have

$$|f_m(x) - f(x)| < \frac{\varepsilon}{2}.$$

Consequently, by the triangle inequality,

$$\left|f_n(x) - f(x)\right| \le \left|f_n(x) - f_n(x)\right| + \left|f_n(x) - f(x)\right|$$

$$< \frac{\varepsilon}{2} + \frac{\varepsilon}{2} = \varepsilon. \qquad \square$$

An infinite series $\sum_{k=1}^{\infty} f_k$ with functions f_k as its terms is handled in an almost obvious way by considering the sequence of its partial sum functions as in the next definition:

4.12 DEFINITION

SERIES OF FUNCTIONS Let $(f_n)_{n \in \mathbb{N}}$ be a sequence of functions, each of which is defined on a set S. For each $n \in \mathbb{N}$ define the **partial sum function** $g_n : S \to \mathbb{R}$ by $g_n(x) = \sum_{k=1}^{n} f_k(x)$ for all $x \in S$. If the sequence of functions $(g_n)_{n \in \mathbb{N}}$ converges pointwise to the limit function g on S, we write $g = \sum_{k=1}^{\infty} f_k$ and we say that the **series of functions** $\sum_{k=1}^{\infty} f_k$ **converges pointwise on** S.

If the convergence of the sequence $(g_n)_{n \in \mathbb{N}}$ of partial sum functions in Definition 4.12 to the function g is uniform, we say that the series of functions $\sum_{k=1}^{\infty} f_k$ **converges uniformly on** S. A Cauchy criterion for this type of convergence is given in Problem 18. The following theorem gives a very useful sufficient (but not necessary!) condition for uniform convergence of a series of functions.

4.13 THEOREM

THE WEIERSTRASS *M*-TEST Let $(f_n)_{n \in \mathbb{N}}$ be a sequence of functions each of which is defined on S and suppose that $(M_n)_{n \in \mathbb{N}}$ is a sequence of nonnegative numbers such that, for all $n \in \mathbb{N}$ and all $x \in S$,

$$\left|f_n(x)\right| \le M_n.$$

Then, if the series $\sum_{k=1}^{\infty} M_k$ converges, it follows that the series of functions $\sum_{k=1}^{\infty} f_k$ converges uniformly on S.

PROOF   Let g_n be the nth partial sum function defined by

$$g_n(x) = \sum_{k=1}^{n} f_k(x)$$

for all $x \in S$, and let $\varepsilon > 0$. By Theorem 2.1, there exists $N \in \mathbb{N}$ such that, for every integer $n \ge N$,

$$\sum_{k=N}^{n} M_k = \left|\sum_{k=N}^{n} M_k\right| < \frac{\varepsilon}{2}.$$

Now suppose that $n, m \geq N$ and $x \in S$. We may assume that $n \geq m \geq N$. (A similar argument takes care of the case in which $m \geq n$.) Then, for all $x \in S$,

$$
\begin{aligned}
\left| g_n(x) - g_m(x) \right| &= \left| \sum_{k=N}^{n} f_k(x) - \sum_{k=N}^{m} f_k(x) \right| \\
&\leq \left| \sum_{k=N}^{n} f_k(x) \right| + \left| \sum_{k=N}^{m} f_k(x) \right| \\
&\leq \sum_{k=N}^{n} |f_k(x)| + \sum_{k=N}^{m} |f_k(x)| \\
&\leq \sum_{k=N}^{n} M_k + \sum_{k=N}^{m} M_k < \frac{\varepsilon}{2} + \frac{\varepsilon}{2} = \varepsilon.
\end{aligned}
$$

Thus, the sequence of functions $(g_n)_{n \in \mathbb{N}}$ satisfies the condition in Theorem 4.11, and so it converges uniformly on S. Therefore, $\sum_{k=1}^{\infty} f_k$ converges uniformly on S. $\square$

On the basis of Definition 4.12, all of the theorems pertaining to sequences of functions may be translated so as to apply to series of functions. The theorems that follow are obtained in this way, and we leave their proofs as exercises. Note that all of these theorems remain true (with only minor rewording) if the summations begin with $k = 0$ (or any other integer).

4.14 THEOREM

> **CONTINUITY OF A SERIES OF FUNCTIONS**   If $(f_n)_{n \in \mathbb{N}}$ is a sequence of functions each of which is continuous on S, and if the series $\sum_{k=1}^{\infty} f_k$ converges uniformly to a function g on S, then g is continuous on S.

PROOF   The proof follows from Corollary 4.7 (Problem 21). $\square$

4.15 THEOREM

> **TERM-BY-TERM INTEGRATION OF A SERIES**   If each term of the series $\sum_{k=1}^{\infty} f_n$ is Riemann integrable on the bounded closed interval $[a, b]$, and if the series converges uniformly to a function g on $[a, b]$, then g is Riemann integrable on $[a, b]$ and
>
> $$ \int_a^b g(x)\,dx = \sum_{k=1}^{\infty} \int_a^b f_k(x)\,dx $$

PROOF   The proof follows from Theorem 4.9 (Problem 23). $\square$

The conclusion of Theorem 4.15 may be written in the form

$$\int_a^b \left[\sum_{k=1}^\infty f_k(x) \right] dx = \sum_{k=1}^\infty \int_a^b f_k(x)\, dx,$$

and again we have a situation in which limit operations commute. Theorem 4.15 is often paraphrased by the statement that a *uniformly convergent series may be integrated term-by-term*. The following theorem gives conditions under which a series may be differentiated term-by-term.

4.16 THEOREM	**TERM-BY-TERM DIFFERENTIATION OF A SERIES** Let $\sum_{k=1}^\infty f_k$ be a series each term of which has a continuous derivative on the bounded closed interval $[a, b]$ and which converges pointwise to a function g on $[a, b]$. Suppose that the series $\sum_{k=1}^\infty f'_k$ converges uniformly on $[a, b]$. Then for all $x \in [a, b]$, $$g'(x) = \sum_{k=1}^\infty f'_k(x).$$

PROOF   The proof follows from Corollary 4.10 (Problem 25).   □

The conclusion of Theorem 4.16 may be written in the form

$$\frac{d}{dx} \sum_{k=1}^n f_k(x) = \sum_{k=1}^\infty \frac{d}{dx} f_k(x)$$

and thus may be regarded as yet another situation in which limit operations commute.

PROBLEM SET 6.4

1.   Let $f_n(x) = 1/(nx + 1)$ for $n \in \mathbb{N}$ and $x \in [0, 1]$. Show that:
(a) $(f_n)_{n \in \mathbb{N}}$ converges pointwise on $[0, 1]$.
(b) $(f_n)_{n \in \mathbb{N}}$ does not converge uniformly on $[0, 1]$.

2.   Let $(f_n)_{n \in \mathbb{N}}$ be a sequence of functions each of which is defined and bounded on the set S and let $f : S \to \mathbb{R}$ be bounded on S. Prove that $f_n \to f$ uniformly on S if and only if

$$\lim_{n \to \infty} \sup_{x \in S} |f_n(x) - f(x)| = 0.$$

3.   Let $f_n(x) = x/(nx + 1)$ for $n \in \mathbb{N}$ and $x \in [0, 1]$. Show that $(f_n)_{n \in \mathbb{N}}$ converges uniformly on $[0, 1]$.

4. Let $(f_n)_{n \in \mathbb{N}}$ be a sequence of functions each of which is defined and bounded on the set S and suppose that $f_n \to f$ uniformly on S. Prove that:
   (a) f is bounded on S.
   (b) The functions in the sequence are **uniformly bounded** on S in the sense that there exists $M \geq 0$ such that $|f_n(x)| \leq M$ for every $n \in \mathbb{N}$ and every $x \in S$.

5. Let $f_n(x) = 1/(1 + x^n)$ for $n \in \mathbb{N}$ and $x \in [0, 1]$. Show that:
   (a) $(f_n)_{n \in \mathbb{N}}$ converges pointwise on $[0, 1]$.
   (b) $(f_n)_{n \in \mathbb{N}}$ does not converge uniformly on $[0, 1]$.
   (c) If $0 < b < 1$, then $(f_n)_{n \in \mathbb{N}}$ converges uniformly on $[0, b]$.

6. Let $f_n(x) = nx(1 - x)^n$ for $n \in \mathbb{N}$ and $x \in [0, 1]$. Show that:
   (a) $f_n \to 0$ pointwise on $[0, 1]$.
   (b) For each $n \in \mathbb{N}$, $\sup_{0 \leq x \leq 1} f_n(x) = [n/(n + 1)]^{n+1}$.
   (c) $(f_n)_{n \in \mathbb{N}}$ does not converge uniformly on $[0, 1]$. [Hint: See Problem 2.]

7. If $f_n \to f$ uniformly on S, $g_n \to g$ uniformly on S, and $A, B \in \mathbb{R}$, prove that $Af_n + Bg_n \to Af + Bg$ uniformly on S.

8. For each $n \in \mathbb{N}$, define $f_n : [0, 1] \to \mathbb{R}$ by $f_n(x) = nx$ for $0 \leq x \leq 1/n$, $f_n(x) = n(2 - nx)$ for $1/n < x \leq 2/n$, and $f_n(x) = 0$ for $2/n < x \leq 1$.
   (a) Sketch the graph of f_n.
   (b) Show that f_n is continuous.
   (c) Show that $f_n \to 0$ pointwise on $[0, 1]$.
   (d) Show that $(f_n)_{n \in \mathbb{N}}$ does not converge uniformly on $[0, 1]$.
   (e) Show that $\lim_{n \to \infty} \int_0^1 f_n(x) \, dx$ is not equal to $\int_0^1 [\lim_{n \to \infty} f_n(x)] \, dx$.

9. In the proof of Theorem 4.6, fill in the details of the argument showing that $|f(x) - f(a)| < \varepsilon$.

10. Fill in the details in the following alternative proof of Dini's theorem (Theorem 4.8): Let $\varepsilon > 0$. For each $n \in \mathbb{N}$, define $g_n = f - f_n$ and $K_n = S \cap (g_n)^{-1}([\varepsilon, \infty))$, noting that, because $(f_n)_{n \in \mathbb{N}}$ is an increasing sequence converging pointwise to f, $g_n = |f - f_n|$ and $K_{n+1} \subseteq K_n$ for all $n \in \mathbb{N}$. Since f and f_n are continuous, g_n is continuous, and therefore K_n is a closed and bounded subset of S. Since $f_n \to f$ pointwise on S, $\bigcap_{n \in \mathbb{N}} K_n = \emptyset$. Therefore, by Problem 11 in Problem Set 4.5, there is an $N \in \mathbb{N}$ such that $K_n = \emptyset$ for all $n \geq N$. Consequently, $f_n \to f$ uniformly.

11. Let $(f_n)_{n \in \mathbb{N}}$ be a sequence of functions each of which is defined on S and let $f : S \to \mathbb{R}$. Define $g_n : S \to \mathbb{R}$ by $g_n(x) = f(x) - f_n(x)$ for each $x \in S$. Prove that $f_n \to f$ uniformly if and only if $g_n \to 0$ uniformly.

12. Let $f_n(x) = |x|^{(n+1)/n}$ for $n \in \mathbb{N}$ and $x \in [-1, 1]$. Show that:
   (a) $f_n(x) \to |x|$ uniformly for $x \in [-1, 1]$.
   (b) f_n has a continuous derivative on $[-1, 1]$.

13. Let $f_n(x) = nx/(nx + 1)$ for $n \in \mathbb{N}$ and $x \in S = (0, 1)$.
   (a) Show that $f_n(x) \leq f_{n+1}(x)$ for all $n \in \mathbb{N}$ and all $x \in S$.
   (b) Show that $\lim_{n \to \infty} f_n(x) = 1$ pointwise for all $x \in S$.

(c) Show that $(f_n)_{n \in \mathbb{N}}$ does not converge uniformly on S.

(d) Explain why these results do not contradict Dini's theorem (Theorem 4.8).

14. Let

$$f_n(x) = \frac{\sin nx}{\sqrt{n}}$$

for $n \in \mathbb{N}$ and $x \in \mathbb{R}$. Show that:

(a) $f_n \to 0$ uniformly on $\mathbb{R}$.

(b) f_n has a continuous derivative on $\mathbb{R}$.

(c) $f_n'(\pi)$ has no limit as $n \to \infty$.

15. In the proof of Theorem 4.9, show that

$$L(f, P) \geq L(f_N, P) - \frac{\varepsilon}{3}.$$

16. By considering the sequence of Riemann integrable functions $(f_n)_{n \in \mathbb{N}}$ in Problem 6, show that uniform convergence is a sufficient, but not a necessary, condition for

$$\lim_{n \to \infty} \int_a^b f_n(x)\, dx = \int_a^b \left[\lim_{n \to \infty} f_n(x)\right] dx$$

as in Theorem 4.9.

17. Given that the condition in Theorem 4.11 is satisfied, prove that, for each $x \in S$, $(f_n(x))_{n \in \mathbb{N}}$ is a Cauchy sequence of real numbers.

18. Let $(f_n)_{n \in \mathbb{N}}$ be a sequence of functions each of which is defined on the set S. Prove that the series $\sum_{k=1}^{\infty} f_k$ converges uniformly on S if and only if it satisfies the following condition: For every $\varepsilon > 0$, there exists $N \in \mathbb{N}$ such that for all $n, m \in \mathbb{N}$ and all $x \in S$, $n \geq m \geq N$ implies that $\left|\sum_{k=m}^{n} f_k(x)\right| < \varepsilon$.

19. Let $f_n(x) = x^n$ for $n \in \mathbb{N}$ and $x \in \mathbb{R}$, let $0 < r < 1$, and define $M_n = r^n$ for all $n \in \mathbb{N}$.

(a) Prove that $|f_n(x)| \leq M_n$ for all $n \in \mathbb{N}$ and all $x \in [-r, r]$.

(b) Use the Weierstrass M-test (Theorem 4.13) to prove that the series $\sum_{k=1}^{\infty} f_k$ converges uniformly on $[-r, r]$.

20. In Problem 19, prove that the series $\sum_{k=1}^{\infty} f_k$ converges pointwise, but not uniformly, on the interval $(-1, 1)$. [Hint: Let g_n be the nth partial sum of the series. If the series were to converge uniformly on $(-1, 1)$, then, by Theorem 4.11, there would exist $n \in \mathbb{N}$ such that, for all $x \in (-1, 1)$, $|x|^n = |g_n(x) - g_{n-1}(x)| < \frac{1}{2}$. Let $x = 2^{-1/n}$ and thus obtain a contradiction.]

21. Prove Theorem 4.14.

22. Let $(f_n)_{n \in \mathbb{N}}$ be a sequence of functions each of which is uniformly continuous on S and suppose that $f_n \to f$ uniformly on S. Prove that f is uniformly continuous on S.

23. Prove Theorem 4.15.

24. Suppose that $f_n \to f$ uniformly on S, $g_n \to g$ on S, and that, for each $n \in \mathbb{N}$, both f_n and g_n are bounded on S. Prove that $f_n g_n \to fg$ uniformly on S.

25. Prove Theorem 4.16.

26. True or false: If $(f_n)_{n \in \mathbb{N}}$ is a sequence of functions that is uniformly convergent on S, and if g is a continuous function such that, for every $n \in \mathbb{N}$, $f_n(S) \subseteq \mathrm{dom}(g)$, then the sequence of functions $(g \circ f_n)_{n \in \mathbb{N}}$ is uniformly convergent on S. If true, prove it; if false, give a counterexample.

27. Let $f_n(x) = x^n/n!$ for integers $n \geq 0$ and all $x \in \mathbb{R}$. If $b > 0$, show that the series $\sum_{k=0}^{\infty} f_k$ is uniformly convergent on $[-b, b]$. [Hint: Use the Weierstrass M-test with $M_n = b^n/n!$, and use the ratio test to establish the convergence of $\sum_{k=0}^{\infty} M_k$.]

28. In Problem 27, show that $\sum_{k=0}^{\infty} f_k$ converges pointwise, but not uniformly, on $\mathbb{R}$.

5

POWER SERIES

Students of elementary calculus, having worked numerous drill problems involving power series, are usually familiar with the basic properties of these series (radius of convergence, term-by-term differentiability and integrability, and so on). However, rigorous proofs of these properties are often beyond the scope of beginning calculus courses. We are now in a position to furnish these proofs.

5.1 DEFINITION

> **POWER SERIES**   Let $a_n \in \mathbb{R}$ for every integer $n \geq 0$ and let $c \in \mathbb{R}$. For each integer $n \geq 0$, define $f_n \colon \mathbb{R} \to \mathbb{R}$ as follows: For all $x \in \mathbb{R}$, $f_0(x) = a_0$ and $f_n(x) = a_n(x - c)^n$ for $n \geq 1$. Then the series of functions $\sum_{k=0}^{\infty} f_k$ is called a **power series** with **coefficients** a_n, $n = 0, 1, 2, \ldots$ and with **center** c.

The power series in Definition 5.1 is usually written in the form

$$\sum_{k=0}^{\infty} a_k(x - c)^k = a_0 + a_1(x - c) + a_2(x - c)^2 + \cdots$$

and interpreted as representing a function

$$f(x) = \sum_{k=0}^{\infty} f_k(x) = \sum_{k=0}^{\infty} a_k(x - c)^k$$

defined on the set I consisting of all numbers x for which the series converges. We note that $c \in I$ and that

$$f(c) = a_0.$$

In what follows, we concentrate our attention on power series with center zero. Such a power series has the form

$$\sum_{k=0}^{\infty} a_k x^k = a_0 + a_1 x + a_2 x^2 + \cdots$$

and is thus an extension of the idea of a polynomial in x. Theorems concerning power series with center zero are easily translated into theorems about power series with an arbitrary center c simply by replacing x by $x - c$.

5.2 LEMMA

> If the power series $\sum_{k=0}^{\infty} a_k x^k$ converges for $x = p \neq 0$ and diverges for $x = q$, then it converges absolutely for all x such that $|x| < |p|$ and it diverges for all x such that $|x| > |q|$.

PROOF   Suppose that $\sum_{k=0}^{\infty} a_k p^k$ converges and that $|x| < |p|$. By Theorem 2.2,

$$\lim_{n \to \infty} a_n p^n = 0,$$

and therefore, by Problem 31 in Problem Set 6.1, there is a number $M \geq 0$ such that, for every integer $k \geq 0$,

$$|a_k p^k| \leq M.$$

Thus, if $|x| < |p|$, we have $|x/p| < 1$, and

$$|a_k x^k| = \left| a_k p^k \left(\frac{x}{p} \right)^k \right| = |a_k p^k| \left| \frac{x}{p} \right|^k \leq M \left| \frac{x}{p} \right|^k$$

for every integer $k \geq 0$. Consequently, if $|x| < |p|$, then the series

$$\sum_{k=0}^{\infty} |a_k x^k|$$

converges by comparison with the convergent geometric series

$$\sum_{k=0}^{\infty} M \left| \frac{x}{p} \right|^k.$$

Now suppose that $|x| > |q|$. If $\sum_{k=0}^{\infty} a_k x^k$ were to converge, then, by what has already been proved (with p replaced by x), it would follow that $\sum_{k=0}^{\infty} a_k q^k$ would converge, contrary to our hypothesis.   □

5.3 COROLLARY

> Let I be the set of all real numbers x for which the power series $\sum_{k=0}^{\infty} a_k x^k$ converges. Then one of the following holds:
>
> (i) $I = \{0\}$.
> (ii) $I = \mathbb{R}$.
> (iii) There exists $R > 0$ such that $(-R, R) \subseteq I \subseteq [-R, R]$.

PROOF   We have already observed that $0 \in I$. It is possible that the power series converges only for $x = 0$, in which case we have $I = \{0\}$ (see Example 5.5 below). It is also possible that it converges for all values of x, in which case we have $I = \mathbb{R}$ (see Example 5.6 below). Assuming, then, that both (i) and (ii) are false, we have to prove that (iii) holds.

Since (i) is false, there exists $p \neq 0$ such that the power series converges for $x = p$. Since (ii) is false, there exists $q \in \mathbb{R}$ such that the series diverges for $x = q$. By Lemma 5.2, $|p|/2 \in I$, so there are positive numbers in I. Also, by Lemma 5.2, if r is a positive number in I, then $r \leq |q|$; hence the set of positive numbers in I is nonempty and bounded above. Let

$$R = \sup\{r \in I \,|\, r > 0\}.$$

Suppose that $x \in (-R, R)$. Then $|x| < R$, so, by Part (i) of Lemma 3.23 in Chapter 3, there exists $r \in I$ with $r > 0$ such that $|x| < r \leq R$. Consequently, by Lemma 5.2, $x \in I$, and it follows that

$$(-R, R) \subseteq I.$$

Suppose that $x \notin [-R, R]$. Then $R < |x|$, so there exists $d \in \mathbb{R}$ such that $R < d < |x|$. Since $R < d$, it follows that $d \notin I$, and therefore that $x \notin I$ by Lemma 5.2. This shows that

$$x \notin [-R, R] \Rightarrow x \notin I,$$

from which we may conclude that

$$I \subseteq [-R, R].$$ $\square$

5.4 DEFINITION

> **INTERVAL AND RADIUS OF CONVERGENCE**   The set I of all real numbers x for which the power series $\sum_{k=0}^{\infty} a_k x^k$ converges is called its **interval of convergence**. If $R > 0$ and $(-R, R) \subseteq I \subseteq [-R, R]$, then R is called the **radius of convergence** of the power series.

It is convenient to extend the notion of the radius of convergence R of a power series $\sum_{k=0}^{\infty} a_k x^k$ to include the other two cases in Corollary 5.3: If the power series converges only for $x = 0$, so that $I = \{0\}$, we define

$R = 0$; if it converges for all real values of x, so that $I = \mathbb{R}$, we define $R = +\infty$.

Both the ratio test and the root test are useful for finding the radius of convergence of a power series. We illustrate the technique in the next three examples.

5.5 Example   Find the radius of convergence of $\sum_{k=0}^{\infty} k^k x^k$.

SOLUTION   If $x \neq 0$, then

$$\lim_{n \to \infty} |n^n x^n|^{1/n} = \lim_{n \to \infty} n|x| = \infty,$$

so the series diverges by the root test (Problem 3). Therefore, $I = \{0\}$ and $R = 0$.   □

5.6 Example   Find the radius of convergence of

$$\sum_{k=0}^{\infty} \frac{x^k}{k!}.$$

SOLUTION   For every $x \neq 0$, we have

$$\lim_{n \to \infty} \left| \frac{x^{n+1}/(n+1)!}{x^n/n!} \right| = \lim_{n \to \infty} \left| \frac{x^{n+1} n!}{x^n (n+1)!} \right| = \lim_{n \to \infty} \left| \frac{x}{n+1} \right|$$

$$= \lim_{n \to \infty} \frac{|x|}{n+1} = 0,$$

so the series converges by the ratio test. Therefore, $I = \mathbb{R}$ and $R = \infty$.   □

5.7 Example   Find the radius of convergence of

$$\sum_{k=0}^{\infty} (-1)^k \frac{k}{3^k} (x-2)^k.$$

SOLUTION   The series converges for $x = 2$. For $x \neq 2$, we have

$$\lim_{n \to \infty} \left| \frac{(-1)^{n+1}(n+1)(x-2)^{n+1}/3^{n+1}}{(-1)^n n(x-2)^n/3^n} \right| = \lim_{n \to \infty} \frac{n+1}{3n} |x-2| = \frac{|x-2|}{3}.$$

Therefore, by the ratio test, the series converges for $|x-2| < 3$ and diverges for $|x-2| > 3$. Thus, the interval of convergence I of the series satisfies

$$(-1, 5) = (2-3, 2+3) \subseteq I \subseteq [2-3, 2+3\} = [-1, 5]$$

and $R = 3$.   □

In Example 5.7, the ratio test gives no information about whether or not the series converges at the two *endpoints* -1 and 5 of the interval

$[-1, 5]$. This is typical, and it turns out that, in general, anything can happen—a power series may converge at both endpoints of its interval of convergence, it may converge at one endpoint and diverge at the other, or it may diverge at both endpoints. As we illustrate in the next two examples, appropriate tests must be applied to each endpoint to determine convergence or divergence.

5.8 Example   Find the interval of convergence I of the power series in Example 5.7.

SOLUTION   We know from Example 5.7 that

$$(-1, 5) \subseteq I \subseteq [-1, 5].$$

Substituting $x = -1$ into the power series, we obtain

$$\sum_{k=0}^{\infty} (-1)^k \frac{k}{3^k} (-1 - 2)^k = \sum_{k=0}^{\infty} (-1)^{2k} k = \sum_{k=0}^{\infty} k,$$

which diverges by the test for divergence (Corollary 2.4). Substituting $x = 5$, we obtain the series

$$\sum_{k=0}^{\infty} (-1)^k \frac{k}{3^k} (5 - 2)^k = \sum_{k=0}^{\infty} (-1)^k k,$$

which diverges for the same reason. Hence,

$$I = (-1, 5).$$ □

5.9 Example   Find the interval of convergence I of the series

$$\sum_{k=1}^{\infty} \frac{x^k}{k}.$$

SOLUTION   Using the ratio test, we find that the radius of convergence is $R = 1$ (Problem 7 with $c = 0$ and $p = 1$), so that

$$(-1, 1) \subseteq I \subseteq [-1, 1].$$

Substituting $x = -1$ into the power series, we obtain the alternating harmonic series, which converges (Example 3.4). Substituting $x = 1$, we obtain the harmonic series, which diverges (Example 2.18). Hence,

$$I = [-1, 1).$$ □

5.10 THEOREM   **UNIFORM CONVERGENCE OF POWER SERIES**   If $r > 0$ and if r is a point in the interior of the interval of convergence I of the power series $\sum_{k=0}^{\infty} a_k x^k$, then the power series converges uniformly on $[-r, r]$.

PROOF   Since $r \in I^o$, there exists $\varepsilon > 0$ such that $(r - \varepsilon, r + \varepsilon) \subseteq I$. Let $p = r + \varepsilon$ and, for each $n \in \mathbb{N}$, let

$$M_n = |a_n| r^n.$$

Since $p \in I$ and $r < p$, the series $\sum_{k=0}^{\infty} M_k$ converges by Lemma 5.2. Also, for $x \in [-r, r]$, we have

$$|a_n x^n| = |a_n|\, |x|^n \leq |a_n| r^n = M_n,$$

and it follows from the Weierstrass M-test (Theorem 4.13) that $\sum_{k=0}^{\infty} a_k x^k$ converges uniformly on $[-r, r]$.   □

In Theorem 5.10 (and in the theorems to follow), we note that, if $I = \{0\}$, then $I^o = \varnothing$, and so the theorem, although it remains true, has no content.

5.11 THEOREM

> **CONTINUITY OF A POWER SERIES** If I is the interval of convergence of the power series $\sum_{k=0}^{\infty} a_k x^k$, then the function f defined for all $x \in I$ by $f(x) = \sum_{k=0}^{\infty} a_k x^k$ is continuous on the interior of I.

PROOF   Let $b \in I^o$, noting that $|b| \in I^o$ by Corollary 5.3 (Problem 9), and choose $r \in I^o$ with $|b| < r$. By Theorem 5.10, the power series converges uniformly to f on $[-r, r]$. Since the function f_n defined by $f_n(x) = a_n x^n$ for $x \in \mathbb{R}$ is continuous, it follows from Theorem 4.14 that f is continuous on $[-r, r]$. But $b \in (-r, r)$, so f is continuous at the point b. Since b was an arbitrary point of I^o, it follows that f is continuous on I^o.   □

5.12 THEOREM

> **TERM-BY-TERM INTEGRATION OF A POWER SERIES** If b is a point in the interior of the interval of convergence I of the power series $\sum_{k=0}^{\infty} a_k x^k$, then
>
> $$\int_0^b \left(\sum_{k=0}^{\infty} a_k x^k \right) dx = \sum_{k=0}^{\infty} \left(\int_0^b a_k x^k \, dx \right) = \sum_{k=0}^{\infty} \frac{a_k}{k+1} b^{k+1}.$$

PROOF   Let $r = |b|$, noting that $r \in I^o$. By Theorem 5.10, the power series converges uniformly for $x \in [-r, r]$, and therefore it converges uniformly for x in the closed interval with 0 and b as endpoints. The conclusion of the theorem now follows immediately from Theorem 4.15.   □

Note that the power series

$$\sum_{k=0}^{\infty} \frac{a_k}{k+1} x^{k+1}$$

is obtained by a term-by-term antidifferentiation of the power series $\sum_{k=0}^{\infty} a_k x^k$. Thus, Theorem 5.12 has the following corollary (Problem 13).

5.13 COROLLARY

If I is the interval of convergence of the power series $\sum_{k=0}^{\infty} a_k x^k$, then I^{o} is contained in the interval of convergence of the power series

$$\sum_{k=0}^{\infty} \frac{a_k}{k+1} x^{k+1},$$

and the latter power series is an antiderivative of the former on the interior of I.

5.14 LEMMA

The two power series

$$\sum_{k=0}^{\infty} a_k x^k \qquad \text{and} \qquad \sum_{k=0}^{\infty} \frac{a_k}{k+1} x^{k+1}$$

have the same radius of convergence.

PROOF

As a consequence of Corollary 5.13, the radius of convergence R^* of the antidifferentiated power series cannot be smaller than the radius of convergence R of the original power series. Suppose then that $R < R^*$ (where we include the possibility that R is finite and $R^* = +\infty$). Choose real numbers s and t such that

$$R < s < t < R^*$$

(with the understanding that there is no restriction on t if $R^* = +\infty$). Then

$$\sum_{k=0}^{\infty} \frac{a_k}{k+1} t^{k+1}$$

converges; hence, by factoring out the constant t, we find that

$$\sum_{k=0}^{\infty} \frac{a_k}{k+1} t^{k}$$

converges. Therefore, the nth term of the latter series approaches zero as $n \to \infty$, and it follows that there exists a positive constant M such that, for all $n \in \mathbb{N}$,

$$\left| \frac{a_n}{n+1} t^n \right| \leq M.$$

Now let $q = s/t$, noting that $0 < q < 1$; hence, by the ratio test, the series $\sum_{k=0}^{\infty} M(k + 1)q^k$ converges. But, for all values of n, we have

$$|a_n s^n| = \left| \frac{a_n}{n + 1} t^n(n + 1) \frac{s^n}{t^n} \right| = \left| \frac{a_n}{n + 1} t^n \right| (n + 1)q^n \leq M(n + 1)q^n,$$

and it follow that the series $\sum_{k=0}^{\infty} a_k s^k$ converges by comparison with $\sum_{k=0}^{\infty} M(k + 1)q^k$, contradicting the fact that $R < q$. $\square$

We note that, if two power series have the same radius of convergence, then the interiors of their intervals of convergence are also the same (Problem 15). Therefore, by Lemma 5.14, the interior of the interval of convergence of a power series is the same as the interior of the interval of convergence of the power series obtained by antidifferentiating it, term-by-term, as in Corollary 5.13. We use this fact in the proof of the following theorem.

5.15 THEOREM

> **TERM-BY-TERM DIFFERENTIATION OF A POWER SERIES** The two power series $\sum_{k=0}^{\infty} a_k x^k$ and $\sum_{k=1}^{\infty} k a_k x^{k-1}$ have the same radius of convergence and, if I is the interval of convergence of $\sum_{k=0}^{\infty} a_k x^k$, then, for every $x \in I^o$,
>
> $$\frac{d}{dx} \sum_{k=0}^{\infty} a_k x^k = \sum_{k=1}^{\infty} k a_k x^{k-1}.$$

PROOF   Let R be the radius of convergence of $\sum_{k=1}^{\infty} k a_k x^{k-1}$. The series obtained by term-by-term antidifferentiation of this series is $\sum_{k=1}^{\infty} a_k x^k$ and, by Lemma 5.14, the latter series also has radius of convergence R. Thus, R is the radius of convergence of $\sum_{k=0}^{\infty} a_k x^k = a_0 + \sum_{k=1}^{\infty} a_k x^k$ and I^o is the interior of the interval of convergence of $\sum_{k=1}^{\infty} k a_k x^{k-1}$. By Corollary 5.13, $\sum_{k=1}^{\infty} a_k x^k$ is an antiderivative of $\sum_{k=1}^{\infty} k a_k x^{k-1}$ on I^o, and therefore, for all $x \in I^o$,

$$\frac{d}{dx} \sum_{k=0}^{\infty} a_k x^k = \frac{d}{dx} \left(a_0 + \sum_{k=1}^{\infty} a_k x^k \right) = \frac{d}{dx} \sum_{k=1}^{\infty} a_k x^k$$

$$= \sum_{k=1}^{\infty} k a_k x^{k-1}. \quad \square$$

If I is an interval and $f(x) = \sum_{k=0}^{\infty} a_k (x - c)^k$ for all $x \in I$, we say that the power series on the right **represents** the function f on I. For instance, by Theorem 1.22, we have

$$\frac{1}{1 - x} = \sum_{k=0}^{\infty} x^k \qquad \text{for } |x| < 1,$$

and so the geometric series on the right represents the function $f(x) = 1/(1 - x)$ on the interval $(-1, 1)$. By starting with a known power series representation and integrating or differentiating term-by-term, it is possible to obtain interesting new power series representations.

5.16 Example   Prove that

$$\frac{1}{(1-x)^2} = \sum_{k=1}^{\infty} kx^{k-1} \qquad \text{for } |x| < 1.$$

SOLUTION   Differentiate both sides of the equation

$$\frac{1}{1-x} = \sum_{k=0}^{\infty} x^k \qquad \text{for } |x| < 1. \qquad \square$$

5.17 Example   Prove that

$$\ln(1+x) = \sum_{k=1}^{\infty} (-1)^{k+1} \frac{x^k}{k} \qquad \text{for } |x| < 1.$$

SOLUTION   By Theorem 1.22, we have

$$\frac{1}{1+t} = \sum_{k=0}^{\infty} (-t)^k = \sum_{k=0}^{\infty} (-1)^k t^k \qquad \text{for } |t| < 1.$$

Therefore, for $|x| < 1$,

$$\ln(1+x) = \int_0^x \frac{dt}{1+t} = \sum_{k=0}^{\infty} (-1)^k \int_0^x t^k \, dt = \sum_{k=0}^{\infty} (-1)^k \frac{x^{k+1}}{k+1}$$

$$= \sum_{k=1}^{\infty} (-1)^{k+1} \frac{x^k}{k},$$

where, in the last step, we have changed the variable of summation. $\square$

Although the next definition is suggested by Definition 3.4 in Chapter 5, it is important not to confuse a Taylor *polynomial* with a Taylor *series*.

5.18 DEFINITION

> **TAYLOR SERIES**   Let $c \in \mathbb{R}$ and suppose that the function f is defined and has derivatives of all orders on some neighborhood of c. Then the power series
>
> $$\sum_{k=0}^{\infty} \frac{f^{(k)}(c)}{k!} (x-c)^k$$
>
> is called the **Taylor series for f at c.**

In Definition 5.18, we do not necessarily mean to suggest that there is any relationship whatsoever between the function f and its Taylor series at c, except that the coefficients of the Taylor series are obtained from f

as indicated. However, we do have the following important result:

5.19 THEOREM

WHEN A POWER SERIES IS A TAYLOR SERIES Suppose that U is an open interval, that $c \in U$, and that the function f is represented on U by a power series with center c. Then the power series is the Taylor series for f at c.

PROOF   Suppose that

$$f(x) = \sum_{k=0}^{\infty} a_k(x - c)^k \qquad \text{for } x \in U.$$

Then, if I is the interval of convergence of the power series, we have $U \subseteq I$. Since U is open, it follows that $U \subseteq I^o$. By Theorem 5.15, f is differentiable on U and

$$f'(x) = \sum_{k=1}^{\infty} ka_k(x - c)^{k-1} \qquad \text{for } x \in U.$$

Applying Theorem 5.15 again, we find that f' is differentiable on U and that

$$f''(x) = \sum_{k=2}^{\infty} k(k - 1)a_k(x - c)^{k-2} \qquad \text{for } x \in U.$$

This process can be continued by mathematical induction, and we find that f has derivatives of all orders on U and that, for all $n \in \mathbb{N}$,

$$f^{(n)}(x) = \sum_{k=n}^{\infty} \frac{k!}{(k - n)!} a_k(x - c)^{k-n} \qquad \text{for } x \in U$$

(Problem 23). By putting $x = c$ in the last equation, we find that

$$f^{(n)}(c) = n!a_n$$

since all terms in the power series except for the first (when $k = n$) contain the factor $x - c$. Hence,

$$a_n = \frac{f^{(n)}(c)}{n!} \qquad \text{for all } n \in \mathbb{N}. \qquad \square$$

5.20 THEOREM

REPRESENTATION OF A FUNCTION BY ITS TAYLOR SERIES Let U be an open interval, let f be a function that is defined and has derivatives of all orders on U, and suppose that there is a positive constant M such that

$$\left| f^{(n)}(x) \right| \leq M$$

holds for all $n \in \mathbb{N}$ and all $x \in U$. Then, if $c \in U$, the Taylor series for f with center c converges to f on U.

PROOF   For each $n \in \mathbb{N}$, define $p_n: U \to \mathbb{R}$ by

$$p_n(x) = \sum_{k=0}^{n} \frac{f^{(k)}(c)}{k!} (x - c)^k$$

for all $x \in U$, noting that p_n is the restriction to U of the nth Taylor polynomial for f at c (Definition 3.4 in Chapter 5) and also that it is the nth partial sum of the Taylor series for f at c (Definition 5.18). Hence, to show that the Taylor series for f at c converges pointwise to f on U, it will be sufficient to prove that

$$\lim_{n \to \infty} |f(x) - p_n(x)| = 0$$

holds for each $x \in U$.

If $x \in U$ and $n \in \mathbb{N}$, then, by the Lagrange form of Taylor's theorem (Theorem 3.6 in Chapter 5), there exists a real number γ between x and c such that

$$f(x) = p_n(x) + \frac{f^{(n+1)}(\gamma)}{(n+1)!} (x - c)^{n+1}.$$

Therefore, by our hypothesis,

$$|f(x) - p_n(x)| = \left| \frac{f^{(n+1)}(\gamma)}{(n+1)!} \right| |x - c|^{n+1} \leq M \frac{|x - c|^{n+1}}{(n+1)!}$$

holds for every $n \in \mathbb{N}$ and every $x \in U$. As a consequence of Example 5.6, the series

$$\sum_{k=0}^{\infty} \frac{|x - c|^k}{k!}$$

converges for every $x \in \mathbb{R}$, and it follows from Theorem 2.2 that

$$\lim_{n \to \infty} \frac{|x - c|^{n+1}}{(n+1)!} = \lim_{n \to \infty} \frac{|x - c|^n}{n!} = 0.$$

Therefore, for each $x \in U$, we have

$$\lim_{n \to \infty} |f(x) - p_n(x)| = 0. \qquad \square$$

5.21 Example   Prove that

$$e^x = \sum_{k=0}^{\infty} \frac{x^k}{k!} \qquad \text{for all } x \in \mathbb{R}.$$

SOLUTION   Let r be any positive number and let $U = (-r, r)$. Define $f(x) = e^x$ for all $x \in \mathbb{R}$, noting that f has derivatives of all orders on $\mathbb{R}$ and that, for any $n \in \mathbb{N}$,

$$f^{(n)}(x) = e^x = f(x)$$

for all $x \in \mathbb{R}$. In particular, since f is an increasing function,

$$|f''(x)| = |e^x| = e^x \le e^r$$

holds for all $x \in U$. Thus, with $M = e^r$ and $c = 0$ in Theorem 5.20, we may conclude that

$$e^x = \sum_{k=0}^{\infty} \frac{f^{(k)}(0)}{k!} (x - 0)^k = \sum_{k=0}^{\infty} \frac{x^k}{k!}$$

holds for all $x \in U = (-r, r)$. But, since we may choose r as large as we please, the last equation holds for all $x \in \mathbb{R}$. $\qquad\square$

5.22 LEMMA

If $\sum_{k=0}^{\infty} a_k x^k$ converges on $[0, 1]$, then it converges uniformly on $[0, 1]$.

PROOF   Let $\varepsilon > 0$ and, for each $n \in \mathbb{N}$, let

$$A_n = \sum_{k=n}^{\infty} a_k.$$

Since, by hypothesis, $\sum_{k=0}^{\infty} a_k$ converges, there exists $N \in \mathbb{N}$ such that, for every $n \in \mathbb{N}$,

$$n \ge N \Rightarrow |A_n| < \frac{\varepsilon}{2}$$

(Problem 31). Hence, for $n \ge N$ and $0 \le x < 1$, the series $\sum_{k=n}^{\infty} A_k x^k$ converges absolutely by comparison with the geometric series

$$\sum_{k=n}^{\infty} \frac{\varepsilon}{2} x^k.$$

Also, as a consequence of Theorem 2.11,

$$A_k - A_{k+1} = a_k$$

holds for every $k \in \mathbb{N}$, and so, for $n \ge N$ and $0 \le x < 1$,

$$\sum_{k=n}^{\infty} a_k x^k = \sum_{k=n}^{\infty} (A_k - A_{k+1}) x^k = \sum_{k=n}^{\infty} (A_k x^k - A_{k+1} x^k)$$

$$= \sum_{k=n}^{\infty} A_k x^k - \sum_{k=n}^{\infty} A_{k+1} x^k$$

$$= A_n x^n + \sum_{k=n}^{\infty} A_{k+1} x^{k+1} - \sum_{k=n}^{\infty} A_{k+1} x^k$$

$$= A_n x^n + \sum_{k=n}^{\infty} A_{k+1} (x^{k+1} - x^k)$$

$$= A_n x^n - x^n (1 - x) \sum_{k=n}^{\infty} A_{k+1} x^{k-n}.$$

Therefore, for $n \geq N$ and $0 \leq x < 1$, we have

$$\left| \sum_{k=n}^{\infty} a_k x^k \right| \leq \frac{\varepsilon}{2} x^n + x^n(1-x) \frac{\varepsilon}{2} \sum_{k=n}^{\infty} x^{k-n}$$

$$= \frac{\varepsilon}{2} \left[x^n + x^n(1-x) \frac{1}{1-x} \right] = \varepsilon x^n < \varepsilon.$$

When $x = 1$, we also have

$$\left| \sum_{k=n}^{\infty} a_k x^k \right| = |A_n| < \frac{\varepsilon}{2} < \varepsilon,$$

and so, for $n \geq N$ and $0 \leq x \leq 1$,

$$\left| \sum_{k=0}^{n-1} a_k x^k - \sum_{k=0}^{\infty} a_k x^k \right| = \left| \sum_{k=n}^{\infty} a_k x^k \right| < \varepsilon.$$

This shows that the partial sums of the series converge uniformly on $[0, 1]$, and the proof is complete. $\square$

As a consequence of Lemma 5.22, we have the following theorem, the proof of which we leave as an exercise (Problem 32).

5.23 THEOREM

> **ABEL'S THEOREM** Suppose that $\sum_{k=0}^{\infty} a_k(x-c)^k$ has a finite positive radius of convergence R. Then, if the series converges for $x = R$ (or for $x = -R$), it converges uniformly on $[c, c+R]$ (or on $[c-R, c]$).

In Problem Set 6.5, we derive some of the consequences of Abel's theorem.

PROBLEM SET 6.5

1. If I is the set of all real numbers for which a power series $\sum_{k=0}^{\infty} a_k(x-c)^k$ converges, prove that one of the following holds: $I = \{c\}$, or $I = \mathbb{R}$, or there exists $R > 0$ such that $(c-R, c+R) \subseteq I \subseteq [c-R, c+R]$.

2. Let $\sum_{k=0}^{\infty} a_k(x-c)^k$ be a power series with radius of convergence R and assume that $a_k \neq 0$ for all integers $k \geq 0$. Suppose that $L = \lim_{n \to \infty} |a_{n+1}/a_n|$, where L is either a nonnegative real number or $+\infty$. If we make the convention (for purposes of this exercise only!) that $1/0 = +\infty$ and $1/(+\infty) = 0$, prove that $R = 1/L$.

3. In Example 5.5, we need a slight extension of the root test in which $\lim_{n \to \infty} |a_n|^{1/n} = +\infty$ implies that $\sum_{k=0}^{\infty} a_k$ diverges. Prove this; in fact, prove the following even more general result: If there is an $N \in \mathbb{N}$ such that $|a_n|^{1/n} > 1$ for all $n \geq N$, then the series diverges.

4. State and prove an extension of the ratio test analogous to the result in Problem 3 for the root test.

5. If r is a constant, find the radius and interval of convergence of the power series $\sum_{k=0}^{\infty} r^k(x - c)^k$.

6. If r is a constant and p is a positive integer, find the radius and interval of convergence of the power series $\sum_{k=0}^{\infty} r^k(x - c)^{pk}$.

7. If p is a constant, find the radius and interval of convergence of the power series

$$\sum_{k=1}^{\infty} \frac{(x - c)^k}{k^p}.$$

8. If $a > b \geq 0$, find the radius and interval of convergence of the power series

$$\sum_{k=0}^{\infty} \frac{x^k}{a^k + b^k}.$$

9. If I is the interval of convergence of the power series $\sum_{k=0}^{\infty} a_k x^k$ and $b \in I^o$, prove that $|b| \in I^o$.

10. If R is the radius of convergence of the power series $\sum_{k=0}^{\infty} a_k(x - c)^k$, if R is positive and finite, and if p is a positive integer, show that $R^{1/p}$ is the radius of convergence of the power series $\sum_{k=0}^{\infty} a_k(x - c)^{pk}$.

11. Let a be a constant real number.
    (a) Show that the series

$$\sum_{k=1}^{\infty} \frac{a(a - 1)(a - 2) \cdots (a - k)}{k!} x^k$$

   converges for $|x| < 1$.
    (b) Use the result of Part (a) to show that, if $|x| < 1$, then

$$\lim_{n \to \infty} \frac{a(a - 1)(a - 2) \cdots (a - n)}{n!} x^n = 0.$$

12. (a) Prove that

$$\frac{1}{1 + x^2} = \sum_{k=0}^{\infty} (-1)^k x^{2k} \qquad \text{for } |x| < 1.$$

   [Hint: Replace x by x^2 in the geometric series representation for $1/(1 + x)$.]
    (b) Use a term-by-term integration of the series in Part (a) to prove that

$$\tan^{-1} x = \sum_{k=0}^{\infty} (-1)^k \frac{x^{2k + 1}}{2k + 1} \qquad \text{for } |x| < 1.$$

   This is called **Gregory's series**.
    (c) If $f(x) = \tan^{-1} x$, find a formula for $f^{(n)}(0)$.

13. Prove Corollary 5.13.

14. (a) Prove the identity $\pi/4 = \tan^{-1}(1/7) + 2\tan^{-1}(1/3)$.
    (b) Use the identity in Part (a) and Gregory's series in Problem 12 to compute the value of π to four decimal places.

15. If two power series with the same center have the same radius of convergence, prove that the interiors of their intervals of convergence are the same.

16. Find power series representations for each of the following. Be sure to indicate the radius of convergence of each power series.

    (a) $\dfrac{1}{(1-x)^2}$ (b) $\dfrac{1}{a-x}$ (c) $\dfrac{1}{1+ax}$ (d) $\dfrac{x}{1-x^2}$

17. (a) Find a power series representation for $\ln(1-x)$.
    (b) Find a power series representation for

    $$\ln\frac{1+x}{1-x}.$$

    [Hint: Combine the results in Part (a) and Example 5.17 and use the fact that $\ln(a/b) = \ln a - \ln b$.]

18. Use the known power series representation of $1/(1-x)$ for $|x| < 1$ to find a function that is represented by each power series. In each case, be sure to indicate the radius of convergence.

    (a) $\displaystyle\sum_{k=0}^{\infty} x^{2k}$ (b) $\displaystyle\sum_{k=1}^{\infty} x^{2k}$ (c) $\displaystyle\sum_{k=1}^{\infty} x^{k+1}$

    (d) $\displaystyle\sum_{k=1}^{\infty} kx^{2k-1}$ (e) $\displaystyle\sum_{k=0}^{\infty} \frac{x^{k+1}}{k+1}$

19. If a function f has a power series expansion with center c on an interval I, prove that this power series is uniquely determined.

20. (a) Find the Taylor series for $\sin x$ at $c = \pi/6$.
    (b) For what values of x does this series represent $\sin x$?

21. Suppose that f is defined and has derivatives of all orders on $\mathbb{R}$ and that, for each positive real number r there exists a positive real number M_r such that $|f^{(n)}(x)| \le M_r$ for all $n \in \mathbb{N}$ and all $x \in (-r, r)$. Prove that the Taylor series for f at c converges to f on all of $\mathbb{R}$.

22. Use the power series representation for e^x in Example 5.21 to find the value of e to four decimal places. [Hint: Let $x = 1$ and use Taylor's theorem in the Lagrange form to keep the error under control.]

23. Under the hypotheses of Theorem 5.19, prove that, for all $n \in \mathbb{N}$ and all $x \in U$,

    $$f^{(n)}(x) = \sum_{k=n}^{\infty} \frac{k!}{(k-n)!} a_k(x-c)^{k-n}.$$

24. Let $f : \mathbb{R} \to \mathbb{R}$ be defined by $f(x) = e^{-1/x^2}$ for $x \neq 0$ and $f(0) = 0$.
    (a) Sketch the graph of f.
    (b) Find $f'(x)$, $f''(x)$, and $f'''(x)$ for $x \neq 0$.
    (c) Prove by induction that, for every positive integer n, there exists a polynomial function P_n such that $f^{(n)}(x) = P_n(1/x)f(x)$ holds for $x \neq 0$.
    (d) Use Part (c) to prove that $\lim_{x \to 0}[f^{(n)}(x)/x] = 0$ for all $n \in \mathbb{N}$.
    (e) Show that f has derivatives of all orders on $\mathbb{R}$ and that $f^{(n)}(0) = 0$ for all $n \in \mathbb{N}$.
    (f) Show that f has no power series representation with center $c = 0$.

25. A Taylor series with center $c = 0$ is called a **Maclaurin series**. Find the Maclaurin series representations for the following, and show that these representations are valid for all $x \in \mathbb{R}$:
    (a) $\sin x$ (b) $\cos x$

26. Find the value of $\int_0^1 \sin x^2 \, dx$ to four decimal places. [Hint: Use the Maclaurin series representation for $\sin x$ (Problem 25), substitute x^2 for x, and integrate term-by-term. Then take sufficiently many terms of the resulting series to obtain the desired result with an error no more than 0.5×10^4.]

27. Let p be a constant and define $f(x) = (1 + x)^p$ for $x > -1$.
    (a) Prove that f satisfies the differential equation $(1 - x)f'(x) = pf(x)$.
    (b) Supposing that f can be represented by a power series $f(x) = \sum_{k=0}^{\infty} a_k x^k$ so that $f'(x) = \sum_{k=1}^{\infty} k a_k x^{k-1}$, substitute these expressions into the differential equation in Part (a), simplify, and equate coefficients on both sides of the resulting equation to conclude that

    $$a_{n+1} = \frac{p - n}{n + 1} a_n \qquad \text{for all } n \in \mathbb{N}.$$

    (c) Show that $a_0 = 1$, and thus conclude that

    $$a_n = \frac{p(p - 1)(p - 2) \cdots (p - n + 1)}{n!} \qquad \text{for all } n \in \mathbb{N}.$$

    The series $\sum_{k=0}^{\infty} a_k x^k$ is called the **binomial series** for $(1 + x)^p$.

28. If p is a positive integer, prove that the binomial series in Problem 27 reduces to a finite sum and that this sum is equal to $(1 + x)^p$ for all $x \in \mathbb{R}$. [Hint: Recall the binomial theorem.]

29. In Problem 27, suppose that $p \neq 0$ and that $p \notin \mathbb{N}$.
    (a) Show that all of the coefficients in the binomial series are nonzero.
    (b) Show that the binomial series has radius of convergence $R = 1$.
    (c) Let g be the function to which the binomial series converges on $(-1, 1)$. Prove that g satisfies the differential equation $(1 + x)g'(x) = pg(x)$ for all $x \in (-1, 1)$.

(d) By computing its derivative, show that $g(x)(1 + x)^{-p}$ is a constant for all $x \in (-1, 1)$, and, by substituting $x = 0$, show that this constant is 1.

(e) Conclude that $(1 + x)^p$ is represented by the binomial series on the interval $(-1, 1)$.

30. Using the result of Problem 29:

(a) Find a power series representation for $(1 + x)^{1/3}$, $|x| < 1$.

(b) Use Part (a) to find the value of $28^{1/3}$ to four decimal places. [Hint: Find the value of $(1 + \frac{1}{27})^{1/3} = (\frac{28}{27})^{1/3} = 28^{1/3}/3$.]

31. Attend to a detail in the proof of Lemma 5.22 by showing that, if $\sum_{k=0}^{\infty} a_k$ converges and $\varepsilon > 0$, then there exists $N \in \mathbb{N}$ such that, for every integer $n \geq N$, $\left| \sum_{k=n}^{\infty} a_k \right| < \varepsilon$.

32. Prove Theorem 5.23. [Hint: Let $t = (x - c)/R$ and consider the power series $\sum_{k=0}^{\infty} a_k t^k$.]

33. If $f(x) = \sum_{k=0}^{\infty} a_k x^k$ for all x in I, the interval of convergence of the power series, prove that f is continuous on I. [Hint: Use Abel's theorem (Theorem 5.23) to take care of continuity at the endpoints (if any) of I.]

34. (a) If $\sum_{k=0}^{\infty} a_k(x - c)^k$ has a finite positive radius of convergence R, and if its interval of convergence is $I = [-R, R]$, prove that the series converges uniformly on I.

(b) True or false: A power series with a finite positive radius of convergence always converges uniformly on its interval of convergence. If true, prove it; if false, give a counterexample.

35. Prove that the sum of the alternating harmonic series is ln 2. [Hint: Combine the result of Example 5.17 with Abel's theorem.]

36. Prove that $\pi = 4(1 - \frac{1}{3} + \frac{1}{5} - \frac{1}{7} + \cdots)$, that is, π is four times the alternating sum of the reciprocals of the odd integers. [Hint: Use Gregory's series (Problem 12) and Abel's theorem.]

HISTORICAL NOTES

Although Archimedes had summed the geometric series, other results on infinite series did not appear in Europe until the fourteenth century, when Nicole Oresme (c. 1330–1382) showed in his *Quaestiones Super Geometriam Euclidis* that the harmonic series diverges. He did it by the method still used in calculus texts today, grouping the terms. Pietro Mengoli (1625–1686) rediscovered this important fact and went on to show that the alternating harmonic series sums to ln 2. Mengoli also showed that the sum of the reciprocals of the triangular numbers converges, a result also shown by Christian Huygens (1629–1695). The next natural series considered was the sum of the reciprocals of the squares of the positive integers. Mengoli failed on this problem, as did a number of others, including the Bernoullis. It was not until 1734–1735 that Euler found a beautiful expression for this series, $\pi^2/6$. At the same time, he found formulas for the sums of the reciprocals of each even power of the positive integers. A second derivation was given in what may be Euler's most beautiful book, his *Introductio in Analysin Infinitorum* of 1748. These sums involve the so-called Bernoulli numbers, although Euler did not point out this connection until 1755. It was one of Euler's great achievements. His methods were not entirely rigorous, but rigorous methods were found later.

Numerous other series were also summed. In 1674, Leibniz showed that the reciprocals of the odd integers, with alternating signs, sum to $\pi/4$. Newton, in his *De Analysi* of 1669, gave power series for $\sin x$, $\cos x$, $\arcsin x$, and e^x, and James Gregory (1638–1675) added the series for $\tan x$ and $\sec x$ that same year. In 1673, Leibniz independently found series for $\sin x$, $\cos x$, and $\arctan x$. So, well before Leibniz's first work of 1684 on calculus, there were derivations of infinite series for a number of the standard functions of mathematics.

Brook Taylor (1685–1731), in his 1715 *Methodus Incrementorum Directa et Inversa*, published a result of his dating from 1712, the theorem on infinite series that bears his name. This provides further evidence, if any were needed, of a metatheorem in mathematics—if a theorem is named for someone, that person was probably not the first to prove it. In this case, Taylor's result had been discovered by Gregory in 1670 and was independently discovered by Leibniz and Bernoulli just a few years later. There is even evidence that it was known in India as early as 1550. Indeed, modern scholarship has turned up evidence of the discovery of a number of mathematical ideas in China, India, and Persia well before they came to be known in Europe.

In the seventeenth century, there seemed to be little concern for convergence of series, and most of the derivations considered the series only formally, really just as extensions of polynomials. Leibniz, however, did show concern for convergence, and provided in 1713 the proof that an alternating series whose terms decrease in absolute value monotonically converges. But even during the eighteenth century, although there was an awareness of the distinction between convergence and divergence, such questions were often ignored. Even Euler was rather casual about using divergent series when they served his purposes.

Lagrange and d'Alembert wrote on questions of convergence and divergence. And Edward Waring (1734–1798), another in the long line of Lucasian professors of mathematics at Cambridge and now remembered mainly for a famous problem in number theory that bears his name, was ahead of his time in working on questions of convergence. He produced what we now call the ratio test, a result often attributed to Cauchy, who discovered it considerably later.

The first rigorous investigation of convergence was given by Gauss in 1812 in his *Disquisitiones Generales Circa Seriem Infinitam*. In 1817,

Bolzano gave a correct condition for convergence of a sequence, but, as mentioned before, this work did not become widely known. In 1821, in his *Cours d'Analyse*, Cauchy gave what is now called the Cauchy criterion. He also asserted that a convergent series of continuous functions is continuous. Abel then pointed out that this could only be concluded if the series is *uniformly* convergent. Weierstrass observed that the condition of uniform convergence was also needed in order to integrate a series term-by-term, an observation also noted by Phillipp L. von Seidel (1821–1896), by Sir George Gabriel Stokes (1819–1903), and, eventually, by Cauchy. And Heine was able to show that the Fourier representation of a continuous function is unique with the condition of uniform convergence.

Mathematicians in the seventeenth and eighteenth centuries rearranged the terms of series without qualms. It was not until 1837 that Dirichlet showed that with absolutely convergent series one may rearrange or group the terms without changing the sum, but with conditionally convergent series such operations may indeed alter the sum. In 1854, Riemann showed that with a suitable rearrangement of terms, the sum could be made to be anything at all.

As mentioned earlier, Euler had extensively used divergent series to obtain results that were indeed correct. Yet in the nineteenth century it was recognized that there were dangers, so the study of divergent series fell into disrepute although Abel and Cauchy expressed doubt about the wisdom of abandoning divergent series since they had been so useful. The study of divergent series became popular again in the twentieth century and attracted the attention of a number of first-class mathematicians, among them George David Birkhoff (1884–1944) of Harvard University and G. H. Hardy (1887–1947) of Cambridge University.

ANNOTATED BIBLIOGRAPHY

Hirshman, I. I., Jr. *Infinite Series* (New York: Holt, Rinehart and Winston), 1962.
   A nice, concise summary of facts about series.

Knopp, Konrad. *Theory and Applications of Infinite Series* (New York: Hafner), n.d.
   This is surely the great book on series, by one of the important mathematicians in analysis in the twentieth century. For those interested in series, it is full of good information and beautiful mathematics.

Pólya, George. *Mathematics and Plausible Reasoning*. Vol. 1. *Induction and Analogy in Mathematics* (Princeton: Princeton University Press), 1954.
   In Chapter 2 of Volume 1 of this series there is an elegant and easily understood treatment of Euler's remarkable derivation of the sum of the reciprocals of the squares, showing that they add to $\pi^2/6$. If you're browsing in this book anyway, you might as well check out the treatment here of Archimedes's anticipation of integral calculus in Chapter 9. A beautifully written account in each case.

MATHEMATICAL INDUCTION

In your precalculus and calculus courses, you may have studied the method of proof known as *mathematical induction*. Since this topic is not always stressed in these courses, you might need a brief review of the method. The idea of mathematical induction is often illustrated by the so-called domino analogy. Imagine a row of dominoes standing on end as in Figure A-1. Assume that the following two conditions hold:

1. The first domino is tipped over.
2. If any domino tips over, it will hit the next domino and tip it over.

Evidently, then, *all* of the dominoes tip over.

By analogy with the row of dominoes, consider a sequence of statements, or propositions,

$$P_1, P_2, P_3, P_4, P_5, P_6, \ldots,$$

each of which could be either true or false. If a proposition P_n is true, imagine that the nth domino tips over. Then the two conditions above are analogous to the following:

1. The first proposition P_1 is true.
2. If any proposition P_n is true, then the next proposition P_{n+1} is also true.

If these two conditions hold, then, by analogy with the dominoes, it appears that *all* of the propositions should be true. The **principle of mathematical induction**, which states that this argument is valid, is proved in Section 4 of Chapter 3 (see Theorem 4.4 on page 194).

To prove by mathematical induction that all propositions in a sequence

$$P_1, P_2, P_3, P_4, P_5, P_6, \ldots$$

are true, begin by specifying the exact meaning of the proposition P_n for each positive integer n. Then carry out the following two steps:

Step 1. Prove that P_1 is true.

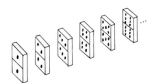

FIGURE A-1

Step 2.   Let n denote an arbitrary, but fixed, positive integer. Assume that P_n is true. On the basis of this assumption, prove that P_{n+1} is true.

Having done this, you are entitled to conclude, by the principle of mathematical induction, that P_n is true for every positive integer $n = 1, 2, 3, 4, 5, 6, \ldots$. This is called a proof **by induction on n**.

In Step 2, the assumption that P_n is true for an arbitrary, but fixed, value of n is called the **induction hypothesis**. When you make the induction hypothesis, you are not really stating that P_n is, in fact, true: you are just *supposing* it is, to see if you can prove that P_{n+1} is true on the basis of this supposition. In the domino analogy, it is as if you were checking to see if the dominoes are close enough together so that if any one of the dominoes, say the nth one, tips over, it will hit the next one and tip it over.

Example 1   If n is a positive integer, prove that the nth odd positive integer is $2n - 1$.

SOLUTION   We make a proof by induction on n. The odd positive integers are 1, 3, 5, 7, 9, and so forth. Note that, by adding 2 to an odd positive integer, we obtain the next odd positive integer. For each positive integer n, let P_n be the proposition stating that the nth odd positive integer is $2n - 1$. We indicate this by writing

P_n: *The nth odd positive integer is $2n - 1$.*

For instance, P_3 says that

The third odd positive integer is $2(3) - 1 = 5$

(which happens to be true). The proposition P_{n+1} is obtained by replacing n by $n + 1$ in the expression for P_n; so P_{n+1} is given by

P_{n+1}: *The $(n + 1)$st odd positive integer is $2(n + 1) - 1$.*

To prove that P_n is true for all positive integers n, we carry out the two steps in the mathematical-induction procedure.

Step 1. P_1 is the proposition given by

P_1: *The 1st odd positive integer is $2(1) - 1$;*

that is, P_1 states that

The first odd positive integer is 1,

which is true.

Step 2.   Let n be an arbitrary, but fixed, positive integer. We assume, for the sake of our induction argument, that P_n is true. Thus, we assume, as our induction hypothesis, that

The nth odd positive integer is $2n - 1$.

Then the next *odd* positive integer after $2n - 1$ is

$$(2n - 1) + 2 = 2n + 2 - 1 = 2(n + 1) - 1;$$

that is,

The $(n + 1)$st odd positive integer is $2(n + 1) - 1$.

This shows that P_{n+1} is true, and completes our proof by induction on n.

□

Example 2   Prove that the sum of the first n odd positive integers is n^2.

SOLUTION   Again, we make a proof by induction on n. For each positive integer n, let P_n be the proposition given by

P_n: *The sum of the first n odd positive integers is n^2.*

By the result of Example 1, the nth odd positive integer is $2n - 1$; so we can rewrite P_n in the equivalent form

P_n: $1 + 3 + 5 + \cdots + (2n - 1) = n^2.$

Then P_{n+1} is given by

P_{n+1}: $1 + 3 + 5 + \cdots + (2n - 1) + [2(n + 1) - 1] = (n + 1)^2;$

that is,

P_{n+1}: $1 + 3 + 5 + \cdots + (2n - 1) + (2n + 1) = (n + 1)^2.$

Step 1. P_1 is the proposition given by

P_1: $1 = 1^2,$

which is clearly true.

Step 2.   Assume, as our induction hypothesis, that P_n is true for some value of n. Thus, we assume that

$$1 + 3 + 5 + \cdots + (2n - 1) = n^2 \tag{1}$$

is true. Our goal is to prove, on the basis of Equation (1), that P_{n+1} is true. A glance at the statement P_{n+1} above suggests that we add $2n + 1$ to both sides of Equation (1) to obtain

$$1 + 3 + 5 + \cdots + (2n - 1) + (2n + 1) = n^2 + (2n + 1). \tag{2}$$

Since

$$n^2 + (2n + 1) = n^2 + 2n + 1 = (n + 1)^2, \tag{3}$$

we can rewrite Equation (2) as

$$1 + 3 + 5 + \cdots + (2n - 1) + (2n + 1) = (n + 1)^2. \tag{4}$$

Thus, P_{n+1} is true, and our proof by induction is complete.   □

After you become more familiar with the technique of proving theorems by induction on n, you can just keep in mind the proposition corresponding to n, rather than denoting it explicitly by P_n. The following example illustrates how this is done:

Example 3   If $f(x)$ is a real-valued function of a real variable, denote the derivative $f'(x)$ by $D_x f(x)$. Using the fact that $D_x x = 1$ and the rule for differentiating products, prove by induction on n that

$$D_x x^n = n x^{n-1}$$

holds for all positive integers n.

SOLUTION   For $n = 1$, the statement to be proved becomes

$$D_x x^1 = 1 \cdot x^{1-1},$$

that is,

$$D_x x = x^0,$$

or

$$D_x x = 1,$$

which is true. Now we assume, as our induction hypothesis, that

$$D_x x^n = n x^{n-1}$$

for some arbitrary, but fixed, value of n. Our goal is to prove that

$$D_x x^{n+1} = (n+1) x^{(n+1)-1}.$$

In other words, we want to prove that

$$D_x x^{n+1} = (n+1) x^n.$$

But, using the facts that $D_x x = 1$ and the induction hypothesis, we have

$$
\begin{aligned}
D_x x^{n+1} &= D_x(x \cdot x^n) \\
&= (D_x x)x^n + x(D_x x^n) \quad \text{(Product rule)} \\
&= 1 \cdot x^n + x(n x^{n-1}) \\
&= x^n + n x^n \\
&= (1 + n)x^n \\
&= (n+1)x^n.
\end{aligned}
$$
□

PROBLEM SET FOR APPENDIX I

In Problems 1–17, use mathematical induction on n to prove that the given proposition is true for all positive integers n.

1. $1 + 2 + 3 + \cdots + n = n(n+1)/2$; that is, the sum of the first n positive integers is $n(n+1)/2$.

2. $1^2 + 2^2 + 3^2 + \cdots + n^2 = n(n+1)(2n+1)/6$; that is, the sum of the first n perfect squares is $n(n+1)(2n+1)/6$.

3. The sum of the first n perfect cubes is $[n(n + 1)/2]^2$.

4. $1 + 5 + 9 + \cdots + (4n - 3) = n(2n - 1)$. [Hint: In the series of numbers to be added, notice that each number is 4 more than its immediate predecessor.]

5. The sum of the squares of the first n odd positive integers is $n(2n - 1)(2n + 1)/3$.

6. $1 \cdot 2 + 2 \cdot 3 + 3 \cdot 4 + \cdots + n(n + 1) = n(n + 1)(n + 2)/3$.

7. If a and b are real numbers, then $(ab)^n = a^n b^n$.

8. $(1 \cdot 2)^{-1} + (2 \cdot 3)^{-1} + (3 \cdot 4)^{-1} + \cdots + [n(n + 1)]^{-1} = n/(n + 1)$.

9. $\cos n\pi = (-1)^n$.

10. De Moivre's theorem: If θ is a real number, then

$$(\cos \theta + i \sin \theta)^n = \cos n\theta + i \sin n\theta.$$

11. If r is a real number and $r \neq 1$, then

$$r + r^2 + r^3 + r^4 + \cdots + r^n = \frac{r(r^n - 1)}{r - 1}.$$

12. Bernoulli's inequality: If h is a real number and $h \geq -1$, then $(1 + h)^n \geq 1 + nh$.

13. $n^3 - n$ is exactly divisible by 3.

14. If $n \geq 2$, then $n^2 + 4 < (n + 1)^2$.

15. $D_x^n(x^{-1}) = (-1)^n n! x^{-(n + 1)}$.

16. A polygonal region of the plane is divided into n polygonal subregions in such a way that two of the subregions overlap, if at all, only on a common edge or a common vertex. Counting an edge only once, even if it is common to two subregions, suppose there are e edges. Counting a vertex only once, even if it is common to two or more subregions, suppose there are v vertices. Then $n + v = e + 1$.

17. There are n people in a room. If every person shakes hands with every other person, show that $n(n + 1)/2$ handshakes take place.

ANSWERS AND HINTS FOR SELECTED ODD-NUMBERED PROBLEMS

1. c, d, i, l, m, and n are propositions; a, b, e, g and o are not. The rest are indeterminate; e.g., for k, there could be a question of, where? When?

3. (a) If P and Q have the same truth values, and if Q and R have the same truth values, then P and R have the same truth values.
    (b) Tautology 15

5. (a) True   (b) False   (c) False   (d) True   (e) True   (f) False

7. (a) $P \wedge (\sim Q)$ (b) $(\sim P) \wedge Q$ (c) $Q \Rightarrow (\sim P)$ (d) $(\sim Q) \Rightarrow P$ (e) $(\sim P) \wedge Q$

9. (a) This equation does not have a solution and I am not stupid.
    (b) Alfie is not studying hard or he is not doing well in his classes.

11. (a) I am not cold.   (b) If I am cold, then it is snowing and I am not going home.
    (c) It is snowing and I am not going home.   (d) I am cold if and only if I am going home.
    (e) If I am going home, then I am cold.   (f) Either I am cold or I am going home, but not both.
    (g) If it is snowing or I am cold, then I am going home.   (h) I am going home and I am not cold.

13. (a) $P \Rightarrow R$ (b) $(P \wedge Q) \Rightarrow R$ (c) $R \Rightarrow Q$ (d) $R \Rightarrow Q$ (e) $R \Rightarrow (P \vee Q)$ (f) $\sim (P \Leftrightarrow Q)$

15.

P	Q	R	$P \vee R$	$Q \vee R$	$P \wedge R$	$Q \wedge R$	$P \wedge (Q \wedge R)$	$P \wedge (Q \vee R)$	$P \vee (Q \wedge R)$
1	1	1	1	1	1	1	1	1	1
0	1	1	1	1	0	1	0	0	1
1	0	1	1	1	1	0	0	1	1
0	0	1	1	1	0	0	0	0	0
1	1	0	1	1	0	0	0	1	1
0	1	0	0	1	0	0	0	0	0
1	0	0	1	0	0	0	0	0	1
0	0	0	0	0	0	0	0	0	0

17.

P	Q	R	$(P \wedge ((\sim Q) \vee R)) \vee Q$						
1	1	1	1	1	0	1	1	1	1
0	1	1	0	0	0	1	1	1	1
1	0	1	1	1	1	1	1	1	0
0	0	1	0	0	1	1	1	0	0
1	1	0	1	0	0	0	0	1	1
0	1	0	0	0	0	0	0	1	1
1	0	0	1	1	1	1	0	1	0
0	0	0	0	0	1	1	0	0	0

19.

P	$\sim P$	$P \Leftrightarrow (\sim(\sim P))$		
1	0	1	1	1
0	1	0	1	0

21.

P	Q	$(P \wedge Q) \Leftrightarrow (Q \wedge P)$						
1	1	1	1	1	1	1	1	1
0	1	0	0	1	1	1	0	0
1	0	1	0	0	1	0	0	1
0	0	0	0	0	1	0	0	0

P	Q	$(P \vee Q) \Leftrightarrow (Q \vee P)$						
1	1	1	1	1	1	1	1	1
0	1	0	1	1	1	1	1	0
1	0	1	1	0	1	0	1	1
0	0	0	0	0	1	0	0	0

P	Q	R	$(P \wedge Q) \wedge R \Leftrightarrow P \wedge (Q \wedge R)$										
1	1	1	1	1	1	1	1	1	1	1	1	1	1
0	1	1	0	0	1	0	1	1	0	0	1	1	1
1	0	1	1	0	0	0	1	1	1	0	0	0	1
0	0	1	0	0	0	0	1	1	0	0	0	0	1
1	1	0	1	1	1	0	0	1	1	0	1	0	0
0	1	0	0	0	1	0	0	1	0	0	1	0	0
1	0	0	1	0	0	0	0	1	1	0	0	0	0
0	0	0	0	0	0	0	0	1	0	0	0	0	0

P	Q	R	$(P \vee Q) \vee R \Leftrightarrow P \vee (Q \vee R)$										
1	1	1	1	1	1	1	1	1	1	1	1	1	1
0	1	1	0	1	1	1	1	1	0	1	1	1	1
1	0	1	1	1	0	1	1	1	1	1	0	1	1
0	0	1	0	0	0	1	1	1	0	1	0	1	1
1	1	0	1	1	1	1	0	1	1	1	1	1	0
0	1	0	0	1	1	1	0	1	0	1	1	1	0
1	0	0	1	1	0	1	0	1	1	1	0	0	0
0	0	0	0	0	0	0	0	1	0	0	0	0	0

23.

P	Q	R	$P \wedge (Q \vee R) \Leftrightarrow (P \wedge Q) \vee (P \wedge R)$												
1	1	1	1	1	1	1	1	1	1	1	1	1	1	1	1
0	1	1	0	0	1	1	1	1	0	0	1	0	0	0	1
1	0	1	1	1	0	1	1	1	1	0	0	1	1	1	1
0	0	1	0	0	0	1	1	1	0	0	0	0	0	0	1
1	1	0	1	1	1	1	0	1	1	1	1	1	1	0	0
0	1	0	0	0	1	1	0	1	0	0	1	0	0	0	0
1	0	0	1	0	0	0	0	1	1	0	0	0	1	0	0
0	0	0	0	0	0	0	0	1	0	0	0	0	0	0	0

P	Q	R	$P \vee (Q \wedge R) \Leftrightarrow (P \vee Q) \wedge (P \vee R)$												
1	1	1	1	1	1	1	1	1	1	1	1	1	1	1	1
0	1	1	0	1	1	1	1	1	0	1	1	1	0	1	1
1	0	1	1	1	0	0	1	1	1	1	0	1	1	1	1
0	0	1	0	0	0	0	1	1	0	0	0	0	0	1	1
1	1	0	1	1	1	0	0	1	1	1	1	1	1	1	0
0	1	0	0	0	1	0	0	1	0	1	1	0	0	0	0
1	0	0	1	1	0	0	0	1	1	1	0	1	1	1	0
0	0	0	0	0	0	0	0	1	0	0	0	0	0	0	0

25.

P	Q	$\sim(P \wedge Q) \Leftrightarrow (\sim P) \vee (\sim Q)$							
1	1	0	1	1	1	1	0	0	0
0	1	1	0	0	1	1	1	1	0
1	0	1	1	0	0	1	0	1	1
0	0	1	0	0	0	1	1	1	1

P	Q	$\sim(P \vee Q) \Leftrightarrow (\sim P) \wedge (\sim Q)$							
1	1	0	1	1	1	1	0	0	0
0	1	0	0	1	1	1	1	0	0
1	0	0	1	1	0	1	0	0	1
0	0	1	0	0	0	1	1	1	1

27.

P	Q	$(P \Rightarrow Q) \Leftrightarrow ((\sim Q) \Rightarrow (\sim P))$						
1	1	1	1 1 1	0	1	0
0	1	0	1 1 1	0	1	1
1	0	1	0 0 1	1	0	0
0	0	0	1 0 1	1	1	1

29.

P	Q	$(P \Rightarrow Q) \Leftrightarrow ((\sim P) \vee Q)$					
1	1	1	1 1 1	0	1 1
0	1	0	1 1 1	1	1 1
1	0	1	0 0 1	0	0 0
0	0	0	1 0 1	1	1 0

P	Q	$(P \Leftrightarrow Q) \Leftrightarrow ((P \Rightarrow Q) \wedge (Q \Rightarrow P))$						
1	1	1	1 1 1	1	1 1 1 1 1 1
0	1	0	0 1 1	0	1 1 0 1 0 0
1	0	1	0 0 1	1	0 0 0 0 1 1
0	0	0	1 0 1	0	1 0 1 0 1 0

31.

P	Q	$(P \wedge Q) \Rightarrow P$			
1	1	1 1 1	1 1
0	1	0 0 1	1 0
1	0	1 0 0	1 1
0	0	0 0 0	1 0

P	Q	$P \Rightarrow (Q \Rightarrow (P \wedge Q))$			
1	1	1 1 1 1	1 1 1
0	1	0 1 1 0	0 0 1
1	0	1 1 0 1	1 0 0
0	0	0 1 0 1	0 0 0

P	Q	$P \Rightarrow (P \vee Q)$		
1	1	1 1	1 1 1
0	1	0 1	0 1 1
1	0	1 1	1 1 0
0	0	0 1	0 0 0

33.

P	Q	$(P \wedge (P \Rightarrow Q)) \Rightarrow Q$			
1	1	1 1 1 1 1	1 1
0	1	0 0 0 1 1	1 1
1	0	1 0 1 0 0	1 0
0	0	0 0 0 1 0	1 0

35. (a) If $2n + 1$ is not an odd integer, then n is not an integer.
(b) If you have not passed Math 211, then you cannot take Math 212.
(c) If $\lim_{n \to \infty} a_n \neq 0$, then the series $\sum_{k=1}^{\infty} a_k$ does not converge.
(d) If a citizen cannot vote, then he or she is not of age 18 or over.

37. 2 is prime. **39.** False; $((-4)^2 > 9$, yet $-4 \not> 3)$.

41. Note that for $n = 1$, there are exactly $2^1 = 2$ truth combinations possible. Assume that for $n = k$ there are 2^k truth combinations. By adding another propositional variable, the number of combinations grows to $2 \cdot 2^k$ since the $(k + 1)$st variable may be either true or false. Thus, since $2 \cdot 2^k = 2^{k+1}$, we conclude that for $k + 1$ propositional variables, there are 2^{k+1} truth combinations.

43. A truth table for such a connective would contain $2^2 = 4$ rows. Arguing as in Problem 41, we see that there are $2^4 = 16$ ways to place 0's and 1's in these 4 rows.

PROBLEM SET 1.2, PAGE 31

1. (a) True   (b) False   (c) False   **3.** (a) True   (b) True   (c) False
5. (a) True   (b) False   (c) True   (d) False
7. (a) $(\forall x)(T(x) \Rightarrow S(x))$, where $T(x)$ means x is a teacher and $S(x)$ means x is a sadist.
(b) There exists at least one teacher who is not a sadist.
9. (a) $(\forall x)(S(x))$, where $S(x)$ means x is a sadist.   (b) There exists at least one teacher who is not a sadist.

11. (a) $(\exists x)(L(x) \wedge (\sim J(x)))$, where $L(x)$ means x is a lawyer and $J(x)$ means x is a judge.
(b) All lawyers are judges.

13. (a) $(\exists x)(\sim J(x))$, where $J(x)$ means x is a judge. (b) All lawyers are judges.

15. (a) $(\exists x)(P(x) \wedge T(x))$, where $P(x)$ means x is prime and $T(x)$ means x is divisible by 3.
(b) There does not exist a prime number that is exactly divisible by 3.

17. (a) $(\forall y)(x > y)$ means that all numbers are less than x.
(b) $(\exists x)(x > y)$ means that there is at least one number greater than y.
(c) $(\exists x)(\forall y)(x > y)$ means that there exists some number that is greater than all numbers.
(d) $(\forall y)(\exists x)(x > y)$ means that for any number, there exists a number greater than that number.

19. (a) $(\forall y)(x + y = 0)$ means that any real number will serve as an additive inverse for x.
(b) $(\exists x)(x + y = 0)$ means that there exists an additive inverse for y.
(c) $(\exists x)(\forall y)(x + y = 0)$ means that there exists some number which serves as an additive inverse for all real numbers.
(d) $(\forall y)(\exists x)(x + y = 0)$ means that for any real number, an additive inverse exists.

21. (a) True (b) True (c) False (d) True (e) True (f) True (g) True (h) True
(i) False (j) True (k) False (l) True

23. (a) $(\exists x)(\forall n)((n > x) \vee (x \geq n + 1))$ means that there exists some number x such that, for every integer n, x is less than n or x is greater than or equal to $n + 1$.
(b) $(\forall x)(\exists n)(n \leq x < n + 1)$ means that for any real number x, there exist consecutive integers n and $n + 1$ such that $n \leq x < n + 1$.

25. (a) $(\forall x)(P(x))$ (b) $(\exists x)(P(x))$

27. (a) $(\forall \varepsilon)(\exists \delta)(|x - a| < \delta \Rightarrow |f(x) - f(a)| < \varepsilon)$ (b) $(\exists \varepsilon)(\forall \delta)(|x - a| < \delta \wedge |f(x) - f(a)| \geq \varepsilon)$

29. $(\forall x)(\forall y)(x > y \Rightarrow f(x) > f(y))$

31. (1) If every person is a politician, then Joe Smith is a politician.
(2) If Joe Smith is a politician, then there is at least one politician.
(3) If every person is a politician, then there is at least one politician. And so on.

33. The statement $(\forall x)(P(x) \wedge Q(x))$ in theorem 11 on page 28 says that for all x, $P(x)$ and $Q(x)$ are simultaneously true; thus, for all x, $P(x)$ is true and for all x, $Q(x)$ is true. Conversely, $[(\forall x)(P(x)) \wedge (\forall x)(Q(x))]$ means that for any choice of x, $P(x)$ is true and for any choice of x, $Q(x)$ is true. Thus, any choice of x will suffice for both $P(x)$ and $Q(x)$ to be true.

35. By the hypothesis, $P(x)$ is true for all x or else $Q(x)$ is true for all x (or both). Suppose, for instance, that $P(x)$ is true for all x. Then, for any x, $P(x) \vee Q(x)$ is certainly true. Likewise, if $Q(x)$ is true for all x, then, for any x, $P(x) \vee Q(x)$ is again true. Hence, $(\forall x)(P(x) \vee Q(x))$ is true.

37. Similar to Problems 33 and 35. **39.** For example, if two athletes, cars, or computers are said to be "equal."

41. (a) $(\forall x)(\forall y)(\forall z)((x = y) \wedge (y = z) \Rightarrow (x = z))$
(b) If a and c are logically identical to a third object b, we have $P(a) \Leftrightarrow P(b)$ and $P(b) \Leftrightarrow P(c)$ for every propositional function $P(x)$. But, from this and tautology 15 in Section 1.1, it follows that $P(a) \Leftrightarrow P(c)$.

43. Note that $2^3 = 8$. Now if $x^3 = 8$ and $y^3 = 8$, then $x^3 - y^3 = 0$; that is, $(x - y)(x^2 + xy + y^2) = 0$. Considering $x^2 + xy + y^2 = 0$ as a quadratic equation in x (for an arbitrary, but fixed, value of y), we have a discriminant $y^2 - 4y^2 = -3y^2$. Thus, unless $y = 0$, this equation has no solution. Since $y^3 = 8$, we cannot have $y = 0$. Hence, $x^2 + xy + y^2 \neq 0$, and it follows that $x = y$.

47. (a) "For all x" means "for each and every x in the universe." If only 1 and 2 belong to the universe, then $(\forall x)(P(x))$ is equivalent to $P(1) \wedge P(2)$.
(b) "For some $x \ldots$" means "there is at least one x in the universe $\ldots$." If only 1 and 2 belong to the universe, then $(\exists x)(P(x))$ is equivalent to $P(1) \vee P(2)$.

49. $\sim(\forall x)(P(x)) \Leftrightarrow \sim(P(1) \wedge P(2)) \Leftrightarrow (\sim P(1)) \vee (\sim P(2)) \Leftrightarrow (\exists x)(\sim P(x))$.
Also, $\sim(\exists x)(P(x)) \Leftrightarrow \sim(P(1) \vee P(2)) \Leftrightarrow (\sim P(1)) \wedge (\sim P(2)) \Leftrightarrow \sim(\forall x)(P(x))$.

PROBLEM SET 1.3, PAGE 42

1.

P	Q	$P \Rightarrow Q$
1	1	1
0	1	1
1	0	0
0	0	1

Note that, in the second row, P is false.

3. If $P \Rightarrow Q$ is justified, then the law of contraposition states that $\sim Q \Rightarrow \sim P$ is justified. Thus, if $\sim Q$ is justified, $\sim P$ is justified by modus ponens.

5. True, by modus ponens.   **7.** False, converse need not be true.   **9.** False, converse need not be true.

11. Yes, by 3.5, Assertion of Hypothesis.   **13.** Yes, by 3.6, Justification of Conclusion.

15. $P \Leftrightarrow \sim(\sim P)$, where P is replaced by $\cos 0 = 1$.

17. $(P \wedge (P \Rightarrow Q)) \Rightarrow Q$, where P is replaced by $\cos 0 = 1$ and Q is replaced by $\sin 0 = 0$.

19. By a truth table check, $[P \wedge (P \Rightarrow Q) \wedge (\sim Q)] \Leftrightarrow [P \wedge (\sim Q)]$. Replace P by $x = 1$ and Q by $y = 2$.

21. Since, by tautology 18 of Section 1.1, $(P \wedge Q) \Rightarrow P$, if $(P \wedge Q)$ is justified, then, by modus ponens, P is justified as well.

23. Use substitution of equals and tautology 18 of Section 1.1.

25. Mathematically, an axiom need be neither self-evident nor recognized as being, in fact, true.

27. Suppose s is the largest positive integer. Then $s + 1 > s$, contradicting the supposition that s is the largest positive integer.

29. Keen is correct, since assuming $P \Rightarrow Q$ is false is tantamount to assuming P is true and Q is false. She must then reach a contradiction in order to show (arguing by contradiction) that $P \Rightarrow Q$ is true.

31. Assume that a second point of intersection Q exists, and show that L can no longer be perpendicular to $\overline{OP}$.

33. The proof that a number at which a function has a relative extremum is a critical number is often an indirect proof.

PROBLEM SET 1.4, PAGE 53

1. (a) $\{1, 2, 3, 4, 5, 6, 7, 8\} = A$ (b) $\{-9, -8, -7, -6, -5, -4, -3, 3, 4, 5, 6, 7, 8, 9\} = B$
(c) $\{3, 4, 5, 6, 7, 8\} = C$ (d) $\{-9, -8, -7, -6, -5, -4, -3, 1, 2, 3, 4, 5, 6, 7, 8, 9\} = D$
(e) $\{4, 8, 12, 16, 20, 24, 28, 32, 36, 40\} = E$ (f) $\{-3, 1, 2, 3\}$

3. $\{-2, 2\}$ **5.** $\varnothing$ **7.** $\{3\}$ **9.** (a) Empty   (b) Nonempty   (c) Empty   (d) Nonempty

11. (a) $M \subseteq N$ (b) $M \subseteq N$ (c) $M \nsubseteq N$ (d) $M \subseteq N$ (e) $M \subseteq N$ (f) $M \nsubseteq N$
(g) $M \nsubseteq N$ (h) $M \nsubseteq N$

13. (a) $A = B$ (b) $A \neq B$ (c) $A = B$ (d) $A \neq B$ (e) $A = B$

15. $(A \subseteq B) \wedge (B \subseteq C) \Leftrightarrow (\forall x)(x \in A \Rightarrow x \in B) \wedge (\forall x)(x \in B \Rightarrow x \in C)$. Thus, by tautology 16 in Section 1.1, the conclusion follows.

17.

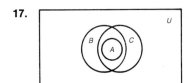

19. $\{a, b, c, d\} = M$ **21.** $\{\varnothing, \{1\}\}$

23. $\{\varnothing, \{1\}, \{2\}, \{1, 2\}\}$ **25.** $\{\varnothing, \{a\}, \{b\}, \{c\}, \{a, b\}, \{b, c\}, \{a, c\}, \{a, b, c\}\}$

27. $\varnothing, \{1\}, \{2\}, \{3\}, \{1, 2\}, \{2, 3\}, \{1, 3\}$ **29.** $\{1\}, \{2\}, \{3\}, \{1, 2\}, \{2, 3\}, \{1, 3\}$
31. False, if the set is nonempty, then $\varnothing$ is proper but trivial.   **33.** Use the result of Problem 15.

PROBLEM SET 1.5, PAGE 66

1. (a) $\{1, 2, 3, 4, 5\}$ (b) $\{3, 4\}$ (c) $\{3, 4\}$ (d) $\{3, 4, 6, 7, 8, 9\}$ (e) $\{1, 2\}$ (f) $\{3, 4, 7\}$
(g) $\{1, 2\}$ (h) $\{6, 7\}$ (i) $\{1, 2, 3, 4, 6\}$ (j) $\{3, 4\}$ (k) $\{3, 4, 6\}$ (l) $\{3, 4, 5, 6\}$
(m) $\{1, 2, 3, 4, 5\}$ (n) $\{1, 2, 3, 4\}$

3. (a)

(b)

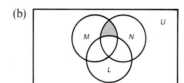

(c)

(d)

(e)

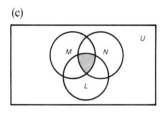 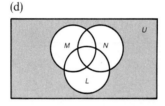 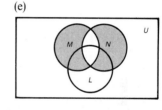

5. (a) $\{5, 6\}$ (b) $\{3\}$ (c) $\{3, 5, 6\}$ (d) $\varnothing$ (e) $\{3, 5, 6\}$ (f) $\varnothing$ (g) $\{2, 3, 4, 5, 6\}$
(h) $\{3\}$ (i) $\{5, 6\}$

7. Utilize appropriate tautology from Section 1.1.   **9.** Utilize appropriate tautologies from Section 1.1.

11. (a) $(M \cap N)' = M' \cup N'$ (b) $(M \cup N)' = M' \cap N'$

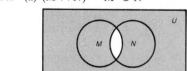 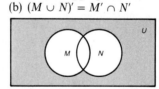

15. (a) By definition, $M \cap N = \{x \,|\, x \in M, x \in N\}$ and $M \setminus N = \{x \,|\, x \in M, x \notin N\}$. Thus, since we cannot have both $x \in N$ and $x \notin N$, we conclude that $(M \cap N) \cap (M \setminus N) = \varnothing$.

(c)

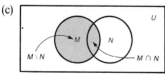

17. False   **19.** (a) $\{1, 4, 5\}$ (b) $\{1, 4, 5\}$ (c) $\varnothing$ (d) $\{1, 2, 3\}$

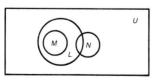

21. Give a pick-a-point proof.   **23.** (a) 2   (b) 0   (c) 1   (d) 1
25. (a) 10   (b) 3   (c) 2   **27.** Note that for $N \subseteq M$, $\#(M \cap N) = \#N$.
29. Use Theorems 5.10 and 5.12.   **31.** (a) $2^m - 1$ (b) $2^m - 2$
33. The 3 sets must decompose U into $2^3 = 8$ nonempty subsets as shown below.

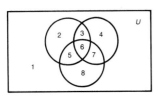

35. Since $H \subseteq H \cup M = G$ and $L \subseteq L \cup N = G$, it follows that $H \cup L \subseteq G$. To prove $G \subseteq H \cup L$, suppose $x \in G$ but $x \notin H \cup L$. Then $x \notin H$ and $x \notin L$. Since $x \in G = H \cup M$ and $x \notin H$, it follows that $x \in M$. Since $x \in G = L \cup N$ and $x \notin L$, it follows that $x \in N$. But, then, $x \in M \cap N = \emptyset$, a contradiction. Therefore, $G \subseteq H \cup L$, so $G = H \cup L$. A similar argument proves that $M \cup N = G$.

PROBLEM SET 1.6, PAGE 77

1. Use $I = \{a, b, c\}$ as the indexing set, and let $M_a = \{$Alabama, Alaska$\}$, $M_b = \{$Michigan, Ohio$\}$, and $M_c = \{$Maine, Texas$\}$.
3. Use $I = \{a, b, c, d, e\}$ as the indexing set, and let $M_a = \{1, 4, 5, 9\}$, $M_b = \{1, 3, 4, 5\}$, $M_c = \{4, 5, 7, 8\}$, $M_d = \{1, 4, 7, 9\}$, and $M_e = \{4, 5, 8, 9\}$.
5. Use $I = \{1, 2, 3, \ldots\} = \mathbb{N}$ as the indexing set, and let $M_i = \{x \cdot i \mid x$ is an integer, for each $i \in I\}$.
7. (a) $\{1, 3\}$ (b) $\{2, 4, 6\}$ (c) $\{4, 6, 12\}$ (d) $\{1, 3, 5, 9\}$ (e) $\{3\}$ (f) $\{1, 2, 3, 4, 5, 6, 9, 12\}$ (g) $\emptyset$
9. (a) $\{1, 2, 3, 4, 5, 7\}$ (b) $\{2\}$ **11.** (a) $\{a, b, c, d, e, f\}$ (b) $\{c, d\}$
13. (a) $\{$red, yellow, green, blue, violet, pink, orange$\}$ (b) $\{$green, blue$\}$
15. (a) $\{$right, left, up, down, front, back$\}$ (b) $\{$right, left$\}$ **17.** (a) Problems 1 and 2   (b) None
19. (a) $[0, \infty)$ (b) $\emptyset$ **21.** (a) $[0, 1]$ (b) $\{0\}$
23. Say $p, q \in I$ with $p \neq q$. Then $\bigcap_{i \in I} M_i \subseteq M_p \cap M_q = \emptyset$.   **25.** $\bigcap \mathscr{E} \subseteq \emptyset$
27. Use pick-a-point process.
29. (a) Let $x \in A$. By hypothesis, $x \in M_i$, $\forall i \in I$. By definition, $x \in \bigcap_{i \in I} M_i$.
     (b) Let $x \in \bigcup_{i \in I} M_i$. By definition, $\exists i \in I$ with $x \in M_i$. By hypothesis, $x \in B$.
31. $\mathscr{E} \in \mathscr{P}(\mathscr{P}(X)) \Leftrightarrow \mathscr{E} \subseteq \mathscr{P}(X)$

CHAPTER 2   PROBLEM SET 2.1, PAGE 88

1. $x = 1, y = 1$ **3.** $(1, 2), (2, 4), (4, 3), (3, 1)$ **5.** $\{(2, 1), (2, 2), (4, 1), (4, 2), (5, 1), (5, 2)\}$
7. $\{(1, 2), (2, 2), (1, 4), (2, 4), (1, 5), (2, 5)\}$ **9.** $\emptyset$
11. $\{(2, 1), (1, 2), (2, 2), (4, 1), (1, 4), (4, 2), (2, 4), (5, 1), (1, 5), (5, 2), (2, 5)\}$
13. $\{(a, c), (a, d), (a, e), (a, f), (b, c), (b, d), (b, e), (b, f)\}$ **15.** $\{(a, c), (a, d), (b, c), (b, d), (a, e), (a, f), (b, e), (b, f)\}$
17. A prism   **19.** A solid right circular cylinder   **21.** 6   **23.** 78   **25.** 216
27. Let $(a, b) \in A \times B$. Then $a \in A$, and thus, $a \in X$. Similarly, $b \in Y$. Thus, $(a, b) \in X \times Y$.

29. (For first half of proof see Example 1.8.) Let $(x, y) \in (A \times C) \cup (B \times C)$. Then $(x, y) \in (A \times C)$ or $(x, y) \in (B \times C)$; that is, $y \in C \wedge (x \in A \vee x \in B)$, which is to say $(x \in A \cup B) \wedge y \in C$. Thus, $(x, y) \in (A \cup B) \times C$.

31. Use pick-a-point procedure.

33.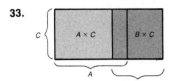

35. 2^{2m}

PROBLEM SET 2.2, PAGE 100

1. (a) Graph consists of the pairs $(2, 4)$ and $(3, 9)$.   (b) Graph consists of the pairs $(2, 3)$, and $(3, 2)$.
(c) Graph consists of the pairs $(2, 4)$, $(2, 3)$, $(3, 9)$, and $(3, 2)$.

3. Suppose $R \subseteq X \times X$. We must show that dom $R \subseteq X$ and codom $R \subseteq X$. Let $x \in$ dom R. Then $\exists y, (x, y) \in R$. Since $R \subseteq X \times X$, it follows that $(x, y) \in X \times X$, so $x \in X$ and $y \in X$. In particular, $x \in X$, so dom $R \subseteq X$. A similar argument shows that codom $R \subseteq X$. The proof that (dom $R \subseteq X$ and codom $R \subseteq X$) $\Rightarrow R \subseteq X \times X$ is just as easy.

5. (a) $\text{dom}(R) = \{1, 2, 3, 7\}$; $\text{codom}(R) = \{4, 5, 6, 7\}$; $R^{-1} = \{(5, 1), (5, 2), (4, 1), (6, 2), (7, 3), (6, 7)\}$
(b) $\text{dom}(R) = \{a, b, c\}$; $\text{codom}(R) = \{b, c, e, g\}$; $R^{-1} = \{(b, c), (g, b), (e, c), (b, b), (e, b), (e, a), (c, a)\}$
(c) $\text{dom}(R) = \{x \,|\, -3 \le x \le 3\}$; $\text{codom}(R) = \{x \,|\, -2 \le x \le 2\}$; $R^{-1} = \{(x, y) \in \mathbb{R} \times \mathbb{R} \,|\, 9x^2 + 4y^2 = 36\}$
(d) $\text{dom}(R) = \{2, 4, 5\}$; $\text{codom}(R) = \{1, 2, 4\}$; $R^{-1} = \{(4, 2), (1, 2), (4, 4), (2, 4), (2, 5)\}$

7. (a) No, since $(4, 3) \notin R$.   (b) No, since $(2, 1) \in R$ and $(1, 2) \in R$ but $1 \ne 2$.

9. (a) Δ_X is reflexive since $\forall x \in$ dom $\Delta_X, (x, x) \in$ $X \times X$ is reflexive since $\forall x \in X, (x, x) \in X \times X$.
(b) Δ_X is symmetric since $\forall (x, y) \in \Delta_X$, we have $x = y$ and thus $(y, x) \in \Delta_X$. $X \times X$ is reflexive since $\forall (x, y) \in X \times X$, we have $x \in X$, $y \in X$, and thus, $(y, x) \in X \times X$.
(c) Δ_X is transitive since, if $(x, y) \in \Delta_X$ and $(y, z) \in \Delta_X$, then $x = y$ and $y = z$. Thus, $x = z$ and $(x, z) \in \Delta_X$. $X \times X$ is transitive since, if $(x, y) \in X \times X$ and $(y, z) \in X \times X$, then $x, y, z \in X$. Thus, $(x, z) \in X \times X$.

11. Yes, since $xRy \wedge yRz \Rightarrow xRz$.

13. (a) See Problem 9.   (b) $\forall x \in \mathbb{R}, x \ge x$. Also, $x \ge y$ and $y \ge z \Rightarrow x \ge z$.

15. Note that any positive integer is a multiple of itself. Also, if x is a multiple of y and y is a multiple of x, then $x = y$. Finally, if y is a multiple of x and z is a multiple of y, then z is a multiple of x.

17. (a) Symmetric   (b) Reflexive, symmetric, transitive, preorder   (c) Reflexive, transitive, preorder
(d) Reflexive, symmetric, transitive, preorder
(e) Reflexive, antisymmetric, transitive, preorder, partial order
(f) Reflexive, symmetric, antisymmetric, transitive, preorder, partial order
(g) Reflexive, symmetric, transitive, preorder   (h) Reflexive, transitive, preorder

19. (a) Use definitions of reflexive and Δ_X.   (b) Use definitions of symmetric and R^{-1}.

21. Note that for $x \in X, (x, x) \in (R \cup \Delta_X)$.

23. If R is reflexive on X, then (by Problem 19a) $\Delta_X \subseteq R$. Therefore, $\Delta_X \subseteq R \cup S$, so (by Problem 19a) $R \cup S$ is reflexive on X.

25. Use pick-a-point.   **27.** Assume $(x, y) \in (X \times X) \setminus R$ and show $(y, x) \in (X \times X) \setminus R$.

29. Use definition of partial order.   **31.** Use definition of partial order.

33. (a) 1   (b) 0   (c) 1   (d) No least element exists.

35. (a) Use definition of partial order.   (b) $\sup\{M, N\} = M \cup N$; $\inf\{M, N\} = M \cap N$

37. (a) Use antisymmetric property to prove uniqueness.   (b) Analogous to Part (a).

PROBLEM SET 2.3, PAGE 114

1. b, c, and d are functions.
3. (a) $\{1, 2, 3, 5, 7\}$ (b) $\{a, b, c, d\}$ (c) c (d) d (e) b (f) c
5. (a) $\text{dom}(f) = \{1, 2, 4, 6\}$, $\text{range}(f) = \{a, b, c\}$ (b) $\text{dom}(g) = \{1, 2, 3, 5\}$, $\text{range}(g) = \{b\}$
   (c) $\text{dom}(h) = \mathbb{R}$, $\text{range}(h) = \{x \mid -1 \le x \le 1\}$ (d) $\text{dom}(F) = \mathbb{R}$, $\text{range}(F) = \{x \mid x \ge 4\}$
   (e) $\text{dom}(G) = \{x \mid x \in \mathbb{R},\ x \ne (\pi/2) + n\pi \text{ for } n \text{ an integer}\}$, $\text{range }(G) = \{x \mid x \le -1 \lor x \ge 1\}$
   (f) $\text{dom}(H) = \{x \mid x \in \mathbb{R},\ x \ne -2\}$, $\text{range}(H) = \{x \mid x \in \mathbb{R},\ x \ne 0\}$
7. $\{(a, y), (b, x), (c, z), (d, y)\}$
9. The function under consideration is $\{(x, y) \mid x \in \mathbb{R} \land y \in \mathbb{R} \land x > -2 \land y = 1/\sqrt{x + 2}\}$.
11. Only Part (a)   13. f and g are injections.
15. If f is injective, then $f(a) = f(b) \Rightarrow a = b$, $\forall a, b \in X$. This means that two arrows pointing to the same element of Y must start at the same element of X. That is, the two arrows are really one. The converse reverses this argument.
17. (a) f is a mapping because f satisfies the definition of function, $\text{dom}(f) = X$, and $\text{range}(f) \subseteq Y$.
   (b) $\text{range}(f) = \{c, d\}$ (c) No   (d) No
19. No; no arrow points to 1 or 4.
21. (a) Injective and surjective   (b) Neither   (c) Injective and surjective   (d) Injective   (e) Neither
23. Since $f(1) \ne f(2) \ne f(3) \ne f(1)$, f is injective. Since $\text{range}(f) = \{a, b, c\} = Y$, f is surjective. Thus, f is a bijection.
25. Show that $\forall a, b \in (0, 1)$, $a/(1 - a) = b/(1 - b) \Rightarrow a = b$ (cross multiply). Let $y \in (0, \infty)$ and show $\exists x \in (0, 1)$ such that $y = x/(1 - x)$ (solve the equation).
27. (a) See Problem 21(d).   (b) Let $f(x) = x^3 + x^2$.   (c) See Problem 21(b) and (e).
   (d) See Problem 21(a) and (c).
29. (a) $f^{-1}(x) = (2 - x)/7$ (b) $g^{-1}(x) = 1/x$ (c) $h^{-1}(x) = (2x - 3)/(3x - 2)$
   (d) $F^{-1}(x) = (x - 1)^2$, $x \ge 1$ (e) $G^{-1}(x) = \ln(x + \sqrt{x^2 + 1})$
31. (b) Define $f : X \to Y$ by $f = \{(x, y) \mid x \text{ is a positive integer} \land y = 2x\}$, and show that f is bijective.
33. For each of the m objects x in X we have n choices for $f(x)$.
35. Show that an injection $f : X \to Y$ corresponds to a permutation of the n elements of Y taken m at a time.

PROBLEM SET 2.4, PAGE 126

1. (a) $\mathbb{R}$ (b) $(g \circ f)(x) = -21x - 6$ (c) $\mathbb{R}$ (d) $(f \circ g)(x) = -21x + 2$
3. (a) $\{x \mid x > -\frac{1}{3}\}$ (b) $(g \circ f)(x) = \sqrt{3x + 1}$ (c) $\{x \mid x \ge -1\}$ (d) $(f \circ g)(x) = 3\sqrt{x + 1}$
5. (a) $\{x \mid x \in \mathbb{R} \land x \ne -3\}$ (b) $(g \circ f)(x) = (3x^2 + 14x + 19)/(x^2 + 6x + 9)$ (c) $\mathbb{R}$
   (d) $(f \circ g)(x) = (x^2 + 3)/(x^2 + 5)$
7. (a) $\mathbb{R}$ (b) $(g \circ f)(x) = 3 \sin^2 x + 2$ (c) $\mathbb{R}$ (d) $(f \circ g)(x) = \sin(3x^2 + 2)$
9. (a) $\mathbb{R}$ (b) $(g \circ f)(x) = 3$ (c) $\mathbb{R}$ (d) $(f \circ g)(x) = 16$
11. (a) $\{(a, b), (b, c), (c, c)\}$ (b) $\{(1, 2), (2, 2), (3, 1)\}$
13. (a) $g(x) = (x + 1)/(x - 1)$; $f(x) = \sqrt{x}$; $\text{dom}(f \circ g) = \{x \mid x > 1 \lor x \le -1\}$
   (b) $g(x) = \sqrt{x}$; $f(x) = \cos^{-1} x$; $\text{dom}(f \circ g) = \{x \mid 0 \le x \le 1\}$
   (c) $g(x) = x^2 - x - 6$; $f(x) = \ln x$; $\text{dom}(f \circ g) = \{x \mid x < 2 \lor x > 3\}$
15. (a) $ad + b = cb + d$ (b) Let $x = b/(1 - a) = d/(1 - c)$.
17. Let $(x, z) \in f \circ (g \circ h)$ and argue that $(x, z) \in (f \circ g) \circ h$.
19. (a) $\text{dom}(g \circ f) = \{x \mid x \in \text{dom}(f) \land f(x) \in \text{dom}(g)\} = \{x \mid x \in X \land f(x) \in Y\} = X$
   (b) $\text{range}(g \circ f) = \{g(y) \mid y = f(x) \land x \in X\} \subseteq \{g(y) \mid y \in Y\} = Z$

21. (a) a r (b) $\{s, t\}$

        b s

        c t

23. (a) Suppose that $(g \circ f): X \to Z$ is injective and that $a, b \in X$ with $f(a) = f(b)$. Then $(g \circ f)(a) = g(f(a)) = g(f(b)) = (g \circ f)(b)$; hence, $a = b$.

    (b) Suppose that $(g \circ f): X \to Z$ is surjective and let $z \in Z$. Then $\exists x \in X$ such that $z = (g \circ f)(x) = g(f(x))$. Therefore, $\exists y \in Y [\text{namely}, y = f(x)]$ such that $z = g(y)$.

25. $g = \{(1, 1), (2, 2), (3, 3)\}$, $g = \{(1, 2), (2, 3), (3, 1)\}$, $g = \{(1, 3), (2, 1), (3, 2)\}$

27. (a) $\{a\}$ (b) $\{b, d\}$ (c) $\{a, b, c, d\}$ (d) $\{1, 3\}$ (e) $\{2, 5\}$ (f) $\{2, 4, 5\}$ (g) $\{4, 6\}$

29. (a) $f(X) = \{f(x) | x \in X\} = \{y | y = f(x) \land x \in X\} = \text{range}(f)$ (b) $f^{-1}(Y) = \{x \in \text{dom}(f) | f(x) \in Y\} = X$

    (c) $f(\varnothing) = \{f(x) | x \in \varnothing\} = \varnothing$ (d) $f^{-1}(\varnothing) = \{x \in X | f(x) \in \varnothing\} = \varnothing$

31. (a) Assume $y \in f(M)$. Then $\exists x$ with $f(x) = y$ and $x \in M$. Thus, $\exists x$ such that $x \in \text{dom}(f)$ and $x \in M$, and thus, $x \in M \cap \text{dom}(f)$ with $f(x) = y \in f(M \cap \text{dom}(f))$. For $y \in f(M \cap \text{dom}(f))$, $\exists x \in M \cap \text{dom}(f)$ with $f(x) = y$. Thus, $y \in f(M)$.

    (b) $x \in f^{-1}(N) \Leftrightarrow x \in \text{dom}(f) \land f(x) \in N \Leftrightarrow x \in \text{dom}(f) \land f(x) \in N \land f(x) \in \text{range}(f) \Leftrightarrow x \in \text{dom}(f) \land f(x) \in N \cap \text{range}(f) \Leftrightarrow x \in f^{-1}(N \cap \text{range}(f))$

    (d) Let $M \subseteq N$ and suppose that $x \in f^{-1}(M)$. Then $f(x) \in M$, so $f(x) \in N$; hence, $x \in f^{-1}(N)$.

    (f) Let $N \subseteq \text{range}(f)$. Suppose $y \in N$. Then $\exists x \in \text{dom}(f)$, $y = f(x)$. Since $y = f(x) \in N$, we have $x \in f^{-1}(N)$, so $y = f(x) \in f(f^{-1}(N))$. Conversely, suppose that $y \in f(f^{-1}(N))$. Then $\exists x \in f^{-1}(N)$ such that $y = f(x)$. But since $x \in f^{-1}(N)$, we have $y = f(x) \in N$.

33. (a) $y \in f(M \cup N) \Leftrightarrow y \in \{f(x) | x \in M \lor x \in N\} \Leftrightarrow y \in \{f(x) | x \in M\} \cup \{f(x) | x \in N\} \Leftrightarrow y \in f(M) \cup f(N)$

    (b) $x \in f^{-1}(M \cup N) \Leftrightarrow f(x) \in M \cup N \Leftrightarrow f(x) \in M \lor f(x) \in N \Leftrightarrow x \in f^{-1}(M) \lor x \in f^{-1}(N) \Leftrightarrow x \in f^{-1}(M) \cup f^{-1}(N)$

    (c) $f^{-1}(M \cap N) = \{x | f(x) \in M \cap N\} = \{x | f(x) \in M \land f(x) \in N\} = \{x | f(x) \in M\} \cap \{x | f(x) \in N\} = f^{-1}(M) \cap f^{-1}(N)$

35. (a) Let $y \in f(M) \setminus f(N)$. Since $y \in f(M)$, $\exists x \in M$, $y = f(x)$. Since $y \notin f(N)$, we cannot have $x \in N$ [else we would have $y = f(x) \in f(N)$]; hence, $x \in M \setminus N$. Therefore, $y \in f(M \setminus N)$.

37. (a) Let $y \in f(\bigcap_{i \in I} M_i)$. Then $\exists x \in \bigcap_{i \in I} M_i$ such that $y = f(x)$. Therefore, because $x \in M$ for every $i \in I$, we have $y = f(x) \in f(M_i)$ for every $i \in I$. Thus, $y \in \bigcap_{i \in I} f(M_i)$.

39. (a) $y \in \pi_2(f \cap \pi_1^{-1}(M)) \Leftrightarrow \exists x, \quad (x, y) \in f \cap \pi_1^{-1}(M) \Leftrightarrow \exists x \in \text{dom}(f), \quad y = f(x) \land x \in M \Leftrightarrow y \in f(M)$

PROBLEM SET 2.5, PAGE 139

1. Partition   **3.** Not a partition (7 belongs to no set in the set of sets).   **5.** $b, d, f,$ and g

7. Follows from using Definition 5.5.

    (b) {Los Angeles, San Francisco}, {Miami, Fort Lauderdale, Daytona Beach}, {Dallas, San Antonio}, {Chicago}

9. $\forall x \in \mathbb{R}, x^2 = x^2$. $\forall x, y \in \mathbb{R}, x^2 = y^2 \Rightarrow y^2 = x^2$. $\forall x, y, z \in \mathbb{R}, x^2 = y^2 \land y^2 = z^2 \Rightarrow x^2 = z^2$.

11. Show that Definition 5.5 is satisfied.   **13.** $\{n\pi + (\pi/2) | n \text{ is an integer}\}$

15. For Part (b), $[\![x]\!]_E = \{x\} \cup \{y | y$ is a brother or sister of $x\}$. For Part (d), $[\![x]\!]_E = \{y | y$ is a triangle and y is congruent to $x\}$. For Part (f), $[\![x]\!]_E = \{x\}$. For Part (g), $[\![x]\!]_E = \{y | y - x$ is divisible by $n\}$.

17. (a) $\{(a, a), (a, d), (d, d), (d, a), (b, b), (b, e), (e, e), (e, b), (c, c)\}$ (b) $\{\{a, d\}, \{b, e\}, \{c\}\}$

19. Let $l, s, m, f, b, d, a,$ and c represent Los Angeles, San Francisco, Miami, Fort Lauderdale, Daytona Beach, Dallas, San Antonio and Chicago, respectively. Then $E = \{(l, l), (l, s), (s, s), (s, l), (m, m), (m, f), (m, b), (f, m), (f, f), (f, b), (b, m), (b, f), (b, b), (d, d), (d, a), (a, a), (a, d), (c, c)\}$.

25. Let $x \in X$. Then $\exists y$ such that xRy. By symmetry, yRx, and by transitivity, $xRy \wedge yRx \Rightarrow xRx$. Thus, if $X = \text{dom}(R)$, the reflexivity condition is not needed. However, if $X \neq \text{dom}(R)$, then the argument above cannot be made.

27. Let $(x, y) \in R$. By symmetry, $(y, x) \in R$, so, by antisymmetry, $x = y$.   **29.** (b) $[\![x]\!]_G = [\![x]\!]_E \cap [\![x]\!]_F$

31. (a) $\{X\}$ is both exhaustive and mutually exclusive.   (b) $E = X \times X$

33. Note that cRa and bRd. Thus, by the transitivity of R, $cRa \wedge aRb \wedge bRd \Rightarrow cRd$.

35. Use the fact that R is antisymmetric.

37. For $G \in \mathscr{E}$, $G \leq G$. For $G, H \in \mathscr{E}$, $G \leq H \wedge H \leq G \Rightarrow H = G$. For $G, H, K \in \mathscr{E}$, $G \leq H \wedge H \leq K \Rightarrow G \leq K$.

39. Use Theorem 5.15.

41. B_2 represents the number of equivalence relations on X if $\#X = 2$. Since the only partitions possible are $\{\{a\}, \{b\}\}$ or $\{X\}$, where $X = \{a, b\}$, we have $B_2 = 2$. Similarly, if $X = \{a, b, c\}$, the only partitions are $\{\{a\}, \{b\}, \{c\}\}, \{\{a, b\}, \{c\}\}, \{\{a\}, \{b, c\}\}, \{\{a, c\}, \{b\}\}$, and $\{X\}$. Thus, $B_3 = 5$. Similarly, $B_4 = 15$, $B_5 = 52$, $B_6 = 203$, $B_7 = 877$, $B_8 = 4140$.

43. The n endings to the lines may be partitioned in B_n different ways. Two line endings are in the same cell of the partition (or the same equivalence class) if and only if they rhyme.

CHAPTER 3   PROBLEM SET 3.1, PAGE 157

1. (a) e (b) e (c) f (d) c (e) c (f) c (g) f (h) f

3. (a) Yes   (b) Yes   (c) No, $\sqrt{x^2 + y^2}$ need not be an integer.   (d) Yes   (e) No, $(\frac{1}{2}) * (-1) \notin S$.

5. (a) $x * (y * z) = x * (y + z + yz) = x + y + z + yz + x(y + z + yz) = x + y + z + yz + xy + xz + xyz$,
   $(x * y) * z = (x + y + xy) * z = x + y + xy + z + (x + y + xy)z = x + y + z + yz + xy + xz + xyz$
   (b) $x * y = x + y + xy = x + y + yx = y * x$
   (c) $x * 0 = x + 0 + x \cdot 0 = x$, $\quad 0 * x = 0 + x + 0 \cdot x = x$

7. (a) $e * (y * z) = y * z = (e * y) * z$, $\quad x * (e * z) = x * z = (x * e) * z$, $\quad x * (y * e) = x * y = (x * y) * e$
   (b) It is only necessary to check all equations of the form $x * (y * z) = (x * y) * z$ for $x, y, z \in S \setminus \{e\}$.

11. (a) No   (b) No   (c) No   (d) No   **13.** $((x * y) * z) * w, x * ((y * z) * w), x * (y * (z * w))$

15. By definition, $x * y * z * w = (x * y * z) * w = ((x * y) * z) * w$.
    Also, $x * ((y * z) * w) = (x * (y * z)) * w = ((x * y) * z) * w = x * y * z * w$,
    $(x * y) * (z * w) = ((x * y) * z) * w = x * y * z * w$, and so forth.

17. (a) $(x * y) * z = x * y = x$, $\quad x * (y * z) = x$ (b) $(S, *)$ is a monoid if and only if $\#S = 1$.
    (c) In view of Part (b), $(S, *)$ is a group if and only if $\#S = 1$.

19. (a) $(x * y) * z = x * (y * z) = a$. (b) $(S, *)$ is a monoid if and only if $\#S = 1$.   (c) Only if $\#S = 1$.

21. (a) If $y * x = e$, then $(x * y) * (x * y) = (x * (y * x)) * y = (x * e) * y = x * y$.
    (b) Let $(S, *)$ be a group and suppose $a \in S$ with $a * a = a$. Then $a * a = a * e$, and so $a = e$ by the left-cancellation law.

23. $z * (x * y) = ((y^-) * (x^-)) * (x * y) = (y^-) * ((x^-) * (x * y)) = (y^-) * (((x^-) * x) * y) = (y^-) * (e * y)$
    $= (y^-) * y = e$

25. By Problem 9, $(S, *)$ is a semigroup. By the table in Example 1.2, e is a neutral element, $a^- = a$, $b^- = b$, and $c^- = c$.

27. $-(-x) = x$ and $-(x + y) = (-y) + (-x)$.

29. An element b appears in the row labeled by a and in the columns labeled by x and y if and only if $a * x = a * y = b$.

31. Suppose $(S, *)$ is a group, $a, x, y \in S$, and $x * a = y * a$. Then $(x * a) * (a^-) = (y * a) * (a^-)$, so $x * (a * (a^-)) = y * (a * (a^-))$, that is, $x * e = y * e$, so $x = y$.

33. See the answer to Problem 29.

35. $a * x = b \Rightarrow (a^-) * (a * x) = (a^-) * b \Rightarrow ((a^-) * a) * x = (a^-) * b \Rightarrow e * x = (a^-) * b \Rightarrow x = (a^-) * b$. Likewise, $y * a = b \Rightarrow y = b * (a^-)$.

37. Let S be the set of all nonnegative integers and let $x * y = x + y$ for all $x, y \in S$.

39. $M * (N * P) = (M \setminus (N \cup P)) \cup (N \setminus (M \cup P)) \cup (P \setminus (M \cup N)) \cup (M \cap N \cap P) = (M * N) * P$, $M * \varnothing = \varnothing * M = M, M * M = \varnothing$.

41. We know by Theorem 4.7 in Chapter 2 that $*$ is associative. The function $\varepsilon : X \to X$ defined by $\varepsilon(x) = x$ for all $x \in X$ serves as a neutral element, and, for $\phi \in \beta(X)$, ϕ^{-1} serves as ϕ^-.

PROBLEM SET 3.2, PAGE 173

1. (a) It has already been shown that $(R, +)$ is a group. Since $(0 \cdot 0) \cdot 0 = 0 = 0 \cdot (0 \cdot 0)$ and since $0 \cdot (0 + 0) = 0 \cdot 0 = 0 = 0 \cdot 0 + 0 \cdot 0$ and, similarly, $(0 + 0) \cdot 0 = 0 \cdot 0 + 0 \cdot 0$, $(R, +, \cdot)$ is a ring
(b) Suppose 0 is a neutral element for $(R, \cdot)$. Then, for $x \in R$, $x = x \cdot 0 = 0$ by Theorem 2.3.
(c) If $1 = 0$, then, by Part (b), $R = \{0\}$, which isn't very interesting.

3.

+	Even	Odd		·	Even	Odd
Even	Even	Odd		Even	Even	Even
Odd	Odd	Even		Odd	Even	Odd

By earlier work, $(R, +)$ is a group. Also, $(R, \cdot)$ is associative since $(\mathbb{R}, \cdot)$ is. Since $(\mathbb{R}, +, \cdot)$ obeys the left- and right-distributive laws, so does $(R, +, \cdot)$. Finally, it is evident that $(R, +)$ and $(R, \cdot)$ are commutative.

5. (a) By the left-distributive law, $(a + b)(c + d) = (a + b)c + (a + b)d$. By the right-distributive law, $(a + b)c + (a + b)d = ac + bc + ad + bd$, and the commutativity of $+$ gives $ac + ad + bc + bd$.
(b) Not true unless $ab = ba$.

7. (a) Compute the tables.   (b)

+	0	1	2		·	0	1	2
0	0	1	2		0	0	0	0
1	1	2	0		1	0	1	2
2	2	0	1		2	0	2	1

(c) 1, 3

9. Yes. Check all conditions in Theorem 2.2.

11. (a) Show that $x(-y) + xy = 0$.
(b) Add $-y$ to both sides of $x + y = z$ and then compute to show that $x = z - y$. Add y to both sides of $x = z - y$ and then compute to show that $x + y = z$.

13. (a) Verify all conditions in Theorem 2.2 by direct computation, taking $\begin{bmatrix} 0 & 0 \\ 0 & 0 \end{bmatrix}$ as the additive neutral element and $\begin{bmatrix} 1 & 0 \\ 0 & 1 \end{bmatrix}$ as the multiplicative neutral element.
(b) Follow the hint.

15. Let $\delta = ad - bc$. If $\delta \neq 0$, show that $\begin{bmatrix} a & b \\ c & d \end{bmatrix}^{-1} = \begin{bmatrix} d/\delta & -b/\delta \\ -c/\delta & a/\delta \end{bmatrix}$. For the converse, suppose $\begin{bmatrix} a & b \\ c & d \end{bmatrix}^{-1}$ exists. Multiply both sides of the equation $\begin{bmatrix} a & b \\ c & d \end{bmatrix} \begin{bmatrix} a & b \\ c & d \end{bmatrix}^{-1} = \begin{bmatrix} 1 & 0 \\ 0 & 1 \end{bmatrix}$ on the left by $\begin{bmatrix} d & -b \\ -c & a \end{bmatrix}$ and conclude that $\begin{bmatrix} \delta & 0 \\ 0 & \delta \end{bmatrix} \begin{bmatrix} a & b \\ c & d \end{bmatrix}^{-1} = \begin{bmatrix} d & -b \\ -c & a \end{bmatrix}$. Now suppose that $\delta = 0$ and obtain a contradiction.

17. (a) $\begin{bmatrix} 1 & 0 \\ 0 & 0 \end{bmatrix} \cdot \begin{bmatrix} 0 & 0 \\ 0 & 1 \end{bmatrix} = \begin{bmatrix} 0 & 0 \\ 0 & 0 \end{bmatrix}$. Consider $\begin{bmatrix} 0 & 1 \\ 0 & 0 \end{bmatrix}$.   (b) Yes.

19. (a) Verify all conditions in Theorem 2.2.
(b) The nonzero polynomials of degree zero.
23. One must show that, in a field, $ab = 0 \Rightarrow a = 0$ or $b = 0$. Suppose that $ab = 0$ but that neither a nor b is zero and multiply by a^{-1}.
25. (a) $x^2 = 0 \Rightarrow x = 0$ (b) $x = x^{-1} \Rightarrow x = 1$

PROBLEM SET 3.3, PAGE 189

1. $(x/y) \in P \Rightarrow (x/y)y^2 \in P$ (since $y \neq 0 \Rightarrow y^2 \in P$)
3. (i) $x \in P \Leftrightarrow x - 0 \in P \Leftrightarrow 0 < x$ (ii) $-x \in P \Leftrightarrow 0 - x \in P \Leftrightarrow x < 0$
(iii) $x < y \Leftrightarrow y - x \in P \Rightarrow y - x \neq 0 \Rightarrow x \neq y$
5. Apply the law of trichotomy to $y - x$.
13. Consider separately the cases $y < z$ and $y = z$.
17. To prove $x + z < y + z \Rightarrow x < y$, use Part (i) of Theorem 3.10 to add $-z$ to both sides.
21. Case analysis **29.** $|x - y| = |x + (-y)| \leq |x| + |-y| = |x| + |y|$. **31.** Case analysis
33. If $|x + y| = |x| + |y|$, then $|x + y|^2 = |x|^2 + 2|x||y| + |y|^2$, so $x^2 + 2xy + y^2 = x^2 + 2|xy| + y^2$, and hence, $xy = |xy|$. For the converse argument, read this argument backwards.
35. (a) Case analysis (b) $(x + y - |x - y|)/(1 + 1)$
37. (b) Yes. Only the properties of an ordered field are used in the proof.

PROBLEM SET 3.4, PAGE 201

1. Since $1 \in \mathbb{R}$ and $1 > 0$, we have $1 \in P$. Also, $\forall n \in \mathbb{R}, n \in P \Rightarrow n > 0 \Rightarrow n + 1 > 0 \Rightarrow n + 1 \in P$.
3. Since $1 \in \mathbb{R}$ and $1 \geq 1$, we have $1 \in J$. Also, $\forall n \in \mathbb{R}, n \in J \Rightarrow n \geq 1 \Rightarrow n + 1 \geq 2 \Rightarrow n + 1 \geq 1 \Rightarrow n + 1 \in J$.
5. By Lemma 4.3, if I is an inductive subset of $\mathbb{R}$, then $\mathbb{N} \subseteq I$. Thus, if $I \subseteq \mathbb{N}$, we have $I = \mathbb{N}$.
7. False. Let $I = \mathbb{N}$ and let $J = \mathbb{N} \cup \{1/2\}$.
9. By Theorem 4.7, $m \leq n$ or $m \geq n + 1$. Since $m > n$, we cannot have $m \leq n$ and so $m \geq n + 1$.
11. Fix $m \in \mathbb{N}$. Then $m \cdot 1 = m \in \mathbb{N}$. Assume the induction hypothesis $mn \in \mathbb{N}$. Then, $m(n + 1) = mn + m \in \mathbb{N}$ by Theorem 4.9.
13. By Theorem 4.11, $\exists m \in \mathbb{N}, m\varepsilon > 1$. Thus, $\varepsilon > 1/m$. Now, if $n > m$, then $1/m > 1/n$, so $\varepsilon > 1/m$.
15. Use Theorem 4.7.
17. Since $n \leq x$ and $x < m + 1$, we have $n < m + 1$. Also, since $m \leq x$ and $x < n + 1$, we have $m < n + 1$. Thus, $m + 1 < n + 2$, so $n < m + 1 < n + 2$ and, from Problem 15, $m + 1 = n + 1$, from which it follows that $m = n$.
19. (i) $m + 1 = n + 1 \Rightarrow m = n$. (ii) $1 = x + 1 \Rightarrow x = 0$, contradicting $0 \notin \mathbb{N}$. (iii) Use Problem 5.
21. Let $b = \sup(M)$ and prove that $b \in M$ by arguing as in the proof of Theorem 4.14.
23. (a) If n is even, then $n = 2k$ for some $k \in \mathbb{N}$. If $n + 1$ were also even, we would have $n + 1 = 2j$ for some $j \in \mathbb{N}$. Then, with $m = j - k$, we would have $1 = 2m, m = 1/2$. Thus, $m > 0$, so $j > k$, and it would follow from Theorem 4.8 that $1/2 = m \in \mathbb{N}$, contradicting Example 4.5.
(b) Suppose there were a number n such that both n and $n + 1$ are odd. Then, by Theorem 4.14, there would be a smallest such n. Since 2 is even, $n \neq 1$, hence, $n - 1 \in \mathbb{N}$ by Lemma 4.6. By our choice of n, we cannot have both $n - 1$ and n odd; hence, $n - 1$ is even, so $\exists k \in \mathbb{N}$ with $n - 1 = 2k$. Then $n + 1 = 2k + 2 = 2(k + 1)$, contradicting the fact that $n + 1$ is odd.
(c) Obvious, since odd means not even.
(d) Let m and n be odd. Without loss of generality, we can suppose $m, n > 1$. By Part (b), both $m - 1$ and $n - 1$ are even, so $\exists j, k \in \mathbb{N}$ such that $m - 1 = 2j$ and $n - 1 = 2k$, it follows that $mn = (2j + 1)(2k + 1) = 2q + 1$, where $q = 2jk + j + k \in \mathbb{N}$. Since $2q$ is even, it follows from Part (a) that mn is odd.
(e) Argue by contradiction, using Part (d).

25. (a) Use Theorem 4.14.   (b) Follow the hint given.

(c) Since $n^2 = 2m^2$ and $n = 2k$, we have $(2k)^2 = 2m^2$, $4k^2 = 2m^2$, and so $m^2 = 2k^2$. Now apply the reasoning in Part (b) to m.

(d) Substituting $m = 2r$ into $m^2 = 2k^2$ and canceling the factor 2 from both sides, we obtain $k^2 = 2r^2$. That $k < n$ follows from $n = 2k$.

PROBLEM SET 3.5, PAGE 212

1. Follows from the facts that $m > 0$ for every $m \in \mathbb{N}$ and $m \leq 0$ for every $m \in \{0\} \cup \{-n \mid n \in \mathbb{N}\}$.

3. First prove $k \in \mathbb{Z} \Rightarrow -k \in \mathbb{Z}$ by a case analysis. Then replace k by $-k$ to obtain $-k \in \mathbb{Z} \Rightarrow -(-k) \in \mathbb{Z}$, from which the converse implication $-k \in \mathbb{Z} \Rightarrow k \in \mathbb{Z}$ follows.

5. Suppose $mn = 1$ with m, $n \in \mathbb{Z}$. We consider the case in which $m > 0$. (Evidently, $m \neq 0$, and the case in which $m < 0$ is handled similarly.) Then $m \in \mathbb{N}$, so $m \geq 1$ by Example 4.5. If $m = 1$, then $n = 1$ and we are done. But if $m > 1$, then $0 < 1/m < 1$, so $0 < n < 1$, contradicting Theorem 5.4.

7. $x^* \cdot 10^n = [\![x \cdot 10^n + 0.5]\!]$, so $x^* \cdot 10^n \leq x \cdot 10^n + 0.5 < 10^n x^* + 1$. Subtracting $x^* \cdot 10^n + 0.5$ from all members of the last inequality, we find that $-0.5 \leq (x - x^*) \cdot 10^n < 0.5$, so that $-(10^{-n})/2 \leq x - x^* < 10^{-n}/2$. Consequently, $|x - x^*| \leq 10^{-n}/2$.

9. (b) Use the result of Problem 5.

13. (a) $h = np + \rho(h)$ and $k = nq + \rho(k)$ for p, q, $\rho(h)$, $\rho(k) \in \mathbb{Z}$ with $0 \leq \rho(h)$, $\rho(k) < n$. Then $h + k = n(p + q) + \rho(h) + \rho(k)$. Also, $\rho(h) + \rho(k) = ns + \rho(\rho(h) + \rho(k))$ for s, $\rho(\rho(h) + \rho(k)) \in \mathbb{Z}$ with $0 \leq \rho(\rho(h) + \rho(k)) < n$. Hence, $h + k = n(p + q + s) + \rho(\rho(h) + \rho(k))$, and so $\rho(h + k) = \rho(\rho(h) = \rho(k))$.

15. $m = 10$, $h = 4$, $k = 5$

17. Following the hint given and the fact that $r = k - cq$, prove that $m \mid r$ and $n \mid r$.

19. Suppose n is the smallest natural number such that $n > 1$ and n cannot be factored as a product of finitely many primes. Then n is not a prime, so we can write $n = ab$ where a, $b > 1$. Since a, $b < n$, we can factor a and b as products of finitely many primes, and so n factors as a product of finitely many primes.

21. $u + (-u) = 0$, $u - u = 0$, $u(1/u) = 1$, $u/u = 1$.

23. If P is the set of positive rational numbers, show that P satisfies conditions (i) through (iv) of Definition 3.1.

25. Since $1 \in F$ and F is closed under addition, it follows from mathematical induction that F contains every natural number n. Since F is closed under negation, it contains all numbers of the form $-n$ for $n \in \mathbb{N}$. Since $0 \in F$, it follows that $\mathbb{Z} \subseteq F$. Since F is closed under division, we have $\mathbb{Q} \subseteq F$.

27. Let $x = \alpha_1 \alpha_2 \ldots \alpha_n . \beta_1 \beta_2 \ldots \beta_m$ be the decimal expansion of x, where α_i, $\beta_j \in \mathbb{Z}$ with $0 \leq \alpha_i$, $\beta_j \leq 9$ for $1 \leq i \leq n$, $1 \leq j \leq m$. Let $y = \alpha_1 \alpha_2 \ldots \alpha_m \beta_1 \beta_2 \ldots \beta_m$, where the α's and β's are the digits in the expansion of y in the decimal system. Then $y \in \mathbb{N}$ and $x = y/10^m$.

29. If $y = 10^n x$, then y is obtained from x by moving the decimal point n places to the right. Then $[\![y]\!]$ is obtained by dropping all digits to the right of the decimal point, and multiplication by 10^{-n} moves the decimal point back to its original position.

CHAPTER 4   PROBLEM SET 4.1, PAGE 230

1. Clearly, a is a lower bound for (a, b). Let c be any other such lower bound. We must prove that $c \leq a$. Suppose not. Then $a < c$, so there exists d such that $a < d < c$ and $d < b$. Hence, $d \in (a, b)$ with $d < c$, contradicting the fact that c is a lower bound for (a, b).

3. Use the fact that $\mathbb{R} \setminus (\mathbb{R} \setminus M) = M$.

5. (a) Suppose, for instance, that $p \leq q$ so that $p = \min(p, q)$. Then $p \in \{p, q\}$ and $p \leq x$ for all $x \in \{p, q\}$, so $p = \inf\{p, q\}$. A similar argument applies to the remaining case in which $q < p$.

7. Use a pick-a-point argument to show that $J \cap K \subseteq (c - \varepsilon, c + \varepsilon)$ and that $(c - \varepsilon, c + \varepsilon) \subseteq J \cap K$.

9. By Lemma 1.7, every bounded open interval is open and, by Theorem 1.8, the union of open sets is open. Conversely, by Theorem 1.6, every open set U is the union of all neighborhoods N of points $u \in U$ such that $N \subseteq U$.

11. Let $(F_j)_{j=1}^n$ be a finite family of closed subsets of $\mathbb{R}$ and let $F = \bigcup_{j=1}^n F_j$. Then, by De Morgan's law, $F' = \bigcap_{j=1}^n (F_j)'$. By Theorem 1.5, each set $(F_j)'$ is open, so F' is open by Theorem 1.11, and it follows that F is closed by Theorem 1.5.

13. (a) For $x \in I = (a, \infty)$, we have $(a, x) \subseteq (a, \infty)$ and so $\bigcup_{x \in I}(a, x) \subseteq (a, \infty)$. Conversely, suppose $p \in (a, \infty)$. Then $a < p$. Select $x \in \mathbb{R}$ with $p < x$. Then $p \in (a, x)$, and it follows that $p \in \bigcup_{x \in I}(a, x)$.

15. Clearly, $\{a\} \subseteq \bigcap_{\varepsilon > 0}(a - \varepsilon, a + \varepsilon)$. Suppose $x \in \bigcap_{\varepsilon > 0}(a - \varepsilon. a + \varepsilon)$ but that $x \ne a$. Let $\varepsilon = \frac{1}{2}|x - a|$. Then $x \notin (a - \varepsilon, a + \varepsilon)$, a contradiction.

17. By the hint, $\mathbb{R} \setminus [a, b]$ is a union of open sets.

19. (a) Let $N = (b - \varepsilon, b + \varepsilon)$ be a neighborhood of b of radius $\varepsilon > 0$. By Part (i) of Lemma 3.23 in Chapter 3, there exists $s \in M$ such that $b - \varepsilon < s \le b$. Since $b \notin M$, it follows that $s \ne b$. Therefore, $s \in (N \setminus \{b\}) \cap M$.

21. (a) Follow the hint.   (b) $a = \inf(M)$ and $a \in M$; $b = \sup(M)$ and $b \in M$.

23. (a) Let $b \in bd(M)$. We claim that $b \in \bar{M}$. If $b \in M$, this is clear, so assume $b \notin M$. Then every neighborhood of b contains a point of M other than b, and so b is an accumulation point of M; hence, $b \in \bar{M}$. Since $b \in bd(M)$, we have $b \in bd(\mathbb{R} \setminus M)$, so, by the argument above, $b \in \overline{\mathbb{R} \setminus M}$. Hence, $b \in \bar{M} \cap \overline{\mathbb{R} \setminus M}$. Conversely, suppose $b \in \bar{M} \cap \overline{\mathbb{R} \setminus M}$. Exactly one of the conditions $b \in M$ or $b \in \mathbb{R} \setminus M$ must hold. Suppose $b \in \mathbb{R} \setminus M$. (The other case is handled similarly.) Since $b \in \bar{M}$, it follows that b is an accumulation point of M; hence, every neighborhood of b contains a point of M. Such a neighborhood also contains b, which is a point of $\mathbb{R} \setminus M$; hence $b \in bd(M)$.

     (b) Part (a) shows that $bd(M)$ is the intersection of closed sets.

25. p is an isolated point of M if and only if there is a neighborhood N of p such that $(N \setminus \{p\}) \cap M = \varnothing$. By elementary set theory and the fact that $p \in M \cap N$, it follows that $N \cap M = \{p\}$.

27. Use Lemma 1.19 and Theorems 4.7 and 5.4 of Chapter 3.

29. $\mathbb{R} \setminus M \subseteq \overline{\mathbb{R} \setminus M}$, and so $\mathbb{R} \setminus (\overline{\mathbb{R} \setminus M}) \subseteq M$. Also, $\overline{\mathbb{R} \setminus M}$ is closed, so $\mathbb{R} \setminus (\overline{\mathbb{R} \setminus M})$ is open. Thus, if $p \in \mathbb{R} \setminus (\overline{\mathbb{R} \setminus M})$, there is a neighborhood N of p such that $N \subseteq \mathbb{R} \setminus (\overline{\mathbb{R} \setminus M}) \subseteq M$. This shows that $\mathbb{R} \setminus (\overline{\mathbb{R} \setminus M}) \subseteq M^o$. Conversely, suppose that $p \in M^o$. Then there is a neighborhood N of p such that $N \subseteq M$. Thus, $(N \setminus \{p\}) \cap (\mathbb{R} \setminus M) \subseteq N \cap (\mathbb{R} \setminus M) = \varnothing$, and it follows that p is not an accumulation point of $\mathbb{R} \setminus M$. Since $p \notin \mathbb{R} \setminus M$, it follows that $p \notin \overline{\mathbb{R} \setminus M}$; hence, we have $p \in \mathbb{R} \setminus (\overline{\mathbb{R} \setminus M})$.

31. (a) If F is a closed subset of $\mathbb{R}$ and $M \subseteq F$, then $\bar{M} \subseteq \bar{F} = F$ by Parts (iii) and (ii) of Theorem 1.17. Hence, $\bar{M}$ is contained in the intersection of all closed sets that contain M. Conversely, suppose that p belongs to every closed set F that contains M. Then, since $M \subseteq \bar{M}$ and $\bar{M}$ is closed, we have $p \in \bar{M}$.

33. By Problem 23a, $bd(bd(M)) = \overline{bd(M)} \cap (\overline{\mathbb{R} \setminus bd(M)}) \subseteq bd(M) = bd(M)$, since $bd(M)$ is closed by Problem 23b.

35. (a) We use the formula in Problem 23a. Then $bd(P \cup Q) = \overline{(P \cup Q)} \cap (\overline{\mathbb{R} \setminus (P \cup Q)}) = (\bar{P} \cup \bar{Q}) \cap ((\overline{\mathbb{R} \setminus P}) \cap (\overline{\mathbb{R} \setminus Q})) \subseteq (\bar{P} \cap (\overline{\mathbb{R} \setminus P})) \cup (\bar{Q} \cap (\overline{\mathbb{R} \setminus Q})) = bd(P) \cup bd(Q)$.

37. True.

39. All but (a) are topological. Although a neighborhood is an open set, it is not possible to tell whether an open set containing the point p has the form $(p - \varepsilon, p + \varepsilon)$ for some $\varepsilon > 0$ merely by knowing which sets are open.

PROBLEM SET 4.2, PAGE 245

1. $(\forall \varepsilon)(\exists \delta)(\forall x)((x \in D \cap 0 < |x - a| \le \delta) \Rightarrow |f(x) - L| < \varepsilon)$.

3. (a) Let $x \in J = (a - \delta, a + \delta)$. Then, by Part (ii) of Lemma 2.2, with $L = a$, $\varepsilon = \delta$, and $y = x$, we have $|x - a| < \delta$. By the triangle inequality, $|x| = |a + (x - a)| \le |a| + |x - a| < |a| + \delta$. Also, $|a| = |x + (a - x)| \le |x| + |a - x| = |x| + |-(x - a)| = |x| + |x - a| < |x| + \delta$, so $|a| - \delta < |x|$.

     (b) By Part (a), $x \in J \Rightarrow |a| - \delta < |x| < |a| + \delta \Rightarrow |x| \in (|a| - \delta, |a| + \delta) = N$.

5. Let $\varepsilon = L - B$. Then $\exists \delta > 0$ such that $\forall x \in \mathrm{dom}(f)$, $|x - a| < \delta \Rightarrow |f(x) - L| < \varepsilon \Rightarrow B = L - \varepsilon < f(x)$. Let $J = (a - \delta, a + \delta)$.

7. Use the facts that every neighborhood of a is an open set and that, if V is an open set and $a \in V$, then there is a neighborhood N of a such that $N \subseteq V$.

9. Assume the hypotheses of Theorem 2.8 and suppose that $L = \lim_{x \to a} g(x)$. By the argument already given in the proof, Condition (i) in Theorem 2.7 is satisfied, Let $U \subseteq \mathbb{R}$ be open and let $L \in U$. Then there exists an open set V with $a \in V$ such that, for all $x \in \mathbb{R}$, $x \in (V \setminus \{a\}) \cap \mathrm{dom}(g) \Rightarrow g(x) \in U$. Let $D = \mathrm{dom}(f)$. Then $\mathrm{dom}(g) = (W \setminus \{a\}) \cap D$, so $(V \setminus \{a\}) \cap \mathrm{dom}(g) = ((V \cap W) \setminus \{a\}) \cap D$. Note that $V \cap W$ is an open set and $a \in V \cap W$. Now show that f satisfies Condition (ii) in Theorem 2.7.

11. To prove Part (i), let $\delta > 0$ be arbitrary. To prove Part (ii), let $\delta = \varepsilon$.

13. Let N be a neighborhood of LL'. By Lemma 2.13, there exist neighborhoods J and K of L and L', respectively, such that $u \in J$, $v \in K \Rightarrow uv \in N$. Also, there exist neighborhoods I_1 and I_2 of a such that $x \in (I_1 \setminus \{a\}) \cap \mathrm{dom}(f) \Rightarrow f(x) \in J$ and $x \in (I_2 \setminus \{a\}) \cap \mathrm{dom}(g) \Rightarrow g(x) \in K$. Let $I = I_1 \cap I_2$, and prove that $x \in (I \setminus \{a\}) \cap \mathrm{dom}(h) \Rightarrow h(x) \in N$.

15. Follow the hint.   **17.** Use mathematical induction on n.

19. The proof is quite similar to the proof of Lemma 2.11.

25. Combine Theorem 2.14 and the result of Problem 23.

PROBLEM SET 4.3, PAGE 259

1. In Example 3.2, let $a = 2$, $\varepsilon = 0.007$. Then the required δ is given by $\delta = \min(1, \varepsilon/(2|a| + 1)) = 0.007/5 = 0.0014$.

3. Use the fact that $|(2x - 3) - (2a - 3)| = 2|x - a|$.

7. Let $N = (f(a) - \varepsilon, f(a) + \varepsilon)$ and $J = (a - \delta, a + \delta)$.

9. (b) We must prove that there exists $\varepsilon > 0$ such that, for all $\delta > 0$, there exists x such that $|x - 0| < \delta$ and $|f(x) - f(0)| \geq \varepsilon$. Note that $f(x) = 1$ for $x \neq 0$ and that $f(0) = 0$. Take $\varepsilon = 1$. We must prove that, for all $\delta > 0$, there exists x such that $|x| < \delta$ and $|f(x)| \geq 1$. But, given any $\delta > 0$, let $x = \delta/2$, and we have $|x| < \delta$ and $|f(x)| = 1 \geq 1$.

(c) Let $a \neq 0$, let $\varepsilon > 0$, and choose $\delta > 0$ sufficiently small that $0 \notin (a - \delta, a + \delta)$. (For example, let $\delta = |a|/2$.) Then, if $|x - a| < \delta$, it follows that $|f(x) - f(a)| = |1 - 1| = 0 < \varepsilon$.

11. For instance, to prove that s is continuous at a, we use Lemma 2.11 and Theorem 3.8.

13. $\lim_{x \to a} s(x) = \lim_{x \to a}(f(x) + g(x)) = \lim_{x \to a} f(x) + \lim_{x \to a} g(x) = f(a) + g(a) = s(a)$.

15. Let $c, d \in \mathbb{R}$ and let N be a neighborhood of $c \star d = g(c \ast d)$. Since g is continuous, there exists a neighborhood W of $c \ast d$ such that, for all $x \in W$, $g(x) \in N$. Because $\ast$ is a continuous binary operation, there exists neighborhoods J and K of c and d, respectively, such that $u \in J$, $v \in K \Rightarrow u \ast v \in W \Rightarrow u \star v = g(u \ast v) \in N$.

17. By Problem 19 in Problem Set 4.2, subtraction is a continuous binary operation. By Example 3.15, the absolute value function is continuous; hence, by Problem 15, ρ is a continuous binary operation.

27. Let f be continuous at a and let $\varepsilon > 0$. Then there exists a neighborhood J of a such that, for $x \in J \cap \mathrm{dom}(f)$, $f(a) - \varepsilon < f(x) < f(a) + \varepsilon$. Thus, $f(a) - \varepsilon$ is a lower bound, and $f(a) + \varepsilon$ is an upper bound for $f(J)$.

29. Let f be continuous at a with $f(a) \neq 0$. Let $\varepsilon = |f(a)|/2$, and choose a neighborhood J of a such that, for $x \in J \cap \mathrm{dom}(f)$, $f(a) - \varepsilon < f(x) < f(a) + \varepsilon$. Prove that $f(J)$ does not intersect the neighborhood $(-\varepsilon, \varepsilon)$ of 0.

31. If H is relatively open in M, then there exists an open set $U \subseteq \mathbb{R}$ such that $H = U \cap M$. Let $F = \mathbb{R} \setminus U$. Then F is closed and $M \setminus H = M \cap F$.

33. Use Theorem 3.19 and Problem 31.

35. Suppose f is continuous. By Problem 33 there is a closed set $F \subseteq \mathbb{R}$ such that $f^{-1}(\bar{M}) = F \cap \mathrm{dom}(f)$. In particular, $f^{-1}(M) \subseteq f^{-1}(\bar{M}) \subseteq F$, and so $\overline{f^{-1}(M)} \subseteq \bar{F} = F$, and therefore, $\overline{f^{-1}(M)} \cap \mathrm{dom}(f) \subseteq$

$F \cap \mathrm{dom}(f) = f^{-1}(\bar{M})$. Conversely, if $\overline{f^{-1}(M)} \cap \mathrm{dom}(f) \subseteq f^{-1}(\bar{M})$ holds for all sets $M \subseteq \mathbb{R}$, then, if F is any closed subset of $\mathbb{R}$, $\overline{f^{-1}(F)} \cap \mathrm{dom}(f) \subseteq f^{-1}(\bar{F}) = f^{-1}(F) \subseteq \overline{f^{-1}(F)} \cap \mathrm{dom}(f)$, which shows that $f^{-1}(F)$ is relatively closed in $\mathrm{dom}(f)$.

37. Use the fact that $\mathbb{N}$ is a discrete subset of $\mathbb{R}$.

39. Let $a \in M$ and let $\varepsilon > 0$. Then there exists $\delta > 0$ such that, for $x \in \mathrm{dom}(f)$, $|x - a| < \delta \Rightarrow |f(x) - f(a)| < \varepsilon$. Consequently, for $x \in \mathrm{dom}(f|_M) = M \subseteq \mathrm{dom}(f)$, we have $|x - a| < \delta \Rightarrow |f|_M(x) - f|_M(a)| < \varepsilon$.

PROBLEM SET 4.4, PAGE 274

1. $\bar{A} \cap B = (A \cup A^*) \cap B = (A \cap B) \cup (A^* \cap B) = A \cap B$ since $A^* \cap B = \varnothing$.

3. Let $A \cap B = \varnothing$. Then $A \subseteq \mathbb{R} \setminus B$, and so $\bar{A} \subseteq \overline{\mathbb{R} \setminus B}$; hence, by Problem 23a in Problem Set 4.1, $A \cap bd(B) = A \cap \bar{B} \cap (\mathbb{R} \setminus B) = A \cap \bar{B}$. Likewise, $bd(A) \cap B = \bar{A} \cap B$. Now use Lemma 4.2.

5. (a) $\varnothing$ cannot be written as the union of two nonempty sets.
(b) A singleton set cannot be written as the union of two nonempty disjoint sets.

7. Prove that both a and b are accumulation points of (a, b) and that, if $c \in (a, b)$, $c \neq a$, and $c \neq b$, then c cannot be an accumulation point of (a, b).

9. If d is an isolated point of D, prove that $A = \{d\}$, $B = D \setminus \{d\}$ provides a separation of D.

11. Check the condition for convexity for each type of set.

13. Assume the hypotheses of Lemma 4.13 and let D be the union of all closed bounded intervals $[a, b]$ such that $c \in [a, b] \subseteq C$. Obviously, $D \subseteq C$. Choose a point $r \in C$ with $r \neq c$. If $r < c$, then $c \in [r, c] \subseteq D$, and, if $c < r$, then $c \in [c, r] \subseteq D$. In any case, $c \in D$. To prove that $C \subseteq D$, let $p \in C$. We have to prove that $p \in D$. We may assume that $p \neq c$. If $p < c$, then $p \in [p, c] \subseteq D$, and if $c < p$, then $p \in [c, p] \subseteq D$. In any case, $p \in D$.

15. Use Theorem 4.11.

17. Let C be a connected component of M. Then there exists $x \in M$ such that C is the union of all connected subsets of M that contain x. Suppose that $C \subseteq K \subseteq M$ and K is connected. Then K is one of the connected subsets of M that contain x, and it follows that $K \subseteq C$, so that $C = K$. Conversely, suppose that C is a maximal connected subset of M. If $C = \varnothing$, then $M = \varnothing$ and there is nothing to prove. Suppose $M \neq \varnothing$. We cannot have $C = \varnothing$ because any singleton subset of M is connected, so the empty set cannot be a maximal connected subset of M. Choose any point $x \in C$ and let D be the connected component of x in M. Since C is a connected subset of M and $x \in C$, we have $C \subseteq D \subseteq M$, and, since C is maximal, $C = D$.

19. (a) If $x \in C_1 \cap C_2$, then both C_1 and C_2 are the connected components of x in M, so $C_1 = C_2$.

21. (a) Let U be a nonempty open subset of $\mathbb{R}$, let $u \in U$, and let C be the connected component of u in U. To prove that C is open, let $c \in C$. Then $c \in U$ and, since U is open, there exists a neighborhood N of c with $N \subseteq U$. By Theorem 4.8, N is connected, and, by Theorem 4.11, $C \cup N$ is connected. But $C \subseteq C \cup N \subseteq U$, so, by Problem 17, $C = C \cup N$, and therefore, $N \subseteq C$.

23. The set A is nonempty, so we may choose a point $a \in A$. Since $A \subseteq f(C)$, we have $a \in f(C)$, so there exists $c \in C$ with $a = f(c)$. But then $c \in C \cap f^{-1}(A) = H$, so $H \neq \varnothing$. A similar argument shows that $K \neq \varnothing$.

25. Follow the hints.

27. Suppose, for instance, that $f(a) < 0 < f(b)$ and that $a < b$. Then $[a, b]$ is connected, so, by Theorem 4.17, there exists $x \in [a, b]$ such that $f(x) = 0$.

29. The polynomial f is continuous, $f(0) = -1$, and $f(1) = 1$. Use Corollary 4.18.

31. Let $f(x) = x^5 - 2x^3 - 9$, and apply Corollary 4.18.

33. Note that $g(0) \geq 0$ and $g(1) \leq 0$. If $g(0) = 0$, or if $g(1) = 0$, then 0 or 1 would be a fixed point of f. Otherwise, apply Corollary 4.18.

35. (a) Define $h:[0, 1] \to [0, 1]$ by $h(x) = [f(a + x(b - a)) - a]/(b - a)$. Prove that h is well-defined and continuous. Conclude by Problem 33 that h has a fixed point. Deduce from this that f has a fixed point.
(b) Consider the function $f:(0, 1) \to (0,1)$ defined by $f(x) = x^2$.

37. (a) Let $f:\mathbb{R} \to \mathbb{R}$ be defined by $f(x) = e^x - 3x$. Then $f(0) = 1 > 0$ and $f(1) = e - 3 < 0$. Use Corollary 4.18.
(b) $f(1) = e - 3 < 0$ and $f(2) = e^2 - 6 > 2.7^2 - 6 > 0$.
(c) No. Use methods of elementary calculus to sketch the graph.
(d) 0.6191 and 1.5121.

39. (a) Use mathematical induction on the number of points in F.
(b) Use Corollary 4.18.

PROBLEM SET 4.5, PAGE 289

1. To prove that $\mathscr{C}$ covers $(0, 1)$, suppose that $x \in (0, 1)$, and let $\varepsilon = (1 - x)/2$. Suppose that $\mathscr{F} = \{(0, 1 - \varepsilon)|\varepsilon \in F\}$ is a finite subcover of $\mathscr{C}$, let ε_0 be the minimum of all the positive numbers in the finite set F, and let $x = 1 - \varepsilon_0/2$. show that $x \in (0, 1)$, but that $x \notin (0, 1 - \varepsilon)$ for $\varepsilon \in F$.

3. Let $\mathscr{C} = \{(n - \frac{1}{2}, n + \frac{1}{2})|n \in \mathbb{N}\}$.

5. Take an open cover $\mathscr{C}$ of the union. For each of the compact subsets, reduce $\mathscr{C}$ to a finite subcover and then take the union of all the resulting subcovers.

7. If $\mathscr{C}$ is a collection of relatively open subsets of M, there is a collection $\mathscr{C}^*$ of open subsets of $\mathbb{R}$ such that $\mathscr{C} = \{M \cap U | U \in \mathscr{C}^*\}$. For each subcollection of $\mathscr{F}$ of $\mathscr{C}$ there is a corresponding subcollection of $\mathscr{F}^*$ of $\mathscr{C}^*$ such that $\mathscr{F} = \{M \cap U | U \in \mathscr{F}^*\}$. Moreover, $\mathscr{F}$ covers M if and only if $\mathscr{F}^*$ covers M.

9. The boundary of a set is closed (Problem 23b in Problem Set 4.1), and a closed subset of a compact set is compact (Lemma 5.6).

11. The argument is almost word for word the same.

13. I is compact, nonempty, and connected; hence, $f(I)$ is compact, nonempty, and connected. Since $f(I)$ is connected and nonempty, it is either a singleton set or an interval. If $f(I)$ is not a singleton set, it is a compact interval; therefore, it is a closed bounded interval.

15. Since dom(f) is compact, it is closed. Since M is closed, $M \cap \text{dom}(f)$ is a closed, hence, compact, subset of dom(f). Hence, $f(M) = f(M \cap \text{dom}(f))$ is compact, and therefore, it is closed.

17. To prove that $f^{-1}:\gamma \to X$ is continuous, it is sufficient to show that, for every closed subset M of $\mathbb{R}$, $(f^{-1})^{-1}(M) = f(M)$ is relatively closed in $Y = \text{dom}(f^{-1})$ (Problem 33 in Problem Set 4.3). By Problem 15, $f(M)$ is closed in $\mathbb{R}$, and therefore it is relatively closed in Y.

21. With the agreement that ε and δ represent positive numbers:
(a) $(\forall x)(\forall \varepsilon)(\exists \delta)(\forall y)((x, y \in D \wedge |x - y| < \delta) \Rightarrow |f(x) - f(y)| < \varepsilon)$
(b) $(\forall \varepsilon)(\exists \delta)(\forall x)(\forall y)((x, y \in D \wedge |x - y| < \delta) \Rightarrow |f(x) - f(y)| < \varepsilon)$
(c) By the theory of quantifiers in predicate calculus:
$(\forall x)(\forall \varepsilon)(\exists \delta)(\forall y)(\ldots) \Rightarrow (\forall \varepsilon)(\forall x)(\exists \delta)(\forall y)(\ldots) \Rightarrow (\forall \varepsilon)(\exists \delta)(\forall x)(\forall y)(\ldots)$.

23. $(\exists \varepsilon)(\forall \delta)(\exists x)(\exists y)(x, y \in D \wedge |x - y| < \delta \wedge |f(x) - f(y)| \geq \varepsilon)$.

25. Follow the hint.

27. Let $\varepsilon > 0$ and let $\delta = \min(1, \varepsilon/3)$. Then, for $x, y \in [0, 1]$ with $|x - y| < \delta$, $|x^2 - y^2| = |x - y||x + y| < \delta|x + y| \leq \delta(|x| + |y|) \leq \delta(|y| + |x - y| + |y|) = \delta(2|y| + |x - y|) \leq \delta(2|y| + 1) \leq \delta(2 + 1) < \varepsilon$.

29. For a given ε, the same δ that works on D will work for every subset of D.

31. False. The function $f:\mathbb{R} \to \mathbb{R}$ is given by $f(x) = x$ for all x is uniformly continuous on $\mathbb{R}$, but the function $g = f \cdot f$ is not (Problem 25).

33. Follow the hint in Part (d).

35. g is uniformly continuous on C by Theorem 5.18; hence, g is uniformly continuous on D by Problem 29.

37. Since f is continuous, $f^{-1}(M)$ is relatively closed in C. Since C is compact, it is closed, so $f^{-1}(M)$ is a closed subset of C. But, a closed subset of a compact set is compact.

PROBLEM SET 4.6, PAGE 304

1. If $\#M = \#N = n$, there are bijections $g:I_n \to M$ and $h:I_n \to N$. Let $f = h \circ g^{-1}$. Conversely, suppose that $f:M \to N$ is a bijection and let $\#M = n$. Then there exists a bijection $g:I_n \to M$. Define $h:I_n \to N$ by $h = f \circ g$, and prove that h is a bijection.

3. (a) Suppose, on the contrary, that $\mathbb{N}$ is a finite set. Then there exists $n \in \mathbb{N}$ and a bijection $f:I_n \to \mathbb{N}$. Then $f(I_n)$ is a nonempty subset of $\mathbb{N}$. By mathematical induction on n, prove that $f(I_n)$ has an upper bound b in $\mathbb{N}$, so that $b + 1 \notin f(I_n)$, and thus conclude that f is not surjective.
   (b) Note that $\{1/n \mid n \in \mathbb{N}\} \subseteq [0, 1]$ and use the result of Part (a).

5. (a) Let $f:L \to L$ be defined by $f(x) = x$ for all $x \in L$.
   (b) If $f:L \to M$ is a bijection, then $f^{-1}:M \to L$ is a bijection.
   (c) If $f:L \to M$ and $g:M \to N$ are bijections, then $g \circ f:L \to N$ is a bijection.

7. Suppose that $n, m \in \mathbb{N}$ and that $f(n) = f(m)$. Show that $f(n) \le 0 \Leftrightarrow n$ is odd and that, $f(n) > 0 \Leftrightarrow n$ is even. Conclude that n and m are either both even or both odd. Supposing that n and m are both even, we have $n/2 = m/2$, so $n = m$. Make a similar argument for the case in which n and m are both odd. Thus, f is injective. To prove that it is surjective, let $k \in \mathbb{Z}$. Show that if $k > 0$, then $f(2k) = k$ and, if $k \le 0$, then $f(1 - 2k) = k$.

9. If X is finite, then $\mathcal{P}(X)$ is finite by Theorem 5.14 in Chapter 1. Conversely, suppose $\mathcal{P}(X)$ is finite, let $\mathcal{S} = \{\{x\} \mid x \in X\}$, show that $\#X = \#\mathcal{S}$, and, using the fact that $\mathcal{P}(X)$ is finite, conclude that $\mathcal{S}$, and therefore also X, is finite.

11. $s_n = 2^{n-1}$ **13.** Follow the hints.

15. The restriction of an injection to a subset of its domain remains an injection. That $g|_J$ is a surjection follows from the fact that $X = g(J)$.

17. (a) If $h:M \to \mathbb{N}$ is an injection, then M is countable by Theorem 6.13. Conversely, suppose M is countable. If M is empty, then the empty set, regarded as a set of ordered pairs, is an injection from M into $\mathbb{N}$. If M is finite and nonempty there exists a bijection $f:I_n \to M$, so $f^{-1}:M \to \mathbb{N}$ is an injection. On the other hand, if M is infinite, then, by Definition 6.7, there exists a bijection $f:\mathbb{N} \to M$, so $f^{-1}:M \to \mathbb{N}$ is a bijection, hence, an injection.
   (b) If $f:\mathbb{N} \to M$ is a surjection, then M is countable by Theorem 6.15. Conversely, suppose M is countable and nonempty. Choose and fix $k \in M$. If M is finite, there exists a bijection $g:I_n \to M$, and we can define a surjection $f:\mathbb{N} \to M$ by $f(m) = g^{-1}(m)$ if $m \in I_n$ and $f(m) = k$ if $m \notin I_n$. On the other hand, if M is infinite, there exists a bijection, hence, a surjection, from $\mathbb{N}$ onto M.

19. That the mapping is an injection follows from the fact that p and q are injections and that two ordered pairs are equal if and only if their corresponding entries are equal.

23. Suppose that $i, k \in J$ and that $D_i \cap D_k \ne \varnothing$. Then there exists $(j, p) \in D_j$ and $(k, q) \in D_k$ such that $(j, p) = (k, q)$. In particular, then, $j = k$.

25. Let $y \in \bigcup_{j \in J} C_j$. Then there exists $j \in J$ such that $y \in C_j$. Thus, $(j, y) \in D_j$ and $f((j, y)) = y$.

27. (a) By definition, I is of one of $[a, b]$, $[a, (a + b)/2]$, or $[(x + b)/2, b]$ with $a \le x < b$. In any case, I is a closed bounded interval.
   (b) In each of the three cases in Part (a), we have $I \subseteq [a, b]$.
   (c) Again, look at each case.

29. By hypothesis, there exist $a, b \in C$ with $a < b$ and, since C is connected, $[a, b] \subseteq C$ by Theorem 4.15. From here on, the argument is almost word for word the same as the proof of Lemma 6.22.

31. Let $J = \{1, 2\}$, $C_1 = A$, $C_2 = B$, and consider the family of sets $(C_j)_{j \in J}$, noting that $M = A \cup B = \bigcup_{j \in J} C_j$. If A and B are both countable, then M is countable by Theorem 6.21.

33. Follow the hint, noting that $\mathbb{R} = \mathbb{Q} \cup (\mathbb{R} \setminus \mathbb{Q})$.

PROBLEM SET 4.7, PAGE 319

1. Use a pick-a-point argument to show that $(x, y) \in h \Rightarrow (x, y) \in f$ or $(x, y) \in g$ or the converse.

3. Let $f(x) = ax + b$ and $g(x) = cx + d$ for all $x \in \mathbb{R}$. Then $(f \circ g)(x) = f(g(x)) = ag(x) + b = a(cx + d) + b = (ac)x + (ad + b)$ for all $x \in \mathbb{R}$.

5. Begin by using the fact that multiplication by -1 reverses inequalities to prove that, for $a, b \in \mathbb{R}$, $\max(a, b) = -\min(-a, -b)$.

7. (a) Use the fact that, for $a \in \mathbb{R}$, $|a| = (a^2)^{1/2}$.
   (b) Begin by proving that, for $a, b \in \mathbb{R}$, $\max(a, b) = (a + b + |a - b|)/2$. To do this, consider separately the two cases $a \le b$ and $a > b$.

9. Let $f(x) = ax + b$ and $g(x) = cx + d$ for all $x \in \mathbb{R}$. Let $A, B \in \mathbb{R}$. Then $(Af + Bg)(x) = Af(x) + Bg(x) = Aax + Ab + Bcx + Bd = (Aa + Bc)x + (Ab + Bd)$ for all $x \in \mathbb{R}$.

11. By using standard results from elementary calculus, we know that a constant multiple of a continuous function is continuous and that the sum of two continuous functions is continuous.

13. The product of continuous functions is continuous and the constant function $u(x) = 1$ for all $x \in X$ is continuous. Therefore, $C(X)$ is closed under multiplication. That it forms a commutative ring with unity u is proved by checking the conditions in Theorem 2.2 of Chapter 3.

15. Prove that the sum and product of polynomial functions is again a polynomial function and that a constant multiple of a polynomial function is a polynomial function, then proceed as in Problem 13.

17. Suppose $a, b \in \text{dom}(f)$ and $a \ne b$. Then either $a < b$, in which case $f(a) < f(b)$, or else $b < a$, in which case $f(b) < f(a)$. In either case, $f(a) \ne f(b)$. Suppose that $c, d \in \text{range}(f)$ with $c < d$. Since $c, d \in \text{range}(f)$, there exist $a, b \in \text{dom}(f)$ such that $f(a) = c$ and $f(b) = d$. We cannot have $a \ge b$, otherwise we would have $c \ge d$, contradicting $c < d$; hence, $a < b$. Therefore, $f^{-1}(c) = a < b = f^{-1}(d)$.

19. Note that, because f is an injection, if $x, y \in \text{dom}(f)$ with $x \ne y$, then either $f(x) < f(y)$ or else $f(y) < f(x)$. Because f is not increasing, there exist $p, q \in C$ such that $p < q$ and $f(q) < f(p)$. Because f is not decreasing, there exist $r, s \in C$ such that $r < s$ and $f(r) < f(s)$. If $f(q) \le f(r)$, let $u = q$; otherwise, let $u = r$. Thus, $f(u) \le f(q)$, $f(r)$, and $f(u) < f(p)$, $f(s)$. In particular, $u \ne p$, s. If $f(p) \le f(s)$, let $v = s$; otherwise, let $v = p$. Thus, $f(p)$, $f(s) \le f(v)$ and $f(q)$, $f(r) < f(v)$. In particular, $v \ne q$, r. Since $f(u) < f(v)$, we cannot have $u = v$. Either $u < v$, or else $v < u$. Suppose first that $u < v$. If $p < u$, we have $p < u < v$ and $f(u) < f(p)$, $f(v)$, and we are done. If $u < p$, we have $u < p < q$ with $f(u)$, $f(q) < f(p)$, and again we are done. In the only remaining case, we have $v < u$. If $r < v$, then we have $r < v < u$ with $f(r)$, $f(u) < f(v)$, and we are done. If $v < r$, we have $v < r < s$ with $f(r) < f(v)$, $f(s)$, and again we are done.

21. Follow the hint.

23. There are two approaches; either make an argument similar to the proof of Lemma 7.4 or let $g(x) = -f(-x)$ and apply Lemma 7.4 to g.

27. If $f = g - h$ where g, h are increasing, then $-f = h - g$. Also, if a is a nonnegative constant, then ag, ah are increasing and $af = ag - ah$. Thus, a constant multiple of a function of bounded variation is again of bounded variation. Prove that the sum of two functions of bounded variation is again of bounded variation by using the fact that the sum of two increasing functions is increasing.

29. Use Theorem 7.7 and the fact that the union of two countable sets is countable.

31. To prove that g is increasing, suppose that $a, b \in C$ with $a < b$. Let $A = \{f(x) \mid x \in C, a < x\}$, $B = \{f(x) \mid x \in C, b < x\}$. By Theorem 7.5, $g(a) = \inf(A)$ and $g(B) = \inf(B)$. Since $a < b$, it follows that $B \subseteq A$, and therefore that $\inf(A) \le \inf(B)$. To prove that $g(a) = \inf_{a<x} g(x) = \lim_{x \to a^+} g(x)$, use Theorem 7.5 to conclude that $g(a) \le \inf_{a<x} g(x) = \lim_{x \to a^+} g(x)$, then argue that $g(a)$ cannot be smaller than $\inf_{a<x} g(x)$.

CHAPTER 5　PROBLEM SET 5.1, PAGE 338

1. Suppose that $w(x) = C$ for all $x \in \mathbb{R}$. Then f approaches zero at a more rapidly than w does if and only if $\lim_{x \to a}(1/C)f(x) = 0$, that is, if and only if $(1/C)\lim_{x \to a} f(x) = 0$.

3. By Lemma 1.2 with $w(x) = x - a$ for all $x \in \mathbb{R}$ and $\mathscr{E}(x) = f(x) - \alpha(x)$ for all $x \in \text{dom}(f)$, the linear function α is a local linear approximation to f at a if and only if there exists a function ε such that (i) $\text{dom}(\varepsilon) = \text{dom}(\mathscr{E}/w) \cup \{a\}$, (ii) $\mathscr{E}(x) = \varepsilon(x)(x - a)$ for all $x \in \text{dom}(\mathscr{E}/w) \setminus \{a\}$, (iii) $\varepsilon(a) = 0$, and (iv) $\lim_{x \to a} \varepsilon(x) = 0$. Note that $\text{dom}(\mathscr{E}/w) = \text{dom}(\mathscr{E}) \setminus \{a\} = \text{dom}(f) \setminus \{a\}$, so that $\text{dom}(\mathscr{E}/w) \cup \{a\} = \text{dom}(f)$.

5. Note that $\text{dom}(w) = \text{dom}(|w|)$ and that $\text{dom}(\mathscr{E}/w) = \text{dom}(\mathscr{E}/|w|)$. We have to prove that $\lim_{x \to a} \dfrac{\mathscr{E}(x)}{w(x)} = 0$ if and only if $\lim_{x \to a} \dfrac{\mathscr{E}(x)}{|w(x)|} = 0$. Write out these two conditions in terms of ε and δ and note that $\left| \dfrac{\mathscr{E}(x)}{w(x)} \right| = \left| \dfrac{\mathscr{E}(x)}{|w(x)|} \right|$.

7. Use the results of Problems 3 and 5.

9. Let $f(x) = C$ for all $x \in \text{dom}(f)$. By Definition 1.3, we have to find a constant m and a function ε with $\text{dom}(\varepsilon) = \text{dom}(f)$, $\varepsilon(a) = 0 = \lim_{x \to a} \varepsilon(x)$, and $f(x) = f(a) + m(x - a) + \varepsilon(x)(x - a)$ for all $x \in \text{dom}(f)$. Let $m = 0$ and define $\varepsilon : \text{dom}(f) \to \mathbb{R}$ by $\varepsilon(x) = 0$ for all $x \in \text{dom}(f)$, and show that these conditions hold.

11. As in Problem 9, let $m = 1$ and define $\varepsilon : \mathbb{R} \to \mathbb{R}$ by $\varepsilon(x) = 0$ for all $x \in \mathbb{R}$.

13. Only straightforward algebraic calculation is required here.

15. Follow the hint.

17. Use mathematical induction on the degree of the polynomial, the result of Problem 15, and the fact that the sum and product of differentiable functions are differentiable.

19. Only straightforward algebraic calculation is required here.

21. Follow the hint.　　**23.** Note that $f/g = f \cdot (1/g)$.

25. If n is a negative integer and $m = |n|$, note that $x^n = (1/x)^m$ and apply the results of Problem 21.

27. Use the result of Problem 17 and Theorem 1.14.

29. If $|x|$ were differentiable at 0, then, by Theorem 1.15, $\lim_{x \to 0}(|x|/x)$ would exist. But $|x|/x = 1$ for $x > 0$ and $|x|/x = -1$ for $x < 0$.

31. Use Corollary 1.17 and consider separately the two cases $x > 0$ and $x < 0$.

33. By the result of Problem 31 and the product rule, $f'(x) = 2|x|$ for all $x \ne 0$. Use Corollary 1.17 to show that $f'(0)$ exists and equals 0.

35. The sum of differentiable functions is differentiable and any constant multiple of a differentiable function is differentiable.

37. If f and g have continuous first derivatives, show that $f + g$ has a continuous first derivative. If f has a continuous first derivative and c is a constant, show that cf has a continuous first derivative.

PROBLEM SET 5.2, PAGE 353

1. Follow the hint.

3. Either make a direct proof, or apply the result in Part (i) of Corollary 2.4 to the function $-f$.

5. Use Lemma 2.6.

7. The absolute maximum, for instance, must occur either at an endpoint of the interval $[a, b]$ or at a point in the open interval (a, b). In the latter case, it occurs at a critical point in (a, b).

9. If f' fails to have a derivative at a, then, by definition, a is a critical point of the domain of f'. On the other hand, if f' is differentiable at a, use Theorem 2.8 applied to f' to conclude that $f''(a) = 0$.

11. Let $|h| \neq 0$ be sufficiently small so that $p + h \in (a, b)$. If $h > 0$, apply Theorem 2.10 to the interval $[p, p + h]$ and, if $n < 0$, apply it to the interval $[p + h, p]$. In either case, conclude that there exists c (depending on h), strictly between p and $p + h$, such that $f'(c)g(p + h) = g'(c)f(p + h)$. Because $g(p) = 0$ and $g'(x) \neq 0$ for x between p and $p + h$, Rolle's theorem implies that $g(p + h) \neq 0$. Therefore, $f'(c)/g'(c) = f(p + h)/g(p + h)$. Now let $h \to 0$, keeping in mind that c is "trapped" between p and $p + h$.

13. (a) $\lim_{x \to 0} \dfrac{\sin x}{x} = \lim_{x \to 0} \dfrac{\cos x}{1} = 1$

(b) The usual derivation of the formula for the derivative of $\sin x$ uses the fact that $\lim_{x \to 0} \dfrac{\sin x}{x} = 1$.

15. (a) 1   (b) 1/2   (c) $-1/7$

17. See any good elementary calculus textbook.

19. Let $f(x) = x^n$. By applying the mean value theorem to the interval $[1, 1 + x]$, we find that $\exists c \in (1, 1 + x)$ such that $(1 + x)^n - 1^n = nc^{n-1}x$. Since $c > 1$, we have $c^{n-1} > 1$, so $(1 + x)^n - 1 > nx$.

21. Apply the mean value theorem to the function $f(x) = \ln x$ on the interval $[1, x]$.

23. (a) Follow the hint.   (b) To make $|f(x) - f(y)| < \varepsilon$, just make sure that $|x - y| < \varepsilon/M$.

25. If f' is continuous on $[a, b]$, then it is bounded on $[a, b]$.

27. Look at the proof of Theorem 2.17.

29. Consider $f(x) = x^3$ on the interval $[-1, 1]$.

31. Assume that Conditions (i) and (ii) of Lemma 2.18 hold, let $b \in J$, and let $\varepsilon > 0$. Then there exist $\delta_1 > 0$ and $\delta_2 > 0$ such that $0 < b - x < \delta_1 \Rightarrow |g(x) - g(b)| < \varepsilon$ and $0 < x - b < \delta_2 \Rightarrow |g(x) - g(b)| < \varepsilon$. Let $\delta = \min(\delta_1, \delta_2)$ and show that $|x - b| < \delta \Rightarrow |g(x) - g(b)| < \varepsilon$. For the converse, note that, for a given $\varepsilon > 0$, the δ that works in the last implication will automatically work in the former two implications.

33. $\text{dom}(f^{-1}) = J$, an interval, and, since $y_0, b \in J$ with $y_0 < b$, it follows that $[y_0, b) \subseteq (-\infty, b) \cap J$. But b is an accumulation point of $[y_0, b)$, and therefore, it is an accumulation point of $(-\infty, b) \cap J$.

35. Follow the hint.

37. If n is even, then $D = [0, \infty)$ and, if n is odd, $D = \mathbb{R}$. Study these two cases separately. Show that, if $x \in D$ and $x \neq 0$, then $f'(x)$ exists and equals rx^{r-1}. Show that, if n is odd and $m > n$, then $f'(0)$ exists.

PROBLEM SET 5.3, PAGE 368

1. Use mathematical induction on n.

3. If $-b_n(x - a)^n$ is expanded by the binomial theorem, then the resulting term of highest degree is $-b_n x^n$, which cancels with the term $b_n x^n$ of highest degree in $p(x)$.

5. This is a straightforward application of mathematical induction.

7. Follow the hint.   **9.** Follow the hint.   **11.** Follow the hint.   **13.** Follow the hint.

15. Follow the hint.   **17.** Fix n and use mathematical induction on k for $k \leq n$.

19. Choose a closed bounded interval $[c, d] \subseteq N$ such that $a \in (c, d)$. Define $F : \text{dom}(f) \to \mathbb{R}$ by $F(x) = f(x) - p(x)$ for all $x \in \text{dom}(f)$. Define $G(x) = (x - a)^{n+1}$ for all $x \in \mathbb{R}$. Note that F and G are continuous on $[c, d]$ and differentiable on (c, d). Also, $G'(x) = (n + 1)(x - a)^n \neq 0$ for all $x \in (c, d)$ with $x \neq a$. Also, $F(a) = G(a) = 0$, so we can apply L'Hospital's rule to the functions F and G using the closed interval $[c, d]$ and the point $a \in (c, d)$.

21. Use Theorem 3.10.

23. Use the chain rule and prove that $(p \circ q)^{(k)}(b) = (f \circ q)^{(k)}(b)$ for all integers $k = 0, 1, 2, \ldots, n$.

25. (a) By the intermediate value theorem, f has at least one zero r in (d, a). Suppose that there were a second such zero, r^*, of f in (d, a). Then, by Rolle's theorem, $f'(c) = 0$ for some c between r and r^*, contradicting the fact that $f'(x) \geq m > 0$ for all $x \in (d, a)$.

   (b) Use Equation (5) in the proof of Theorem 3.11 and the fact that both $f'(a)$ and $f''(c)$ are positive to find that $b - r > 0$.

27. Let $f(x) = x^2 - c$ for $x \in \mathbb{R}$. By taking $a > 0$ to be a first approximation to a zero of f, Newton's method suggests that $b = a - f(a)/f'(a) = a - (a^2 - c)/2a = (a + c/a)/2$ is a better approximation to a zero of f.

29. (a) 0.93969262

   (b) Let $\theta = 20° = \pi/9$ radian. Then $\cos 2\theta = \cos^2 \theta - \sin^2 \theta$, $\sin 2\theta = 2 \sin \theta \cos \theta$, and $\cos 3\theta = \cos(\theta + 2\theta) = \cos \theta \cos 2\theta - \sin \theta \sin 2\theta$. By combining these equations and using the fact that $\sin^2 \theta = 1 - \cos^2 \theta$, we find that $\cos 3\theta = 4 \cos^3 \theta - 3 \cos \theta$. But $\cos 3\theta = \cos(\pi/3) = 1/2$, and it follows that $\cos \theta$ is a root of the polynomial $8x^3 - 6x - 1$.

31. (a) Apply Newton's method to f.

   (b) Given that $0 < a < 2c^{-1}$, show that $|a - c^{-1}| < c^{-1}$. Multiply the last inequality by $|1 - ac|$ and thus deduce that $|b - c^{-1}| < |a - c^{-1}|$.

PROBLEM SET 5.4, PAGE 385 _____

1. (a) Given $x \in [a, b]$, let i be the smallest integer k such that $x \leq x_k$, and conclude that $x \in [x_{k-1}, x_k]$.

   (b) Suppose for instance, that $i < j$. If $j = i + 1$, show that x_i is the only point that belongs to both $[x_{i-1}, x_i]$ and $[x_{j-1}, x_j]$; otherwise, show that these two intervals are disjoint.

   (c) Expand the sum and show that all terms except y_n and y_0 cancel in pairs.

   (d) Use the result of Part (c).

3. (a) If $|f(x)| \leq M$ for all $x \in [p, q]$, then $|f(x)| \leq M$ for all $x \in [r, s]$.

   (b) For every $x \in [r, s]$ we have $f(x) \leq \sup_{p \leq x \leq q} f(x)$; hence, $\sup_{r \leq x \leq s} f(x) \leq \sup_{p \leq x \leq q} f(x)$.

   (c) Either argue directly, or apply the result in Part (b) to $-f$.

5. (a) If P and Q are partitions of $[a, b]$, then P and Q are finite subsets of $[a, b]$, $a, b \in P$, and $a, b \in Q$. Thus, $P \cup Q$ is a finite subset of $[a, b]$ and $a, b \in P \cup Q$. Since $P, Q \subseteq P \cup Q$, it follows that $P \cup Q$ is a refinement of both P and Q.

   (b) $P, Q \subseteq R$ if and only if $P \cup Q \subseteq R$.   (c) Set-theoretic containment is transitive.

7. (a) By Lemma 4.6, we have $m(b - a) \leq L(f, P)$ for every partition P of $[a, b]$, and it follows that $m(b - a) \leq \sup_P L(f, P) = \underline{\int} f$.

   (b) The argument is similar to that in Part (a).

   (c) By Lemma 4.9, $\underline{\int} f \leq \overline{\int} f$ and, by Definition 4.8, for any partition P of $[a, b]$, $L(f, P) \leq \underline{\int} f$ and $\overline{\int} f \leq U(f, P)$.

9. (a) If $|f(x)| \leq M$ for all $x \in [a, b]$, then $|-f(x)| = |f(x)| \leq M$ for all $x \in [a, b]$.

   (b) Follow the hint.   (c) Follow the same hint.   (d)–(e) Use the results of Parts (a) and (b).

11. Follows immediately from the results in Problem 9.

13. (a) If $|f(x)| \leq M$ for all $x \in [a, b]$, then $|kf(x)| \leq kM$ for all $x \in [a, b]$.

   (b) Prove that $\inf_{x_{i-1} \leq x \leq x_i} kf(x) = km_i$ for $i = 1, 2, \ldots, n$ and use this to show that $L(kf, P) = kL(f, P)$.

   (c) Argue as in Part (b) or else apply Part (b) to $-f$.

   (d) Prove that $\sup_P kL(f, P) = k \cdot \sup_P L(f, P)$.

   (e) Argue as in Part (d) or else apply Part (d) to $-f$.

15. Suppose that $u > v$ and let $\varepsilon = u - v$.

17. If δ and n are positive numbers, then $(b - a)/\delta < n$ implies that $(b - a)/n < \delta$.

19. If ε and n are positive numbers, then $(b - a)[f(b) - f(a)]/\varepsilon < n$ implies that $(b - a)/n < \varepsilon/[f(b) - f(a)]$.

21. Since f is bounded on $[a, b]$, it follows that g is bounded on $[c, d]$. Let $\varepsilon > 0$. By Theorem 4.13, there exists a partition P of $[a, b]$ such that $U(f, P) - L(f, P) < \varepsilon$. Let $Q = P \cup \{c, d\}$, noting that Q is a partition of $[a, b]$ and that Q refines P. Use Corollary 4.5 to conclude that $U(f, Q) - L(f, Q) < \varepsilon$. Let $R = Q \cap [c, d]$. Prove that R is a partition of $[c, d]$ and show that $U(g, R) - L(g, R) \leq U(f, Q) - L(f, Q) < \varepsilon$.

23. Since f is bounded above, and since it is bounded below by 0, then it is bounded and, in Lemma 4.9, we have $0 \leq m$ and $0 < b - a$.

27. Follow the hint.   **29.** Follow the hint.   **31.** Follow the hint.

PROBLEM SET 5.5, PAGE 400

1. Follow the hint.

3. For each i, $m_i \Delta x_i \leq f(x_i^*) \Delta x_i \leq M_i \Delta x_i$.

5. Supposing that $\lambda \neq \lambda'$, let $\varepsilon = |\lambda - \lambda'|/2$, choose a δ for λ and a δ' for λ', and choose a partition P with $\|P\| < \min(\delta, \delta')$.

7. Follow the hint.   **9.** Follow the hint.

11. For each $i \in C'$, choose $j_i \in \{1, 2, 3, \dots, m - 1\} = I_{m-1}$ such that $y_{j_i} \in (x_{i-1}, x_i)$ and define $\phi: C' \to I_{m-1}$ by $\phi(i) = j_i$ for every $i \in C'$. Note that ϕ is an injection because, if $i \neq k$, then the intervals (x_{i-1}, x_i) and (x_{k-1}, x_k) are disjoint. Therefore, $\#C' \leq I_{m-1} = m - 1$.

13. If $x_i^* \neq c$ for $i = 1, 2, 3, \dots, n$, then $S(f, P, x^*) - S(g, P, x^*) = 0$. Thus, suppose that $x_k^* = c$ for some $k = 1, 2, \dots, n$. If $c \notin P$, or if $c = a$, or if $c = b$, then k is uniquely determined and $f(x_i^*) \Delta x_i = g(x_i^*) \Delta x_i$ holds for all $i = 1, 2, \dots, n$ except for $i = k$. If $c \in P$, but $c \neq a$ and $c \neq b$, then it is possible that $x_i^* = c$ for exactly two successive values $i = k$ and $i = k + 1$.

15. Argue as in the proof of Lemma 5.12.

17. Use mathematical induction and Lemma 5.16.

19. Consider these cases: (i) $a = b = c$; (ii) $a = b \neq c$; (iii) $a = c \neq b$; (iv) $a \neq b = c$; (v) $a < b < c$; (vi) $a < c < b$; (vii) $b < a < c$; (viii) $b < c < a$; (ix) $c < a < b$; (x) $c < b < a$. Show that these ten cases exhaust all possibilities, and prove that $\int_a^c f(x)\,dx = \int_a^b f(x)\,dx + \int_b^c f(x)\,dx$ in each case.

21. We sketch the argument for the case in which $d \in (a, b)$. Slight modifications of this argument take care of the cases in which $d = a$ or $d = b$. Let B be a positive number such that $|f(x)| \leq B$ for all $x \in [a, b]$. Let $\varepsilon > 0$ and let $\eta = \min(b - d, d - a, \varepsilon/(12B))$, noting that $d - \eta, d + \eta \in (a, b)$. Let $g = f|_{[a, d-\eta]}$ and $h = f|_{[d+\eta, b]}$. Then g and h are continuous, hence, integrable, on $[a, d - \eta]$ and $[d + \eta, b]$, respectively. By Theorem 4.13, there are partitions P and Q of $[a, d - \eta]$ and $[d + \eta, b]$, respectively, such that $U(g, P) - L(g, P) < \varepsilon/3$ and $U(h, Q) - L(h, Q) < \varepsilon/3$. Let $R = P \cup \{d - \eta, d + \eta\} \cup Q$, note that R is a partition of $[a, b]$, prove that $U(f, R) - L(f, R) = U(g, P) - L(g, P) + [\sup_{d-\eta \leq x \leq d+\eta} f(x) - \inf_{d-\eta \leq x \leq d+\eta} f(x)][(d + \eta) - (d - \eta)] + U(h, Q) - L(h, Q) < \varepsilon/3 + (2B)(2\eta) + \varepsilon/3 < \varepsilon/3 + \varepsilon/3 + \varepsilon/3 = \varepsilon$. Use Theorem 4.13 again to conclude that f is integrable on $[a, b]$.

23. Follow the hint.

25. This is just a way of rephrasing the result of Problem 23.

PROBLEM SET 5.6, PAGE 414

1. Use Problems 11 and 13 in Problem Set 5.4

3. $\int_a^b f(x)\,dx = A(b - a)$, and $A(b - a)$ is the area of the rectangle.

5. Use Theorem 6.5, the triangle inequality, Theorem 6.3, and Theorem 6.1.

7. The result is obvious if $a = b$. If $b < a$, apply Theorem 6.7 to the interval $[b, a]$ and use the fact that $\int_a^b f(x)\,dx = -\int_b^a f(x)\,dx$.

9. Follow the hint.

11. If $g(x) \leq 0$ for all x, use Theorem 6.9 with g replaced by $-g$.

13. If $g(x) \neq 0$ for $x \in [a, b]$, then, by the intermediate value theorem, either $g(x) > 0$ for all $x \in [a, b]$ or $g(x) < 0$ for all $x \in [a, b]$. In the former case, use Theorem 6.9; in the latter case, use Problem 11.

15. (a) Note that $g(1) = 1$, so $\sup_{0 \leq x \leq 1} g(x) = 1$.
    (b) There are at most $[\![2/\varepsilon]\!]$ positive integers n with $n \leq 2/\varepsilon$ and, for each such n, there are at most n positive integers m with $m \leq n$.

17. (a) Follow the hint.
    (b) By the Archimedian property of $\mathbb{R}$, there exists an integer K with $2N/\varepsilon < K$ and, since there are only K positive integers $p \leq K$, and there are infinitely many primes, there must be a prime $p > K$.

19. Since $P \cap A = \{0, 1\}$, it follows that $[x_{j-1}, x_j] \cap P = (x_{j-1}, x_j) \cap P$ for $j = 2, 3, \ldots, p-1$, $[0, x_1] \cap P = [0, x_1) \cap P$, and $[x_{p-1}, 1] \cap P = (x_{p-1}, 1] \cap P$. But the intervals $[0, x_1), (x_2, x_3), \ldots, (x_{p-2}, x_{p-1}), (x_{p-1}, 1]$ are pairwise disjoint; hence, we cannot have $a_j = a_k$ unless $j = k$.

21. (a) Since $S \subseteq [-K, K]$, it follows that $\sup(S) \leq K$ and that $-K \leq \inf(S)$; hence, $\sup(S) - \inf(S) \leq K + K = 2K$.
    (b) Follow the hint.

23. (a) We have $m \leq y \leq M$ and $-M \leq -y' \leq -m$, so $m - M \leq y - y' \leq M - m$, that is, $-(M - m) \leq y - y' \leq M - m$.

25. (a) Suppose, on the contrary, that $M - m > \eta$ and let $\varepsilon = (M - m) - \eta$. Choose $s \in S$ such that $s < m + \varepsilon/2$ and choose $s' \in S$ such that $M - \varepsilon/2 < s'$. Then $\varepsilon = (M - m) - \eta \leq (M - m) + s - s' = (M - s') + (s - m) < \varepsilon/2 + \varepsilon/2 = \varepsilon$, a contradiction.

27. Use Theorem 6.12 with g replaced by f and with f replaced by the square root function.

29. Use Theorem 6.12 with g replaced by f and with f replaced by the function $h(x) = 1/x$ for $x \neq 0$.

PROBLEM SET 5.7, PAGE 428

1. Assume that f is integrable on $[c, a]$ for every $c \in I$ with $c < a$ and on $[a, b]$ for every $b \in I$ with $a < b$, then, by Problem 29 in Problem Set 5.4, f is integrable on $[c, b]$ for every $c, b \in I$ with $c < a < b$. Now suppose that $p, q \in I$ with $p < q$. We have to prove that f is integrable on $[p, q]$. If $p = a$, or $q = a$, or $p < a < q$, then we are done. If $a < p < q$, then, since f is integrable on $[a, q]$, it is integrable on $[p, q]$ by Problem 21 in Problem Set 5.4. A similar argument applies in the only remaining case, that in which $p < q < a$.

3. Suppose, for instance, that b is a right endpoint of the interval I. Then there is an $\eta > 0$ such that $[b - \eta, b] \subseteq I$. Since f is integrable on $[b - \eta, b]$, it is bounded on $[b - \eta, b]$ and so there exists $K \in \mathbb{R}$ such that $|f(t)| \leq K$ for all $t \in [b - \eta, b]$. Let $\delta = \min(\varepsilon/K, \eta)$ and prove that, for $x \in I$ with $|x - b| < \delta$, $x \in (b - \eta, b]$. From this, prove that $|F(x) - F(b)| < \varepsilon$.

5. Suppose that $0 < |h| < \delta$ and that t is between c and $c + h$. If $h \geq 0$, then $c \leq t \leq c + h$, so $|t - c| = t - c \leq h = |h| < \delta$. If $h < 0$, then $c + h \leq t \leq c$, and a similar argument applies.

7. (a) For $x \neq 0$, F is differentiable and $F'(x) = 2x \sin(1/x) - \cos(1/x)$. Also,
$$F'(0) = \lim_{h \to 0} \frac{h^2 \sin(1/h) - 0}{h} = \lim_{h \to 0} h \sin(1/h) = 0, \text{ since } |\sin(1/h)| \leq 1 \text{ for all } h \neq 0.$$
    (b) For $x \neq 0$, $f(x) = 2x \sin(1/x) - \cos(1/x)$ and $\lim_{x \to 0} 2x \sin(1/x) = 0$, but $\cos(1/x)$ has no limit as $x \to 0$; hence $f(x)$ can have no limit as $x \to 0$.
    (c) Follow the hint.

9. (a) By Example 7.9, $(d/dx) \ln x = 1/x > 0$ for $x > 0$, so ln is a strictly increasing function on $(0, \infty)$ by Theorem 2.17.
    (b) $\ln 1 = \int_1^1 (1/t) \, dt = 0$ by Part (iii) of Definition 5.18.

11. Replace f in Theorem 7.14 by f', noting that f is an antiderivative of f' on $[a, b]$.

13. f' is not defined at 0 and it is unbounded on $[-1, 0) \cup (0, 1]$.

15. The mean value of f' on $[a, b]$ is given by $A = \dfrac{1}{b-a} \int_a^b f'(x)\, dx = \dfrac{f(b) - f(a)}{b - a}$. By Theorem 2.11, there exists $c \in (a, b)$ such that $f(b) - f(a) = f'(c)(b - a)$; hence, $A = f'(c)$.

17. Use the fact that $\int_x^b f(t)\, dt = -\int_b^x f(t)\, dt$.

19. If n is even, let $F(x) = x^{n+1}/(n + 1)$. If n is odd, let $F(x) = x^n|x|/(n + 1)$.

21. Follow the hint. **23.** Follow the hint. **25.** 5956/35 **27.** (a) 8/3 (b) 8/3

CHAPTER 6 PROBLEM SET 6.1, PAGE 448

1. If $(s_n)_{n \in \mathbb{N}}$ is eventually in A, there exists $N \in \mathbb{N}$ such that, for all $n \in \mathbb{N}$, $n \geq N \Rightarrow s_n \in A$. Thus, by contra-position, $s_n \notin A \Rightarrow n < N$, and so $\# \{n \in \mathbb{N} \mid s_n \notin A\} \leq N - 1$. Conversely, suppose that $\# \{n \in \mathbb{N} \mid s_n \notin A\} \leq M$. Then, for all $n \in \mathbb{N}$, $n \geq M + 1 \Rightarrow s_n \in A$.

3. Use the result of Problem 1, Definition 1.1, and the fact that $|s_n - L| < \varepsilon$ if and only if $s_n \in (L - \varepsilon, L + \varepsilon)$.

5. Let m, n, N denote natural numbers. Then, if $n \geq N \Rightarrow |a_n - L| < \varepsilon$, it follows that $n \geq N \Rightarrow |a_{n+k} - L| < \varepsilon$. Conversely, if $n \geq N \Rightarrow |a_{n+k} - L| < \varepsilon$, it follows that $m \geq N + k \Rightarrow |a_m - L| \leq \varepsilon$.

7. Use mathematical induction.

9. Either make a similar argument or else apply what has already been proved to the sequence $(-s_n)_{n \in \mathbb{N}}$.

11. Use mathematical induction to prove that $r^n \geq r^{n+1}$ for all $n \in \mathbb{N}$, and then use the result in Part (b) of Problem 7.

13. (b) Combine Lemma 2.11 in Chapter 4 with Lemma 1.4.

15. Follow the hint. **17.** Follow the hint.

19. Let $|a_n| < A$ for every $n \in \mathbb{N}$ and let $\varepsilon > 0$. Choose $N \in \mathbb{N}$ such that, for all $n \in \mathbb{N}$, $n \geq N \Rightarrow |b_n - 0| < \varepsilon/A$. Then show that $n \geq N \Rightarrow |a_n b_n - 0| < \varepsilon$.

21. Prove that $\{n \in \mathbb{N} \mid s_n \in A\}$ is a finite set if and only if $(s_n)_{n \in \mathbb{N}}$ is not frequently in A.

23. If $(s_n)_{n \in \mathbb{N}}$ is eventually in A, there exists $N \in \mathbb{N}$ such that, for all $n \in \mathbb{N}$, $n \geq N \Rightarrow s_n \in A$. Let $k \in \mathbb{N}$ and let $n = \max(k, N)$. Then $n \geq k$ and $s_n \in A$.

25. Let $s_n = b_n - a_n$ for every $n \in \mathbb{N}$, noting that $s_n \geq 0$ and that $\lim_{n \to \infty} s_n = \lim_{n \to \infty} b_n - \lim_{n \to \infty} a_n$. Prove that $\lim_{n \to \infty} s_n \geq 0$. [Hint: Suppose that $\lim_{n \to \infty} s_n < 0$ and let $\varepsilon = -\frac{1}{2} \lim_{n \to \infty} s_n$.]

27. Every neighborhood of L is an open set and every open set containing L contains a neighborhood of L.

29. Suppose that $(s_n)_{n \in \mathbb{N}}$ satisfies the given condition and let $\varepsilon > 0$. Then there exists $N \in \mathbb{N}$ such that, for every integer $n \geq N$, $|s_n - s_N| < \varepsilon/2$. Consequently, if $n, m \in \mathbb{N}$ with $n, m \geq N$, then $|s_n - s_m| \leq |s_n - s_N| + |s_m - s_N| < \varepsilon$. The converse argument is even easier.

31. Suppose that $\lim_{n \to \infty} s_n = L$. Choose $N \in \mathbb{N}$ such that, for all $n \in \mathbb{N}$, $n \geq N \Rightarrow |s_n - L| < 1$. Then $n \geq N \Rightarrow L - 1 < s_n < L + 1$. Let $A = \min\{L - 1, s_1, s_2, \ldots, s_{N-1}\}$ and let $B = \max\{L + 1, s_1, s_2, \ldots, s_{N-1}\}$. Then, for all $n \in \mathbb{N}$, $A \leq s_n \leq B$.

33. Let $a_1 = s_1$ and, for $n \in \mathbb{N}$ with $n > 1$, let $a_n = s_n - s_{n-1}$.

35. Follow the hint.

37. If $a_1 = a$ and if $a_{n+1} = a_n r$ holds for all $n \in \mathbb{N}$, use mathematical induction to prove that $a_k = ar^{k-1}$ for all $k \in \mathbb{N}$.

39. (a) Let $\varepsilon = |r| - 1$, so that $\varepsilon > 0$ and $|r| = 1 + \varepsilon$. Prove by mathematical induction that $(1 + \varepsilon)^n \geq 1 + n\varepsilon$ for all $n \in \mathbb{N}$ and thus use the Archimedian property of $\mathbb{R}$ to conclude that the sequence $(r^n)_{n \in \mathbb{N}}$ is unbounded. Then use the result of Problem 31. [For an alternative proof, suppose that $\lim_{n \to \infty} r^n = L$, argue that $L > 0$, conclude that $\lim_{n \to \infty} (1/r)^n = 1/L \neq 0$, and thus contradict the result of Example 1.10 (with r replaced by $1/r$).]

PROBLEM SET 6.2, PAGE 461

1. Follow the hint.
3. Use Theorem 2.2, the fact that the cosine is continuous, and the fact that $\pi/n \to 0$ as $n \to \infty$.
5. The converse of Theorem 2.2 is not a theorem.
7. (a) Apply Theorem 1.11 to the sequences of partial sums.
9. Supposing that $\sum_{k=1}^{\infty}(a_k + b_k)$ converges, then the fact that $\sum_{k=1}^{\infty}(-a_k)$ converges would imply that $\sum_{k=1}^{\infty} b_k$ converges.
11. In Part (i), the sequence of partial sums is bounded and increasing. In Part (ii), the sequence of partial sums is increasing and convergent. Since it is convergent, it is bounded (Problem 31 in Problem Set 6.1); hence, by Theorem 1.9, its limit is the supremum of all its terms. In particular, each term is less than or equal to this limit.
13. Term-by-term, the series are identical.
15. $\displaystyle\sum_{k=1}^{\infty} \frac{1}{\ln(k+1)}$ 17. Follow the hint.   19. Follow the hint.
21. We have $0 \le a_n/b_n \le M$ for all $n \in \mathbb{N}$; hence, $0 \le a_n \le Mb_n$ for all $n \in \mathbb{N}$. But $\sum_{k=1}^{\infty} Mb_n = M \sum_{k=1}^{\infty} b_n$ is a convergent series, so $\sum_{k=1}^{\infty} a_k$ converges by the comparison test.
23. Follow the hint.
25. We have $f(x) \le f(1)$ for all $x \in [1, 2]$; hence, by Corollary 6.4 in Chapter 5, $\int_1^2 f(x)\, dx \le f(1) = \sum_{k=1}^{1} f(k) = f(1) + \int_1^1 f(x)\, dx$.
27. Suppose that $L = \lim_{n \to \infty} \int_1^n f(x)\, dx$ exists. Then the partial sums of the series $\sum_{k=1}^{\infty} f(k)$ form an increasing sequence that is bounded above by $f(1) + L$; hence, the series converges and its sum does not exceed $f(1) + L$. Also, $L = \lim_{n \to \infty} \int_1^{n+1} f(x)\, dx$, and so $L \le \sum_{k=1}^{\infty} f(k)$. The converse is proved in a similar way.
29. Follow the hint.
31. Use the fact that $\int \ln x\, dx = \ln x - x$.
33. Use the integral test and integration by parts.

PROBLEM SET 6.3, PAGE 475

1. Modify the given proof by replacing appropriate inequalities by strict inequalities.
3. Proceed as in Problem 1.
5. Let $\varepsilon > 0$ and choose positive integers N and M such that, for all $n \in \mathbb{N}$, $n \ge N \Rightarrow |s_{2n} - L| < \varepsilon$ and $n \ge M \Rightarrow |s_{2n-1} - L| < \varepsilon$. Let $K = \max(2N, 2M - 1)$ and suppose that $k \in \mathbb{N}$ with $k \ge K$. If k is even, then $k/2 \ge N$, and it follows that $|s_k - L| < \varepsilon$. If k is odd, then $(k + 1)/2 \ge M$, and it follows again that $|s_k - L| < \varepsilon$.
7. We must use n terms, where $1/(2n + 2)! \le 0.5 \times 10^{-4}$, that is, $20{,}000 < (2n + 2)!$. By trial and error, we find that the smallest number k for which $20{,}000 < k!$ is $k = 8$; hence, we can take $n = 3$. Thus, to four decimal places, the sum of the series is approximately $(1/2!) - (1/4!) + (1/6!) \approx 0.4597$.
9. Use the fact that $\left|\sum_{k=m}^{M+q} a_k\right| \le \sum_{k=m}^{M+q} |a_k| = \left|\sum_{k=m}^{M+q} |a_k|\right|$.
11. Follow the hint, then use comparison with a p-series for $p = 3$.
13. (a) Let $L = \lim_{n \to \infty} b_n$ and suppose that $L < r$. Choose $N \in \mathbb{N}$ such that, for all $n \in \mathbb{N}$, $n \ge N \Rightarrow |b_n - L| < (r - L)/2$.
15. For instance, $(n + 1)^{-2}/n^{-2} = [n/(n + 1)]^2 \to 1$ as $n \to \infty$.

17. Use mathematical induction on j.

19. If $L > 1$, there exists $N \in \mathbb{N}$ such that, for all $n \in \mathbb{N}$, $n \geq N \Rightarrow a_{n+1} > a_n$. By mathematical induction, it follows that $n \geq N \Rightarrow a_n \geq a_N > 0$, from which we conclude that $\lim_{n \to \infty} a_n \neq 0$, so the series diverges by Corollary 2.4.

21. (a) Diverges by Corollary 2.4.

(b) Converges absolutely by application of the integral test to $\sum_{k=1}^{\infty} [(\ln k)/k^2]$.

(c) Converges absolutely by the root test.

(d) Converges conditionally by the alternating series test and application of the integral test to $\sum_{k=2}^{\infty} [1/(k \ln k)]$.

23. (a) Converges by ratio test.   (b) Converges by ratio test. [Note that $\lim_{n \to \infty} [n/(n+1)]^n = 1/e < 1$.]

(c) Converges by ratio test.

25. Use Lemma 3.9 with $c = 1$.

27. Apply what has already been proved to $(-b)_{n \in \mathbb{N}}$.

29. Begin by proving that $\sum_{k=1}^{\infty} kr^k$ converges.

31. Follow the hint.   **33.** Follow the hint.   **35.** Follow the hint.

PROBLEM SET 6.4, PAGE 491

1. (a) For each $x \in [0, 1]$ with $x \neq 0$, $\lim_{n \to \infty} f(x) = 0$, whereas $\lim_{n \to \infty} f_n(0) = 1$.

(b) Suppose that the convergence is uniform. Then there exists $N \in \mathbb{N}$ such that, for all $n \in \mathbb{N}$ and for all x with $0 < x \leq 1$, $n \geq N \Rightarrow 1/(nx + 1) < 1/2$; that is, $1 < nx$. In particular, $1 < Nx$ holds for all such x, which produces a contradiction upon setting $x = 1/2N$.

3. The sequence of functions converges uniformly to the constant zero function. To prove this, let $\varepsilon > 0$ and choose $N \in \mathbb{N}$ with $N > 1/\varepsilon$. Then, assuming that $n \in \mathbb{N}$ with $n \geq N$ and that $0 \leq x \leq 1$, we have $1 < n\varepsilon$, so $x < n\varepsilon x < n\varepsilon x + \varepsilon = (nx + 1)\varepsilon$, and it follows that $x/(nx + 1) < \varepsilon$.

5. (a) For each $x \in [0, 1]$, we have $\lim_{n \to \infty} [1/(1 + x^n)] = f(x)$, where $f(x) = 1$ for $x < 1$ and $f(1) = 1/2$.

(b) The function f is discontinuous on $[0, 1]$, so the convergence cannot be uniform by Corollary 4.7.

(c) Let $\varepsilon > 0$. With no loss in generality, we may assume that $0 < \varepsilon < 1$. Since $0 < b < 1$, there is an $N \in \mathbb{N}$ such that, for all $n \in \mathbb{N}$, $n \geq N \Rightarrow b^n < \varepsilon/(1 - \varepsilon)$. [Recall that $b^n \to 0$ as $n \to \infty$.] Therefore, if $0 < x < b$, we have $x^n < \varepsilon/(1 - \varepsilon)$, from which it follows that $|1/(1 + x^n) - 1| < \varepsilon$.

7. First take care of the case in which $A = 0$ or $B = 0$. Then, assuming that $A, B \neq 0$, let $\varepsilon > 0$ and choose $N \in \mathbb{N}$ such that, for all $n \in \mathbb{N}$ and all $x \in S$, $n \geq N \Rightarrow |f_n(x) - f(x)| < \varepsilon/2A$ and $|g_n(x) - g(x)| < \varepsilon/2B$.

9. By the triangle inequality, note that, if a, b, c, d are any four real numbers, then $|a - d| = |(a - c) + (c - d)| \leq |a - c| + |c - d| = |(a - b) + (b - c)| + |c - d| \leq |a - b| + |b - c| + |c - d|$.

11. Note that $|g_n(x) - 0| < \varepsilon$ if and only if $|f_n(x) - f(x)| < \varepsilon$.

13. (a) $nx(nx + x + 1) = n^2x^2 + nx^2 + nx < n^2x^2 + nx^2 + nx + x = (nx + 1)(nx + x)$, from which it follows that $f_n(x) < f_{n+1}(x)$ holds for all $x \in S = (0, 1)$.

(b) For each $x \in (0, 1)$, we have $\lim_{n \to \infty} f_n(x) = \lim_{n \to \infty} (x/[x + (1/n)]) = x/x = 1$.

(c) If the convergence were uniform, there would exist $N \in \mathbb{N}$ such that $|f_N(x) - 1| < 1/2$ for all $x \in S = (0, 1)$. Let $x = 1/2N$ and derive a contradiction.

(d) $S = (0, 1)$ is not compact.

15. The argument is similar to that already given to show that $U(f, P) \leq U(f_N, P) + \varepsilon/3$, except that inequalities must be reversed.

17. Follows directly from the definition of a Cauchy sequence.

19. (a) $|f_n(x)| = |x^n| = |x|^n$ and $x \in [-r, r]$ implies that $|x| \leq r$, so that $|x|^n \leq r^n = M_n$.

(b) The series $\sum_{k=1}^{\infty} M_n$ converges because it is a geometric series with ratio $r < 1$.

21. Consider the sequence of partial sums of the series. Use the fact the the the sum of finitely many continuous functions is continuous and the result of Corollary 4.7.

23. Consider the sequence of partial sums of the series. Use the fact that the sum of finitely many integrable functions is integrable and that the integral of such a sum is the sum of the integrals, and then invoke Theorem 4.9.

25. Consider the sequence of partial sums of the series. Use the fact that the derivative of the sum of finitely many differentiable functions is the sum of the derivatives, and invoke Corollary 4.10.

27. Follow the hint.

PROBLEM SET 6.5, PAGE 506

1. Apply the result of Corollary 5.3 with x replaced by $x - c$.

3. Use Corollary 2.4

5. If $r = 0$, the series converges for all $x \in \mathbb{R}$. Suppose $r \neq 0$. Then the radius of convergence is $R = 1/|r|$ and the interval of convergence is $(c - R, c + R)$.

7. The radius of convergence is $R = 1$. The series converges for $x = c + 1$ if and only if $p > 1$. The series converges for $x = c - 1$ if and only if $p > 0$.

9. If I consists of a single point, then $I^o = \varnothing$. If $I = \mathbb{R}$, then $I^o = \mathbb{R}$. If $0 < R \in \mathbb{R}$ is the radius of convergence of the series, then $I^o = (-R, R)$, so $b \in I^o \Rightarrow |b| \in I^o$.

11. (a) Use the ratio test. (b) Use Corollary 2.4.

13. Let $x \in I^o$. Then, by Theorem 5.12, $\sum_{k=1}^{\infty}[a_k/(k+1)]x^k$ converges and its sum is $\int_0^x [\sum_{k=1}^{\infty} a_k t^k]\,dt$. In particular, if J is the interval of convergence of $\sum_{k=1}^{\infty}[a_k/(k+1)]x^k$, we have $I^o \subseteq J$. That $\sum_{k=1}^{\infty}[a_k/(k+1)]x^k$ is an antiderivative of $\sum_{k=0}^{\infty} a_k x^k$ on I^o now follows from the Fundamental Theorem of Calculus in the form of Corollary 7.8 in Chapter 5.

15. Consider the three possible cases $R = 0, 0 < R \in \mathbb{R}$, and $R = +\infty$.

17. (a) In Example 5.17, replace x by $-x$. (b) Follow the hint.

19. Use Theorem 5.19.

21. Use Theorem 5.20.

23. Use mathematical induction.

25. $\sin x = \sum_{k=0}^{\infty}(-1)^k[x^{2k+1}/(2k+1)!]$, $\cos x = \sum_{k=0}^{\infty}(-1)^k[x^{2k}/(2k)!]$

27. (a) $f'(x) = p(1 + x)^{p-1}$, so $(1 + x)f'(x) = pf(x)$. (b) Follow the instructions.
(c) Note that $f(0) = 1$ and use mathematical induction.

29. (a) Look at the numerator of the expression for a_n and use the fact that a product is zero if and only if one of the factors is zero.
(b) Use the ratio test. (c) Differentiate term-by-term. (d) Follow the instructions.
(e) Use the result of Part (d).

31. Use Theorem 2.11 to show that $\lim_{n \to \infty}[\sum_{k=n}^{\infty} a_k] = 0$.

33. Follow the hint. **35.** Follow the hint.

INDEX